Acoustic Echo and Noise Control

Acoustic Echo and Noise Control

A Practical Approach

Eberhard Hänsler
Gerhard Schmidt

A JOHN WILEY & SONS, INC., PUBLICATION

The cover of this book was designed by Professor Helmut Lortz, Darmstadt, Germany.

Published by John Wiley & Sons, Inc., Hoboken, New Jersey.
Published simultaneously in Canada.

Library of Congress Cataloging-in-Publication Data:

Hänsler, E. (Eberhard)
Acoustic echo and noise control: a practical approach / Eberhard Haensler, Gerhard Schmidt.
p. cm. — (Adaptive and learning systems for signal processing, communications, and control)
Includes bibliographical references and index.
ISBN 0-471-45346-3 (cloth)
1. Echo suppression (Telecommunications) 2. Acoustical engineering. I. Schmidt, Gerhard. II. Title. III. Series.

TK5102.98.H36 2004
621.382'24—dc22 2004044054

Printed in the United States of America.

10 9 8 7 6 5 4 3 2 1

Contents

List of Figures

List of Tables

Preface

The motivation to write this book originated after a 20-year long engagement in the problems of acoustic echoes and noise control at the Signal Theory Group at Darmstadt University of Technology. About 20 Ph.D. students were involved in various projects on these topics. The authors now intend to present a concise documentation of the results of this work embedded into the state of the art.

The work of the Signal Theory Group spanned the entire range of scientific and development work: theoretical considerations, computer and hardware simulations, and implementation of realtime demonstrators. Testing ideas in real environments at real time turned out to be an extremely useful tool to judge results gained by formal analysis and computer simulations and to create new ideas.

The organization of this book somewhat reflects this working mode; we start with presenting the basic algorithms for filtering, for linear prediction, and for adaptation of filter coefficients. We then apply these methods to acoustic echo cancellation and residual echo and noise suppression. Considerable space is devoted to the estimation of nonmeasurable quantities that are, however, necessary to control the algorithms. Suitable control structures based on these quantities are derived in some detail.

Worldwide knowledge of problems of echo and noise control has increased enormously. Therefore, it was necessary to limit the contents of the book. The main emphasis is put on single-channel systems where—to the opinion of the authors—a certain completeness has been reached. Multichannel systems provide additional options for improved solutions. They receive, currently, increased attention in research and development laboratories. The book deals with the basic ideas.

Implementation issues of acoustic echo and noise control systems are, beyond doubt, just as important as the topics mentioned above. They are, however, not covered in detail in this text.

The readers of this book should have a basic knowledge of linear system theory and of digital signal processing as it is presented, for example, in undergraduate courses. The authors hope that all—theoreticians and practitioners alike—are able to take advantage of the material of this book and learn more about this exciting area of digital signal processing.

Darmstadt and Ulm, Germany

Eberhard Hänsler
Gerhard Schmidt

Acknowledgments

Hidden behind the authors of a book on such an active area as acoustic echo and noise control are a large number of colleagues who helped with intensive discussions, constructive criticism, and constant encouragement. To mention all by name would inevitably mean to forget some. We express our sincere thanks to all of them.

There is, however, no rule without exception. Special thanks go to all former and current research assistants of the Signal Theory Group of Darmstadt University of Technology. Their work, documented in numerous presentations, papers, and Ph.D. dissertations, considerably contributed to the progress of acoustic echo and noise control. Their names can be found in the references to this book.

Furthermore, we have to offer our thanks to all members of the Temic audio research group.

We also have to thank the German Science Foundation for the support of some of the graduate students within the frame of a Graduate College on Intelligent Systems for Information Technology and Control Engineering.

The International Workshop on Acoustic Echo and Noise Control (IWAENC) biennially gathers researchers and developers from many countries to meet their colleagues and to present new ideas. The authors participated in these events and enjoyed the intensive discussions.

We are especially indebted to Professor Simon Haykin for his constant encouragement to write this book and his help in planning it.

Finally, we would like to express our thanks for the friendly assistance of the editors of John Wiley & Sons.

Eberhard Hänsler
Gerhard Schmidt

Abbreviations and Acronyms

AD	Analog/digital
AGC	Automatic gain control
AP	Affine projection
AR	Autoregressive
DA	Digital/analog
DFT	Discrete Fourier transform
DSP	Digital signal processor
ERLE	Echo-return loss enhancement
ES	Exponentially weighted stepsize
ETSI	European Telecommunication Standards Institute
FAP	Fast affine projection
FFT	Fast Fourier transform
FIR	Finite impulse response
FTF	Fast transversal filter
GSM	Global system for mobile communications
IDFT	Inverse discrete Fourier transform
IEEE	Institute of Electrical and Electronics Engineers

IFFT	Inverse fast Fourier transform
IIR	Infinite impulse response
INR	Input-to-noise ratio
IP	Internet Protocol
ITU	International Telecommunication Union
LCMV	Linearly constraint minimum variance
LEM	Loudspeaker–enclosure–microphone (system)
LMS	Least mean square
MAC	Multiply and accumulate
MELP	Mixed excitation linear predictor
MFLOPS	Million floating-point operations per second
MIPS	Million instructions per second
MSE	Mean square error
NLMS	Normalized least mean square
PARCOR	Partial correlation (coefficient)
QMF	Quadrature mirror filterbank
RLS	Recursive least squares
SFTF	Short-time Fourier transform
SNR	Signal-to-noise ratio
VAD	Voice activity detection
XLMS	Extended least mean square

Part I

Basics

1

Introduction

1.1 SOME HISTORY

The commercial use of the telephone started in 1876. This was one year after *Alexander Graham Bell* filed a patent application for a telephone apparatus (see Fig. 1.1). Efforts to transmit voice by electric circuits, however, date further back. Already in 1854 *Charles Bourseul* described a transmission method. *Antonio Meucci* set up a telephone system in his home in 1855. *Philipp Reis* demonstrated his telephone in 1861. One has also to mention *Elisha Gray* who tragically filed his patent application for a telephone just 2 hours later than Alexander Graham Bell.

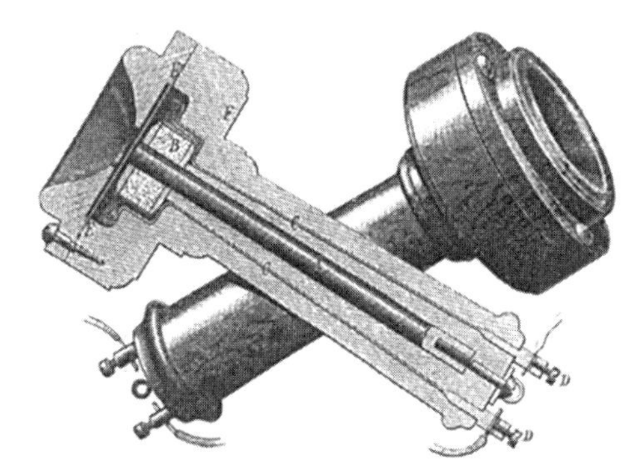

Fig. 1.1 "Bell's new telephone" (with permission from [1]).

In the very early days of the telephone conducting a phone call meant to have both hands busy; one was occupied to hold the loudspeaker close to the ear and the other hand to position the microphone in front of the mouth [1] (see Fig. 1.2). This troublesome way of operation was due to the lack of efficient electroacoustic converters and amplifiers. The inconvenience, however, guaranteed optimal conditions: a high signal-to-(environmental) noise ratio at the microphone input, a perfect coupling between loudspeaker and the ear of the listener, and—last but not least—a high attenuation between the loudspeaker and the microphone. The designers of modern speech communication systems still dream of getting back those conditions.

Fig. 1.2 The "hands-busy" telephone (with permission from [1]).

In a first step one hand had been freed by mounting the telephone device, including the microphone at a wall; further on, only one hand was busy holding the loudspeaker (see Fig. 1.3). In a next step the microphone and the loudspeaker were combined in a handset (see Fig. 1.4). Thus, still one hand was engaged. This basically remained the state of the art until today. How to provide means for hands-free telephone communication will be the focus of this book.

Early attempts to allow telephone calls with a loudspeaker and a microphone at some distance in front of the user had to use analog circuits. In 1957 Bell System introduced a so called *Speakerphone*. By limiting the gain of the communication loop it was—to a certain extent—able to prevent singing [41]. Follow-on developments used voice controlled switches to suppress acoustical echoes and to stabilize the electroacoustical loop [42]. At the same time, however, the telephone connection degraded to a half-duplex loop making natural conversations difficult. The introduction of a "center clipper" may be considered as a last step along this line [15]. This

Fig. 1.3 Wallmounted telephone from 1881 (with permission from the Museumsstiftung Post und Telekommunikation Frankfurt am Main).

Fig. 1.4 Desk telephone from 1897 (with permission from the Museumsstiftung Post und Telekommunikation Frankfurt am Main).

nonlinear device suppresses small amplitudes. Thus, it extinguishes small echoes. Moreover, small speech signals are erased, as well.

The invention of the *least mean square algorithm* in 1960 [237], the application of *adaptive transversal filters* [150, 217], and the availability of *digital circuits* with increasing processing power opened new paths to acoustic echo and noise control.

It was clear from the beginning that modeling a loudspeaker–enclosure–microphone system by an adaptive filter formed a much more demanding task than the modeling of electric systems as used for line echo canceling. Nevertheless, at least laboratory models using adaptive filters were investigated in the 1970s [180]. It took at least two more decades of breathtaking progress in digital technology until commercial applications of adaptive filters for acoustic echo and noise control became feasible.

The problem of acoustic echoes and environmental noise is by no way confined to hands-free telephones. Public announce systems, audio and video conference systems, systems to interview a person that stays in a remote location are just a few examples. Suppression of noise additively mixed with a useful signal can be considered as a "standalone" problem. Originally solutions to these problems used analog technology. New digital solutions led to a considerable increase of their functionality.

Recent advances in digital processing technology caused hardware limitations in the implementation of algorithms to disappear. Nevertheless, providing a fully natural hands-free communication free of background noise between two users of a speech communication system still keeps researchers and industrial developers busy.

1.2 OVERVIEW OF THE BOOK

With this book the authors intend to introduce the reader to the basic problems of acoustic echo and noise control and approaches to their solution. The emphasis in the title of the book is on *acoustic*. "Ordinary" echo and noise control, for instance, on transmission lines, "principally" follows the same theory. The side conditions set by the acoustic environment, however, are responsible for the need of considerably modified solutions.

The book is organized in five main parts: "Basics," "Algorithms," "Acoustic Echo and Noise Control," "Control and Implementation Issues," and finally "Outlook and Appendixes."

In Part I the environment of acoustic echo and noise control is explained. Applications besides the hands-free telephone are briefly reviewed. The signals involved are speech signals and noise. We describe the properties of speech signals and of noise as they are found in the applications treated in this book. Emphasis is placed on office and car noise. We then describe the mechanism generating acoustic echoes and depict an office where experiments have been conducted. The results of these experiments will serve as examples later in the book. The cancellation of acoustic echoes requires filters exhibiting long impulse responses. In spite of this necessity, the application of FIR filters is justified in the last section of this first part.

Part II is devoted to general descriptions of algorithms used for filtering, prediction, and adaptation. It begins with a discussion of error criteria for various applications. Chapter 5 gives a derivations of the discrete-time Wiener filter. Differentiating the mean squared error toward the filter coefficients confirms the principle of orthogonality and leads to the Wiener solution, also called the "normal equation". Specifying the autocorrelation matrix and the cross-correlation vector appropriately takes directly to the linear predictor. The Durbin algorithm solves the normal equation recursively by increasing the order of the predictor by steps of one at a time.

Algorithms for adaptive filters occupy the major fraction of this part. We start with the normalized least mean square (NLMS) algorithm as the most robust and inexpensive procedure. Control parameters are defined and stability as well as convergence are analyzed. The affine projection (AP) algorithm adds memory to the NLMS algorithm by considering also old input signal vectors. Compared to the NLMS algorithm the AP algorithm requires matrix multiplications and the inversion of a low order matrix. Reordering calculations, however, leads to a fast version of this algorithm.

We then move to the recursive least squares (RLS) algorithm. To run this procedure requires the inversion of a matrix of the order of the number of filter coefficients. There are fast versions. In the context of acoustic echo cancellation, however, they cause stability problems. Therefore, the RLS algorithm seems (not yet) well suited for this application and fast versions of it are not described in this book. The derivation of the Kalman algorithm ends this part of the book. Because of the necessary high order of the state vector, this algorithm is hardly applicable for canceling acoustical echoes or suppressing noise in the full band. In subbands, however, the order of the state vector reduces to a manageable size and applications for noise suppression are promising.

In Part III we turn to echo canceling and residual echo and noise suppression. We begin with classical methods of echo canceling like adaptive line enhancement, frequency shift, loss control, and center clipping. Applying the latter ones alone does not lead to systems with satisfying speech quality. Where applied, they are solutions for the time being because of the lack of sufficient processing power. Loss control, however, is required where a minimum echo attenuation has to be guaranteed. In those cases loss control acts as "emergency brake" during times where the adaptive filters are misadjusted.

We continue with a description of architectures for echo canceling systems. Block processing and subband structures offer considerable reductions of computational complexity at the prize of increased memory size and signal delay. They are applicable only where a noticeable delay is tolerable. Even if the book focuses on singlechannel systems, sections are devoted to stereophonic acoustic echo cancellation and multichannel systems. Because of the linear correlation of the signals in the right and the left stereo channels, the echo canceling filters are not uniquely defined. In order to converge to the "true" solution, nonlinear transformations are introduced in current solutions.

Suppression of residual echoes and noise requires a frequency analysis of the noisy signal. We light up this task from different viewpoints. Special emphasis has to be on analysis procedures introducing a low signal delay. The suppression of residual echoes and background noise is first treated separately and then combined. Where more than one input channel is allowed, beamforming provides an effective method for echo and noise suppression. The fundamentals of this technique are explained in Chapter 11.

No one algorithm for echo and noise control that is "left alone" performs satisfactory in a real environment. In Part IV we describe the control of such algorithms and the estimation of the control parameters. Most of the information given here has been published only in papers and conference proceeding. So, this is the first concise presentation of this material. In Chapter 13 we describe estimation procedures of parameters and states of the system, like short-term signal power or the presence of double talk, that are specific for echo and noise control systems. We derive pseudooptimal control parameters and give examples for entire control architectures.

In case of echo and noise suppression the estimation of the noise constitutes the central task. We discuss schemes of various complexities, and we also give hints for the dimensioning of the spectral floor and the overestimation factor. A description of the control of beamformers and a short review of implementation issues ends this part of the book.

The final part, Part V, contains an outlook and two appendixes. Appendix A discusses the properties of subband impulse responses. Special emphasis is put on the fact that subband impulse responses are noncausal. The design of filters and filterbanks used in acoustic echo and noise control systems is the topic of Appendix B.

2

Acoustic Echo and Noise Control Systems

2.1 NOTATION

Throughout this book the notation will follow the same basic rules. Lowercase letters will be used for scalar quantities in the (discrete) time domain; uppercase letters will denote quantities in the frequency domain. Lowercase boldface letters will indicate column vectors; uppercase boldface letters will describe matrices. The discrete-time variable will be n[1]. The Greek letter ω always refers to the angular frequency, the letter Ω indicates the normalized angular frequency. A "hat," as in $\widehat{x}$, is used for estimated quantities. In Table 2.1 the most frequently used symbols as well as their meanings are listed.

Supplementary, a list of abbreviations and acronyms—located at the beginning of this book—contains frequently used terms.

2.2 APPLICATIONS

In this chapter we briefly describe applications where acoustic echo and noise control problems arise. Each application exhibits individual constraints that may require methods especially tailored for it. Procedures successful in other applications may not be applicable.

[1]The authors do hope that the readers forgive them the notation $n(n)$, which refers to a noise signal at discrete time n.

Table 2.1 Notation

Meaning	Symbol
Signals	
Background noise	$b(n), b(t)$
Desired signal	$d(n), d(t)$
Output signal of the adaptive filter	$\hat{d}(n)$
Error signal	$e(n)$
Undistorted error signal	$e_{\mathrm{u}}(n) = e(n) - n(n)$
Sum of all local signals	$n(n)$
Local speech signal	$s(n), s(t)$
Excitation signal	$x(n)$
Microphone signal	$y(n)$
System parameters	
Impulse response vector of the unknown system	$\boldsymbol{h}(n)$
Impulse response vector of the adaptive filter	$\widehat{\boldsymbol{h}}(n)$
Wiener solution	$\widehat{\boldsymbol{h}}_{\mathrm{opt}}$
System mismatch vector	$\boldsymbol{h}_{\Delta}(n) = \boldsymbol{h}(n) - \widehat{\boldsymbol{h}}(n)$
Control parameters	
Regularization parameter	$\Delta,\ \Delta(n)$
Stepsize	$\mu,\ \mu(n)$
Statistical quantities	
Autocorrelation and cross-correlation	$r_{xx}(n), r_{xy}(n)$
Power spectral densities	$S_{xx}(\Omega), S_{xy}(\Omega)$

Characteristic features of systems that require echo and noise control are governed by the volume of the enclosure, the required bandwidth, the tolerable delay, and, if the feature is optional, the extra cost. The reverberation time in an enclosure is proportional to its volume [128]. Thus, it is responsible for the length of the echo and the complexity of the echo canceling filter. The bandwidth of the transmission determines the minimum sampling frequency that directly enters into the processing demands. For telephone systems the standard sampling frequency is 8 kHz. For telecommunication studios, a higher frequency such as 16 kHz is desirable. Upper bounds for the delay introduced by echo and noise control measures are set by physical and psychological requirements and by regulatory bodies. If speaker and listeners are in the same enclosure, such as in an auditorium equipped with a public address system or in a car, that has a car interior communication system, the speaker can be heard directly and via the loudspeakers. In that case only a very small delay is tolerable. Telephone systems, on the other hand, have to fulfill requirements set by international standardization institutes (see Section 3.3). Bounds on delay introduced by front-end processing recommended by them may restrict the application of block processing methods. Last but not least, where echo and noise control is an add-on feature of a communication system the extra cost must be in the range customers are willing to pay. Unfortunately, this can be very low and, therefore, prohibits the application of sophisticated algorithms.

In the following subsections we give a few details of specific applications of echo and noise control systems.

2.2.1 Hands-Free Telephone Systems

Hands-free telephones started with just a loss control unit (see Fig. 2.1) that allowed a half-duplex communication. A full-duplex hands-free system requires an echo canceling filter in parallel to the loudspeaker–enclosure–microphone (LEM) system. The filter has to be matched to this system that is highly time-variable. A microphone array can improve the ratio of the local speech signal to the echo from the loudspeaker and the local noise. Local noise and echoes remaining due to filter mismatch are suppressed by a filter within the outgoing signal path.

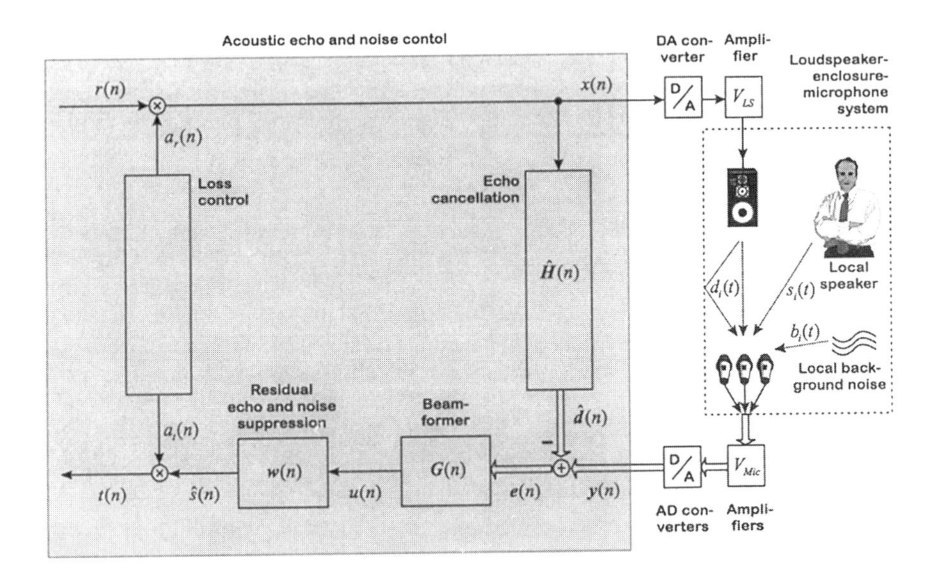

Fig. 2.1 Structure of a monophonic hands-free telephone system.

Hands-free telephone systems are the major focus of this book and will be described in great detail in the following chapters.

2.2.2 Car Interior Communication Systems

In limousines and vans communication between passengers in the front and in the rear may be difficult. Driver and front passengers speak toward the windshield. Thus, they are hardly understandable for those sitting behind them. In the rear-to-front and left-to-right directions, the acoustic loss is smaller. This can be measured by placing an artificial mouth loudspeaker[2] at the driver's seat and torsos with earmicrophones [124] at the front passenger's seat and at the backseats, respectively. In Fig. 2.2 the frequency responses measured between the driver's mouth and the left ear of the front passenger, respectively the left ear of the rear passenger behind the front passenger are depicted. On average the acoustic loss is 5–15 dB larger to the backseat passenger (compared to the front passenger).

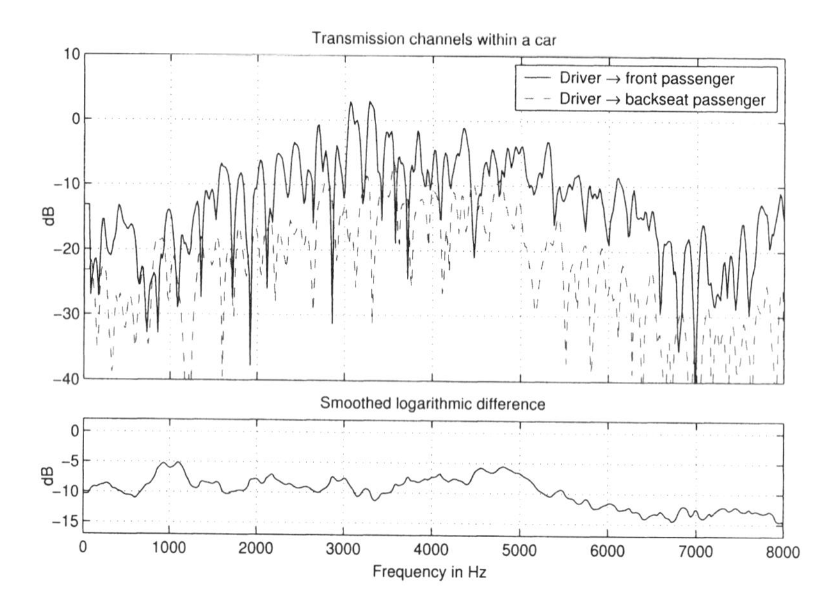

Fig. 2.2 Frequency responses of different (driver to front passenger and driver to left rear passenger) communication directions within a car are measured. Compared to the driver-to-front passenger direction, the rear passengers hear the drivers signal with an attenuation of about 5–15 dB.

Figure 2.3 sketches a car communication system [148, 175] aimed to support front-to-rear conversations. The special challenge in developing systems for this task consists in designing a system that exhibits at most 10 ms delay [108]. Signals from

[2]This is a loudspeaker which has (nearly) the same radiation pattern as the human speech apparatus.

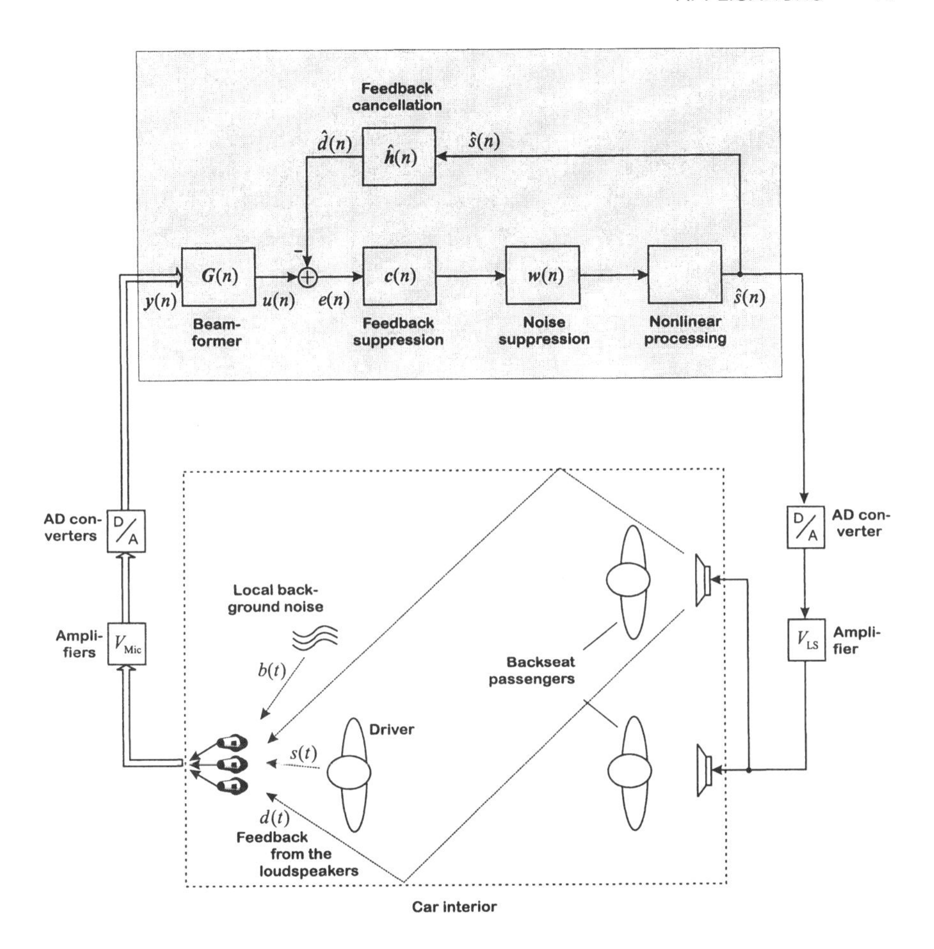

Fig. 2.3 Structure of a car interior communication system.

the loudspeakers delayed for more than that will be perceived as echoes. Since driver and front passenger are at well-defined locations, fixed microphone arrays can point toward each of them requiring fixed beamformers only. This allows to start with the feedback compensation after the beamformer. The feedback cancellation turns out to be extremely difficult since the adaptation of a feedback canceling filter is disturbed by the strongly correlated speech input signal. Feedback suppression and noise suppression can improve the system. A device with nonlinear characteristic attenuates large signal amplitudes. Also an adaptive notch filter can reduce maxima of the closed-loop transfer function. Thus, the howling margin is improved.

2.2.3 Public Address Systems

The problems here are similar to those in car interior communication systems. The enclosure, however, is much larger. Consequently the echo is considerably longer.

Loudspeakers are placed at several locations in the auditorium in order to provide a uniform audibility at all seats. Listeners in front rows should hear the speaker directly and via (more than one) loudspeaker at about the same loudnesses and with about the same delays. A microphone array at the lectern should be sensitive toward the speaker and insensitive in the directions of the loudspeakers.

In addition to feedback cancellation and feedback suppression, a frequency shift can be applied to prevent howling (see Section 8.2). Also adaptive notch filters to attenuate maxima of the closed-loop transfer function are applied.

2.2.4 Hearing Aids

Compared with the examples already mentioned hearing aids are "a world of their own." They are powered by very small batteries. This fact prohibits—at least at the time this book is written—the use of standard signal processing circuits. Devices featuring extremely low power consumption are especially designed for hearing aids. Figure 2.4 shows a behind-the-ear system. An ear hook connects the loudspeaker to the outer auditory canal. It provides a tight fit and, thus, reduces acoustical feedback to the microphone. Completely-in-the-canal hearing aids are so small that they can be worn in the outer ear. Figure 2.5 shows the architecture of an completely-in-the-canal system.

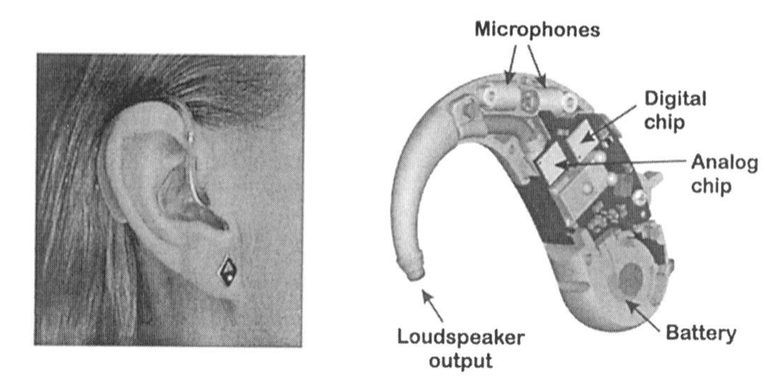

Fig. 2.4 Behind-the-ear hearing aid (with permission from [212, 213]).

Hearing aids provide amplifications of more than 60 dB. Thus, feedback developing through the ventilation openings and the cabinet is extremely critical. In the forward direction the signal is split in up to 32 subbands by a filterbank. In order to ensure short delays, the filter output signals are not subsampled. Thus, there is no synthesis filterbank necessary. The subband signals are amplified according to the needs of the hearing impaired bearer.

In behind-the-ear systems noise reduction is achieved by two microphones that are located close to each other and by differential processing of their outputs [103]. If it is required by the acoustical environment, they can be switched to an omnidirectional characteristic. Because there is no free sound field inside the ear canal, a different noise reduction method has to be applied in completely-in-the-canal sys-

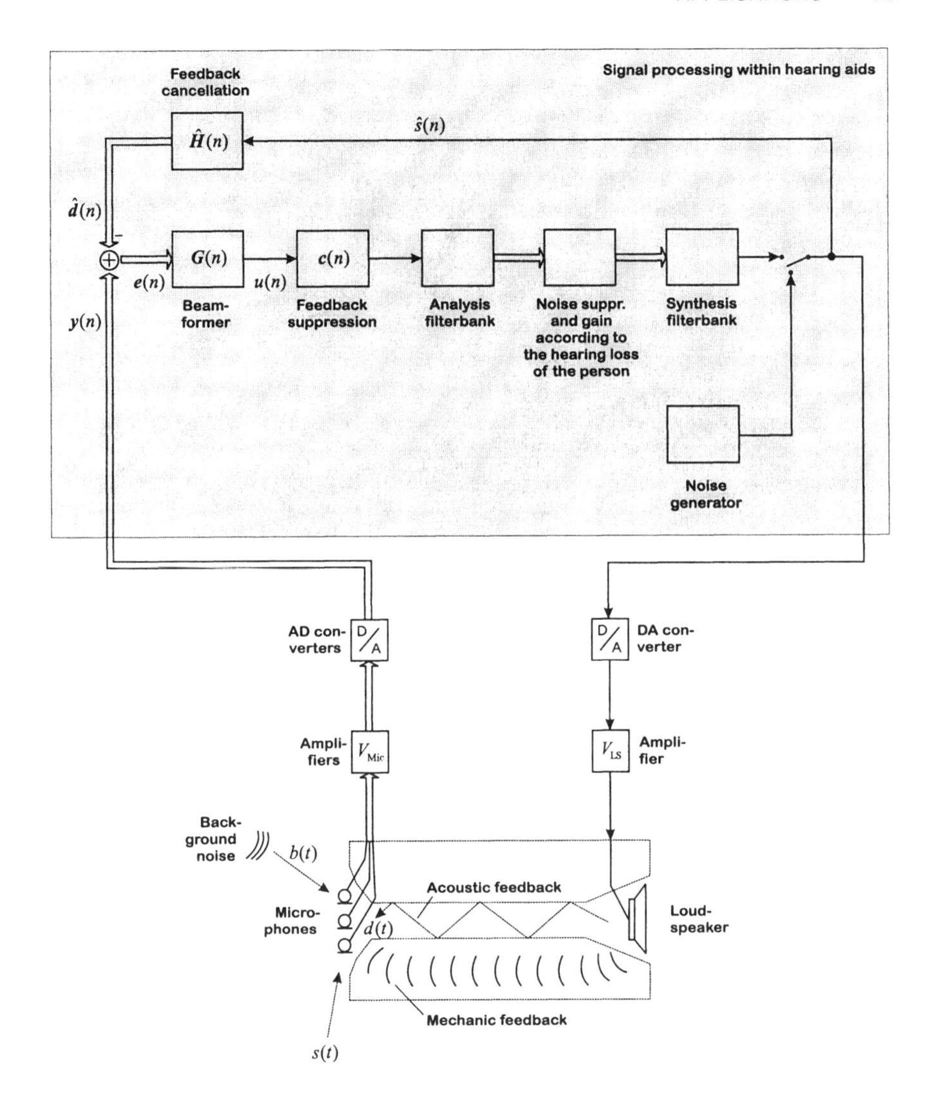

Fig. 2.5 Architecture of a hearing aid.

tems. It uses the filterbank outputs and estimates their amplitude modulation depths. If the modulation depth of a subband signal is low, noise is assumed and the band is attenuated. Improved processing capabilities will also allow spectral subtraction techniques (see Chapter 14) in the future.

In speech pauses the microphone output is switched off and an electronically generated noise is inserted. Since this is not correlated with the microphone output signals, it is very well suited for identifying the acoustical/mechanical feedback loop.

3

Fundamentals

In this chapter we give a short overview of the major types of signals one has to deal with in acoustic echo and noise control systems. Furthermore, we describe the basic mechanism of the origin of acoustic echoes and justify the use of transversal filters for echo suppression. We also briefly review standards formulated by the International Telecommunication Union (ITU) and by the European Telecommunication Standards Institute (ETSI).

3.1 SIGNALS

The two main classes of signals involved in acoustic echo and noise control systems are *speech* and *background noise*. For someone who deals with (adaptive) algorithms, speech certainly belongs to the most difficult signals she or he has to handle. Its properties are far away from the properties of white noise that is popular for designing and analyzing algorithms. As background noise we will discuss noise in a car and noise in an office. Speech signals are treated in the literature in more detail (e.g.,by Rabiner and Juang [193]) than background noise. Therefore, we will be brief on speech and elaborate on background noise.

3.1.1 Speech

Speech might be characterized by nearly periodic (voiced) segments, by noiselike (unvoiced) segments, and by pauses (silence). The envelope of a speech signal exhibits a very high dynamics. If a sampling frequency of 8 kHz (telephony) is used, the mean spectral envelope may range over more than 40 dB [126]. If higher sampling rates are implemented (e.g., for teleconferencing systems), the variations increase even further. Parameters derived from a speech signal may be considered valid only during an interval of about 20 ms [49, 193]. Properties of a speech signal are illustrated in Fig. 3.1. It shows a 5-second time sequence sampled at 8 kHz (upper diagram), the average power spectral density (center), and the result of a time–frequency analysis (lower diagram).

Due to the frequent changes of the properties of speech signals, the (long-term) power spectral density is not sufficient to characterize the spectral behavior of speech. The time–frequency—also called "visible speech"—diagram clearly shows the sequences of voiced and unvoiced sections and of pauses. During voiced intervals the *fundamental* or *pitch frequency* and the harmonics bare the major part of the energy.

A spectral feature of vowels are peaks in the spectral envelope caused by resonances in the vocal tract. These resonance frequencies shape ("form") the spectral envelope. Thus they are called *formant frequencies*. Their change over time characterizes the *intonation* of a word or sentence. The pitch frequency depends on the speaker. Mean pitch frequencies are about 120 Hz for male speakers and about 215 Hz for female speakers. In Fig. 3.2 three 60-ms segments of speech (as well as the entire sequence) are depicted in the time and in the frequency domain. The two diagrams in the center show two vowels: *a* from *wonderful* and *i:* from *be*. In the last diagram the sibilant sequence *sh* from *she* is depicted. The terms *a*, *i:*, and *sh* are written in phonetic transcription.

The differences between voiced and unvoiced speech are also clearly visible in the power spectral density. While voiced speech has a comblike spectrum, unvoiced speech exhibits a nonharmonic spectral structure. Furthermore, the energy of unvoiced segments is located at higher frequencies. The rapidly changing spectral characteristics of speech motivate the utilization of signal processing in subbands or in the frequency domain. These processing structures allow a frequency-selective power normalization leading to a smaller eigenvalue spread [111] and therefore to faster convergence of adaptive filters excited and controlled by such signals.

Sampling frequencies range from 8 kHz in telephone systems up to about 96 kHz in high-fidelity systems. Even in the case of 8 kHz sampling frequency, consecutive samples are highly correlated. The normalized autocorrelation coefficient

$$a_1 = \frac{\mathrm{E}\{\, s(n)\, s(n+1)\,\}}{\mathrm{E}\{\, s^2(n)\,\}} = \frac{s_{ss}(1)}{s_{ss}(0)} \tag{3.1}$$

of neighboring samples reaches values in the range of 0.8–0.95. Because of the periodic components and the pauses, short-time autocorrelation matrices very often become singular. Therefore, speech signals belong to the class of nonpersistent signals. Thus, special precautions are necessary to prevent instability of algorithms that use—directly or indirectly—the inverse of the autocorrelation matrix.

The *probability density function* $f_s(s)$ of the amplitudes of speech signals $s(n)$ is characterized by a marked peak for zero amplitudes and an exponential decay for large amplitudes. Voiced sounds mainly contribute to large values, whereas unvoiced sounds are responsible for the peak at zero amplitude [48]. Analytic approximations

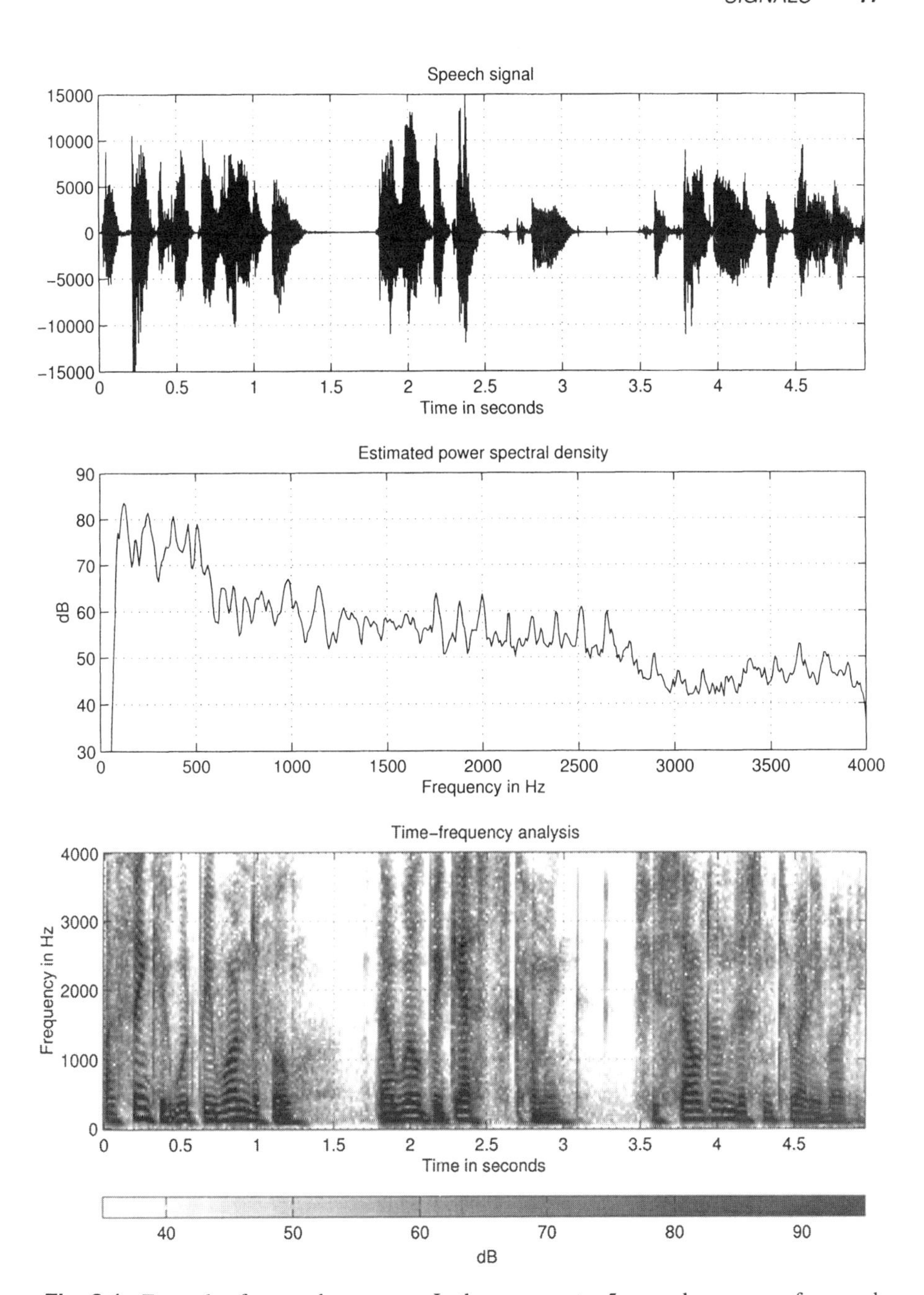

Fig. 3.1 Example of a speech sequence. In the upper part a 5-second sequence of a speech signal is depicted. The signal was sampled at 8 kHz. The second diagram shows the mean power spectral density of the entire sequence (periodogram averaging). In the lowest part a time–frequency analysis is depicted. Dark colors represent areas with high energy; light colors mark areas with low energy.

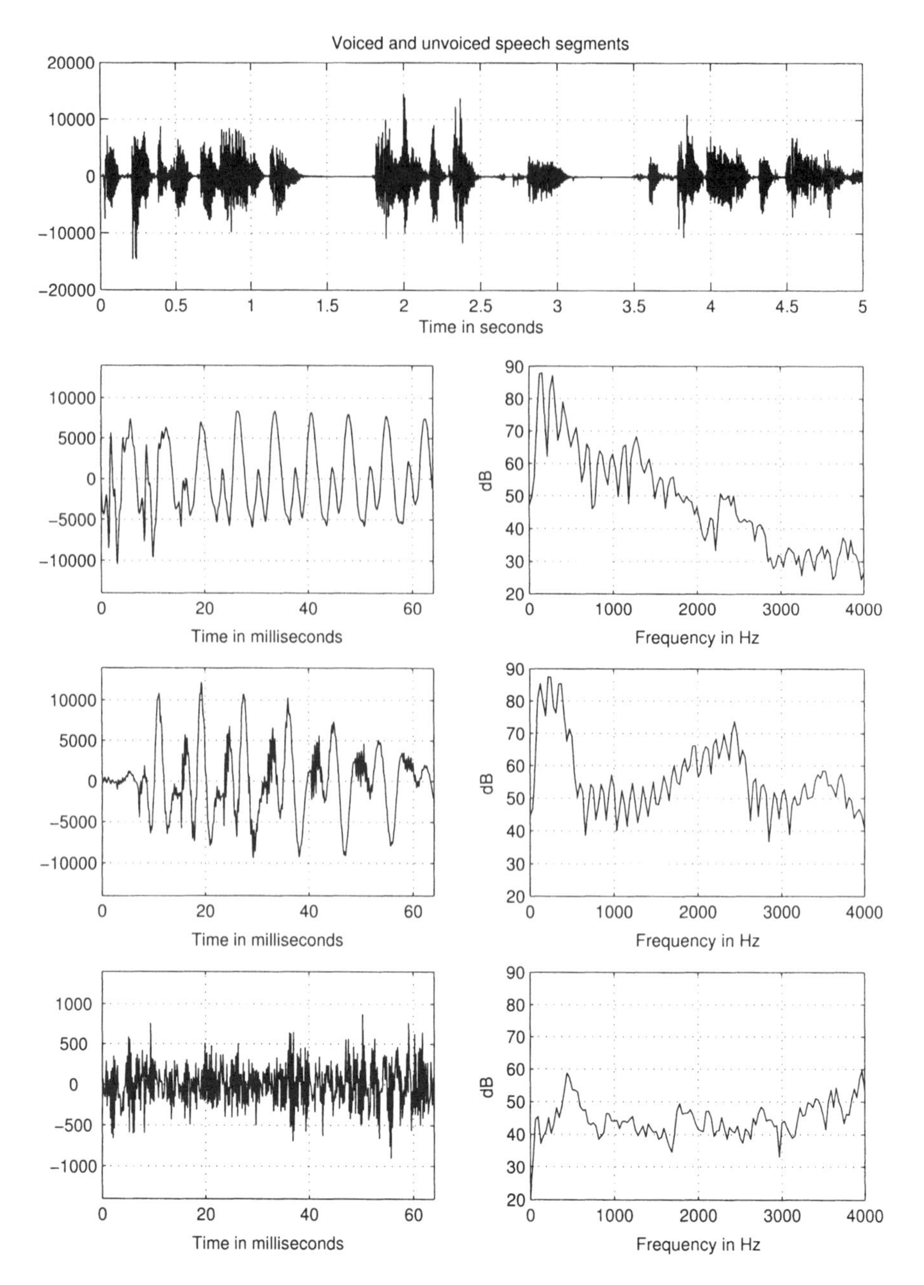

Fig. 3.2 Voiced and unvoiced speech segments. In the upper part a 5-second sequence of a speech signal is depicted. The following three diagram pairs show 60-ms sequences of the speech signal (left) as well as the squared spectrum of the sequences (right). First two vowels (*a* from *wonderful* and *i:* from *be*) are depicted. Finally, an unvoiced (sibilant) sequence (*sh* from *she*) is shown.

for the probability density function are the *Laplacian probability density function*

$$f_{s,1}(s) = \frac{1}{\sqrt{2}\sigma_s} \, e^{-\sqrt{2}\frac{|s|}{\sigma_s}} , \tag{3.2}$$

and the (two-sided) *Gamma probability density function*

$$f_{s,2}(s) = \frac{\sqrt[4]{3}}{2\sqrt{2\,\pi\,\sigma_s}} \, \frac{1}{\sqrt{|s|}} \, e^{-\frac{\sqrt{3}}{2}\frac{|s|}{\sigma_s}} , \tag{3.3}$$

where σ_s is the standard deviation of the signal, that is assumed to be stationary and zero mean in this context [23, 196, 242, 243]. It depends on the speaker which approximation provides the better fit [242].

In order to compare these functions with the well-known Gaussian density

$$f_g(s) = \frac{1}{\sqrt{2\,\pi}\,\sigma_s} \, e^{-\frac{s^2}{2\,\sigma_s^2}} , \tag{3.4}$$

the densities are transformed using

$$\bar{s} = \frac{s}{\sigma_s} \tag{3.5}$$

as normalized amplitude. It follows for Eq. 3.2

$$f_{\bar{s},1}(\bar{s}) = \frac{1}{\sqrt{2}} \, e^{-\sqrt{2}\,|\bar{s}|} , \tag{3.6}$$

and for Eq. 3.3 that

$$f_{\bar{s},2}(\bar{s}) = \frac{\sqrt[4]{3}}{2\sqrt{2\,\pi}} \, \frac{1}{\sqrt{|\bar{s}|}} \, e^{-\frac{\sqrt{3}}{2}|\bar{s}|} . \tag{3.7}$$

The Gaussian density for normalized amplitudes reads

$$f_{\bar{g}}(\bar{s}) = \frac{1}{\sqrt{2\,\pi}} \, e^{-\frac{\bar{s}^2}{2}} . \tag{3.8}$$

Figure 3.3 shows the three functions. It becomes obvious that especially for small amplitudes the approximations of the density of speech differ considerably from the Gaussian density.

In order to be able to handle the wide range of speech amplitudes the densities in Eqs. 3.2 and 3.3 are transformed into functions of normalized logarithmic amplitudes

$$\bar{s}_{\log} = 20\,\log_{10}|\bar{s}| = 20\,\log_{10}\frac{|s|}{\sigma_s} . \tag{3.9}$$

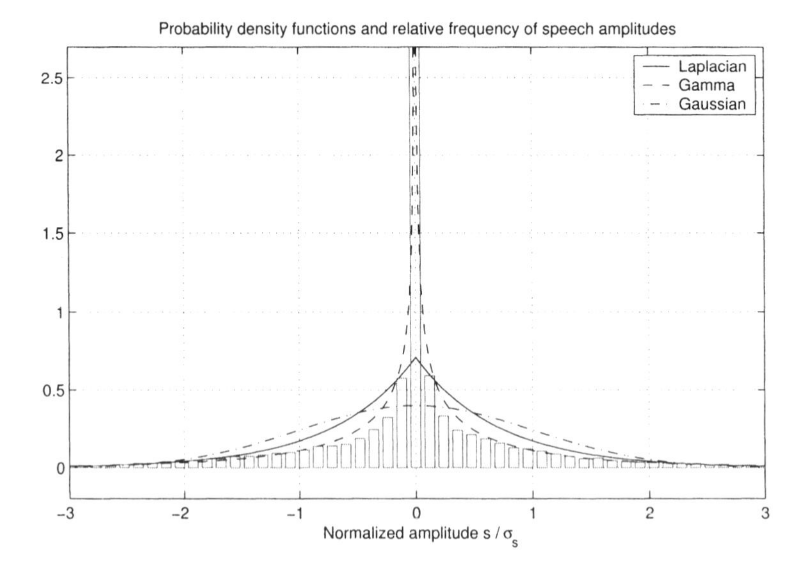

Fig. 3.3 Normalized Gaussian probability density function and normalized approximations for speech signals. The bars represent the measured relative frequency of the amplitudes of a speech sequence.

After a few manipulations the transformed probability density functions read

$$f_{\bar{s}_{\log},1}(\bar{s}_{\log}) = \frac{\sqrt{2}\,\ln 10}{20}\, 10^{\frac{\bar{s}_{\log}}{20}}\, e^{-\sqrt{2}\,10^{\frac{\bar{s}_{\log}}{20}}} \tag{3.10}$$

and

$$f_{\bar{s}_{\log},2}(\bar{s}_{\log}) = \frac{\sqrt[4]{3}}{\sqrt{2\,\pi}}\,\frac{\ln 10}{20}\, 10^{\frac{\bar{s}_{\log}}{40}}\, e^{-\frac{\sqrt{3}}{2}\,10^{\frac{\bar{s}_{\log}}{20}}}\,. \tag{3.11}$$

Both hold for $\bar{s}_{\log} > 0$. Figure 3.4 shows these functions compared with a measured histogram of speech amplitudes.

The assumption of a Gaussian probability density function turns out to be an applicative model for the *logarithm* of the normalized power spectral density $\Theta_{ss}(\Omega)$:

$$\Theta_{ss}(\Omega) = 10\,\log_{10}\frac{S_{ss}(\Omega)}{\sigma_s^2}\,, \tag{3.12}$$

where $S_{ss}(\Omega)$ is the power spectral density of the speech signal at frequency Ω, $S_{ss}(\Omega) > 0$ and σ_s^2 denotes the average power of this signal.[1] With this notation the

[1] If $S_{ss}(\Omega)$ describes a "short-term" power spectral density, this quantity also has a time-index n: $S_{ss}(\Omega, n)$. The time index is omitted here.

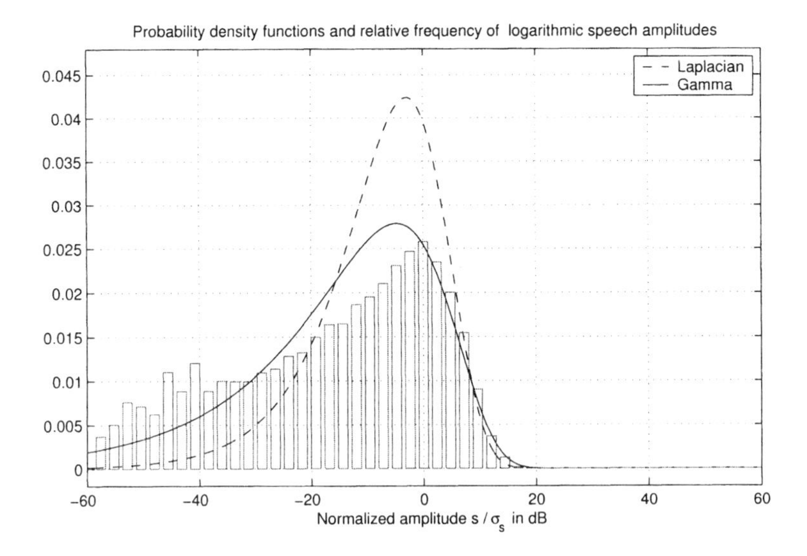

Fig. 3.4 Approximations of the probability density function for logarithmic normalized speech amplitudes and histogram of logarithmic normalized measured speech signal.

assumed density function reads

$$f_{\Theta_{ss}(\Omega)}(\Theta) = \frac{1}{\sqrt{2}\,\pi\,\kappa_{\Theta_{ss}(\Omega)}} \exp\left[-\frac{(\Theta - m_{\Theta_{ss}(\Omega)})^2}{2\,\kappa^2_{\Theta_{ss}(\Omega)}}\right], \tag{3.13}$$

where $m_{\Theta_{ss}(\Omega)}$ stands for the mean of the logarithmic normalized power spectral density at frequency Ω and $\kappa_{\Theta_{ss}(\Omega)}$ denotes the standard deviation of $\Theta_{ss}(\Omega)$. If we normalize once more using

$$\bar{\Theta}_{ss}(\Omega) = \frac{\Theta_{ss}(\Omega)}{\kappa_{\Theta_{ss}(\Omega)}}, \tag{3.14}$$

the probability density function finally reads

$$f_{\bar{\Theta}_{ss}(\Omega)}(\bar{\Theta}) = \frac{1}{\sqrt{2}\,\pi}\, e^{-\dfrac{(\bar{\Theta} - m_{\bar{\Theta}_{ss}(\Omega)})^2}{2}}, \tag{3.15}$$

with

$$m_{\bar{\Theta}_{ss}(\Omega)} = \frac{m_{\Theta_{ss}(\Omega)}}{\kappa_{\Theta_{ss}(\Omega)}}. \tag{3.16}$$

In Fig. 3.5 we show the probability density function (see Eq. 3.13) of the logarithmic normalized power spectral density of speech and histograms calculated from measured signals for selected frequencies.

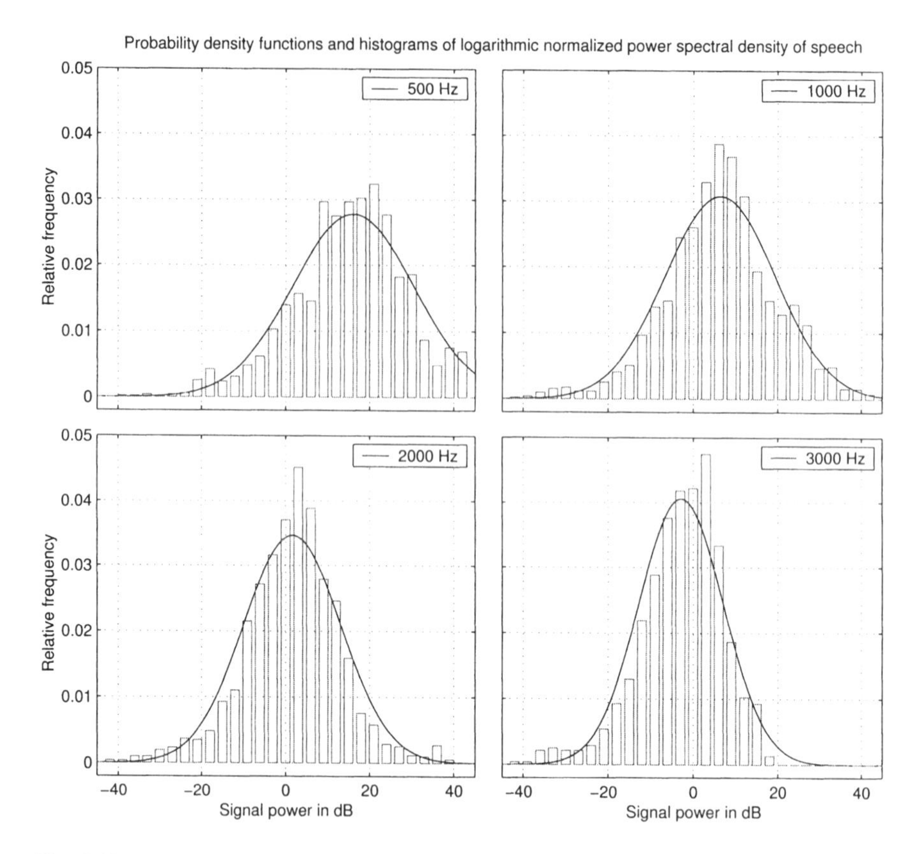

Fig. 3.5 Probability density function of the logarithmic normalized power spectral density of speech and histograms calculated from measured signals for selected frequencies.

Recent results demonstrate that the Laplacian (Eq. 3.2) or the gamma (Eq. 3.3) probability density provide close approximations also of the densities of the real and the imaginary parts of the DFT coefficients of speech signals calculated from frames shorter than 100 ms. Therefore, noise reduction procedures based on these densities prove to be superior to those assuming the Gaussian density [162].

3.1.2 Noise

Among the many types of acoustical noise in human environments, office and car noise are of special interest in connection with echo and noise control. Figure 3.6 shows estimates of power spectral densities of noises measured in a car at various speeds and in different offices. The sound level in a car can vary over 40 dB depending on the speed of the car. In an office the loudness varies depending on the size of the office, the furniture, and the equipment that is in use. However, the noise level in an office is comparable only with the level in cars driving at low speeds.

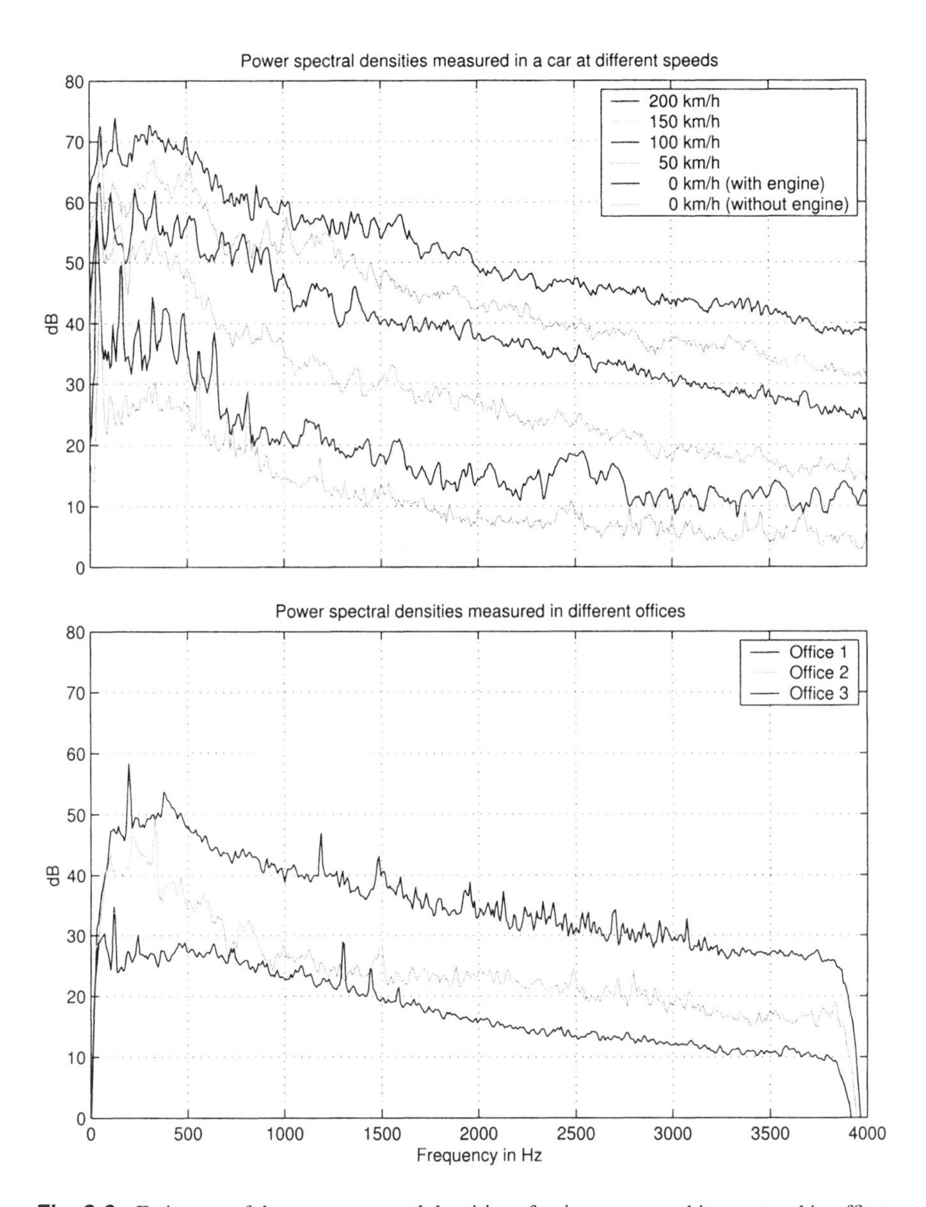

Fig. 3.6 Estimates of the power spectral densities of noises measured in a car and in offices.

3.1.2.1 Office Noise

The main sources of noise in an office are on one hand the ventilators of computers, printers, or facsimile machines and on the other hand (telephone) conversations of colleagues sharing the office.

The first category can be considered as (long-term) stationary. Figure 3.7 gives an example of the noise produced by a computer fan. It shows the time signal, the estimated power spectral density, and the time–frequency analysis. The latter confirms the stationarity of this type of noise. It also shows the harmonic components. With respect to the level of this noise, one has to be aware that its sources are often very close to microphones. This holds especially in IP-based[2] applications that are constantly gaining importance. The slow decay of the power spectral density with increasing frequency may be considered as typical for this type of office noise.

Conversational noise, of course, has speech character and is therefore—even if its level is lower than that of ventilator noise—much harder to attenuate by a noise reduction unit.

3.1.2.2 Car Noise

Car noise results from a large number of sources. The main components are engine noise, wind noise, rolling noise, and noise from various devices (e.g., ventilators) inside the passenger compartment. We will discuss these components in a little bit more detail, because their specific properties will become important later in this book when we present methods for noise reduction.

3.1.2.2.1 Engine Noise

The main energy of the noise produced by the engine of the car is concentrated at frequencies below 500 Hz. The power spectral density shows maxima at half of the engine speed and at integer multiples of it. This holds for 4-stroke engines. Figure 3.8 gives examples of engine noise. The upper diagram shows the result of a time–frequency analysis of an engine during acceleration and during deceleration. The center and lower diagrams depict the time–frequency analyses and the time signals of noises from engines running at constant speed: a gasoline engine at 2100 revolutions per minute (rpm) (center) and a diesel engine at 3500 rpm (lower diagram). In the first case the harmonics are 17.5 Hz apart from each other, in the second case their distance is 29.2 Hz. If it comes to the application of noise reduction methods it is important that the information of the exact engine speed can be read from the data bus system of the car.

3.1.2.2.2 Wind Noise

The wind noise depends on the design of the body of a car. Modern designs reduce the aerodynamic resistance as much as possible in order to lower the gasoline consumption of the car. Therefore wind noise contributes remarkably only at high speeds. Figure 3.9 shows two results of time–frequency analyses and power spectral densities of wind noise at two different speeds of a car. The measurements for these

[2]IP stands for Internet Protocol.

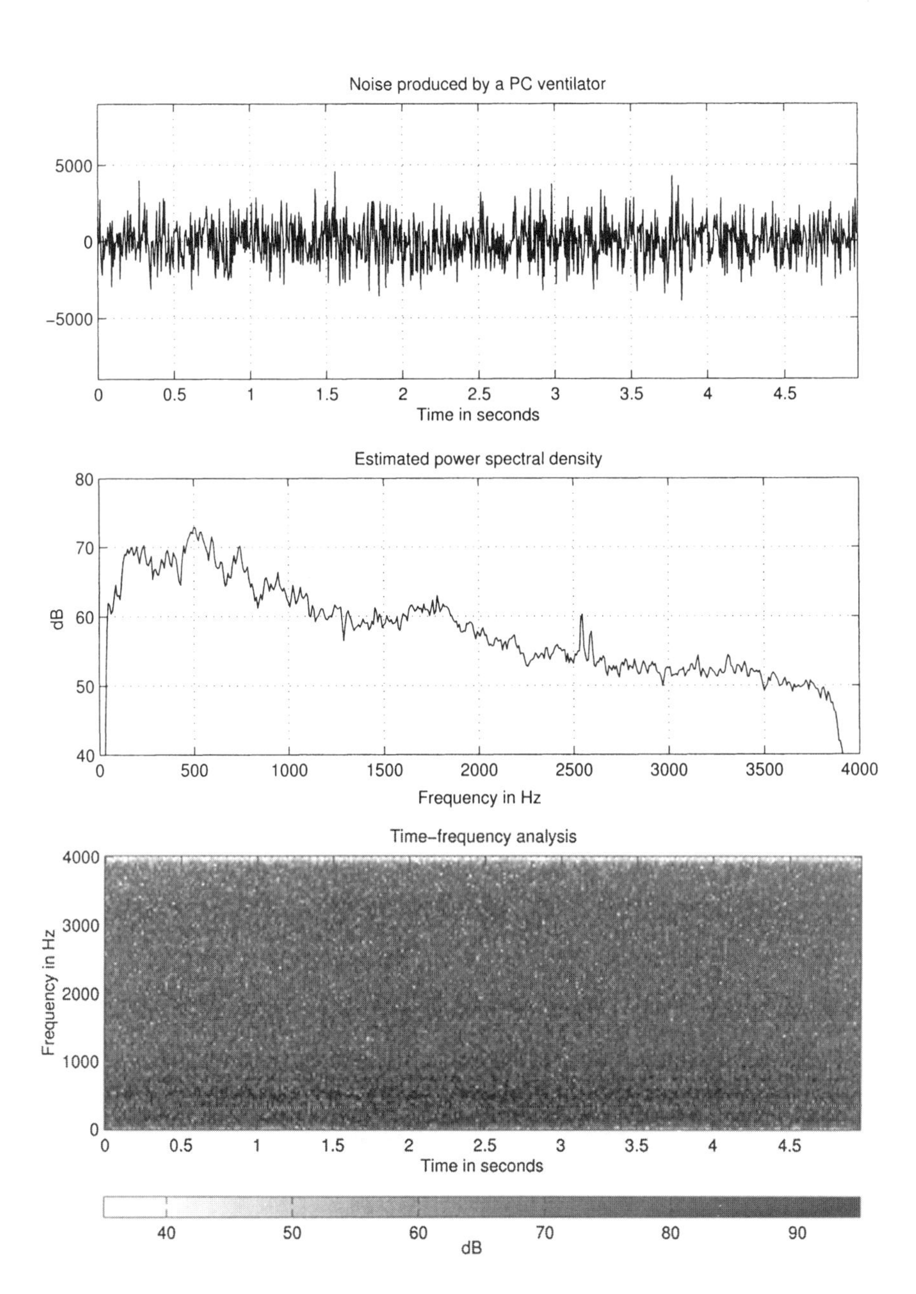

Fig. 3.7 Noise produced by a PC ventilator: time signal (upper diagram), estimated power spectral density (center), and time–frequency analysis (lower diagram). The sampling frequency is 8 kHz.

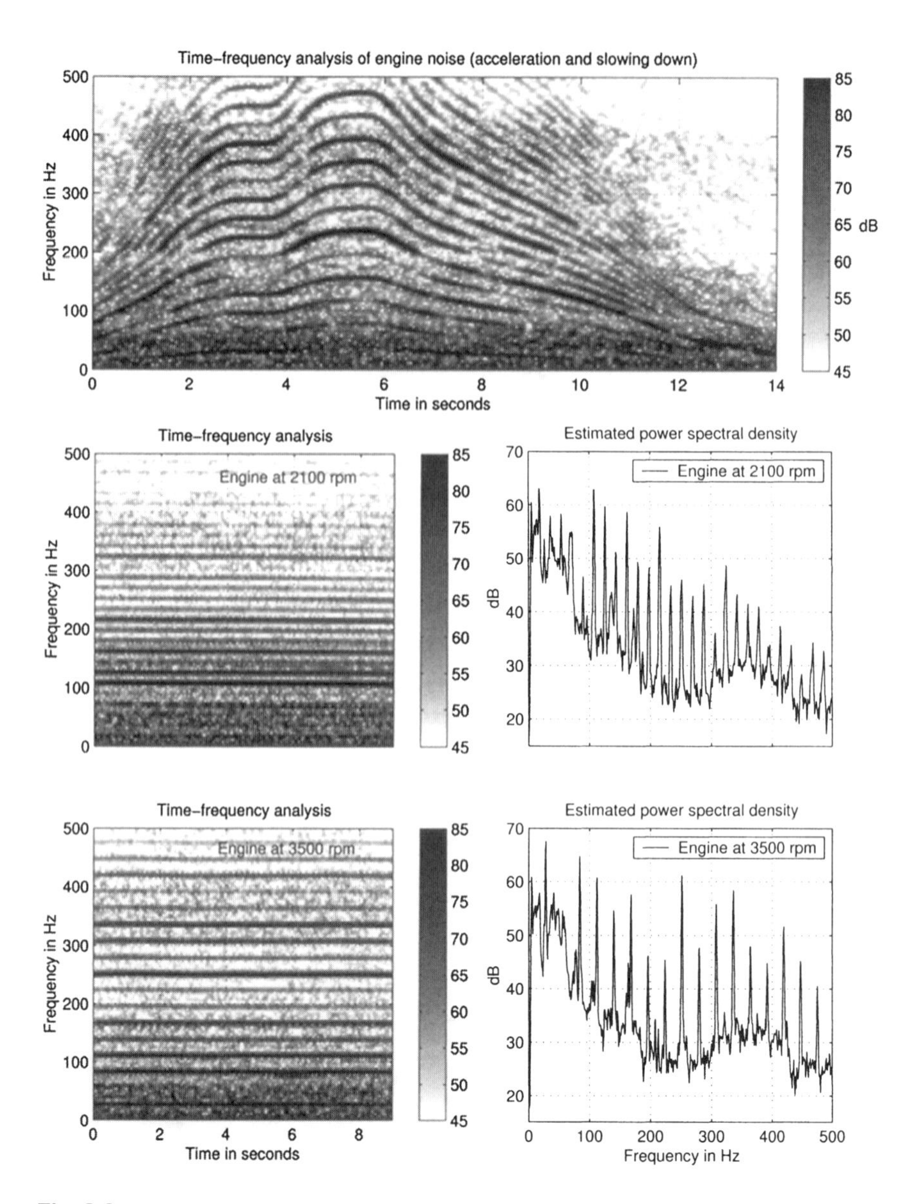

Fig. 3.8 Examples of engine noise: time–frequency analysis during acceleration and during deceleration (upper diagram), time–frequency analyses, and the time signals of noises from engines running at constant speed; gasoline engine at 2100 rpm (center) and diesel engine at 3500 rpm (lower diagram).

diagrams have been performed in a wind tunnel. Thus, the noise was free of noise from other sources.

Wind noise occupies a wider frequency range than engine noise. Since changes in the speed of a car are much slower than that of the engine wind noise is "more stationary" than engine or rolling noise.

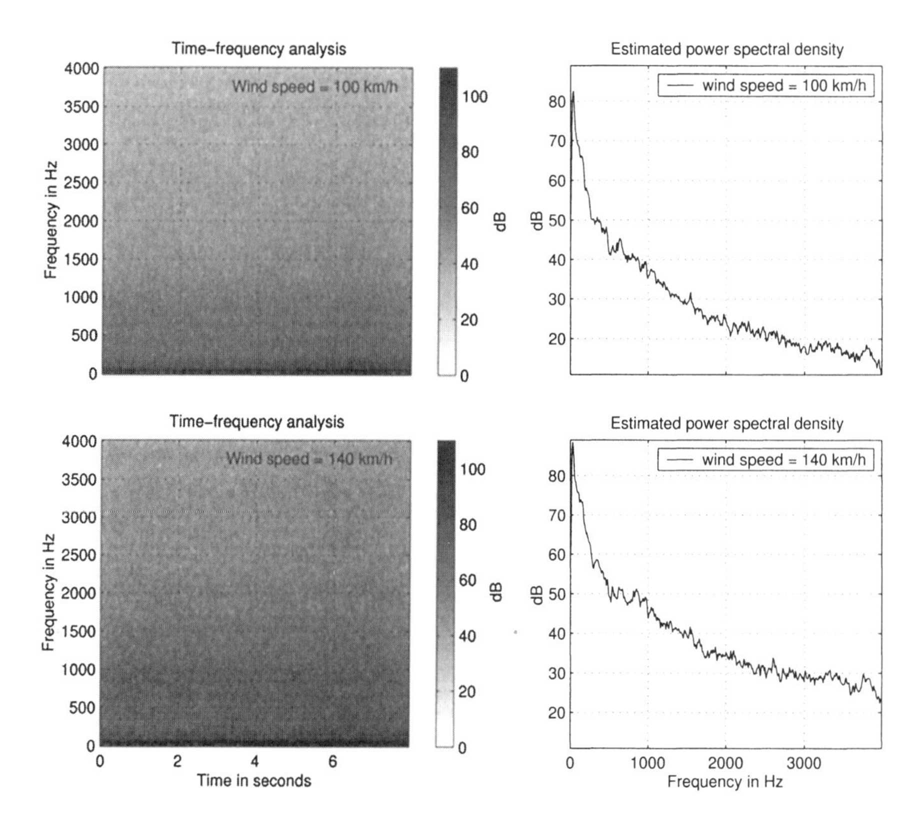

Fig. 3.9 Time–frequency analyses and power spectral densities of wind noise at two different speeds of a car: 100 km/h (upper diagrams) and 140 km/h (lower diagrams).

3.1.2.2.3 Rolling Noise

Rolling noise arises when the tires contact the road surface. It strongly depends on the type of the tires and on the road surface. The tread patterns are designed such that periodic noise at higher frequencies is avoided. Changes of the road surface cause sudden noise changes as shown in Fig. 3.10. During the measurements of these diagrams the car engine was idling to minimize the engine noise. Changes in the power spectral density are up to 15 dB and are mainly below 1000 Hz.

As very coarse approximation one can assume that—in case of nonchanging road surface—the power of the rolling noise increases exponentially with the speed of the car [189]:

$$P_{\text{RN}}(v) = P_{\text{RN}}(v_0)\, e^{K_\text{v}\,(v-v_0)}\,, \tag{3.17}$$

where v is the speed of the car and K_v a constant depending on the type of the tires and the surface of the road. Plotted in decibels (dB), this results in a linear increase.

3.1.2.2.4 *Internal Noise Sources*

Besides the sources for car noise already discussed in this chapter, there are a number of additional sources in the passenger compartment. As an example, we mention ventilators. In unfavorable situations they may blow directly toward close-by microphones.

We show an example of ventilator noise in Fig. 3.11. The upper diagram depicts the time signal when the fan is switched from "low" to "high" through intermediate steps. The result of the time–frequency analysis of the noise is given in the center diagram, whereas the power spectral densities for "low" and "high" are shown at the bottom diagram. At position "low" the ventilator noise may be masked by other noise sources. At "high," however, ventilator noise cannot be neglected.

3.1.3 Probability Density of Spectral Amplitudes of Car Noise

For the real part $R(e^{j\Omega})$ and the imaginary part $I(e^{j\Omega})$ of the (short-term) Fourier transform $B(e^{j\Omega})$ of car noise $b(n)$

$$B\left(e^{j\Omega}\right) = R\left(e^{j\Omega}\right) + j\,I\left(e^{j\Omega}\right)\,, \tag{3.18}$$

Gaussian probability density functions prove to be good approximations [233]:

$$f_{R(\Omega)}(a) = \frac{1}{\sqrt{2\,\pi}\,\sigma_{R(e^{j\Omega})}}\, e^{-\frac{a^2}{2\,\sigma^2_{R(e^{j\Omega})}}}\,, \tag{3.19}$$

$$f_{I(\Omega)}(b) = \frac{1}{\sqrt{2\,\pi}\,\sigma_{I(e^{j\Omega})}}\, e^{-\frac{b^2}{2\,\sigma^2_{I(e^{j\Omega})}}}\,, \tag{3.20}$$

where $\sigma_{R(e^{j\Omega})}$ and $\sigma_{I(e^{j\Omega})}$ are the standard deviations of $R(e^{j\Omega})$ and $I(e^{j\Omega})$, respectively. Consequently, for

$$\sigma_{R(e^{j\Omega})} = \sigma_{I(e^{j\Omega})}\,, \tag{3.21}$$

and

$$|B\left(e^{j\Omega}\right)|^2 = R^2\left(e^{j\Omega}\right) + I^2\left(e^{j\Omega}\right)\,, \tag{3.22}$$

$$\sigma^2_{|B(e^{j\Omega})|^2} = 2\,\sigma^2_{R(e^{j\Omega})} = 2\,\sigma^2_{I(e^{j\Omega})}\,, \tag{3.23}$$

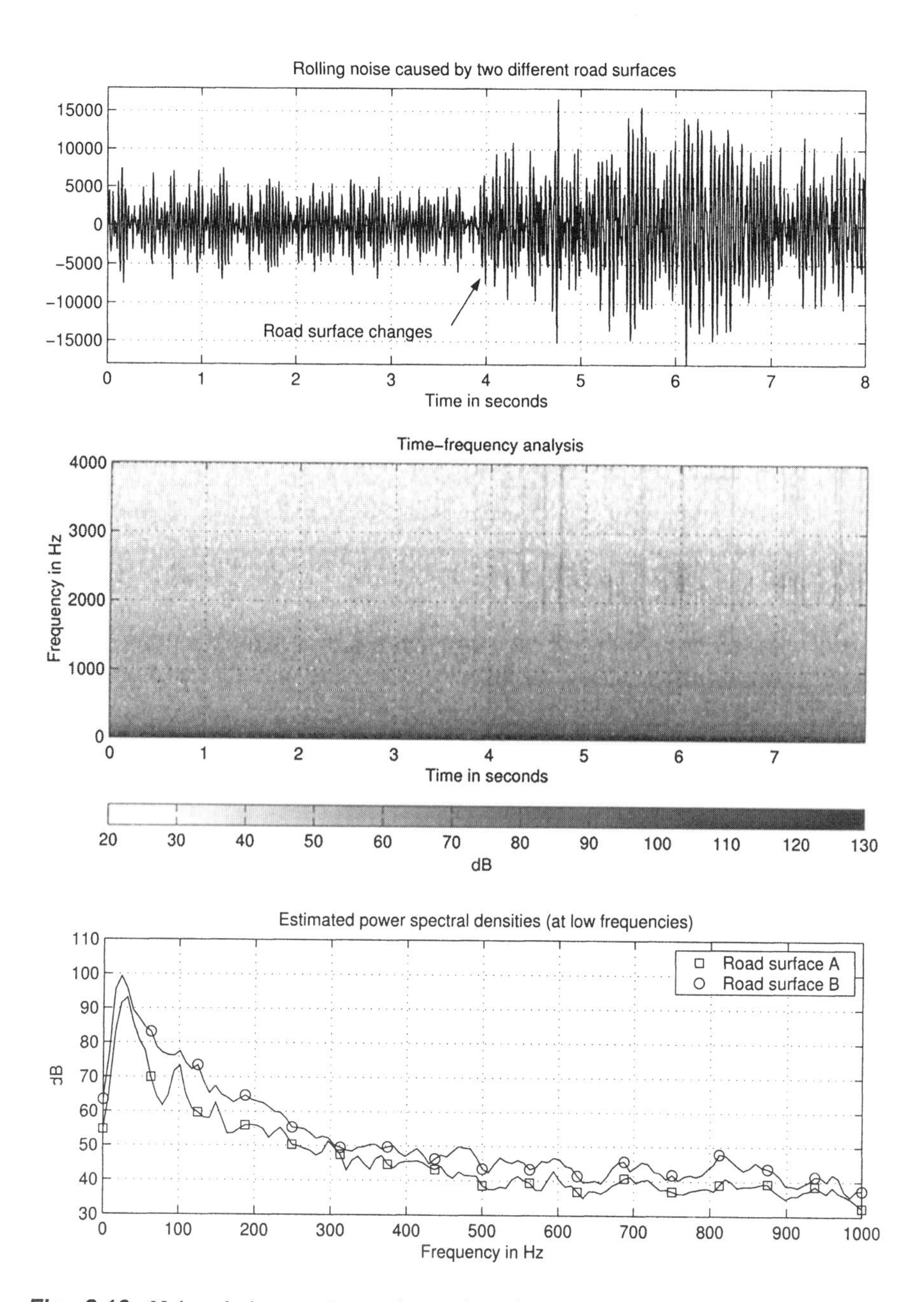

Fig. 3.10 Noise during a change in road surface: time function (upper diagram), time–frequency analysis (center), and power spectral densities (lower diagram).

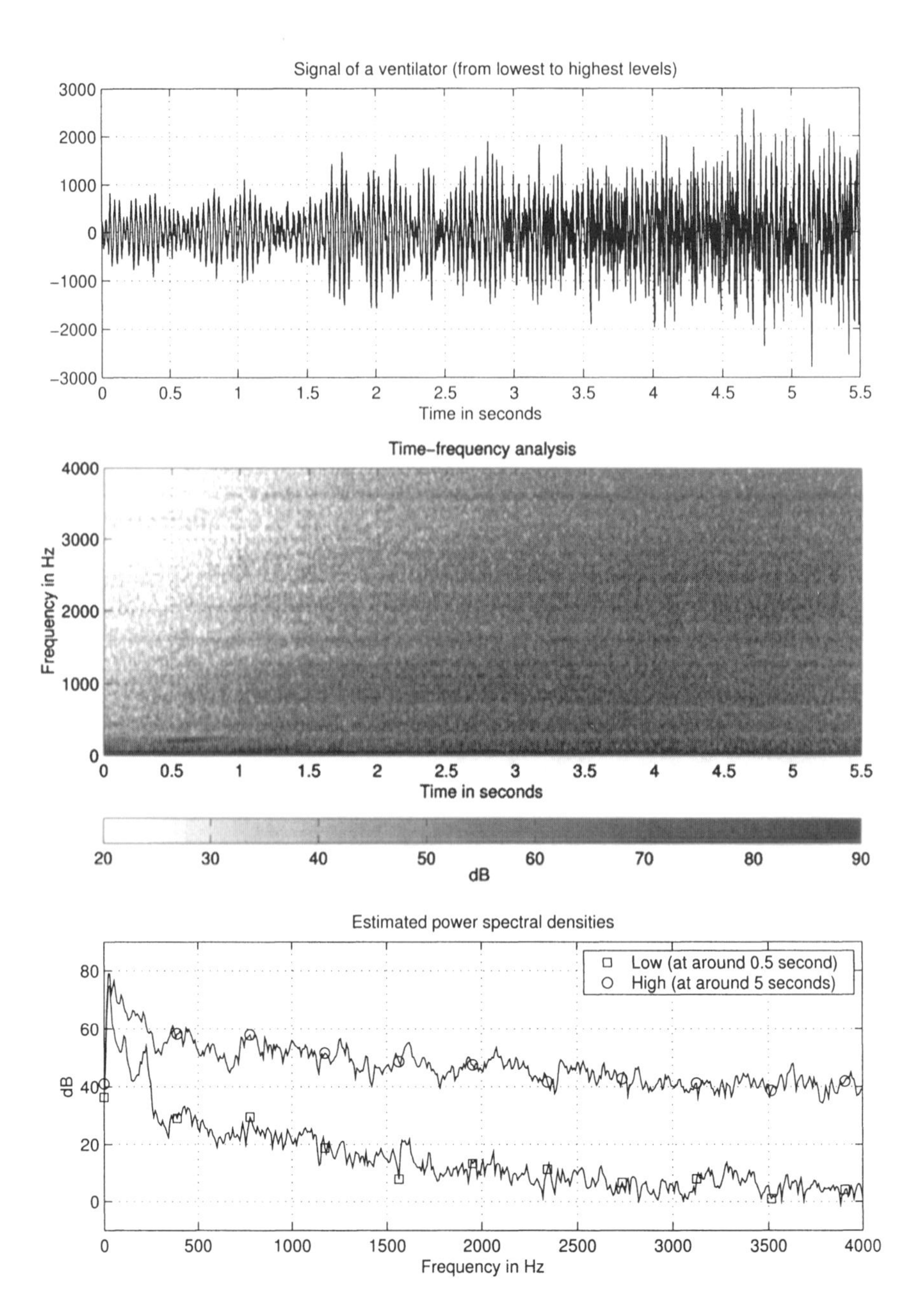

Fig. 3.11 Noise caused by a ventilator switched in steps from "low" to "high": time signal (upper diagram), time–frequency analysis (center diagram), and power spectral densities (bottom diagram).

the probability density function of the absolute value $|B(e^{j\Omega})|$ of the noise spectrum is approximated by a *Rayleigh density*:

$$f_{|B(e^{j\Omega})|}(b) = \begin{cases} \dfrac{2\,b}{\sigma^2_{|B(e^{j\Omega})|^2}}\; e^{-\dfrac{b^2}{\sigma^2_{|B(e^{j\Omega})|^2}}}, & \text{for } b \geq 0\,, \\ 0\,, & \text{else}\,. \end{cases} \tag{3.24}$$

The power spectral density $S_{bb}(\Omega)$—estimated by $S_{bb}(\Omega) = |B(e^{j\Omega})|^2$—then obeys a χ^2 *density* with 2 degrees of freedom:

$$f_{S_{bb}(\Omega)}(S) = \begin{cases} \dfrac{1}{\sigma^2_{|B(e^{j\Omega})|^2}}\; e^{-\dfrac{S}{\sigma^2_{|B(e^{j\Omega})|^2}}}, & \text{for } S \geq 0\,, \\ 0\,, & \text{else}\,. \end{cases} \tag{3.25}$$

With respect to the wide range of amplitudes of the power spectral density a logarithmic transformation may be useful:

$$\Theta_{bb}(\Omega) = 10\,\log_{10} \frac{S_{bb}(\Omega)}{\sigma^2_{|B(e^{j\Omega})|^2}}\,. \tag{3.26}$$

Then, the probability density function transforms to

$$f_{\Theta_{nn}(l)}(\Theta) = \frac{\ln 10}{10}\, 10^{\Theta/10}\; e^{-10^{\frac{\Theta}{10}}}\,. \tag{3.27}$$

In Fig. 3.12 the probability density function (see Eq. 3.27) of the logarithmic normalized power spectral density of car noise and histograms calculated from measured noise signals for selected frequencies are shown.

As a summary of the discussion of noise in a car one can state that besides stationary components it also contains highly nonstationary elements. Consequently, a noise control method can be efficient only if it can cope with both types of noise signals.

3.2 ACOUSTIC ECHOES

Loudspeaker(s) and microphone(s) in the same enclosure are connected by an acoustical path. In the following section we give a few facts from room acoustics about this path and we discuss how to model it by an electronic filter.

3.2.1 Origin of Acoustic Echoes

In a loudspeaker–enclosure–microphone (LEM) system the loudspeaker and the microphone are connected by an acoustical path formed by a *direct connection* (if both

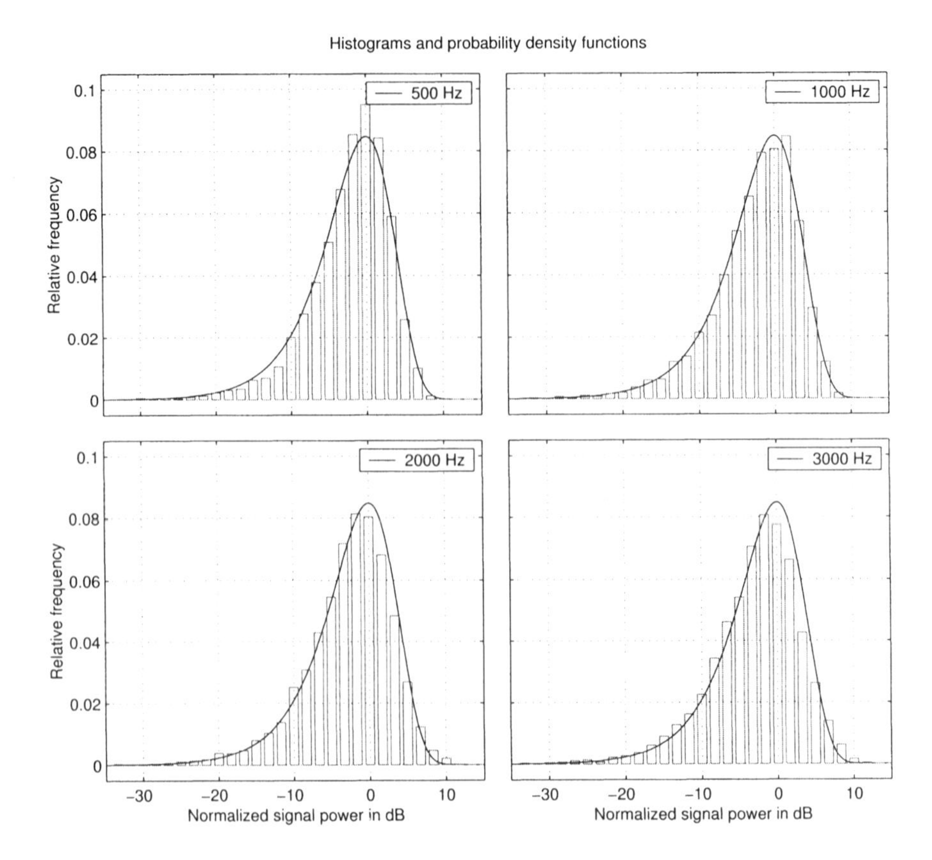

Fig. 3.12 Probability density functions of the logarithmic normalized power spectral density of car noise and histograms calculated from measured noise signals for selected frequencies.

can "see" each other) and in general a large number of *reflections* at the boundaries of the enclosure. For low sound pressure and no overload of the converters, this system may be modeled with sufficient accuracy as a linear system. The echo signal $d(n)$ can be described as the output of a convolution of the (causal) time-variant impulse response of the LEM system $h_i(n)$ and the excitation signal $x(n)$ (see Fig. 3.13):

$$d(n) = \sum_{i=0}^{\infty} h_i(n)\, x(n-i)\,, \tag{3.28}$$

where we assume for the moment that the impulse response is real.[3] The reason for having a time and a coefficient index for the LEM impulse response will become clearer at the end of this section. Signals from local speakers $s(n)$ and background

[3]For complex impulse responses, see Eq. 5.2 and the remarks in surrounding text there.

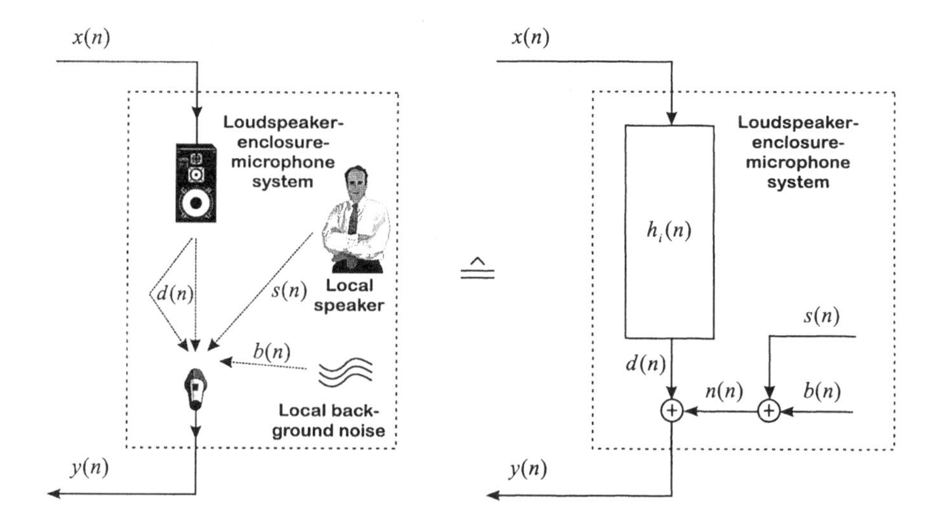

Fig. 3.13 Model of the loudspeaker–enclosure–microphone (LEM) system (see also Fig. 2.1).

noise $b(n)$ are combined to a local signal

$$n(n) = s(n) + b(n). \tag{3.29}$$

Adding the echo signal $d(n)$ and the local signals results in the microphone signal, given by

$$y(n) = d(n) + s(n) + b(n) = d(n) + n(n). \tag{3.30}$$

The time-varying impulse response $h_i(n)$ of an LEM system can be described by a sequence of delta impulses delayed proportionally to the geometrical length of the related path and the inverse of the sound velocity. The amplitudes of the impulses depend on the reflection coefficients of the boundaries and on the inverse of the path lengths. As a first order approximation one can assume that the impulse response decays exponentially. A measure for the degree of this decay is the *reverberation time* T_{60}. It specifies the time necessary for the sound energy to drop by 60 dB after the sound source has been switched off [136]. Depending on the application, it may be possible to design the boundaries of the enclosure such that the reverberation time is small, resulting in a short impulse response. Examples are telecommunication studios. For ordinary offices the reverberation time T_{60} is typically in the order of a few hundred milliseconds. For the interior of a passenger car this quantity is a few tens of milliseconds long.

To give an example, Fig. 3.14 shows the floor plan of the office used for most of the measurements and experiments described in this book. Its size measures about 18 m^2 and its volume is about 50 m^3. The floor is covered by carpet. There are

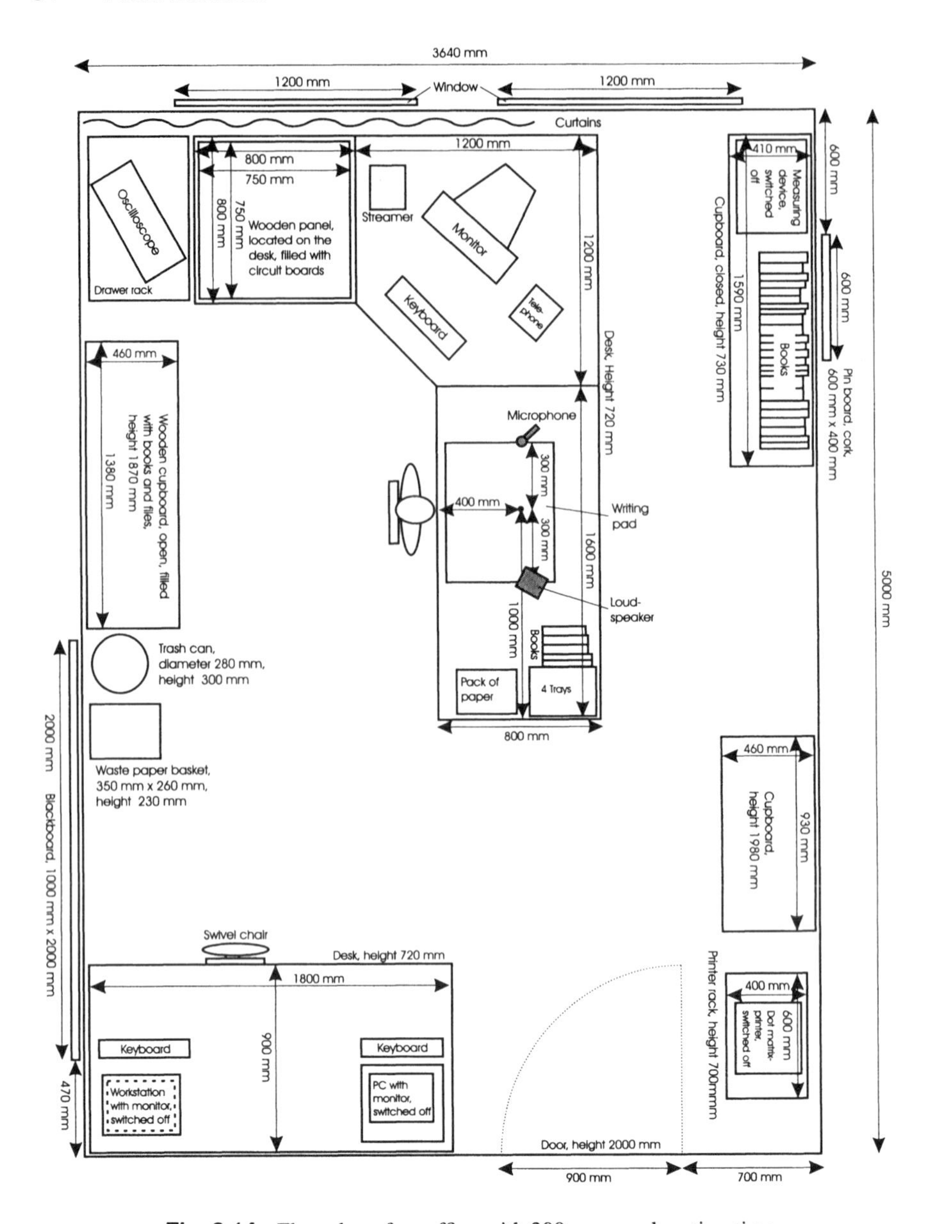

Fig. 3.14 Floorplan of an office with 300 ms reverberation time.

curtains in front of the windows. The ceiling consists of plaster board. All these materials exhibit attenuations that increase with frequency. The walls of the office are acoustically hard. The reverberation time of this office is approximately 300 ms. The microphone and the loudspeaker on the main desk are placed according to the ITU-T recommendations [120].

The impulse responses of LEM systems are highly sensitive to any changes such as the movement of a person within it. This is explained by the fact that, assuming a sound velocity of 343 m/s and 8 kHz sampling frequency, the distance traveled between two sampling instants is 4.3-cm. Therefore, a 4.3-cm change in the length of an echo path, the move of a person by only a few centimeters, shifts the related impulse by one sampling interval. Thus, the impulse response of an LEM system is timevariant. For this reason, we have chosen the notation $h_i(n)$ with filter coefficient index i and time index n for the impulse response in Eq. 3.28.

3.2.2 Electronic Replica of LEM Systems

From a control engineering point of view, acoustic echo cancellation constitutes a system identification problem. However, the system to be identified—the LEM system—is highly complex: Its impulse response exhibits up to several thousand sample values noticeably different from zero and it is time varying at a speed mainly according to human movements. The question of the optimal structure of the model of an LEM system and therefore also the question of the structure of the echo cancellation filter has been discussed intensively. Since a long impulse response has to be modeled by the echo cancellation filter, a recursive (IIR) filter seems best suited at first glance. At second glance, however, the impulse response exhibits a highly detailed and irregular shape. An office impulse response and the absolute value of its frequency response are shown in Fig. 3.15. To achieve a sufficiently good match, the model must offer a large number of adjustable parameters. Therefore, an IIR filter does not show an advantage over a nonrecursive (FIR) filter [142, 164]. The even more important argument in favor of an FIR filter is its guaranteed stability during adaptation.

In the following we assume the echo cancellation filter to have an FIR structure of order $N-1$. In this case the output of the echo cancellation filter $\widehat{d}(n)$, which is an estimate of the echo signal $d(n)$, can be described as a vector product of the real impulse response of the adaptive filter and the excitation vector:

$$\widehat{d}(n) = \sum_{i=0}^{N-1} \hat{h}_i(n)\, x(n-i) = \hat{\boldsymbol{h}}^{\mathrm{T}}(n)\, \boldsymbol{x}(n). \tag{3.31}$$

The vector $\boldsymbol{x}(n)$ consists of the last N samples of the excitation signal

$$\boldsymbol{x}(n) = \left[x(n),\, x(n-1),\, \ldots,\, x(n-N+1)\right]^{\mathrm{T}}, \tag{3.32}$$

and the filter coefficients $\hat{h}_i(n)$ have been combined to a column vector

$$\hat{\boldsymbol{h}}(n) = \left[\hat{h}_0(n),\, \hat{h}_1(n),\, \ldots,\, \hat{h}_{N-1}(n)\right]^{\mathrm{T}}. \tag{3.33}$$

A measure to express the effect of an echo cancellation filter is the *echo-return loss enhancement* (ERLE):

$$ERLE(n) = \frac{\mathrm{E}\left\{d^2(n)\right\}}{\mathrm{E}\left\{\left(d(n) - \widehat{d}(n)\right)^2\right\}}, \tag{3.34}$$

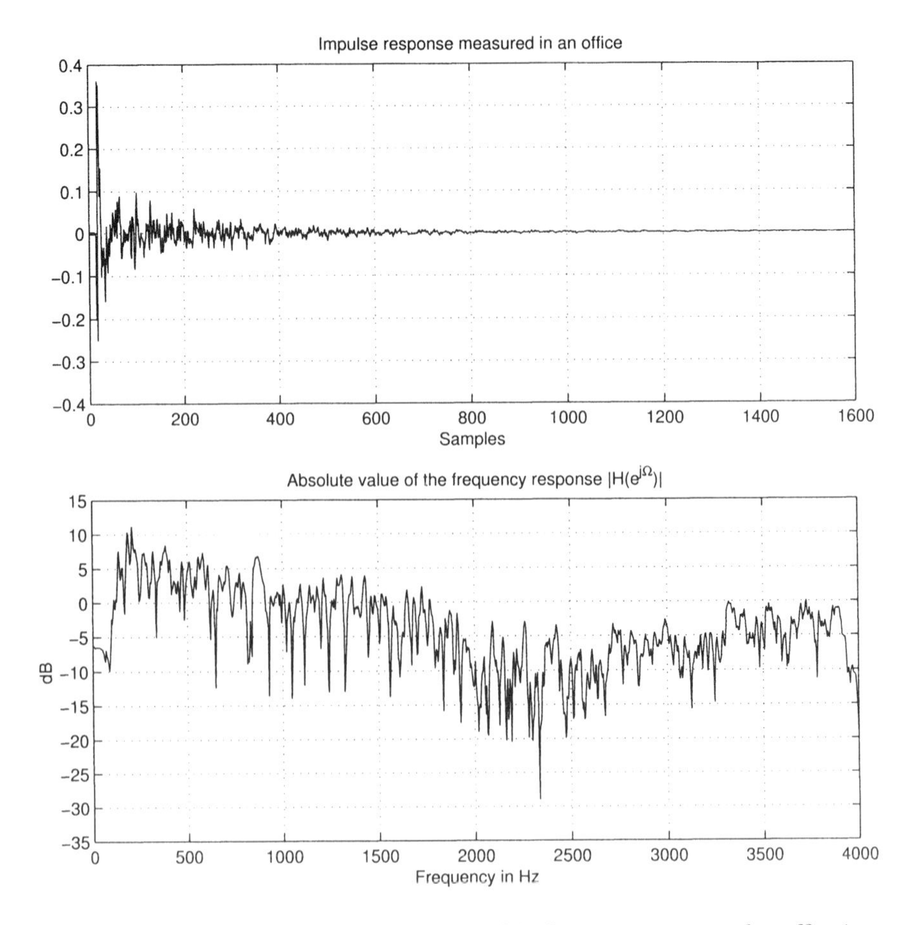

Fig. 3.15 Impulse response and absolute value of the frequency response of an office (sampling frequency $f_s = 8000$ Hz).

where the echo $d(n)$ is equal to the microphone output signal $y(n)$ in case the loudspeaker is the only signal source within the LEM system; thus the local speech signal $s(n)$ and the local noise $b(n)$ are zero. Assuming, for simplicity, a stationary white input signal $x(n)$, the ERLE can be expressed as

$$ERLE(n) = \frac{\mathrm{E}\left\{x^2(n)\right\} \sum\limits_{i=0}^{\infty} h_i^2(n)}{\mathrm{E}\left\{x^2(n)\right\} \left(\sum\limits_{i=0}^{\infty} h_i^2(n) - 2 \sum\limits_{i=0}^{N-1} h_i(n)\, \hat{h}_i(n) + \sum\limits_{i=0}^{N-1} \hat{h}_i^2(n) \right)}. \tag{3.35}$$

An upper bound for the efficiency of an echo cancellation filter of length N can be calculated by assuming a perfect match of the first N coefficients of the adaptive filter with the LEM system,

$$\hat{h}_i(n) = h_i(n) \text{ for } 0 \leq i < N. \tag{3.36}$$

In this case Eq. 3.35 reduces to

$$ERLE_{\max}(n, N) = \frac{\sum\limits_{i=0}^{\infty} h_i^2(n)}{\sum\limits_{i=N}^{\infty} h_i^2(n)}. \tag{3.37}$$

Figure 3.16 shows the impulse responses of LEM systems measured in a car (top), in an office (second diagram), and in a lecture room (third diagram). The microphone signals have been sampled at 8 kHz. It becomes obvious that the impulse responses of an office and of the lecture room exhibit amplitudes noticeably different from zero even after 1000 samples, that is to say after 125 ms. In comparison, the impulse response of the interior of a car decays faster due to the smaller volume of this enclosure.

The bottom diagram in Fig. 3.16 shows the upper bounds of the ERLE achievable with transversal echo cancellation filters of length N. An attenuation of only 30 dB needs filter lengths of about 1900 for the lecture room, 800 for the office and about 250 for the car.

3.3 STANDARDS

The operation of a worldwide network is only feasible on the basis of commonly accepted standards. For the telephone network regulations are formulated by the *International Telecommunication Union* (ITU) as well as by the *European Telecommunication Standards Institute* (ETSI). In addition, industry has set up rules and requirements for specific applications.

3.3.1 Standards by ITU and ETSI

ITU and ETSI set up requirements for for the maximally tolerable front-end delay and the minimal acceptable echo suppression of echo and noise control systems connected to the public telephone network.

3.3.1.1 Delay and Attenuation

For ordinary telephones the additional delay introduced by echo and noise control must not exceed 2 ms [120]. For mobile telephons up to 39 ms additional delay are allowed [64]. It is obvious that these are severe restrictions for the types of algorithms usable for any front-end processing. Especially in case of ordinary telephones, computationally efficient frequency-domain procedures cannot be applied.

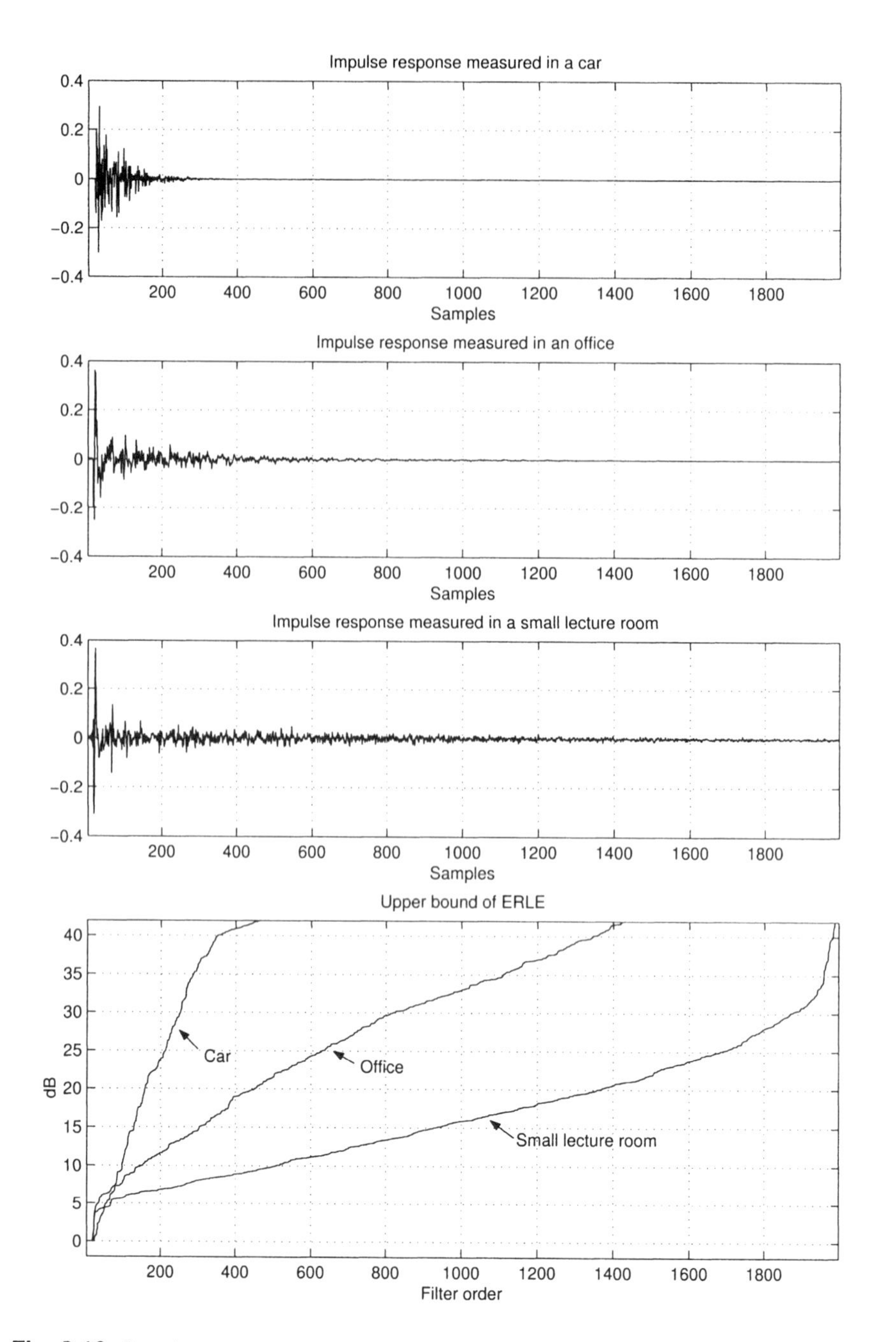

Fig. 3.16 Impulse responses measured in a car, in an office, and in a small lecture room (sampling frequency = 8 kHz). The bottom diagram shows the maximal achievable echo attenuation in dependence of the filter order of an adaptive filter placed in parallel to the LEM system (with permission from [100]).

As the level of the echo is concerned standards require an attenuation of at least 45 dB in case of single talk. During double talk (or during strong background noise) this attenuation can be lowered to 30 dB. In these situations the residual echo is masked at least partially by the double talk and/or the noise. Figure 3.17 shows the result of a study [208] that explains the dependence between echo delay and the acceptable level of the echo attenuation. The three curves mark the attenuation at which 50%, 70%, or 90% of a test audience were not annoyed by residual echoes.

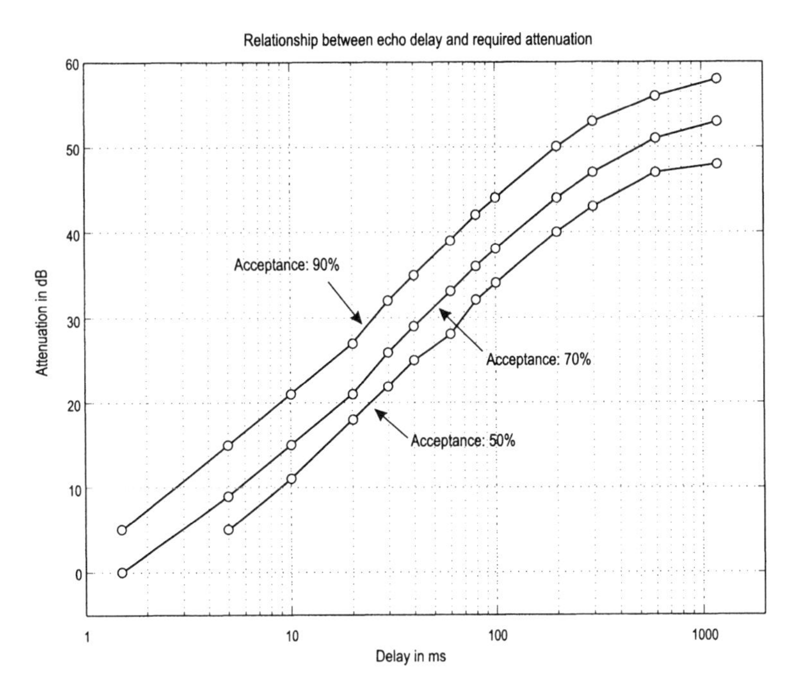

Fig. 3.17 Relationship between echo delay and required attenuation. (Based on a study of AT&T, published in [208].)

3.3.1.2 Test Signals

For measurement purposes in telecommunication systems an artificial voice signal is recommended by ITU [122]. It is mathematically defined such, that it models human speech. The long- and short-term spectra, the probability density function of speech signals, the voiced/unvoiced structure, and the syllabic envelope for male and female speech are emulated.

For communication devices that contain speech activated circuits a "composite source signal" has been defined [79, 123]. It consists of three sections: a 50-ms-long voiced signal taken from artificial voice [122] intended to activate speech detectors in the system, a pseudonoise signal of about 200 ms duration during which measurements can be taken, and a pause that is long enough to set the system back into its quiescent state (see Fig. 3.18). The composite source signal can be repeated several times with alternating polarities.

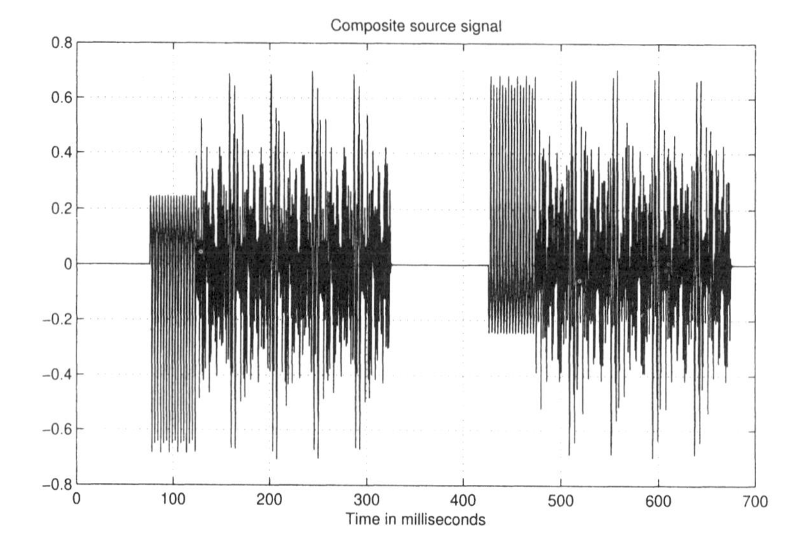

Fig. 3.18 Composite source signal: original and repetition with opposite polarity.

Part II

Algorithms

4

Error Criteria and Cost Functions

In this chapter we make some general comments on error criteria and cost functions. We emphasize the *mean square error* as the most popular error function for the design of adaptive systems, the *minimax* criterion for the design of prototype filters for filterbanks, and the *cepstral distance* used very often in speech processing systems.

Choosing the error and a cost function that perform a weighting of the error represents the first step toward the design of an algorithm to solve an optimization problem. The algorithm that finally implements the solution depends crucially on these choices.

A number of decisions have to be made during the design of an error criterion, such as

- Which information on the problem to be solved is available, and how should it be used?
- How should the cost function weight the error?
- Should the error criterion be deterministic or stochastic?

The error criterion finally used for a specific application always requires a compromise between the closeness to reality and the (foreseeable) signal processing complexity of the solution to be implemented.

4.1 ERROR CRITERIA FOR ADAPTIVE FILTERS

Of pivotal importance is the information that is available and thus may enter the criterion. There may be, for instance, some prior knowledge about the probability density functions of the signals involved or at least of their moments of certain orders. In most cases—and specifically in the application considered here—these quantities, however, are timevariant and, consequently, have to be estimated continuously or at least periodically.

We will split the entire cost function into an outer and an inner one. In general, the outer cost function $F(\cdots)$ depends on the inner cost function $f(\cdots)$, the true (or desired) value $d(n)$ of a quantity, and on its estimate $\widehat{d}(n)$. Minimizing the cost function leads to an optimal set of (e.g., filter) coefficients $\widehat{h}_{i,\text{opt}}$:

$$F\Big(f(\ldots),\, d(n),\, \widehat{d}(n)\Big)\Big|_{\widehat{\boldsymbol{h}}\,=\,\widehat{\boldsymbol{h}}_{\text{opt}}} \longrightarrow \min\,. \tag{4.1}$$

As an often used simplification only the difference of the desired and the estimated value—the error $e(n)$—enters the cost function:

$$F\Big(f(\ldots),\, e(n)\Big) = F\Big(f(\ldots),\, \big(d(n) - \widehat{d}(n)\big)\Big)\,. \tag{4.2}$$

Beside the recent error signal $e(n)$ also past instants can enter the cost function:

$$F\Big(f(\ldots),\, e(n),\, e(n-1),\, \ldots\Big)\Big|_{\widehat{\boldsymbol{h}}\,=\,\widehat{\boldsymbol{h}}_{\text{opt}}} \longrightarrow \min\,. \tag{4.3}$$

In addition, the error depends on time. Consequently, the cost function has not only to depend on amplitudes but also on the time of their occurrences.

For the design and optimization of linear systems mostly the square of the error modulus is used for the inner cost function:

$$f_{\text{s}}\big(e(n)\big) = |e(n)|^2\,. \tag{4.4}$$

Compared to a linear dependency, the square amplifies large errors and attenuates small ones. Large errors may occur in unusual situations like double talk in acoustic echo control applications. With an adaptation control proportional to the squared error, the coefficients of an adaptive filter would undergo large displacements in critical situations. A more robust adaptation control is achieved by gradually reducing the slope of the error function or bounding it to a maximal value [14].

Besides the square of the error magnitude, the magnitude itself is also applied as an inner cost function in some adaptive algorithms (e.g.,, the sign error LMS algorithm [50]):

$$f_{\text{m}}\big(e(n)\big) = |e(n)|\,. \tag{4.5}$$

A common technique to handle the time dependency is to use a windowed sum of the inner cost functions applied to the error signals. A *rectangular time window* means that all errors within this window are considered equally important:

$$F_{\text{rec}}\Big(f(\ldots),\, e(n+W_1),\, \ldots,\, e(n+W_0)\Big) = \sum_{i=W_0}^{W_1} f\big(e(n+i)\big)\,. \tag{4.6}$$

The boundaries W_0 and W_1 of the window may be positive, zero, or negative. For $W_0 < 0$ and $W_1 \leq 0$, only samples of the past enter the cost function. If $W_0 = -n_0$ and $W_1 = +n_0$, the window is symmetric and also future values enter the criterion. In this case a delay of at least n_0 samples results for reasons of causality. A window showing an amplitude that decays toward the past emphasizes present and nearby past data. Especially popular is the *exponentially decaying window*:

$$F_{\text{exp}}\Big(f(\ldots),\, e(n),\, \ldots,\, e(-\infty)\Big) \;=\; \sum_{i=0}^{\infty} \lambda^i f\big(e(n-i)\big) \qquad (4.7)$$

$$\text{with } 0 \ll \lambda < 1\,.$$

If the error criterion is a sum of the exponentially weighted inner cost functions, moving the window one step ahead requires just one multiplication by λ:

$$\begin{aligned} F_{\text{exp}}\Big(f(\ldots),\, e(n+1),\, \ldots,\, e(-\infty)\Big) & \\ = \; & \sum_{i=0}^{\infty} \lambda^i f\big(e(n+1-i)\big) \\ = \; & f\big(e(n+1)\big) + \lambda \sum_{i=0}^{\infty} \lambda^i f\big(e(n-i)\big) \\ = \; & f\big(e(n+1)\big) + \lambda\, F_{\text{exp}}\Big(f(\ldots),\, e(n),\, \ldots,\, e(-\infty)\Big)\,. \end{aligned} \qquad (4.8)$$

No storage of old data is necessary in this case. The decay toward the past is controlled by the factor λ called the *forgetting factor*. Choosing a forgetting factor very close to one results in a long memory. For an adaptive system, this may be desirable if short lacks of proper input data have to be overcome and/or smooth results are desired. On the other hand, an excessively long memory may lead the system to freeze and keep it from tracking environmental changes.

The application of a *stochastic* or a *deterministic* data model is often overemphasized. In the first case the expectation operator is used as the outer cost function:

$$F_{\text{mean}}\Big(f(\ldots),\, e(n)\Big) = \mathrm{E}\left\{\, f\big(e(n)\big) \,\right\}\,. \qquad (4.9)$$

The stochastic error criterion used most of the time is the *mean square error*:

$$F_{\text{mean}}\Big(f_{\text{s}}(\ldots),\, e(n)\Big) = \mathrm{E}\left\{\, |e(n)|^2 \,\right\}\,. \qquad (4.10)$$

Its application requires a priori knowledge about the first and second order ensemble averages (i.e., the first- and second-order moments) of the data involved. In practical applications these quantities have to be estimated from a finite-length record of one sample function of each signal. In those cases, therefore, the statistical model turns into a deterministic one where the expectation is replaced by a time-average running, for instance, from $W_0 = -I_0$ to $W_1 = 0$:

$$\begin{aligned} F_{\text{ls}}\Big(f_{\text{s}}(\ldots),\, e(n),\, \ldots\Big) &= \lim_{I_0\to\infty} \frac{1}{I_0+1} \sum_{i=0}^{I_0} f_{\text{s}}\big(e(n-i)\big) \\ &= \lim_{I_0\to\infty} \frac{1}{I_0+1} \sum_{i=0}^{I_0} |e(n-i)|^2 . \end{aligned} \tag{4.11}$$

In addition, a window may be introduced. For practical cases the sum is restricted to a finite range and proper scaling replaces the division by the number of terms within the summation.

4.2 ERROR CRITERIA FOR FILTER DESIGN

In contrast to cost functions for adaptive filters, error criteria for the design of fixed filters (lowpass, highpass, etc.) are usually specified in the frequency domain. First, a desired (ideal) frequency response $H_{\text{id}}(e^{j\Omega})$ is specified. For an ideal half-band filter, for example, the desired frequency response, would be[1]

$$H_{\text{id, hb}}\big(e^{j\Omega}\big) = \begin{cases} 1\,, & \text{for } |\Omega| < \frac{\pi}{2}\,, \\ 0\,, & \text{else}\,. \end{cases} \tag{4.12}$$

In most cases the frequency response of the desired filter cannot be achieved with a finite amount of filter coefficients. For this reason an error spectrum, which is the difference between the desired frequency response and a frequency response of a filter with a predetermined structure (e.g., FIR of order N)

$$E\big(e^{j\Omega}\big) = H_{\text{id}}\big(e^{j\Omega}\big) - H\big(e^{j\Omega}\big)\,, \tag{4.13}$$

is defined. The coefficients of the filter $H(e^{j\Omega})$ are calculated by minimizing the error spectrum in some sense. As in the design of error criteria for adaptive filters the cost functions for the design of fixed filters can be divided into outer and inner parts. The weighted magnitude of the error spectrum

$$f_{\text{wm}}\Big(E\big(e^{j\Omega}\big)\Big) = W(\Omega)\, \big|E\big(e^{j\Omega}\big)\big| \tag{4.14}$$

and the weighted squared magnitude of the error spectrum

$$f_{\text{ws}}\Big(E\big(e^{j\Omega}\big)\Big) = W(\Omega)\, \big|E\big(e^{j\Omega}\big)\big|^2 \tag{4.15}$$

are often used as inner cost functions. The weighting function $W(\Omega)$ is usually specified by constants in the passband and stopband of the filter and by zero in the

[1]The frequency response is specified here only for the interval $-\pi < \Omega \leq \pi$. For all other frequencies the frequency response is continued with 2π periodicity.

transition band, respectively. For the example of a half-band filter the weighting function would be

$$W_{\text{hb}}(\Omega) = \begin{cases} a\,, & \text{for } |\Omega| < \Omega_0\,, \\ 0\,, & \text{for } \Omega_0 \le |\Omega| \le \Omega_1\,, \\ b\,, & \text{for } |\Omega| < \Omega_1\,. \end{cases} \tag{4.16}$$

The constants a and b should be nonnegative, and in most cases the error spectrum in the stopband is weighted much larger than in the passband ($b \gg a \ge 0$). The transition band of a half-band filter—specified by Ω_0 and Ω_1—is usually located symmetrically around $\frac{\pi}{2}$.

A very popular outer cost function is the maximum of the inner cost function

$$F_{\text{max}}\Big(f(\ldots),\, E\left(e^{j\Omega}\right)\Big) = \max\Big\{f\Big(E\left(e^{j\Omega}\right)\Big)\Big\}\,. \tag{4.17}$$

Because a (weighted) maximum is minimized when this cost function is applied to $f_{\text{wm}}(\cdots)$

$$\max\Big\{W\left(e^{j\Omega}\right)\,\big|E\left(e^{j\Omega}\right)\big|\Big\}\bigg|_{H(e^{j\Omega})\,=\,H_{\text{opt}}(e^{j\Omega})} \longrightarrow \min\,, \tag{4.18}$$

the optimization criterion is called the *minimax* criterion. For maximally narrow FIR lowpass filters this optimization can be solved analytically [53]. The resulting filters, called *Dolph–Chebyshev* filters, will be discussed in more detail in Appendix B, where they have been used for the design of prototype lowpass filters for filterbanks. For other filter types a powerful iterative procedure, called the *Remez exchange algorithm* [203], exists. In order to design a half-band filter in FIR structure this algorithm has been used as an example. The result for an FIR filter with 66 coefficients is depicted in Fig. 4.1.

The transfer functions of filters designed with the minimax criterion will alternate from maximum to minimum in the passband and stopband. The variations from the desired frequency response at these maxima and minima will be equal. For this reason they are also called *equiripple* filters.

Besides applying the maximum for the outer cost function a variety of other functions exist. If, for example, the inner cost function $f_{\text{ws}}(\ldots)$ is integrated over an interval of 2π

$$\begin{aligned} F_{\text{int}}\Big(f_{\text{ws}}(\ldots),\, E\left(e^{j\Omega}\right)\Big) &= \int\limits_{\Omega=0}^{2\pi} f_{\text{ws}}\big(E\left(e^{j\Omega}\right)\big)\, d\Omega \\ &= \int\limits_{\Omega=0}^{2\pi} W(\Omega)\,\big|E\big(e^{j\Omega}\big)\big|^2\, d\Omega \end{aligned} \tag{4.19}$$

and an FIR filter structure is desired the resulting filters are called *eigenfilters* or *prolate sequences* [230] [depending on the utilized weighting function $W(\Omega)$].

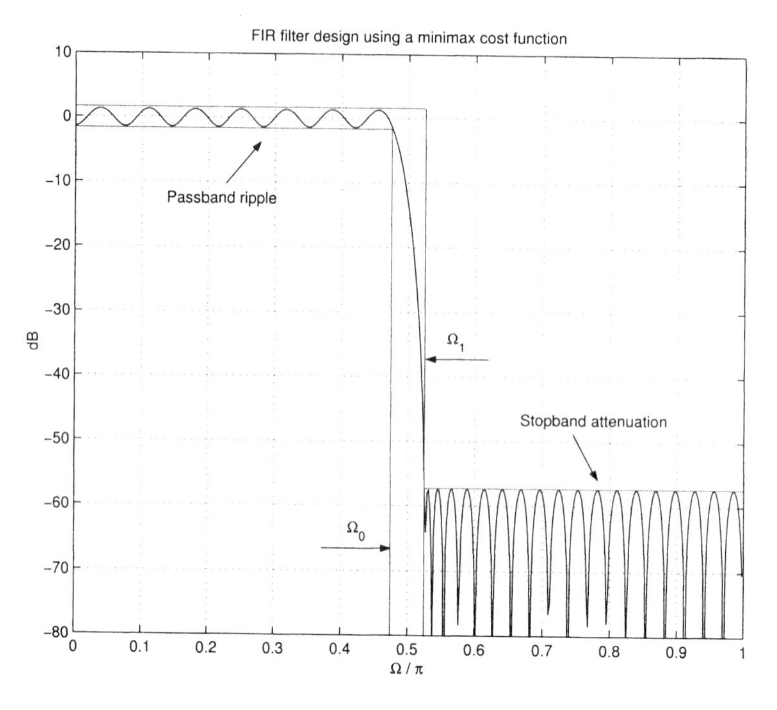

Fig. 4.1 Half-band filter in FIR structure, designed with the Remez exchange algorithm.

Mentioning all typical cost functions used for the design of nonadaptive filters would go far beyond the scope of this book. We referred in this chapter only to those that are important for the design of certain types of polyphase filterbanks. These filterbanks are often used in acoustic echo and noise control. For a more detailed description of filter design the interested reader is referred to the appropriate literature, such as the book by Schlichthärle [203].

4.3 ERROR CRITERIA FOR SPEECH PROCESSING AND CONTROL PURPOSES

If speech signals are involved, the *cepstral distance*,[2] turnes out to be a useful criterion for the decision, regardless of whether two signals are close to each other [87, 193]. In acoustic echo control this problem arises, for example, in cases where double talk has to be detected or denied.

To determine the complex cepstrum of a sequence needs three steps: (1) to take the Fourier transform, (2) to take the logarithm of this transform, and (3) to calculate the inverse Fourier transform of the logarithm. If the sequence contains physical

[2]The term "cepstrum" was formed by reversing the first four letters in the word "spectrum."

values it has to be normalized first since the logarithm exists only for numbers and not for physical quantities.

To give an example we assume $r_{ss}(n)$ to be the autocorrelation sequence of a stationary discrete-time speech process. Then the power spectral density is defined by the Fourier transform

$$S_{ss}(\Omega) = \sum_{n=-\infty}^{+\infty} r_{ss}(n)\, e^{-j\Omega n}\,, \tag{4.20}$$

where Ω is a normalized frequency. $S_{ss}(\Omega)$ is real, nonnegative, and periodic with $2\,\pi$. If the speech process is (assumed to be) *real*, then $r_{ss}(n)$ is real, too, and $r_{ss}(n) = r_{ss}(-n)$. Consequently, also $S_{ss}(\Omega) = S_{ss}(-\Omega)$ in this special case.

Because of the periodicity of $S_{ss}(\Omega)$, the cepstrum can be expressed by a Fourier series

$$\log \frac{S_{ss}(\Omega)}{S_0} = \sum_{n=-\infty}^{+\infty} c_n\, e^{-j\Omega n}\,, \tag{4.21}$$

where S_0 is a constant with the same dimension as $S_{ss}(\Omega)$. The cepstral coefficients c_n can be calculated by

$$c_n = \frac{1}{2\,\pi} \int_{-\pi}^{+\pi} \log \frac{S_{ss}(\Omega)}{S_0}\, e^{j\Omega n}\, d\Omega\,. \tag{4.22}$$

In case of real sequences $S_{ss}(\Omega) = S_{ss}(-\Omega)$ and consequently $c_n = c_{-n}$. If we assume that c_n and $\widehat{c}_n$ are the cepstral coefficients of two different speech processes, then

$$\sum_{n=-\infty}^{\infty} |c_n - \widehat{c}_n|^2 \tag{4.23}$$

represents the squared cepstral distance of the two processes. The coefficients c_0 and $\widehat{c}_0$ are related to the mean powers of these processes. The properties of the spectral envelope are contained in the low-order coefficients. In applications where speech signals are to be compared with each other, it is sufficient to consider this envelope. For real signals the cepstral distance is expressed by a sum running from 1 up to the order of $N \approx 20$ only. If the speech process can be modeled by an autoregressive process the cepstral coefficients can be calculated directly by the coefficients h_i of an all-pole (prediction error) filter (see Eq. 6.5).

Finally, a remark on *stationarity* seems appropriate. Like the random process itself, stationarity is a mathematical definition that in real applications always means an approximation. Using statistical models this assumption is necessary in most cases in order to arrive at manageable results. To come closer to reality and still being able to use the simplifications based on stationarity, the concept of *short-term stationarity* is introduced. It means that results gained by stationary models are valid for some (mostly short) periods of time and that, for instance, optimizations have to be repeated periodically using changed input properties.

For this reason the predictor coefficients $h_i(n)$ and the cepstral coefficients $c_i(n)$, respectively, are computed periodically every 5–20 ms. The autocorrelation functions that are required for computing the predictor and the cepstral parameters are estimated on a basis of a few hundred samples, temporally located around the current sample. Cepstral coefficients $c_i(n)$ and $\widehat{c}_i(n)$ are computed for both signals, which should be compared. The difference of each pair of coefficients

$$e_i(n) = c_i(n) - \widehat{c}_i(n) \tag{4.24}$$

usually enters a magnitude square cost function $f_s(\cdots)$. The outer cost function sums all squared magnitudes of the differences starting at index $i = 1$ up to $i = N - 1$ (with $N \approx 20$):

$$\begin{aligned} F_{\text{cd}}\Big(f_s(\ldots),\, e_1(n),\, \ldots,\, e_{N-1}(n)\Big) &= \sum_{i=1}^{N-1} |c_i(n) - \widehat{c}_i(n)|^2 \\ &= \sum_{i=1}^{N-1} |e_i(n)|^2 \,. \end{aligned} \tag{4.25}$$

The cepstral distance $F_{\text{cd}}(\cdots)$, as well as other typical cost functions used in speech processing [193], is very similar to the cost functions presented in Section 4.1 for adaptive filters, with one exception—the cost function is not applied directly to the signals, but to (nonlinear) transformed versions of them.

In order to show the advantages of applying cepstral transforms as a preprocessor for the inner cost function, a simulation example is presented in Fig. 4.4. For detecting double talk in hands-free telephone systems the short-term spectral envelope of the output of the echo cancellation filter $\widehat{d}(n)$ is compared to the short-term spectral envelope of the microphone signal $y(n)$ (see Fig. 4.3).

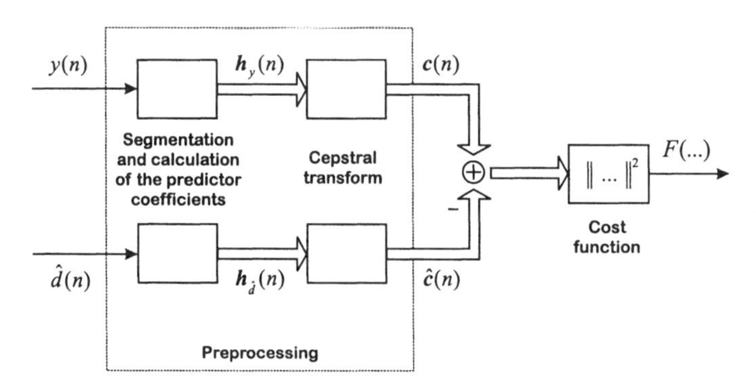

Fig. 4.2 Cepstral transformation as a preprocessor for cost functions applied in speech processing.

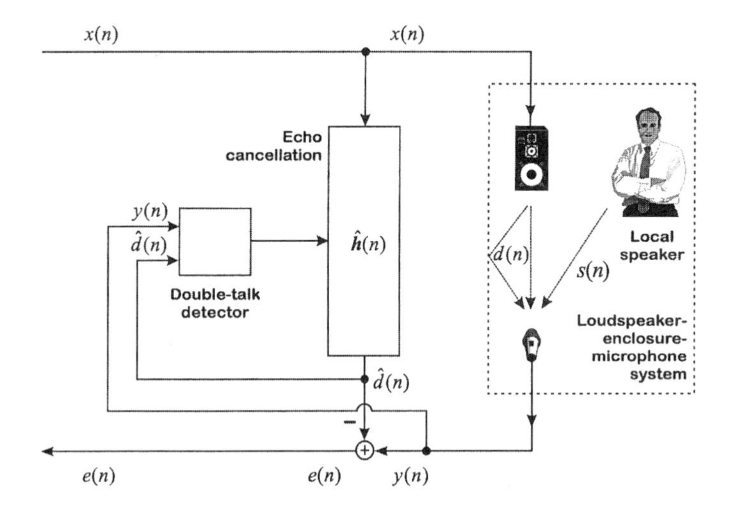

Fig. 4.3 Structure of a single-channel hands-free telephone system with a double-talk detector based on cepstral differences.

The spectral envelopes should be nearly equal if no local speech is present in the microphone signal and the echo cancellation filter $\widehat{\boldsymbol{h}}(n)$ has converged at least moderately. In the upper two diagrams of Fig. 4.4 the remote excitation signal $x(n)$ as well as the signal of a local speaker $s(n)$ are depicted. A double-talk situation starts after 1.5 seconds and lasts about 2 seconds. The resulting microphone signal—consisting of a reverberant version of the excitation signal as well as the local speech—is depicted in the third diagram.

When applying the cost function 4.25 directly to the differences of the predictor coefficients $\boldsymbol{h}_y(n) - \boldsymbol{h}_{\widehat{d}}(n)$, large values appear also during periods of single talk (see fourth diagram in Fig. 4.4). If the predictor coefficients are transformed into cepstral coefficients before using the cost function 4.25 the results are more pronounced. Large peaks occur only within the period of double talk (see lowest diagram).

A number of other criteria have been suggested to measure the closeness of two signals, especially in case of speech recognition [193]. In the following chapters the error criteria discussed in the last three sections will be applied. Further details will be given where required.

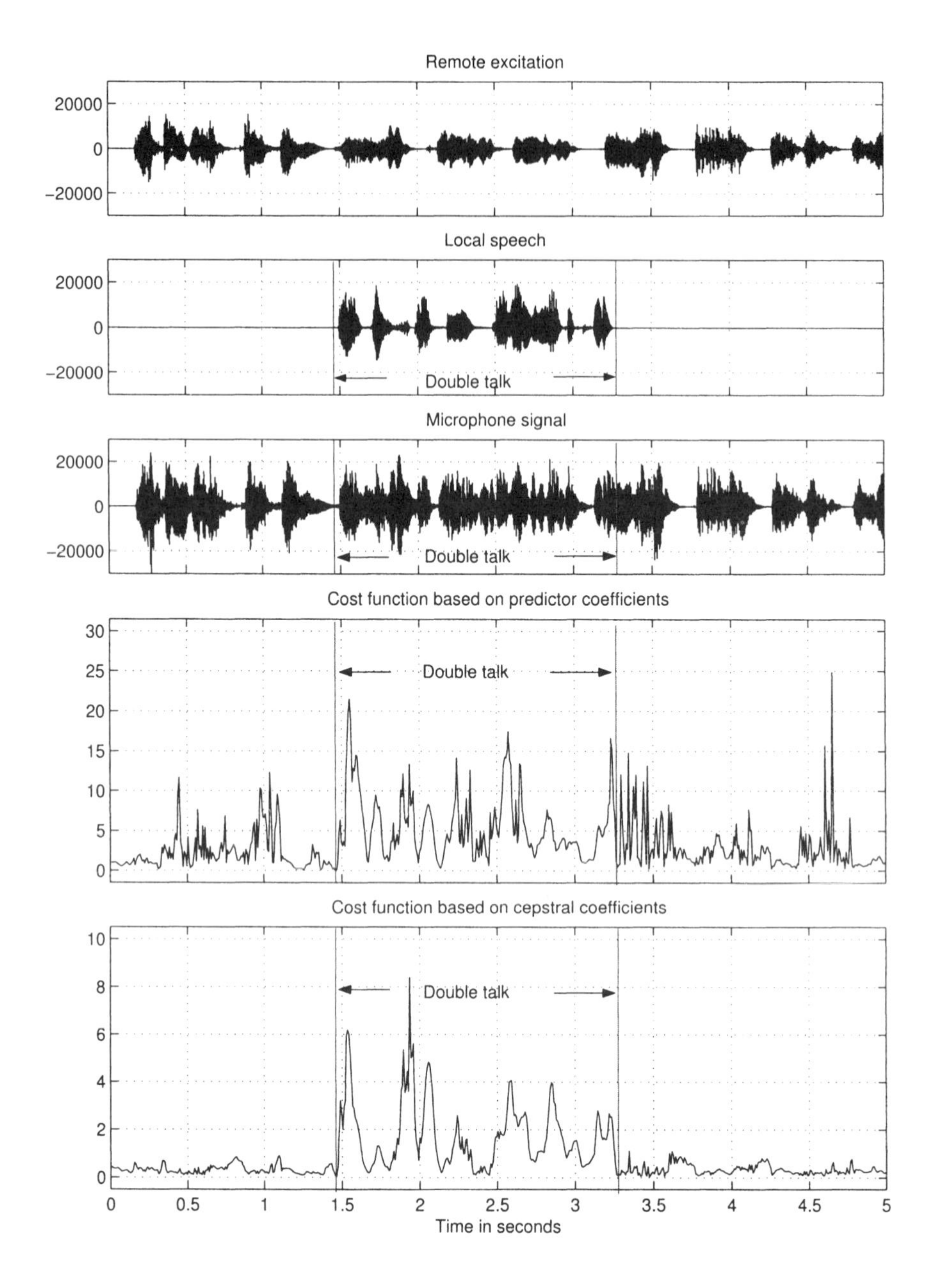

Fig. 4.4 Double-talk detection in hands-free telephone systems. The upper two diagrams show the remote excitation signal and the local speech signal, respectively. The resulting microphone signal is depicted in the third diagram. When applying the cost function 4.25 directly to the predictor coefficients, large values occur also during remote single talk (depicted in the fourth diagram). Better results are achieved when the cepstral transform is applied. The output of this cost function is depicted in the lowest part of the figure.

5

Wiener Filter

The Wiener filter as developed by *Norbert Wiener* in the early 1940s [241] may be considered as a first application of stochastic signal models for filter optimization. Characteristic to such an approach is that some a priori knowledge about the signals has to be available. The filter is a solution to a classical communication problem: An additively disturbed signal has to be separated from noise.

Wiener assumed that the filter input signals can be modeled by stationary stochastic processes and that their power spectral densities are known. He found a linear filter that optimally reconstructs the input signal or a linear transformation thereof from an input that is additively disturbed by noise. The original Wiener solution was derived for time-continuous quantities. The filter is described by its (infinitely) long impulse response or the Fourier transform of it. Solutions are available for a noncausal and a causal filter. In both cases, the observation intervals are infinitely long [178, 225].

According to the class of problems we consider in this book, we assume discrete-time signals and a linear discrete-time filter, as well. Further, we assume an FIR filter [111]. Both assumptions reduce the complexity of the derivation considerably.

5.1 TIME-DOMAIN SOLUTION

For the following we model all signals by stationary complex random processes. The input $y(n)$ to the (for the moment time-invariant) filter with impulse response

h_i, $i = 0, \ldots, N-1$,[1] is an additive mixture of some *useful* signal $s(n)$ and noise $n(n)$ (see Fig. 5.1):

$$y(n) = s(n) + n(n) \,. \tag{5.1}$$

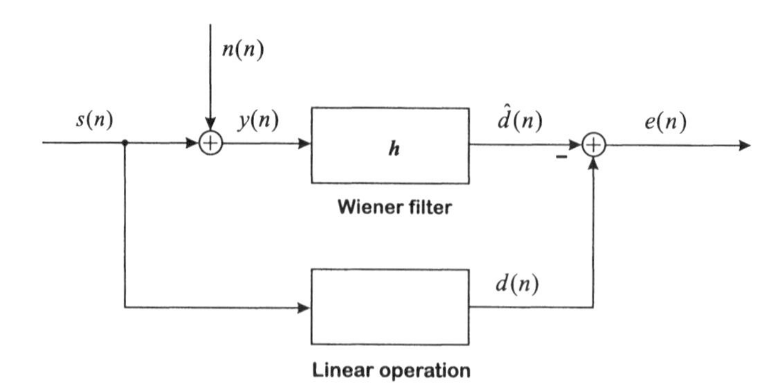

Fig. 5.1 Wiener filter.

The complex impulse response h_i of the FIR filter is of length N. The filter output signal is then given by the convolution of h_i and $y(n)$

$$\widehat{d}(n) = \sum_{i=0}^{N-1} h_i^* \, y(n-i) \,, \tag{5.2}$$

where the asterisk denotes complex conjugation.[2] Combining the sample values of the impulse response into a column vector $\boldsymbol{h}$

$$\boldsymbol{h} = [\, h_0, \, h_1, \, h_2, \, \ldots \, , \, h_{N-1} \,]^{\mathrm{T}} \,, \tag{5.3}$$

and also assembling the N most recent samples of $y(n)$ into a vector $\boldsymbol{y}(n)$

$$\boldsymbol{y}(n) = [\, y(n), \, y(n-1), \, y(n-2), \, \ldots \, , \, y(n-N+1) \,]^{\mathrm{T}} \,, \tag{5.4}$$

the filter output can be written as an inner product

$$\widehat{d}(n) = \boldsymbol{h}^{\mathrm{H}} \, \boldsymbol{y}(n) = \boldsymbol{y}^{\mathrm{T}}(n) \, \boldsymbol{h}^* \,, \tag{5.5}$$

where $\boldsymbol{h}^{\mathrm{H}}$ indicates the transpose of the conjugate complex of the vector $\boldsymbol{h}$.

[1] We write h_i and leave the argument in parentheses for the time dependency.
[2] The definition of the convolution as given by Eq. 5.2 is not unique in the literature. Some authors write

$$\widehat{d}(n) = \sum_{i=0}^{N-1} h_i \, y(n-i) \,,$$

meaning that the impulse response is conjugate complex to the one used in Eq. 5.2.

The desired output $d(n)$ of the filter can be derived from $s(n)$ by any linear operation. If $d(n) = s(n-l)$, $l \geq 0$, an (ordinary) filter is designed. In case of $d(n) = s(n+l)$, $l \geq 1$, the optimization results in a predictor with noisy input. As optimization criterion the expectation of the squared absolute value of the error $e(n)$

$$e(n) = d(n) - \widehat{d}(n) \tag{5.6}$$

is applied:

$$\mathrm{E}\left\{|e(n)|^2\right\} = \mathrm{E}\left\{e(n)\, e^*(n)\right\} . \tag{5.7}$$

This quantity is time-independent since we assume stationary processes. In order to optimize the complex filter coefficients

$$h_i = h_{\mathrm{r},\,i} + j\, h_{\mathrm{i},\,i} \,, \quad \text{for} \quad i = 0,\, \ldots,\, N-1 \,, \tag{5.8}$$

we take the partial derivatives with respect to the real and the imaginary parts of each coefficient:

$$\begin{aligned} &\frac{\partial\, \mathrm{E}\left\{|e(n)|^2\right\}}{\partial\, h_{\mathrm{r},\,i}} + j\, \frac{\partial\, \mathrm{E}\left\{|e(n)|^2\right\}}{\partial\, h_{\mathrm{i},\,i}} \\ &= \mathrm{E}\left\{ \frac{\partial\, e(n)}{\partial\, h_{\mathrm{r},\,i}}\, e^*(n) + \frac{\partial\, e^*(n)}{\partial\, h_{\mathrm{r},\,i}}\, e(n) + \frac{\partial\, e(n)}{\partial\, h_{\mathrm{i},\,i}}\, j\, e^*(n) + \frac{\partial\, e^*(n)}{\partial\, h_{\mathrm{i},\,i}}\, j\, e(n) \right\} \\ &\quad \text{for } i = 0,\, \ldots,\, N-1 \,. \end{aligned} \tag{5.9}$$

Using Eqs. 5.6 and 5.2, the four partial derivatives result in

$$\frac{\partial\, e(n)}{\partial\, h_{\mathrm{r},\,i}} = -y(n-i) \,, \qquad \frac{\partial\, e^*(n)}{\partial\, h_{\mathrm{r},\,i}} = -y^*(n-i) \,, \tag{5.10}$$

$$\frac{\partial\, e(n)}{\partial\, h_{\mathrm{i},\,i}} = j\, y(n-i) \,, \qquad \frac{\partial\, e^*(n)}{\partial\, h_{\mathrm{i},\,i}} = -j\, y^*(n-i) \,. \tag{5.11}$$

These terms combine to

$$\begin{aligned} \frac{\partial\, \mathrm{E}\left\{|e(n)|^2\right\}}{\partial\, h_{\mathrm{r},\,i}} + j\, \frac{\partial\, \mathrm{E}\left\{|e(n)|^2\right\}}{\partial\, h_{\mathrm{i},\,i}} &= -2\, \mathrm{E}\left\{ y(n-i)\, e^*(n) \right\} \,, \\ &\text{for } i = 0,\, \ldots,\, N-1 \,. \end{aligned} \tag{5.12}$$

The minimal mean square error $\mathrm{E}\{|e(n)|^2\}_{\min}$ is achieved by setting the derivatives equal to zero

$$\mathrm{E}\left\{ y(n-i)\, e^*_{\text{opt}}(n) \right\} = 0 \,, \quad \text{for } i = 0,\, \ldots,\, N-1, \tag{5.13}$$

where $e_{\text{opt}}(n)$ denotes the error in case of the optimal filter. Equation 5.13 can also be derived from the *principle of orthogonality* [111]. It states that for an optimal system the input signal and the related error $e_{\text{opt}}(n)$ are orthogonal. *Input signal* in this context means those sample values of $y(n)$ that contribute to the sample value of the filter output that, in turn, leads to error sample $e_{\text{opt}}(n)$.

Extending Eq. 5.13 by Eqs. 5.6 and 5.2 leads to

$$\mathrm{E}\left\{ y(n-i) \left[d^*(n) - \sum_{k=0}^{N-1} h_{\text{opt},k}\, y^*(n-k) \right] \right\} = 0 \,, \quad \text{for } i = 0, \ldots, N-1 \,. \tag{5.14}$$

Exchanging the expectation with linear operations finally results in

$$r_{yd}(-i) - \sum_{k=0}^{N-1} h_{\text{opt},k}\, r_{yy}(k-i) = 0 \,, \quad \text{for } i = 0, \ldots, N-1 \,, \tag{5.15}$$

where we introduced two abbreviations:

$$r_{yd}(i) = \mathrm{E}\{ y(n+i)\, d^*(n) \} \,, \tag{5.16}$$

$$r_{yy}(k-i) = \mathrm{E}\{ y(n-i)\, y^*(n-k) \} \,. \tag{5.17}$$

The function $r_{yd}(i)$ denotes the cross-correlation function between the filter input signal, and the desired filter output for lag i, $r_{yy}(k-i)$ stands for the autocorrelation function of the filter input signal for lag $k-i$. In case the filter input $y(n)$ and the desired output $d(n)$ are orthogonal, the cross-correlation function is zero and the optimal filter produces no output.

Equation 5.15 represents a system of N simultaneous linear equations called *Wiener–Hopf equations*.

To arrive at a more compact notation we define an autocorrelation and a cross-correlation vector:

$$\boldsymbol{r}_{yy}(n) = [\, r_{yy}(n),\, r_{yy}(n-1),\, r_{yy}(n-2),\, \ldots\,,\, r_{yy}(n-N+1)\,]^{\mathrm{T}} \,, \tag{5.18}$$

$$\boldsymbol{r}_{yd}(n) = [\, r_{yd}(n),\, r_{yd}(n-1),\, r_{yd}(n-2),\, \ldots\,,\, r_{yd}(n-N+1)\,]^{\mathrm{T}} \,. \tag{5.19}$$

Finally, we introduce the autocorrelation matrix $\boldsymbol{R}_{yy}$:

$$\begin{aligned} \boldsymbol{R}_{yy} &= \mathrm{E}\{ \boldsymbol{y}(n)\, \boldsymbol{y}^{\mathrm{H}}(n) \} \\ &= \begin{bmatrix} r_{yy}(0) & r_{yy}(1) & \cdots & r_{yy}(N-1) \\ r_{yy}^*(1) & r_{yy}(0) & \cdots & r_{yy}(N-2) \\ \vdots & \vdots & \ddots & \vdots \\ r_{yy}^*(N-1) & r_{yy}^*(N-2) & \cdots & r_{yy}(0) \end{bmatrix} . \end{aligned} \tag{5.20}$$

Using this notation the system of equations, Eq. 5.15 can be expressed by one matrix equation:

$$\boldsymbol{R}_{yy}\ \boldsymbol{h}_{\text{opt}} = \boldsymbol{r}_{yd}(0) \,. \tag{5.21}$$

Thus, we arrived at the *Wiener solution* that, in other connection, is also called the *normal equation*. It plays a key role in linear system optimization. Assuming that the inverse of the autocorrelation matrix $\boldsymbol{R}_{yy}$ exists, the vector of the optimal filter coefficients is given by

$$\boldsymbol{h}_{\text{opt}} = \boldsymbol{R}_{yy}^{-1}\, \boldsymbol{r}_{yd}(0)\ . \tag{5.22}$$

Figure 5.2 shows an example of the performance of a Wiener filter. The filter input $y(n)$ is an additive mixture of a sine wave and a sample function from white noise. Both have zero mean; their variances are $\sigma_s^2 = 0.5$ and $\sigma_n^2 = 0.02$, respectively. The desired filter output $d(n)$ is equal to $s(n)$. The filter length N is equal to 32. The lower part of the figure shows the modulus of the frequency response of the filter over a normalized frequency $\Omega\,/\pi = 2\,f\,/f_s$, where f_s is the sampling frequency of the signals.

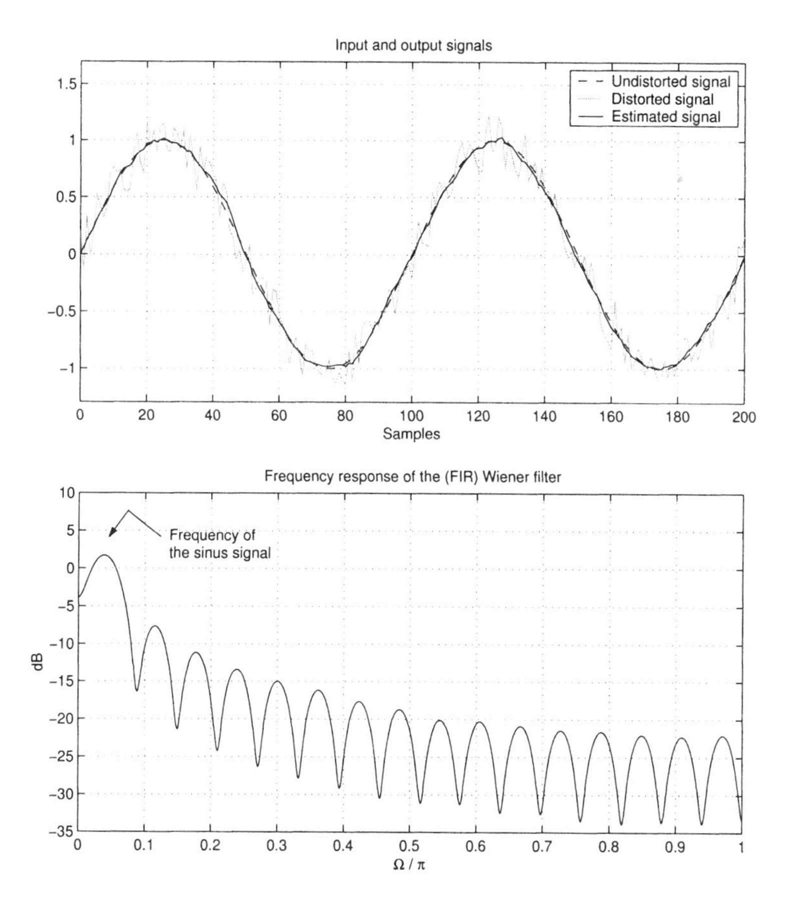

Fig. 5.2 Example of a Wiener filter. Filter input—sine wave $s(n)$ plus white noise $n(n)$; desired filter output—$\widehat{d}(n) = \widehat{s}(n)$.

The autocorrelation matrix $\boldsymbol{R}_{yy}$ has a number of special properties that allow for efficient inversion algorithms (see Chapter 6). Its most important features are that it is *Hermitian*

$$\boldsymbol{R}_{yy} = \boldsymbol{R}_{yy}^{\mathrm{H}} \tag{5.23}$$

and that it is a *Toeplitz matrix*, meaning that all elements along the main diagonal are equal and that this property also holds for elements on other diagonals parallel to the main diagonal.

Analyzing the mean square error leads to some further insight into the Wiener solution. We insert Eq. 5.5 into Eq. 5.6:

$$e(n) = d(n) - \boldsymbol{h}^{\mathrm{H}}\, \boldsymbol{y}(n) = d(n) - \boldsymbol{y}^{\mathrm{T}}(n)\, \boldsymbol{h}^* \,. \tag{5.24}$$

Then Eq. 5.7 reads as follows:

$$\begin{aligned} \mathrm{E}\left\{|e(n)|^2\right\} &= \mathrm{E}\{\, (d(n) - \boldsymbol{h}^{\mathrm{H}}\, \boldsymbol{y}(n))\, (d^*(n) - \boldsymbol{y}^{\mathrm{H}}(n)\, \boldsymbol{h})\, \} \\ &= r_{dd}(0) - \boldsymbol{h}^{\mathrm{H}}\, \boldsymbol{r}_{yd}(0) - \boldsymbol{r}_{yd}^{\mathrm{H}}(0)\, \boldsymbol{h} + \boldsymbol{h}^{\mathrm{H}}\, \boldsymbol{R}_{yy}\, \boldsymbol{h} \,, \end{aligned} \tag{5.25}$$

where the scalar quantity $r_{dd}(0)$ describes the autocorrelation of the desired output signal at lag 0. We now insert the optimal filter vector according to Eq. 5.22 in order to get the minimum mean square error:

$$\begin{aligned} \mathrm{E}\left\{|e(n)|^2\right\}_{\min} &= r_{dd}(0) - \boldsymbol{r}_{yd}^{\mathrm{H}}(0)\, \boldsymbol{h}_{\mathrm{opt}} \\ &= r_{dd}(0) - \boldsymbol{h}_{\mathrm{opt}}^{\mathrm{H}}\, \boldsymbol{r}_{yd}(0) \\ &= r_{dd}(0) - \boldsymbol{r}_{yd}^{\mathrm{H}}(0)\, \boldsymbol{R}_{yy}^{-1}\, \boldsymbol{r}_{yd}(0) \,, \end{aligned} \tag{5.26}$$

where the manipulation from line 1 to line 2 is allowed since $\boldsymbol{r}_{yd}^{\mathrm{H}}(0)\, \boldsymbol{h}_{\mathrm{opt}}$ is scalar and real. Inserting Eq. 5.26 into Eq. 5.25 leads after some manipulations to

$$\mathrm{E}\left\{|e(n)|^2\right\} = \mathrm{E}\left\{|e(n)|^2\right\}_{\min} + (\boldsymbol{h} - \boldsymbol{h}_{\mathrm{opt}})^{\mathrm{H}}\, \boldsymbol{R}_{yy}\, (\boldsymbol{h} - \boldsymbol{h}_{\mathrm{opt}}) \,. \tag{5.27}$$

This result shows that the mean square error forms a quadratic surface with the minimum $\mathrm{E}\{|e(n)|^2\}_{\min}$ at $\boldsymbol{h} = \boldsymbol{h}_{\mathrm{opt}}$. This fact (subsequently) justifies the optimization by a simple gradient technique. Figure 5.3 shows the error surfaces and the contour lines for (left) a white and (right) a colored noise and a filter with two coefficients. The optimal Wiener filter is $\boldsymbol{h}_{\mathrm{opt}} = [4,\ 4]^{\mathrm{T}}$.

5.2 FREQUENCY-DOMAIN SOLUTION

As given by Eq. 5.22, the calculation of the optimal impulse response requires the inversion of the autocorrelation matrix. If we extend N to infinity, the Wiener–Hopf

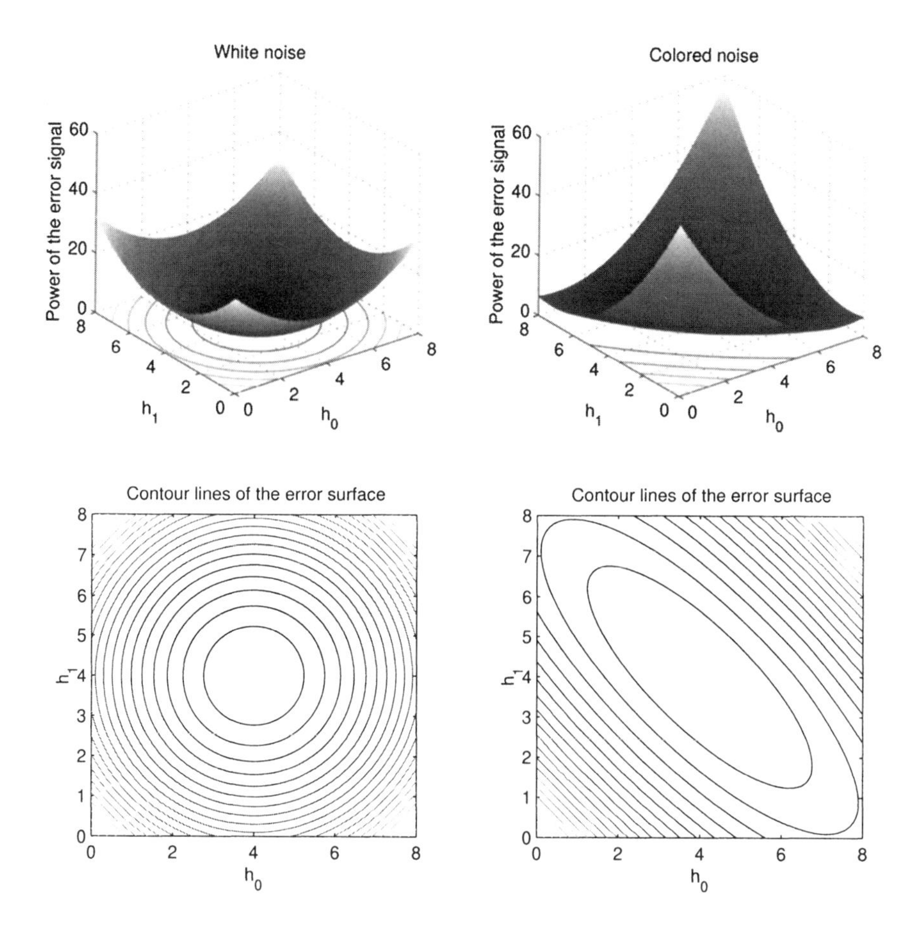

Fig. 5.3 Error surfaces for (left) white and (right) colored noise and a filter with two coefficients.

equation (Eq. 5.15) can be solved by Fourier transformation. The resulting filter is noncausal. The Fourier transformation for discrete-time functions is defined by

$$F(\Omega) = \sum_{n=-\infty}^{+\infty} f(n)\, e^{-j\,\Omega\, n} \,. \tag{5.28}$$

Transforming the correlation functions leads to *power spectral densities* [111]:

$$S_{yy}(\Omega) = \sum_{n=-\infty}^{+\infty} r_{yy}(n)\, e^{-j\,\Omega\, n} \,, \tag{5.29}$$

$$S_{yd}(\Omega) = \sum_{n=-\infty}^{+\infty} r_{yd}(n)\, e^{-j\,\Omega\, n} \,, \tag{5.30}$$

where $S_{yy}(\Omega)$ describes the auto power spectral density of the filter input $y(n)$ and $S_{yd}(\Omega)$ stands for the cross power spectral density between $y(n)$ and the desired filter output $d(n)$. If we finally introduce the transfer function $H_{\text{opt}}(\Omega)$ of the optimal filter

$$H_{\text{opt}}(\Omega) = \sum_{n=-\infty}^{+\infty} h_{\text{opt},\,n}\, e^{-j\,\Omega\, n}\,, \tag{5.31}$$

the Fourier transform of Eq. 5.14 reads as follows:

$$S_{yd}^{*}(\Omega) - H_{\text{opt}}(\Omega)\, S_{yy}(\Omega) = 0\,. \tag{5.32}$$

Assuming that $S_{yy}(\Omega) \neq 0$ for all Ω, we find the following transfer function of the noncausal Wiener filter:

$$H_{\text{opt}}(\Omega) = \frac{S_{yd}^{*}(\Omega)}{S_{yy}(\Omega)}\,. \tag{5.33}$$

We can simplify this expression by assuming that the *useful* signal $s(n)$ and the noise $n(n)$ are orthogonal and that the desired filter output $d(n)$ equals $s(n)$. In this case the power spectral densities are given by

$$S_{yy}(\Omega) = S_{ss}(\Omega) + S_{nn}(\Omega)\,, \tag{5.34}$$

$$S_{yd}(\Omega) = S_{ss}(\Omega)\,. \tag{5.35}$$

The transfer function of the Wiener filter then reads

$$H_{\text{opt}}(\Omega) = \frac{S_{ss}(\Omega)}{S_{ss}(\Omega) + S_{nn}(\Omega)} = \frac{1}{1 + \dfrac{S_{nn}(\Omega)}{S_{ss}(\Omega)}}\,, \tag{5.36}$$

showing that this filter attenuates the input signal according to the ratio of the power spectral densities of the signal and the noise.

6

Linear Prediction

Prediction in general means to forecast signal samples that are not yet available ("forward prediction") or to reestablish already forgotten samples ("backward prediction"). With this capability predictors play an important role in signal processing wherever it is desirable, for instance, to reduce the amount of data to be transmitted or stored. Examples for the use of predictors are encoders for speech or video signals.

On the basis of the derivation of the Wiener filter (see Chapter 5), we are able to regard the linear predictor as a special case. To make things easy, one assumes that the noise component $n(n)$ of the input signal is zero. The desired prediction filter output $d(n)$ equals the input signal $s(n)$ shifted by L sampling intervals:

$$d(n) = s(n + L) \,. \tag{6.1}$$

For $L > 0$ one talks about an L-step *forward predictor*. In the simplest case L is equal to 1. A *backward predictor* is characterized by $L < 1 - N$, where N is the length of the (FIR) prediction filter here called the *order* of the predictor. To predict speech signals, for instance, usual predictor orders are between 8 and 16.

Subtracting the predictor output signal $\widehat{d}(n)$ from the signal $d(n)$ to be predicted leads to the prediction error signal $e(n)$. Figure 6.1 shows a one-step forward *prediction error filter*.

Assuming a prediction filter of length N (see Eq. 5.3) the prediction error is given by

$$e(n) = s(n) - \sum_{i=0}^{N-1} h_i^* \, s(n - L - i) \;, \tag{6.2}$$

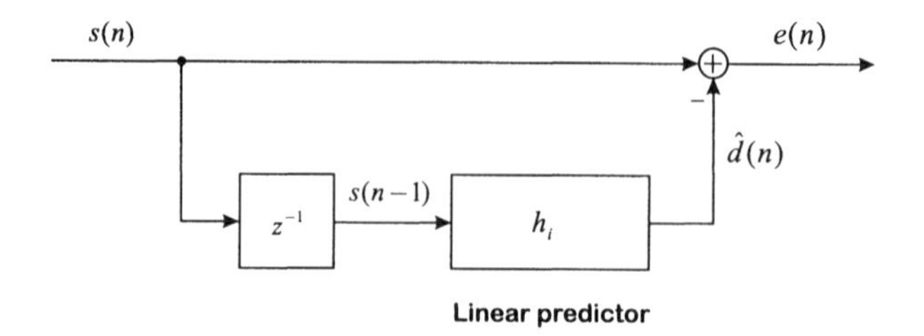

Fig. 6.1 One-step forward prediction error filter.

The filter coefficients are considered optimal if the mean square prediction error (see Eq. 5.7) is minimal.

For the special case of a one-step forward prediction error filter ($L = 1$), the prediction error can be written as

$$e(n) = s(n) - \sum_{i=1}^{N} h_{i-1}^* \, s(n-i) \, . \tag{6.3}$$

The prediction error filter, therefore, is an *all-zero* filter (transversal filter) with the transfer function

$$A(z) = 1 - \sum_{i=1}^{N} h_{i-1}^* \, z^{-i} \, . \tag{6.4}$$

If the filter coefficients are optimized for a speech input, the *all-pole* filter

$$\frac{1}{A(z)} = \frac{1}{1 - \sum\limits_{i=1}^{N} h_{i-1}^* \, z^{-i}} \tag{6.5}$$

excited by sample functions of a white stochastic process can serve as a model for speech production.

The effect of a prediction error filter is demonstrated in Fig. 6.2, where the power spectral densities of a speech signal and of the related prediction error signal—also called *decorrelated speech signal*—are shown. For this example a one-step forward predictor of order 16 is used. Comparing both curves the whitening effect of the prediction error filter becomes obvious. Therefore, it is also called *decorrelation* or *whitening filter*. Filters like this are used, for example, to speed up the convergence of adaptive algorithms that suffer from correlated input signals.

6.1 NORMAL EQUATIONS

Again, we model the predictor input $s(n)$ by a stationary discrete-time random process. The normal equation for an L-step predictor follows from Eq. 5.21 by specifying the autocorrelation matrix $\boldsymbol{R}_{yy}$ (see Eq. 5.20) and the cross-correlation vector

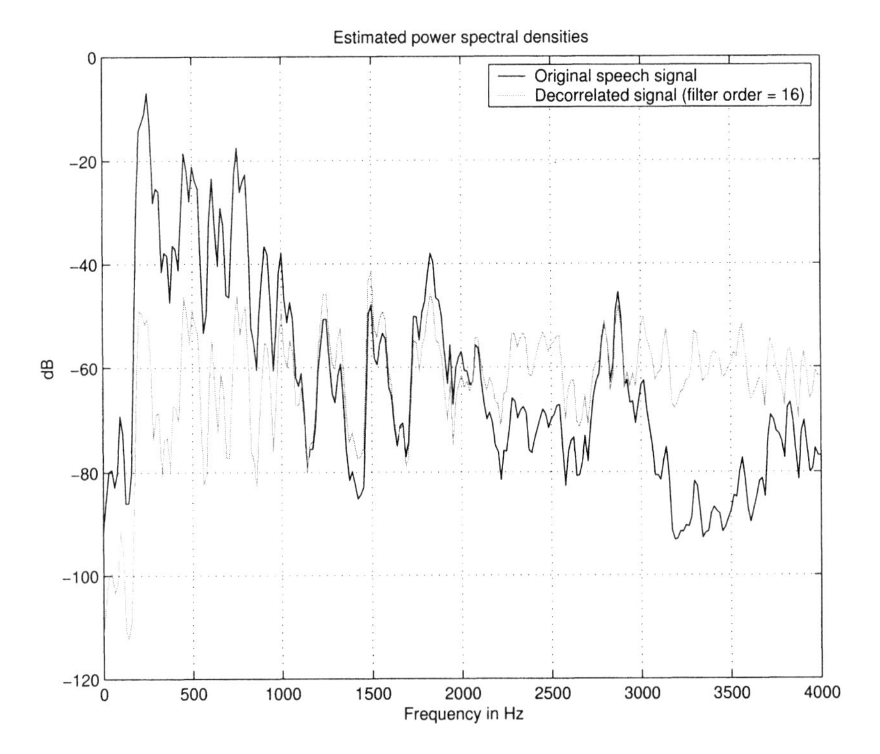

Fig. 6.2 Power spectral densities of a speech signal and the related decorrelated signal using a one-step forward predictor of order 16.

$\boldsymbol{r}_{yd}(n)$ (see Eq. 5.19) according to the requirements here. On the basis of Eq. 6.1, we replace $r_{yy}(n)$ in Eq. 5.20 by $r_{ss}(n)$

$$\begin{aligned} \boldsymbol{R}_{yy} &= \boldsymbol{R}_{ss} = \mathrm{E}\left\{ \boldsymbol{s}(n)\, \boldsymbol{s}^{\mathrm{H}}(n) \right\} \\ &= \begin{bmatrix} r_{ss}(0) & r_{ss}(1) & \dots & r_{ss}(N-1) \\ r_{ss}^*(1) & r_{ss}(0) & \dots & r_{ss}(N-2) \\ \vdots & \vdots & \ddots & \vdots \\ r_{ss}^*(N-1) & r_{ss}^*(N-2) & \dots & r_{ss}(0) \end{bmatrix} \end{aligned} \tag{6.6}$$

and $\boldsymbol{r}_{yd}(n)$ in Eq. 5.19 by $\boldsymbol{r}_{ss}(n-L)$:

$$\begin{aligned} \boldsymbol{r}_{yd}(n) &= \boldsymbol{r}_{ss}(n-L) \\ &= \left[r_{ss}(n-L),\, r_{ss}(n-L-1),\, \dots,\, r_{ss}(n-L-N+1) \right]^{\mathrm{T}}. \end{aligned} \tag{6.7}$$

Thus, the normal equation reads

$$\boldsymbol{R}_{ss}\ \boldsymbol{h}_{\text{opt}} = \boldsymbol{r}_{ss}(-L)\,. \tag{6.8}$$

Assuming that the inverse of $\boldsymbol{R}_{ss}$ exists, the vector of the optimal prediction filter coefficients is given by

$$\boldsymbol{h}_{\text{opt}} = \boldsymbol{R}_{ss}^{-1}\, \boldsymbol{r}_{ss}(-L)\,. \tag{6.9}$$

This result clearly shows that the prediction filter delivers a nonzero output only if the signal samples $y(n-k),\ \ k = 0,\ \dots\ ,\ N-1$ available to the filter and the signal sample $y(n+L)$ to be predicted are correlated. In case of a (zero-mean) white noise only $r_{ss}(0)$ is different from zero resulting in a diagonal autocorrelation matrix $\boldsymbol{R}_{ss}$. The vector $\boldsymbol{r}_{ss}(n)$ has a nonzero element only for $n = 0$, showing that the predictor delivers a zero output.

A stationary random process is not a realistic model for speech. As an approximation, one usually assumes that speech signals keep their properties in intervals of about 20 ms duration. Consequently, a prediction filter for speech signals has to be updated according to this timeframe. Therefore, an efficient algorithm for the inversion of the autocorrelation matrix $\boldsymbol{R}_{ss}$ is crucial for the application of a predictor.

In Fig. 6.3 we show an example for the efficiency of a prediction error filter optimized according to the long-term statistical properties of a speech input—this accords with a stationary input model—and a filter that is updated periodically resulting from a "short-term stationary" model.

6.2 LEVINSON–DURBIN RECURSION

The *Levinson–Durbin algorithm* determines the prediction filter coefficients by a recursive procedure. It starts with a predictor of order 1 and increases the order in steps of 1 until the final order N is reached. It makes use of the special properties of the autocorrelation matrix $\boldsymbol{R}_{ss}$ and results in considerable savings of operations for the calculation of the predictor coefficients compared to a direct inversion of the matrix.

For the following considerations we restrict ourselves to a one-step forward and an N-step backward predictor ($L = -N$). "Predicting N steps backwards" means reconstructing the sample of the input signal that has just "dropped out" of the memory of the prediction filter of length N.

Since the algorithm increases the filter order step by step, we have to refine our notation such that it also monitors the predictor order. We also have to distinguish between the coefficients of a *forward* and a *backward* predictor.

Using the extended notation, one obtains the following for a *one-step forward predictor* ($L = 1$) of order k from Eq. 6.8:

$$\boldsymbol{R}_{ss}^{(k)}\ {}^{\mathrm{f}}\boldsymbol{h}_{\text{opt}}^{(k)} = \boldsymbol{r}_{ss}^{(k)}(-1)\,. \tag{6.10}$$

In the notation $\boldsymbol{r}_{ss}^{(k)}(-1)$ the exponent (k) marks the order of the related predictor and the argument (-1) defines the argument of the first element of the vector. For the *k-step backward predictor* ($L = -k$), the normal equation reads

$$\boldsymbol{R}_{ss}^{(k)}\ {}^{\mathrm{b}}\boldsymbol{h}_{\text{opt}}^{(k)} = \boldsymbol{r}_{ss}^{(k)}(k)\,. \tag{6.11}$$

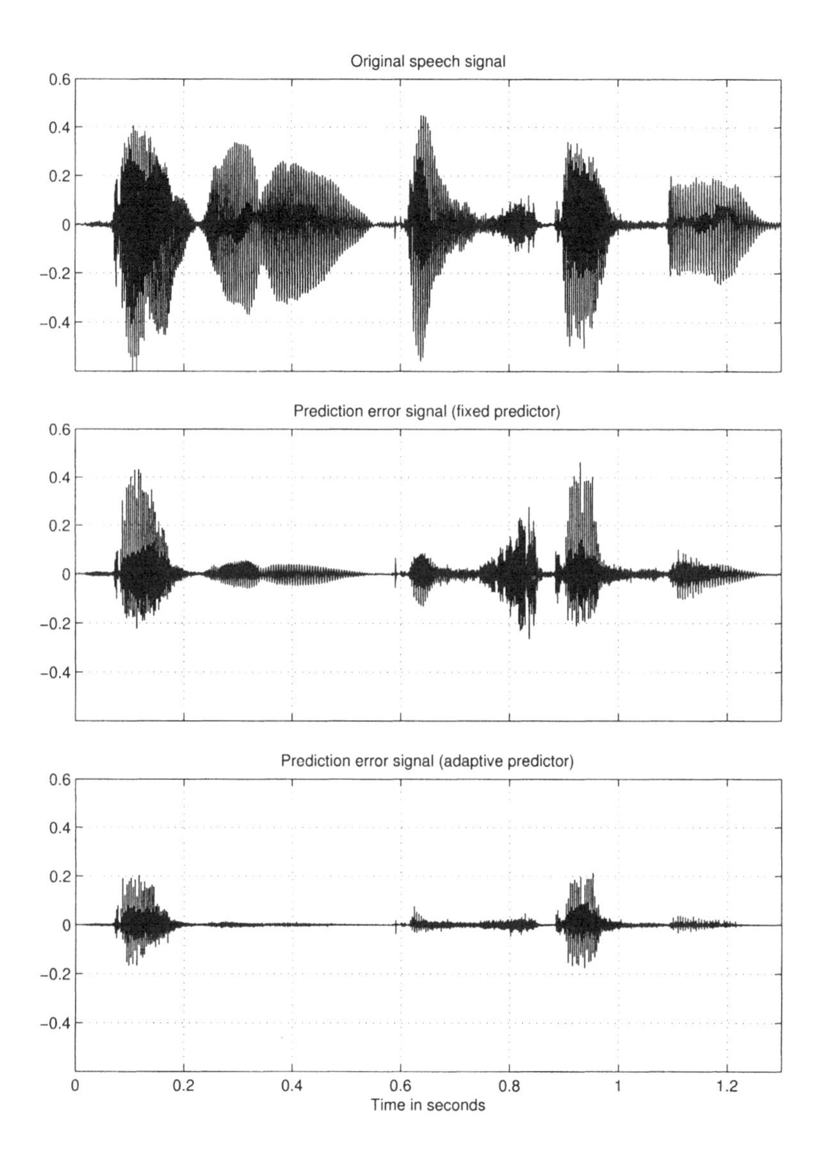

Fig. 6.3 Original speech signal and prediction error signals using a one-step predictor of order 16. The "fixed predictor" is optimized according to the long-term properties of the speech input signal. The"adaptive predictor" is adjusted every 64 samples based on the properties of the input within a symmetric 256-sample window.

The autocorrelation matrix $\boldsymbol{R}_{ss}^{(k)}$ remains unchanged in both cases. The correlation vectors at the right hand sides of Eqs. 6.10 and 6.11 read:

$$\boldsymbol{r}_{ss}^{(k)}(-1) \quad = \quad [\, r_{ss}(-1), r_{ss}(-2), r_{ss}(-3), \ldots, r_{ss}(-k) \,]^{\mathrm{T}}, \qquad (6.12)$$

$$\boldsymbol{r}_{ss}^{(k)}(k) \quad = \quad [r_{ss}(k),\, r_{ss}(k-1),\, r_{ss}(k-2),\, \ldots\,,\, r_{ss}(1)]^{\mathrm{T}} \,. \tag{6.13}$$

From the definition 5.17 of the autocorrelation function, it follows that

$$\boldsymbol{r}_{ss}(k) = \boldsymbol{r}_{ss}^{*}(-k) \,. \tag{6.14}$$

Regarding, further, the symmetry of the autocorrelation matrix $\boldsymbol{R}_{ss}$, it becomes clear that ${}^{\mathrm{b}}\boldsymbol{h}_{\mathrm{opt}}^{(k)}$ contains the conjugate complex elements of ${}^{\mathrm{f}}\boldsymbol{h}_{\mathrm{opt}}^{(k)}$ in reverse order:

$${}^{\mathrm{b}}h_{\mathrm{opt},l}^{(k)} = {}^{\mathrm{f}}h_{\mathrm{opt},k-1-l}^{(k)^{*}} \qquad \text{for } l = 0,\, \ldots\,,\, k-1 \,. \tag{6.15}$$

This relation is a consequence of the assumption of a stationary input process.

To start a recursion from the order $k-1$ to the order k, we separate the last column and the last row from the kth-order autocorrelation matrix $\boldsymbol{R}_{ss}^{(k)}$:

$$\boldsymbol{R}_{ss}^{(k)} = \left[\begin{array}{ccccc|c} r_{ss}(0) & r_{ss}(1) & \ldots & r_{ss}(k-2) & & r_{ss}(k-1) \\ r_{ss}^{*}(1) & r_{ss}(0) & \ldots & r_{ss}(k-1) & & r_{ss}(k-2) \\ \vdots & \vdots & \ddots & \ldots & & \vdots \\ r_{ss}^{*}(k-2) & r_{ss}^{*}(k-3) & \ldots & r_{ss}(0) & & r_{ss}(1) \\ --- & --- & \ldots & --- & & --- \\ r_{ss}^{*}(k-1) & r_{ss}^{*}(k-2) & \ldots & r_{ss}^{*}(1) & & r_{ss}(0) \end{array}\right] . \tag{6.16}$$

Thus, the autocorrelation matrix can be written as follows:

$$\boldsymbol{R}_{ss}^{(k)} = \left[\begin{array}{c|c} \boldsymbol{R}_{ss}^{(k-1)} & \boldsymbol{r}_{ss}^{(k-1)}(k-1) \\ --- & --- \\ \boldsymbol{r}_{ss}^{(k-1)^{\mathrm{H}}}(k-1) & r_{ss}(0) \end{array}\right] . \tag{6.17}$$

If we, further, partition the vector for the coefficients of the forward predictor and of the right hand side of the normal equation 6.10,

$${}^{\mathrm{f}}\boldsymbol{h}_{\mathrm{opt}}^{(k)} = \begin{bmatrix} {}^{\mathrm{f}}h_{\mathrm{opt},0}^{(k)} \\ {}^{\mathrm{f}}h_{\mathrm{opt},1}^{(k)} \\ \vdots \\ {}^{\mathrm{f}}h_{\mathrm{opt},k-2}^{(k)} \\ --- \\ {}^{\mathrm{f}}h_{\mathrm{opt},k-1}^{(k)} \end{bmatrix} , \tag{6.18}$$

$$\boldsymbol{r}_{ss}^{(k)}(-1) = \begin{bmatrix} r_{ss}(-1) \\ r_{ss}(-2) \\ \vdots \\ r_{ss}(1-k) \\ --- \\ r_{ss}(-k) \end{bmatrix} = \begin{bmatrix} \boldsymbol{r}_{ss}^{(k-1)}(-1) \\ --- \\ r_{ss}(-k) \end{bmatrix} , \tag{6.19}$$

we can write the upper $k-1$ rows of Eq. 6.10 as follows:

$$\boldsymbol{R}_{ss}^{(k-1)} \begin{bmatrix} {}^{\mathrm{f}}h_{\mathrm{opt},0}^{(k)} \\ {}^{\mathrm{f}}h_{\mathrm{opt},1}^{(k)} \\ \vdots \\ {}^{\mathrm{f}}h_{\mathrm{opt},k-2}^{(k)} \end{bmatrix} = \boldsymbol{r}_{ss}^{(k-1)}(-1) - \boldsymbol{r}_{ss}^{(k-1)}(k-1)\, {}^{\mathrm{f}}h_{\mathrm{opt},k-1}^{(k)} \,. \tag{6.20}$$

Multiplying from left with the inverse of the autocorrelation matrix leads to

$$\begin{bmatrix} {}^{\mathrm{f}}h_{\mathrm{opt},0}^{(k)} \\ {}^{\mathrm{f}}h_{\mathrm{opt},1}^{(k)} \\ \vdots \\ {}^{\mathrm{f}}h_{\mathrm{opt},k-2}^{(k)} \end{bmatrix} = \boldsymbol{R}_{ss}^{(k-1)^{-1}} \boldsymbol{r}_{ss}^{(k-1)}(-1) - \boldsymbol{R}_{ss}^{(k-1)^{-1}} \boldsymbol{r}_{ss}^{(k-1)}(k-1)\, {}^{\mathrm{f}}h_{\mathrm{opt},k-1}^{(k)} \,. \tag{6.21}$$

Looking at Eq. 6.21, a number of conclusions can be drawn:

- The first term on the right side contains the coefficients of the forward predictor ${}^{\mathrm{f}}\boldsymbol{h}_{\mathrm{opt}}^{(k-1)}$ of order $k-1$.
- The first two factors of the second term represent the backward predictor ${}^{\mathrm{b}}\boldsymbol{h}_{\mathrm{opt}}^{(k-1)}$ of order $k-1$.
- The third factor of the second term shows the last element of the forward predictor of order k that is missing on the left side of Eq. 6.21.

Using abbreviations, one can write

$$\begin{bmatrix} {}^{\mathrm{f}}h_{\mathrm{opt},0}^{(k)} \\ {}^{\mathrm{f}}h_{\mathrm{opt},1}^{(k)} \\ \vdots \\ {}^{\mathrm{f}}h_{\mathrm{opt},k-2}^{(k)} \end{bmatrix} = {}^{\mathrm{f}}\boldsymbol{h}_{\mathrm{opt}}^{(k-1)} - {}^{\mathrm{b}}\boldsymbol{h}_{\mathrm{opt}}^{(k-1)}\, {}^{\mathrm{f}}h_{\mathrm{opt},k-1}^{(k)} \,. \tag{6.22}$$

This equation specifies the update of the predictor coefficients when stepping from order $k-1$ to order k. The relation is called *Levinson–Durbin recursion* [141, 57].

The final incredient is an equation for the last (the "new") element of the predictor coefficient vector. It can be calculated from the last row of the normal equation (see Eq. 6.10) together with Eqs. 6.17, 6.18, and 6.19:

$$\left[\; \boldsymbol{r}_{ss}^{(k-1)^{\mathrm{H}}}(k-1) \;\middle|\; r_{ss}(0) \;\right] \begin{bmatrix} {}^{\mathrm{f}}h_{\mathrm{opt},0}^{(k)} \\ {}^{\mathrm{f}}h_{\mathrm{opt},1}^{(k)} \\ \vdots \\ {}^{\mathrm{f}}h_{\mathrm{opt},k-2}^{(k)} \\ --- \\ {}^{\mathrm{f}}h_{\mathrm{opt},k-1}^{(k)} \end{bmatrix} = r_{ss}(-k)\,. \tag{6.23}$$

Solving Eq. 6.23 for ${}^{\mathrm{f}}h_{\mathrm{opt},k-1}^{(k)}$ leads to

$$ {}^{\mathrm{f}}h_{\mathrm{opt},k-1}^{(k)} = \frac{1}{r_{ss}(0)} \left[r_{ss}(-k) - \left[\, \boldsymbol{r}_{ss}^{(k-1)^{\mathrm{H}}}(k-1) \, \right] \left[\begin{array}{c} {}^{\mathrm{f}}h_{\mathrm{opt},0}^{(k)} \\ {}^{\mathrm{f}}h_{\mathrm{opt},1}^{(k)} \\ \vdots \\ {}^{\mathrm{f}}h_{\mathrm{opt},k-2}^{(k)} \end{array} \right] \right] . \tag{6.24} $$

Inserting the recursion 6.22, and, again, solving for ${}^{\mathrm{f}}h_{\mathrm{opt},k-1}^{(k)}$ leads to

$$ {}^{\mathrm{f}}h_{\mathrm{opt},k-1}^{(k)} = \frac{r_{ss}(-k) - \boldsymbol{r}_{ss}^{(k-1)^{\mathrm{H}}}(k-1)\, {}^{\mathrm{f}}\boldsymbol{h}_{\mathrm{opt}}^{(k-1)}}{r_{ss}(0) - \boldsymbol{r}_{ss}^{(k-1)^{\mathrm{H}}}(k-1)\, {}^{\mathrm{b}}\boldsymbol{h}_{\mathrm{opt}}^{(k-1)}} . \tag{6.25} $$

The coefficient ${}^{\mathrm{f}}h_{\mathrm{opt},k-1}^{(k)}$ is called the *reflection coefficient* or the *PARCOR coefficient* (for "partial correlation coefficient") [84].

To complete the algorithm, an initial condition for $k = 1$ is necessary:

$$ {}^{\mathrm{f}}h_{\mathrm{opt},0}^{(1)} = \boldsymbol{R}_{ss}^{(1)^{-1}} r_{ss}(-1) = \frac{r_{ss}(-1)}{r_{ss}(0)} . \tag{6.26} $$

The last relation follows from the fact that the autocorrelation matrix of order 1 is just the scalar $r_{ss}(0)$.

The Levinson–Durbin algorithm for the coefficients of a forward predictor is thus complete. The initial condition 6.26 is followed by the calculation of the PARCOR coefficient according to Eq. 6.25 and the application of the Levinson–Durbin recursion 6.22.

To understand the recursion better, we calculate the output signal of a one-step forward prediction filter of order k:

$$ \widehat{d}(n) = \sum_{i=0}^{k-1} {}^{\mathrm{f}}h_{\mathrm{opt},i}^{(k)^*} \, s(n-1-i) . \tag{6.27} $$

Next we split the sum such that we can insert the recursion 6.22:

$$ \widehat{d}(n) = \sum_{i=0}^{k-2} {}^{\mathrm{f}}h_{\mathrm{opt},i}^{(k)^*} \, s(n-1-i) + {}^{\mathrm{f}}h_{\mathrm{opt},k-1}^{(k)^*} \, s(n-k) . \tag{6.28} $$

Inserting the recursion leads to

$$ \begin{aligned} \widehat{d}(n) &= \sum_{i=0}^{k-2} {}^{\mathrm{f}}h_{\mathrm{opt},i}^{(k-1)^*} \, s(n-1-i) \\ &\quad + {}^{\mathrm{f}}h_{\mathrm{opt},k-1}^{(k)^*} \left[s(n-k) - \sum_{i=0}^{k-2} {}^{\mathrm{b}}h_{\mathrm{opt},i}^{(k-1)^*} \, s(n-1-i) \right] . \end{aligned} \tag{6.29} $$

Increasing the prediction filter order from $k-1$ to k means using one more sample—$s(n-k)$—of the predictor input to generate the predictor output. However, as the second term of Eq. 6.29 shows, only that part of this sample, which cannot be backwards predicted, enters the prediction filter output (see Fig. 6.4).

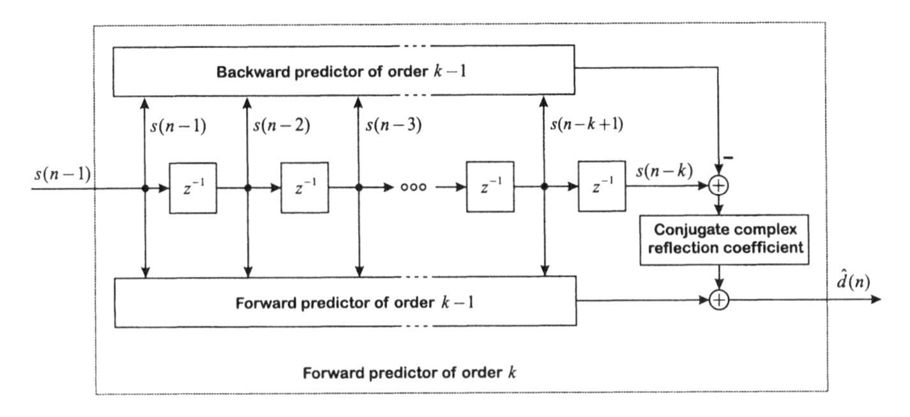

Fig. 6.4 Increasing the predictor order from $k-1$ to k.

In order to develop a recursive expression for the mean square error $\mathrm{E}\{|e^{(k)}(n)|^2\}_{\min}$ of a forward prediction filter of order k, we start with Eq. 5.26 and tailor it for a forward predictor of this order:

$$\begin{aligned}\mathrm{E}\left\{\left|e^{(k)}(n)\right|^2\right\}_{\min} &= r_{ss}(0) - \boldsymbol{r}_{ss}^{(k)^{\mathrm{H}}}(-1)\,{}^{\mathrm{f}}\boldsymbol{h}_{\mathrm{opt}}^{(k)} \\ &= r_{ss}(0) - \boldsymbol{r}_{ss}^{(k-1)^{\mathrm{H}}}(-1)\begin{bmatrix} {}^{\mathrm{f}}h_{\mathrm{opt},0}^{(k)} \\ {}^{\mathrm{f}}h_{\mathrm{opt},1}^{(k)} \\ \vdots \\ {}^{\mathrm{f}}h_{\mathrm{opt},k-2}^{(k)} \end{bmatrix} - r_{ss}^{*}(-k)\,{}^{\mathrm{f}}h_{\mathrm{opt},k-1}^{(k)}\,. \end{aligned} \tag{6.30}$$

Now we are ready to insert the recursion equation 6.22:

$$\begin{aligned}\mathrm{E}\left\{\left|e^{(k)}(n)\right|^2\right\}_{\min} &= r_{ss}(0) - \boldsymbol{r}_{ss}^{(k-1)^{\mathrm{H}}}(-1)\left[{}^{\mathrm{f}}\boldsymbol{h}_{\mathrm{opt}}^{(k-1)} - {}^{\mathrm{b}}\boldsymbol{h}_{\mathrm{opt}}^{(k-1)}\,{}^{\mathrm{f}}h_{\mathrm{opt},k-1}^{(k)}\right] \\ &\quad - r_{ss}^{*}(k)\,{}^{\mathrm{f}}h_{\mathrm{opt},k-1}^{(k)}\,. \end{aligned} \tag{6.31}$$

The first two terms on the right side of this equation, however, represent the prediction error for the order $k-1$:

$$\begin{aligned}\mathrm{E}\left\{\left|e^{(k)}(n)\right|^2\right\}_{\min} &= \mathrm{E}\left\{\left|e^{(k-1)}(n)\right|^2\right\}_{\min} \\ &\quad + \left[\boldsymbol{r}_{ss}^{(k-1)^{\mathrm{H}}}(-1)\,{}^{\mathrm{b}}\boldsymbol{h}_{\mathrm{opt}}^{(k-1)} - r_{ss}^{*}(-k)\right]{}^{\mathrm{f}}h_{\mathrm{opt},k-1}^{(k)}\,. \end{aligned} \tag{6.32}$$

The term in brackets on the right side of this equation can be transformed into the conjugate complex of the numerator of Eq. 6.25. On the other hand, the denominator of Eq. 6.25 equals $\mathrm{E}\left\{\left|e^{(k-1)}(n)\right|^2\right\}_{\min}$. Therefore, Eq. 6.32 finally reads

$$\begin{aligned} &\mathrm{E}\left\{\left|e^{(k)}(n)\right|^2\right\}_{\min} \\ &\quad = \mathrm{E}\left\{\left|e^{(k-1)}(n)\right|^2\right\}_{\min}\left[1 - |{}^{\mathrm{f}}h^{(k)}_{\mathrm{opt},\,k\,-\,1}|^2\right] . \end{aligned} \tag{6.33}$$

This result gives additional insight into the properties of the reflection coefficient ${}^{\mathrm{f}}h^{(k)}_{\mathrm{opt},k\,-\,1}$. Since the mean square prediction error is nonnegative, the following condition holds:

$$|{}^{\mathrm{f}}h^{(k)}_{\mathrm{opt},\,k}|^2 \leq 1 \quad \text{for } k = 0, \ldots, N-1 . \tag{6.34}$$

Equation 6.33 can also serve as stopping criterion for the Levinson–Durbin algorithm. If the reflection coefficient according to Eq. 6.25 falls below a specified (small) value, increasing the order of the predictor may be terminated. In Table 6.1 the Levinson–Durbin algorithm is summarized.

Figure 6.5 shows the inverse of the normalized power spectral density of a predictor input signal

$$10 \log \frac{S_0}{|S_{ss}(\Omega)|} ,$$

where S_0 is a normalizing constant, and the modulus of the prediction error filter

$$20 \log |H^{(k)}_{\mathrm{opt}}(\Omega)| ,$$

for increasing orders k, both in dB. As can be seen in Fig. 6.5, the transfer functions approach closer and closer the inverse power spectral density of the input. Thus, the power spectral densities of the prediction error given by

$$S_{ss}(\Omega)\, |H^{(k)}_{\mathrm{opt}}(\Omega)|^2$$

come closer to a constant, meaning that they approach white noise.

Other filter structures can be used to implement a predictor. In case of a *lattice filter*, the reflection coefficients are immediately applied in each stage of the filter [111].

Table 6.1 Summary of the Levinson–Durbin recursion

Algorithmic part	Corresponding equations
Initialization	
Forward predictor	${}^{\mathrm{f}}h_{\mathrm{opt},0}^{(1)} = \frac{r_{ss}(-1)}{r_{ss}(0)}$
Backward predictor	${}^{\mathrm{b}}h_{\mathrm{opt},0}^{(1)} = {}^{\mathrm{f}}h_{\mathrm{opt},0}^{(1)*}$
Recursion	
Reflexion coefficient	${}^{\mathrm{f}}h_{\mathrm{opt},k-1}^{(k)} = \frac{r_{ss}(-k) - \boldsymbol{r}_{ss}^{(k-1)^{\mathrm{H}}}(k-1)\,{}^{\mathrm{f}}\boldsymbol{h}_{\mathrm{opt}}^{(k-1)}}{r_{ss}(0) - \boldsymbol{r}_{ss}^{(k-1)^{\mathrm{H}}}(k-1)\,{}^{\mathrm{b}}\boldsymbol{h}_{\mathrm{opt}}^{(k-1)}}$
Forward predictor	$\begin{bmatrix} {}^{\mathrm{f}}h_{\mathrm{opt},0}^{(k)} \\ {}^{\mathrm{f}}h_{\mathrm{opt},1}^{(k)} \\ \vdots \\ {}^{\mathrm{f}}h_{\mathrm{opt},k-2}^{(k)} \end{bmatrix} = {}^{\mathrm{f}}\boldsymbol{h}_{\mathrm{opt}}^{(k-1)} - {}^{\mathrm{b}}\boldsymbol{h}_{\mathrm{opt}}^{(k-1)}\,{}^{\mathrm{f}}h_{\mathrm{opt},k-1}^{(k)}$
Backward predictor	${}^{\mathrm{b}}h_{\mathrm{opt},l}^{(k)} = {}^{\mathrm{f}}h_{\mathrm{opt},k-1-l}^{(k)*}$ for $l = 0, \ldots, k-1$
Termination condition	
Num. inaccuracies	If $\left\|{}^{\mathrm{f}}h_{\mathrm{opt},k-1}^{(k)}\right\| > 1$, take the coefficients of the previous iteration and stop the recursion
Final order	If $k = N$, take the recent coefficients and stop the recursion

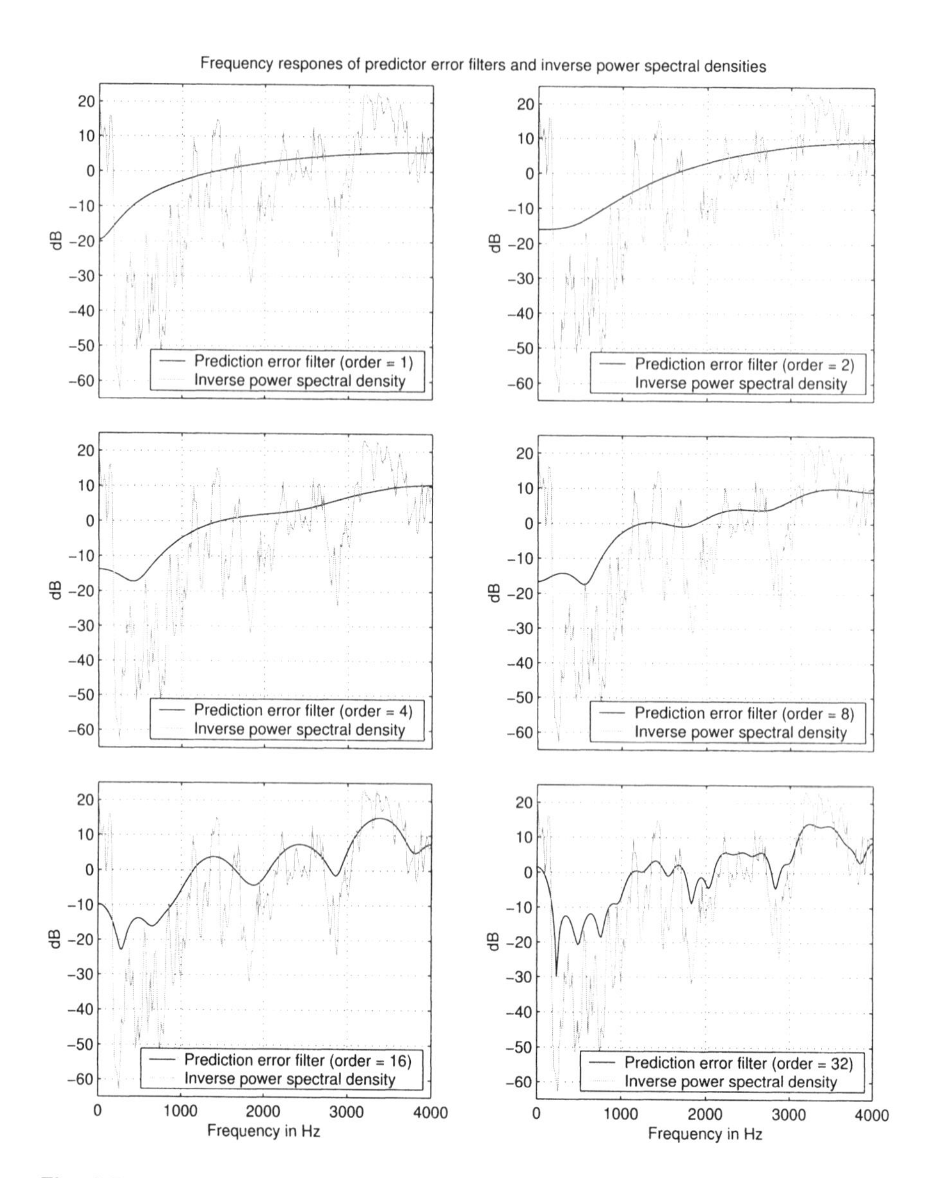

Fig. 6.5 Inverse spectral power density of the predictor input and absolute values of the predictor transfer function for increasing predictor orders.

7

Algorithms for Adaptive Filters

Adaptive filters come in two flavors: those employed for *system identification* and those used for *system equalization*. In the first group the filter parallels an unknown system (see Fig. 7.1). Both use the same input signal $x(n)$ and the filter is adapted such that both output signals $y(n)$ and $\widehat{d}(n)$ are as close as possible. If a minimum is reached, one assumes that the impulse response $\widehat{\boldsymbol{h}}(n)$ of the filter is also close to the impulse response $\boldsymbol{h}(n)$ of the system. It must be mentioned, however, that this is not necessarily true. The adaptive filter performs a *signal* identification. To achieve a *system* identification at the same time, a *persistent* input signal [111] is necessary. This means—loosely expressed—that all frequencies or all modes of the unknown system have to be excited.

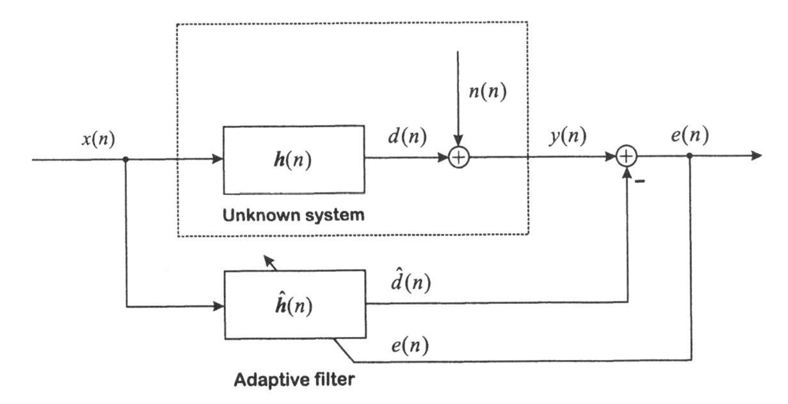

Fig. 7.1 Adaptive filter for system identification.

In order to show this performance, we use an example (see Fig. 7.2). The excitation signal $x(n)$ is colored noise; the disturbing signal $n(n)$ is white noise with a signal-to-disturbing signal ratio of about 30 dB. During the first 1000 iterations the excitation signal has the power spectral density, which is shown in the left part of the second diagram in Fig. 7.2. In the lower part of this figure the short-term power of the undistorted error signal $d(n) - \widehat{d}(n)$ is depicted.

After 1000 iterations the excitation signal is changed such that its power spectral density now has maxima where it had minima before. During the first 1000 iterations the unknown system was well identified only at those frequencies where the excitation signal had most of its energy. Therefore, after the change the power of the error signal increases because the system now is excited at different frequencies.

Also in Fig. 7.2 the squared norm of the system mismatch vector $\boldsymbol{h}_\Delta(n)$

$$\boldsymbol{h}_\Delta(n) = \boldsymbol{h}(n) - \widehat{\boldsymbol{h}}(n)\,, \tag{7.1}$$

called the *system distance*,

$$\|\boldsymbol{h}_\Delta(n)\|^2 = \|\boldsymbol{h}(n) - \widehat{\boldsymbol{h}}(n)\|^2 = [\,\boldsymbol{h}(n) - \widehat{\boldsymbol{h}}(n)\,]^{\mathrm{H}}\,[\,\boldsymbol{h}(n) - \widehat{\boldsymbol{h}}(n)\,]\,, \tag{7.2}$$

is monitored in the lowest diagram. This quantity does not depend directly on the properties of the excitation signal. Consequently, it is still quite large at the end of the first adaptation cycle and continues to decrease during the second cycle.

Only in case of white excitation $x(n)$ and no disturbing signal $n(n)$ are the normalized error power and the expected system distance equal:

$$\begin{aligned}\frac{\mathrm{E}\{\,|e(n)|^2\,\}}{\mathrm{E}\{\,|x(n)|^2\,\}}&\\ &= \frac{[\,\boldsymbol{h}(n) - \widehat{\boldsymbol{h}}(n)\,]^{\mathrm{H}}\,\mathrm{E}\{\,\boldsymbol{x}(n)\,\boldsymbol{x}^{\mathrm{H}}(n)\,\}\,[\,\boldsymbol{h}(n) - \widehat{\boldsymbol{h}}(n)\,]}{\mathrm{E}\{\,|x(n)|^2\,\}}\\ &= \frac{\boldsymbol{h}_\Delta^{\mathrm{H}}(n)\,\mathrm{E}\{\,\boldsymbol{x}(n)\,\boldsymbol{x}^{\mathrm{H}}(n)\,\}\,\boldsymbol{h}_\Delta(n)}{\mathrm{E}\{\,|x(n)|^2\,\}}\,.\end{aligned} \tag{7.3}$$

In case of white noise $\mathrm{E}\{\,\boldsymbol{x}(n)\,\boldsymbol{x}^{\mathrm{H}}(n)\,\}$ equals a diagonal matrix with $\mathrm{E}\{\,|x(n)|^2\,\}$ on its main diagonal. Therefore, Eq. 7.3 finally results in

$$\frac{\mathrm{E}\{\,|e(n)|^2\,\}}{\mathrm{E}\{\,|x(n)|^2\,\}} = \boldsymbol{h}_\Delta^{\mathrm{H}}(n)\,\boldsymbol{h}_\Delta(n)\,. \tag{7.4}$$

For *system equalization* the unknown system (e.g., a transmission channel) and the adaptive filter are series-connected (see Fig. 7.3). The (possibly) noisy output signal $y(n)$ forms the input of the adaptive filter. Since the original input $x(n)$ is not available at the location of the filter, special means are necessary to provide the desired output signal $d(n)$. In case of channel equalization the transmission starts with a test signal that is known to the receiver. After a successful adaptation the error signal $e(n)$ should be small and the transfer function of the filter models the *inverse* of the

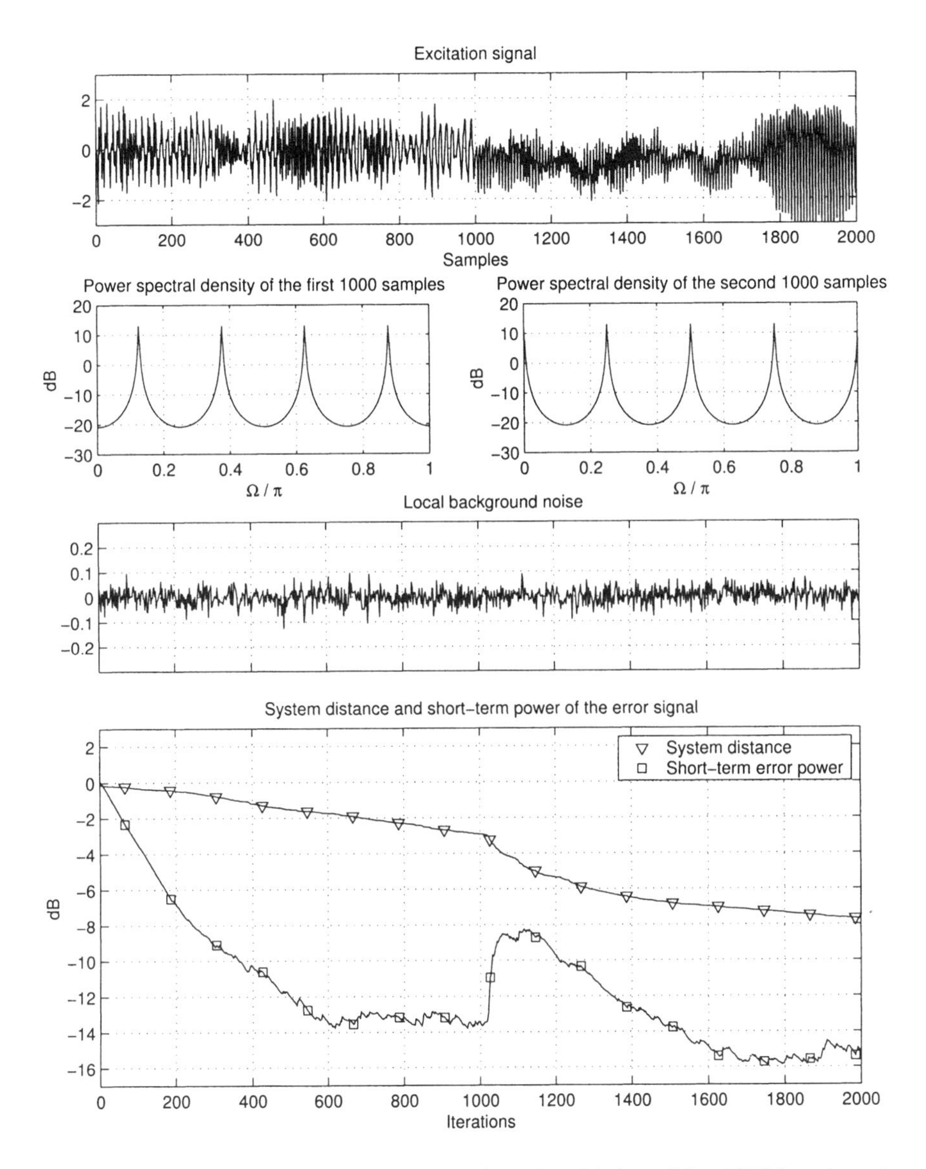

Fig. 7.2 Convergence of an adaptive filter with colored input. After 1000 iterations the excitation signal is changed. The power spectral density of the new signal has maxima where the spectral density of the original signal had its minima.

transfer function of the unknown system. As in the case of system identification, however, a persistent input signal is required.

In the following, we will address the system identification problem [66] only and refer to Fig. 7.1 for the general setup and the notation. We assume that the system, which should be identified, can be modeled with sufficient accuracy as a linear finite

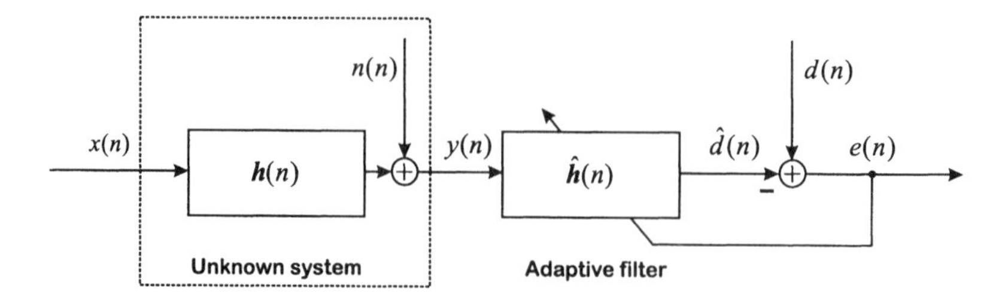

Fig. 7.3 Adaptive filter for system equalization.

impulse response (FIR) filter with complex impulse response $h_i(n)$. Compared to the derivation of the Wiener filter (see Chapter 5), here the impulse response is assumed to be time-varying. It can be written as a vector (see also Eq. 5.3)

$$\boldsymbol{h}(n) = \left[\, h_0(n),\, h_1(n),\, \ldots,\, h_{N-1}(n) \,\right]^{\mathrm{T}} , \tag{7.5}$$

where the subscript $i \in \{0, \ldots, N-1\}$ addresses the ith coefficient of the impulse response at time index n.

The output $y(n)$ of the unknown system consists of the desired signal $d(n)$ and an additional signal $n(n)$, which, in case of acoustic echo cancellation, represents an additive mixture of the local speech signal $s(n)$ and the background noise $b(n)$:

$$n(n) = s(n) + b(n) \,. \tag{7.6}$$

Both together [i.e., $n(n)$] act like noise for the filter adaptation. The desired signal $d(n)$ can be written as a convolution of the (unknown) impulse response $h_i(n)$ and the excitation signal $x(n)$ (see also Eqs. 5.2 and 5.5):

$$\begin{aligned} d(n) &= \sum_{i=0}^{N-1} h_i^*(n)\, y(n-i) \\ &= \boldsymbol{h}^{\mathrm{H}}(n)\, \boldsymbol{x}(n) = \boldsymbol{x}^{\mathrm{T}}(n)\, \boldsymbol{h}^*(n) \,. \end{aligned} \tag{7.7}$$

For the second line of Eq. 7.7 vector notation is also introduced for the excitation signal:

$$\boldsymbol{x}(n) = \left[\, x(n),\, x(n-1),\, \ldots,\, x(n-N+1) \,\right]^{\mathrm{T}} . \tag{7.8}$$

Before we analyze their convergence and tracking performance, we will introduce four specific adaptive algorithms that differ in the error criteria used and in their numerical complexity. We give only brief derivations and refer the reader to the huge amount of literature on this topic (e.g., [116, 240, 167, 152, 8, 227, 111]). Specifically, four adaptive algorithms will be discussed:

- Normalized least mean square (NLMS) algorithm
- Affine projection (AP) algorithm
- Recursive least squares (RLS) algorithm
- Kalman algorithm

The four algorithms are listed in the order of rising numerical complexity. Inherent to the latter is an increased speed of convergence. This, however, does not necessarily lead to better tracking performance.

7.1 THE NORMALIZED LEAST MEAN SQUARE ALGORITHM

In general, the update equation for the filter coefficients of an adaptive filter shows the following structure:

$$\widehat{\boldsymbol{h}}(n+1) = \widehat{\boldsymbol{h}}(n) + \boldsymbol{\Delta}\widehat{\boldsymbol{h}}(n) . \tag{7.9}$$

Referring to Fig. 7.1, the error $e(n)$ before the filter update [the *a priori error* $e(n|n)$] is given by

$$e(n) = e(n|n) = y(n) - \widehat{\boldsymbol{h}}^{\mathrm{H}}(n)\, \boldsymbol{x}(n) . \tag{7.10}$$

The *a posteriori error* $e(n|n+1)$, that is, the error after the filter update but without new input data, is given by

$$\begin{aligned} e(n|n+1) &= y(n) - \widehat{\boldsymbol{h}}^{\mathrm{H}}(n+1)\, \boldsymbol{x}(n) \\ &= y(n) - \widehat{\boldsymbol{h}}^{\mathrm{H}}(n)\, \boldsymbol{x}(n) - \boldsymbol{\Delta}\widehat{\boldsymbol{h}}^{\mathrm{H}}(n)\, \boldsymbol{x}(n) \\ &= e(n|n) - \boldsymbol{\Delta}\widehat{\boldsymbol{h}}^{\mathrm{H}}(n)\, \boldsymbol{x}(n) . \end{aligned} \tag{7.11}$$

A successful filter update requires that the absolute value of the a posteriori error $|e(n|n+1)|$ be at least not larger than the modulus of the a priori error $|e(n|n)|$

$$|e(n|n+1)| \leq |e(n|n)| \tag{7.12}$$

or equivalently

$$|e(n|n+1)|^2 \leq |e(n|n)|^2 . \tag{7.13}$$

By evaluating $|e(n|n+1)|^2$,

$$\begin{aligned} |e(n|n+1)|^2 &= \Big[e(n|n) - \boldsymbol{\Delta}\widehat{\boldsymbol{h}}^{\mathrm{H}}(n)\, \boldsymbol{x}(n)\Big] \Big[e(n|n)^* - \boldsymbol{x}^{\mathrm{H}}(n)\boldsymbol{\Delta}\widehat{\boldsymbol{h}}(n)\Big] \\ &= |e(n|n)|^2 - e(n|n)\, \boldsymbol{x}^{\mathrm{H}}(n)\boldsymbol{\Delta}\widehat{\boldsymbol{h}}(n) \\ &\quad - \boldsymbol{\Delta}\widehat{\boldsymbol{h}}^{\mathrm{H}}(n)\, \boldsymbol{x}(n)\, e^*(n|n) \\ &\quad + \boldsymbol{\Delta}\widehat{\boldsymbol{h}}^{\mathrm{H}}(n)\, \boldsymbol{x}(n)\, \boldsymbol{x}^{\mathrm{H}}(n)\, \boldsymbol{\Delta}\widehat{\boldsymbol{h}}(n) , \end{aligned} \tag{7.14}$$

condition 7.13 reads as follows:

$$\begin{aligned} e(n|n)\,\boldsymbol{x}^{\mathrm{H}}(n)\boldsymbol{\Delta}\widehat{\boldsymbol{h}}(n) \quad &+ \quad e^*(n|n)\,\boldsymbol{\Delta}\widehat{\boldsymbol{h}}^{\mathrm{H}}(n)\,\boldsymbol{x}(n) \\ &- \quad \boldsymbol{\Delta}\widehat{\boldsymbol{h}}^{\mathrm{H}}(n)\,\boldsymbol{x}(n)\,\boldsymbol{x}^{\mathrm{H}}(n)\,\boldsymbol{\Delta}\widehat{\boldsymbol{h}}(n) \geq 0\,. \end{aligned} \tag{7.15}$$

To find a rule for $\boldsymbol{\Delta}\widehat{\boldsymbol{h}}(n)$ that satisfies this condition, we assume

$$\boldsymbol{\Delta}\widehat{\boldsymbol{h}}(n) = \mu\, e^*(n|n)\,\frac{\boldsymbol{M}(n)\,\boldsymbol{x}(n)}{\boldsymbol{x}^{\mathrm{H}}(n)\,\boldsymbol{M}(n)\,\boldsymbol{x}(n)}\,, \tag{7.16}$$

where μ is a *stepsize factor* and $\boldsymbol{M}(n)$ is an arbitrary positive definite square matrix with $\boldsymbol{M}(n) = \boldsymbol{M}^{\mathrm{H}}(n)$. Inserting Eq. 7.16 into Eq. 7.15 leads to a condition on the range of the stepsize factor μ:

$$\mu\,[2-\mu] \geq 0 \quad \text{or} \quad 0 \leq \mu \leq 2 \tag{7.17}$$

In the simplest case the matrix $\boldsymbol{M}(n)$ equals a unit matrix $\boldsymbol{I}$, and one gets for the update:

$$\boldsymbol{\Delta}\widehat{\boldsymbol{h}}(n) = \mu\, e^*(n)\,\frac{\boldsymbol{x}(n)}{\boldsymbol{x}^{\mathrm{H}}(n)\,\boldsymbol{x}(n)}\,. \tag{7.18}$$

This is the filterupdate term for the *normalized least mean square algorithm* (NLMS algorithm), where we returned to the use of $e(n)$ instead of $e(n|n)$. Reordering terms, the complete update equation, finally, reads

$$\widehat{\boldsymbol{h}}(n+1) = \widehat{\boldsymbol{h}}(n) + \frac{\mu}{\boldsymbol{x}^{\mathrm{H}}(n)\,\boldsymbol{x}(n)}\,\boldsymbol{x}(n)\,e^*(n)\,. \tag{7.19}$$

The NLMS algorithm is summarized in Table 7.1.

It should be mentioned that since the excitation signal $x(n)$ and consequentially the error $e(n)$ are modeled by random processes, the filter coefficients $\widehat{\boldsymbol{h}}(n)$ are random processes, too.

The scalar denominator $\boldsymbol{x}^{\mathrm{H}}(n)\,\boldsymbol{x}(n)$ of Eq. 7.19 means a normalization of the stepsize according to the power of the input signal vector $\boldsymbol{x}(n)$ such that the boundaries of the stepsize factor μ become independent of this quantity. This is especially important for speech inputs that show a highly fluctuating short-term power. Furthermore, the coefficient update vector $\boldsymbol{\Delta}\widehat{\boldsymbol{h}}(n)$ points in the direction of the input signal vector $\boldsymbol{x}(n)$, which may not be the direction toward the minimum of a cost function. If $x(n)$ describes a speech signal, consecutive vectors $\boldsymbol{x}(n)$ are strongly correlated, meaning that their directions differ only slightly from each other. Consequently, convergence to the optimal filter coefficient is slow since many adaptation steps are necessary to reach the optimal point in the N-dimensional space of the filter coefficients. This is the price to be paid for the low computational complexity of the NLMS algorithm that does not involve a matrix operation. A matrix multiplication, however, would be required in order to turn the direction of a vector.

A second important feature of the NLMS algorithms also contributing to the low computational complexity is that it uses only "contemporary" data, that is the algorithm has no memory.

Table 7.1 Summary of the NLMS algorithm

Algorithmic part	Corresponding equations
Filtering	$\widehat{d}(n) = \widehat{\boldsymbol{h}}^{\mathrm{H}}(n)\,\boldsymbol{x}(n)$
Error signal	$e(n) = y(n) - \widehat{d}(n)$
Update of the norm	$\|\boldsymbol{x}(n)\|^2 = \|\boldsymbol{x}(n-1)\|^2 - \lvert x(n-N)\rvert^2 + \lvert x(n)\rvert^2$
Filter update	$\widehat{\boldsymbol{h}}(n+1) = \widehat{\boldsymbol{h}}(n) + \mu\,\dfrac{\boldsymbol{x}(n)\,e^*(n)}{\|\boldsymbol{x}(n)\|^2}$

7.1.1 Control Structures

Because of its robustness and its modest numerical complexity the NLMS algorithm may be considered as the workhorse of acoustic echo cancellation. Since, however, it has to work in an extremely "hostile" environment additional control measures are required to achieve a satisfactorily performance. Main obstacles for the fast convergence of the filter coefficients are the properties of speech signals and the signals combined in $n(n)$ (see Fig. 7.1) that are locally generated and that enter the algorithm like noise.

In this chapter we will discuss two possibilities for controlling the adaptation process: the already introduced stepsize factor μ (see Eq. 7.16) and increasing the denominator of the update term by a *regularization* Δ

$$\widehat{\boldsymbol{h}}(n+1) = \widehat{\boldsymbol{h}}(n) + \frac{\mu}{\boldsymbol{x}^{\mathrm{H}}(n)\,\boldsymbol{x}(n)\,+\Delta}\,\boldsymbol{x}(n)\,e^*(n) \tag{7.20}$$

(please compare with Eq. 7.19). Because of the scalar normalization within the NLMS update, both forms of control can easily be exchanged (see Section 13.1.3). Nevertheless, their practical implementations often differ. For this reason we will deal with both methods here. Their differences and their relation to different processing structures will be explained in the next paragraphs.

A further extension of control methods will be the replacement of the *scalar stepsize* parameter μ by a *stepsize matrix* $\mathrm{diag}\{\boldsymbol{\mu}\}$. To simplify our discussions, however, we will first assume time independent control parameters μ, $\boldsymbol{\mu}$, and Δ:

$$\begin{aligned}\widehat{\boldsymbol{h}}(n+1) &= \widehat{\boldsymbol{h}}(n) + \begin{bmatrix} \mu_0 & 0 & \dots & 0 \\ 0 & \mu_1 & & 0 \\ \vdots & & \ddots & \vdots \\ 0 & 0 & \dots & \mu_{N-1} \end{bmatrix} \frac{\boldsymbol{x}(n)\, e^*(n)}{\|\boldsymbol{x}(n)\|^2 + \Delta} \\ &= \widehat{\boldsymbol{h}}(n) + \mathrm{diag}\left\{\boldsymbol{\mu}\right\} \frac{\boldsymbol{x}(n)\, e^*(n)}{\|\boldsymbol{x}(n)\|^2 + \Delta}, \end{aligned} \tag{7.21}$$

where we consider only a diagonal stepsize matrix. Its elements on the main diagonal can be combined to a stepsize vector. In case of a diagonal stepsize matrix the computational complexity of the NLMS algorithm increases from $2N$ to $3N$. The impacts of this additional degree of freedom depend on the used processing structure (see Section 9.1). For the application of network echo cancellation, for instance, the impulse response concentrates its energy in certain areas. If these areas can be detected the adaptation can be enabled only in these regions. Examples of such an impulse response and the associated stepsize vector are depicted in Fig. 7.4. Thus, we exploit prior knowledge of the system in order to improve the performance of the adaptation algorithm.

In acoustic echo cancellation one knows in advance that the impulse response of the LEM system decays exponentially. According to this property a stepsize matrix $\mathrm{diag}\{\boldsymbol{\mu}\}$ with decaying main diagonal elements has been proposed [157].

We will return to time-dependent control parameters later. *Regularization* is usually applied in algorithms with matrix normalization (e.g., the AP or the RLS algorithm) for numerical stabilization. There the main diagonal of the normalization matrix is increased according to the regularization parameter.

In Chapter 13 we will derive pseudooptimal control parameters for the NLMS algorithm. The results, however, can also be implemented with algorithms using an inverse matrix in their update term.

7.1.2 Stability Analysis

In this section convergence and stability issues for LMS adaptive filters are discussed. In the literature several convergence analyses and LMS derivations (e.g., [50, 68, 111, 240]) can be found. Most of them are based on eigenvalue considerations of the autocorrelation matrix of the excitation signal.

The aim of this section is to derive a basic understanding and a general overview about the convergence properties in the presence of local noise. In particular, the dependence of the speed of convergence and the final misadjustment on the control parameters μ and Δ are investigated. For a better understanding, we will treat only scalar, time-invariant control parameters in this section. Furthermore, only system identification problems according to Eykhoff [66] will be addressed here.

If the state of convergence of an adaptive filter is to be judged, the mean square error is not the optimal criterion. In the main part of this section the convergence speed and the final misadjustment of adaptive filters are investigated. Only the case of white

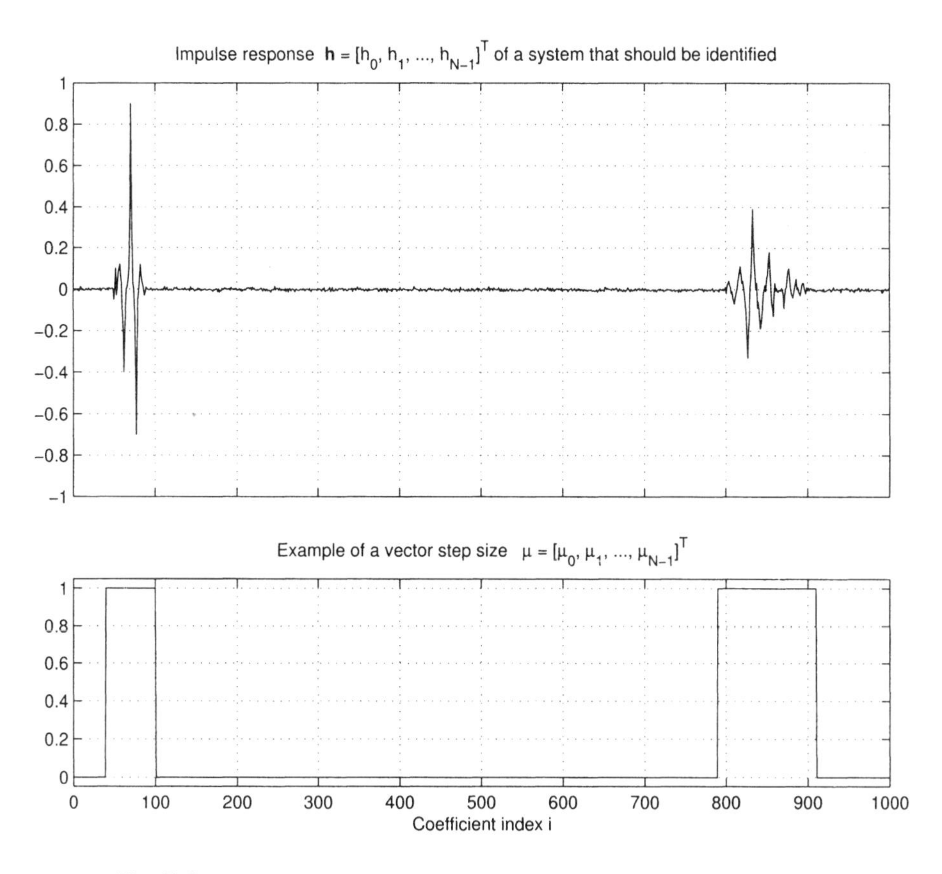

Fig. 7.4 Example of a vector stepsize control (with permission from [102]).

excitation will be mentioned. Even if this "laboratory" case is far away from real implementations, a basic understanding for the relationship of both quantities and the control parameters can be derived. The adaptation process can be separated into two parts; because of the adaptation itself (appropriate control parameters assumed) the system mismatch (see Eq. 7.2) performs a contracting mapping. Unfortunately local noise is disturbing the contraction. Depending on the signal-to-noise ratio the mapping can even be expanding. Both influences can be controlled using the stepsize and the regularization parameter. This connection will be investigated in Sections 7.1.2.1–7.1.2.4.

At the end of this section we will see that in the presence of local noise fast initial convergence, good tracking behavior, and small steady-state error can be achieved only if time-varying control parameters are used. The derivation of optimal, time varying control parameters will be the objective of the Chapters 12–15.

7.1.2.1 Convergence in the Presence of Local Noise

Using the definition of the system mismatch vector (see Eq. 7.1) and assuming that the system is time-invariant

$$\boldsymbol{h}(n+1) = \boldsymbol{h}(n)\,, \tag{7.22}$$

the equation for the NLMS filter update (see Eq. 7.19) can be used to derive an iteration of the system mismatch vector:

$$\boldsymbol{h}_\Delta(n+1) \;=\; \boldsymbol{h}_\Delta(n) - \mu\,\frac{e^*(n)\,\boldsymbol{x}(n)}{\|\boldsymbol{x}(n)\|^2+\Delta}\,. \tag{7.23}$$

The aim of the adaptive algorithm should be to minimize the expected squared norm of the system mismatch vector. Using Eq. 7.23 the squared norm can be computed recursively:

$$\begin{aligned}
\|\boldsymbol{h}_\Delta(n+1)\|^2 \;&=\; \boldsymbol{h}_\Delta^{\mathrm{H}}(n+1)\;\boldsymbol{h}_\Delta(n+1)\\
&=\; \boldsymbol{h}_\Delta^{\mathrm{H}}(n)\;\boldsymbol{h}_\Delta(n) - \mu\,\frac{e(n)\,\boldsymbol{x}^{\mathrm{H}}(n)\,\boldsymbol{h}_\Delta(n)}{\|\boldsymbol{x}(n)\|^2+\Delta}\\
&\quad -\mu\,\frac{\boldsymbol{h}_\Delta^{\mathrm{H}}(n)\,\boldsymbol{x}(n)\,e^*(n)}{\|\boldsymbol{x}(n)\|^2+\Delta} + \mu^2\,\frac{|e(n)|^2\,\boldsymbol{x}^{\mathrm{H}}(n)\,\boldsymbol{x}(n)}{\left(\|\boldsymbol{x}(n)\|^2+\Delta\right)^2}\\
&=\; \|\boldsymbol{h}_\Delta(n)\|^2 - \mu\,\frac{e(n)\,e_u^*(n)}{\|\boldsymbol{x}(n)\|^2+\Delta}\\
&\quad -\mu\,\frac{e_u(n)\,e^*(n)}{\|\boldsymbol{x}(n)\|^2+\Delta} + \mu^2\,\frac{|e(n)|^2\;\|\boldsymbol{x}(n)\|^2}{\left(\|\boldsymbol{x}(n)\|^2+\Delta\right)^2}\,.
\end{aligned} \tag{7.24}$$

The error signal $e(n)$ consists of its undistorted part $e_{\mathrm{u}}(n)$ and the local noise $n(n)$:

$$e(n) \;=\; e_{\mathrm{u}}(n) \;+\; n(n) \;=\; \boldsymbol{h}_\Delta^{\mathrm{H}}(n)\,\boldsymbol{x}(n) \;+\; n(n)\,. \tag{7.25}$$

Using this definition and assuming further that $x(n)$ and $n(n)$ are modeled by statistically independent and zero-mean random processes ($m_x = 0$, $m_n = 0$), the expected squared norm of the system mismatch vector can be written as

$$\begin{aligned}
\mathrm{E}\left\{\|\boldsymbol{h}_\Delta(n+1)\|^2\right\} \;&=\; \mathrm{E}\left\{\|\boldsymbol{h}_\Delta(n)\|^2\right\} - 2\,\mu\,\mathrm{E}\left\{\frac{\boldsymbol{h}_\Delta^{\mathrm{H}}(n)\,\boldsymbol{x}(n)\,\boldsymbol{x}^{\mathrm{H}}(n)\,\boldsymbol{h}_\Delta(n)}{\|\boldsymbol{x}(n)\|^2+\Delta}\right\}\\
&\quad +\mu^2\,\mathrm{E}\left\{\frac{\left(\boldsymbol{h}_\Delta^{\mathrm{H}}(n)\,\boldsymbol{x}(n)\,\boldsymbol{x}^{\mathrm{H}}(n)\,\boldsymbol{h}_\Delta(n) \;+\; |n(n)|^2\right)\;\|\boldsymbol{x}(n)\|^2}{\left(\|\boldsymbol{x}(n)\|^2+\Delta\right)^2}\right\}\,.
\end{aligned} \tag{7.26}$$

For a large filter length $N \gg 1$ the squared norm of the excitation vector can be approximated by a constant that is equal to N times the variance of the signal:

$$\|\boldsymbol{x}(n)\|^2 \;\approx\; N\,\sigma_x^2\,. \tag{7.27}$$

Inserting this approximation into Eq. 7.26 leads to

$$\begin{aligned} \mathrm{E}\Big\{\|\boldsymbol{h}_\Delta(n+1)\|^2\Big\} &\approx \mathrm{E}\Big\{\|\boldsymbol{h}_\Delta(n)\|^2\Big\} + \frac{\mu^2\,N\,\sigma_x^2}{(N\,\sigma_x^2+\Delta)^2}\,\sigma_n^2 \\ &+ \left(\frac{\mu^2\,N\,\sigma_x^2}{(N\,\sigma_x^2+\Delta)^2} - \frac{2\,\mu}{N\,\sigma_x^2+\Delta}\right)\mathrm{E}\left\{\boldsymbol{h}_\Delta^{\mathrm{H}}(n)\boldsymbol{x}(n)\boldsymbol{x}^{\mathrm{H}}(n)\boldsymbol{h}_\Delta(n)\right\}. \end{aligned} \tag{7.28}$$

Next we assume statistical independence of the excitation vector $\boldsymbol{x}(n)$ and the system error vector $\boldsymbol{h}_\Delta(n)$ as well as white excitation $x(n)$. Then the last expectation term in approximation 7.28 can be simplified:

$$\mathrm{E}\left\{\boldsymbol{h}_\Delta^{\mathrm{H}}(n)\boldsymbol{x}(n)\boldsymbol{x}^{\mathrm{H}}(n)\boldsymbol{h}_\Delta(n)\right\} = \sigma_x^2\,\mathrm{E}\Big\{\|\boldsymbol{h}_\Delta(n)\|^2\Big\}. \tag{7.29}$$

If the excitation signal is not white, it is possible to prewhiten or decorrelate the incoming excitation signal before passing it to the adaptive algorithm [111]. Especially in the field of acoustic, hybrid, and network echo cancellation, decorrelation filters are widely applied [70, 245, 247]. Decorrelation filters are predictor error filters whose coefficients are matched to the correlation properties of the input signal $x(n)$.

There are various possibilities for placing the decorrelation filters within the system. For identification purposes, it would be sufficient to simply filter the excitation signal. But if the excitation signal should not be changed before passing it into the unknown system, preprocessing is not allowed. One can move the decorrelation filter into the filter branch and also across the unknown system (see Fig. 7.5). Since the filter is still in the signal path, one has to apply an inverse filter after the calculation of the error value. To invert the linear prediction, one would have to use a recursive filter and apply the same coefficients as in the prediction filter. Since a linear prediction error filter is minimum phase, its inverse is always causal and therefore stable [111]. Further details on decorrelation filters are presented in Section 9.1.1.1.

If the results of Eq. 7.29 are inserted into approximation 7.28, we get

$$\begin{aligned} \mathrm{E}\Big\{\|\boldsymbol{h}_\Delta(n+1)\|^2\Big\} \approx &\left(1 + \frac{\mu^2\,N\,\sigma_x^4}{(N\,\sigma_x^2+\Delta)^2} - \frac{2\,\mu\,\sigma_x^2}{N\,\sigma_x^2+\Delta}\right)\mathrm{E}\Big\{\|\boldsymbol{h}_\Delta(n)\|^2\Big\} \\ &+ \frac{\mu^2\,N\,\sigma_x^2}{(N\,\sigma_x^2+\Delta)^2}\,\sigma_n^2\,. \end{aligned} \tag{7.30}$$

The first term in approximation 7.30 shows the contraction due to the undistorted adaptation process. The factor

$$A(\mu,\Delta,\sigma_x^2,N) = 1 + \frac{\mu^2\,N\,\sigma_x^4}{(N\,\sigma_x^2+\Delta)^2} - \frac{2\,\mu\,\sigma_x^2}{N\,\sigma_x^2+\Delta} \tag{7.31}$$

will be called *contraction parameter* and should always be smaller than one. If the control parameters are chosen within the intervals $0 < \mu < 2$ and $0 \leq \Delta < \infty$,

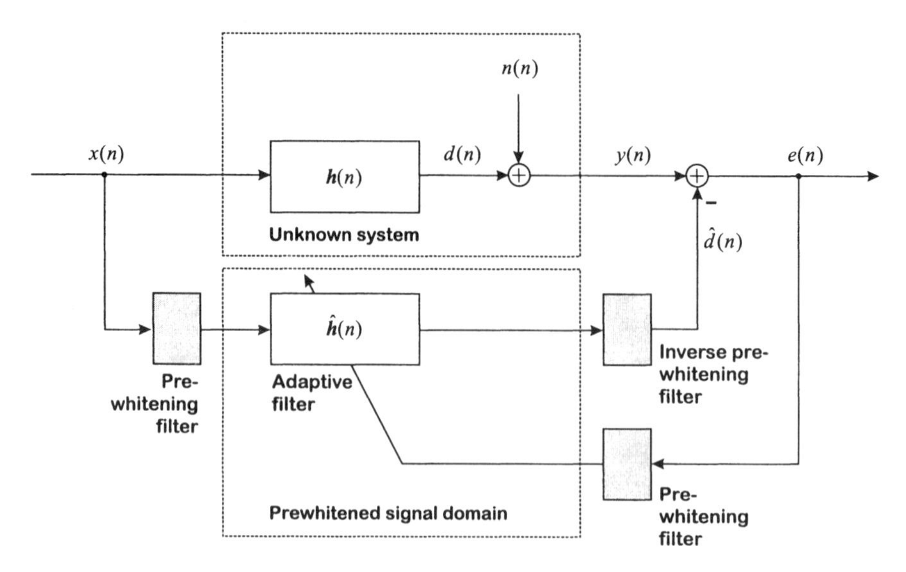

Fig. 7.5 Prewhitening structure with time-invariant decorrelation and inverse decorrelation filters.

this restriction is always fulfilled. The second term in approximation 7.30 describes the influence of the local noise. This signal disturbs the adaptation process. After introducing the abbreviation

$$B(\mu, \Delta, \sigma_x^2, N) = \frac{\mu^2 N \sigma_x^4}{(N \sigma_x^2 + \Delta)^2}, \tag{7.32}$$

which we call the *expansion parameter*, approximation 7.30 can be written in a shorter form:

$$\mathrm{E}\left\{\|\boldsymbol{h}_\Delta(n+1)\|^2\right\} \approx \underbrace{A(\mu, \Delta, \sigma_x^2, N)}_{\substack{\text{Contraction} \\ \text{parameter}}} \mathrm{E}\left\{\|\boldsymbol{h}_\Delta(n)\|^2\right\} + \underbrace{B(\mu, \Delta, \sigma_x^2, N)}_{\substack{\text{Expansion} \\ \text{parameter}}} \frac{\sigma_n^2}{\sigma_x^2}. \tag{7.33}$$

The contraction and the expansion parameters are quantities with dimension one, and both are dependent on the control variables μ and Δ as well as on the filter length N and the excitation power σ_x^2. If the influence of the local noise should be eliminated completely the expansion parameter $B(\mu, \Delta, \sigma_x^2, N)$ should be zero. This can be achieved by setting the stepsize μ to zero or the regularization parameter Δ to infinity. Unfortunately these choices would lead to a contraction parameter $A(\mu, \Delta, \sigma_x^2, N) = 1$. In this case the filter will not be adapted. In Fig. 7.6 the value of the contraction and expansion parameters for a filter length of $N = 100$ and an input power of $\sigma_x^2 = 1$ are depicted.

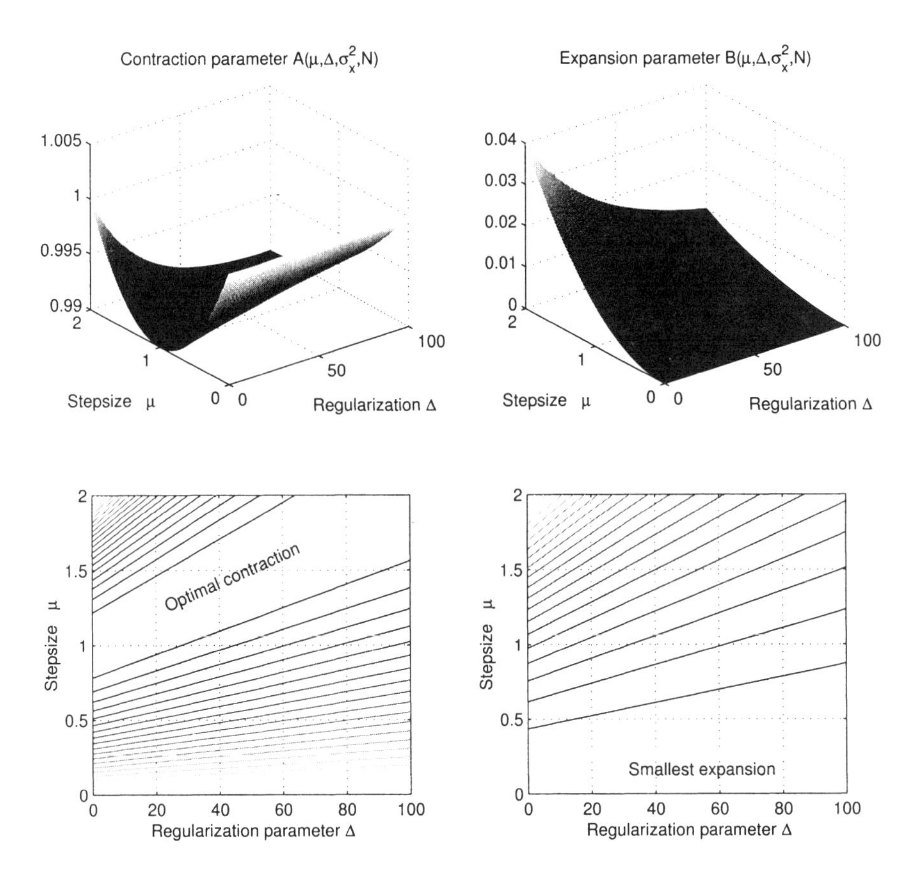

Fig. 7.6 Contraction–expansion parameter. The upper two diagrams show surface plots of both parameter functions for fixed excitation power $\sigma_x^2 = 1$ and a fixed filter length of $N = 100$. In the lower two diagrams appropriate conture plots are depicted (with permission from [102]).

For a fast convergence a compromise between fastest contraction

$$A(\mu, \Delta, \sigma_x^2, N) \ \rightarrow \ \min \tag{7.34}$$

and no influence of the local noise

$$B(\mu, \Delta, \sigma_x^2, N) \ = \ 0 \tag{7.35}$$

has to be found. The optimal compromise will be dependent on the convergence state of the filter. In Section 13.1 this question will be answered. We will first investigate the convergence of the filter for fixed control parameters. We can therefore solve the difference approximation 7.33:

$$\mathrm{E}\left\{\|\boldsymbol{h}_\Delta(n)\|^2\right\} \approx \left(A(\mu,\Delta,\sigma_x^2,N)\right)^n \mathrm{E}\left\{\|\boldsymbol{h}_\Delta(0)\|^2\right\} + B(\mu,\Delta,\sigma_x^2,N)\,\frac{\sigma_n^2}{\sigma_x^2}\sum_{m=0}^{n-1}\left(A(\mu,\Delta,\sigma_x^2,N)\right)^m . \quad (7.36)$$

Due to the local noise, the system distance will not be reduced to zero. Depending on the control parameters, the filter order, and the signal-to-noise ratio, the influence of contraction and expansion will equilibrate. For $|A(\mu,\Delta,\sigma_x^2,N)| < 1$, the system distance will converge for $n \to \infty$ toward

$$\lim_{n\to\infty} \mathrm{E}\left\{\|\boldsymbol{h}_\Delta(n)\|^2\right\} \approx B(\mu,\Delta,\sigma_x^2,N)\,\frac{\sigma_n^2}{\sigma_x^2}\,\frac{1}{1-A(\mu,\Delta,\sigma_x^2,N)} = \frac{\mu\, N\,\sigma_n^2}{(2-\mu)\,N\,\sigma_x^2 + 2\,\Delta} . \quad (7.37)$$

For an adaptation with $\mu = 1$ and $\Delta = 0$, the final system distance will be equal to the inverse of the signal-to-noise ratio:

$$\lim_{n\to\infty} \mathrm{E}\left\{\|\boldsymbol{h}_\Delta(n)\|^2\right\}\Big|_{\mu=1,\,\Delta=0} = \frac{\sigma_n^2}{\sigma_x^2} . \quad (7.38)$$

Finally two simulation examples are presented in Fig. 7.7 to indicate the validity of approximation 7.36. In both cases white noise was chosen as excitation and local noise with a signal-to-noise ratio of 30 dB. In the first simulation the parameter set $\mu_1 = 0.7$ and $\Delta_1 = 400$ has been used. According to approximation 7.37, a final misadjustment of about -35 dB should be achieved. For the second parameter set, the values are $\mu_2 = 0.4$ and $\Delta_2 = 800$. With this set a final misadjustment of about -39 dB is achievable.

In the lowest diagram of Fig. 7.7 the measured system distance as well as its theoretic progression are depicted. The theoretic and the measured curves are overlaying mostly—the maximal (logarithmic) difference is about 3 dB. The predicted final misadjustments coincide very well with the measured ones.

7.1.2.2 Convergence in the Absence of Local Noise

In the last section the theoretical progression of the system distance was investigated. The steady-state performance and its relationship to the control parameters have been of particular interest. In this section we will analyze a second important characteristic of the adaptation process: the speed of convergence.

The speed of convergence is of special importance at the start of an adaptation and after changes of the system $\boldsymbol{h}(n)$. In these situations we will assume that the power of the excitation signal as well as the power of the error signal are much larger than the power of the local noise:

$$\mathrm{E}\left\{|x(n)|^2\right\} \gg \mathrm{E}\left\{|n(n)|^2\right\}, \quad (7.39)$$

$$\mathrm{E}\left\{|e(n)|^2\right\} \gg \mathrm{E}\left\{|n(n)|^2\right\}. \quad (7.40)$$

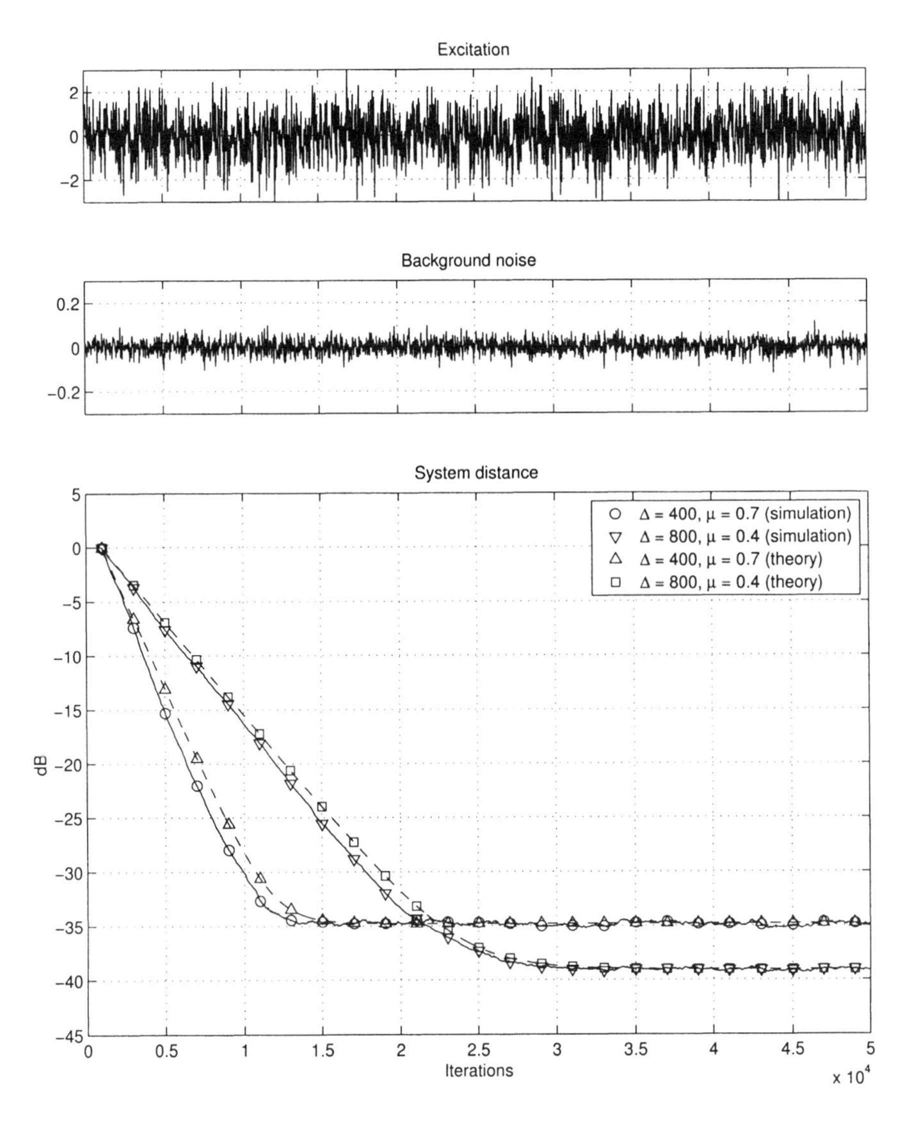

Fig. 7.7 Simulation examples and theoretic convergence. In order to validate approximation 7.36, two simulations are presented. White noise was used as excitation (depicted in the upper diagram) and local noise (presented in the second diagram) with a signal-to-noise ratio of 30 dB. In the last diagram the measured system distance as well as its theoretic progression are depicted (with permission from [102]).

Due to these assumptions, the influence of the local noise can be neglected. Therefore the recursive computation of the expected system distance can be simplified to

$$\mathrm{E}\left\{\|\boldsymbol{h}_\Delta(n+1)\|^2\right\}\Big|_{\sigma_n^2=0} \approx A(\mu,\Delta,\sigma_x^2,N)\,\mathrm{E}\left\{\|\boldsymbol{h}_\Delta(n)\|^2\right\} . \tag{7.41}$$

The average system distance forms a contracting series as long as

$$|A(\mu, \Delta, \sigma_x^2, N)| \; < \; 1\,. \tag{7.42}$$

We will assume a nonnegative stepsize ($\mu \geq 0$) as well as a nonnegative regularization ($\Delta \geq 0$). In this case we can transform condition 7.42 into

$$\mu \; < \; 2 \; + \; \frac{2\,\Delta}{N\,\sigma_x^2}\,. \tag{7.43}$$

Here the additional assumption $N > 2\mu$ was used. The fastest convergence will be achieved if the contraction parameter is minimal:

$$\left.\frac{\partial A(\mu, \Delta, \sigma_x^2, N)}{\partial \mu}\right|_{\mu_{\text{opt}}} \; = \; \left.\frac{\partial A(\mu, \Delta, \sigma_x^2, N)}{\partial \Delta}\right|_{\Delta_{\text{opt}}} \; = \; 0\,. \tag{7.44}$$

After inserting the definition of the contraction parameter (see Eq. 7.31) and equating both derivatives we finally get maximal speed of convergence if the stepsize and the regularization parameter are related as

$$\mu \; = \; 1 \; + \; \frac{\Delta}{N\,\sigma_x^2}\,. \tag{7.45}$$

In Fig. 7.8 the convergence plane for the undistorted adaptation ($\sigma_n^2 = 0$) is depicted in the range $0 \leq \mu \leq 4$ and $0 \leq \Delta \leq 100$. The depicted contraction area is based on a filter length of $N = 100$ and an excitation power of $\sigma_x^2 = 1$. Optimal contraction according to 7.45 can be achieved on the line that cuts the convergence plane into halves.

For the special case $\mu = 1$ and $\Delta = 0$, the recursive computation of the expected system distance in the absence of local noise (see Eq. 7.41) can be simplified to

$$\mathrm{E}\left\{\|\boldsymbol{h}_\Delta(n+1)\|^2\right\}\Big|_{\sigma_n^2=0} \; \approx \; \left(1 - \frac{1}{N}\right)\, \mathrm{E}\left\{\|\boldsymbol{h}_\Delta(n)\|^2\right\}\Big|_{\sigma_n^2=0}\,. \tag{7.46}$$

In this case maximal speed of convergence is achieved. Especially at the beginning of system identifications (initial convergence) where the condition $\sigma_e^2 \gg \sigma_n^2$ is fulfilled, system designers can compare the performance (convergence speed) of their system to this theoretical threshold. In practice, test runs in which all system parameters are accessible are often performed before the control system "starts its work." In these test or pilot runs the time is measured until the system distance has decreased by 10 dB. If the control parameters are adjusted as proposed above ($\mu = 1$, $\Delta = 0$), this time can be computed as follows:

$$\begin{aligned}
\mathrm{E}\left\{\|\boldsymbol{h}_\Delta(n + n_{10\,\text{dB}})\|^2\right\} \; &= \; \frac{1}{10}\,\mathrm{E}\left\{\|\boldsymbol{h}_\Delta(n)\|^2\right\} \\
&\approx \; \left(1 - \frac{1}{N}\right)^{n_{10\,\text{dB}}} \mathrm{E}\left\{\|\boldsymbol{h}_\Delta(n)\|^2\right\}.
\end{aligned} \tag{7.47}$$

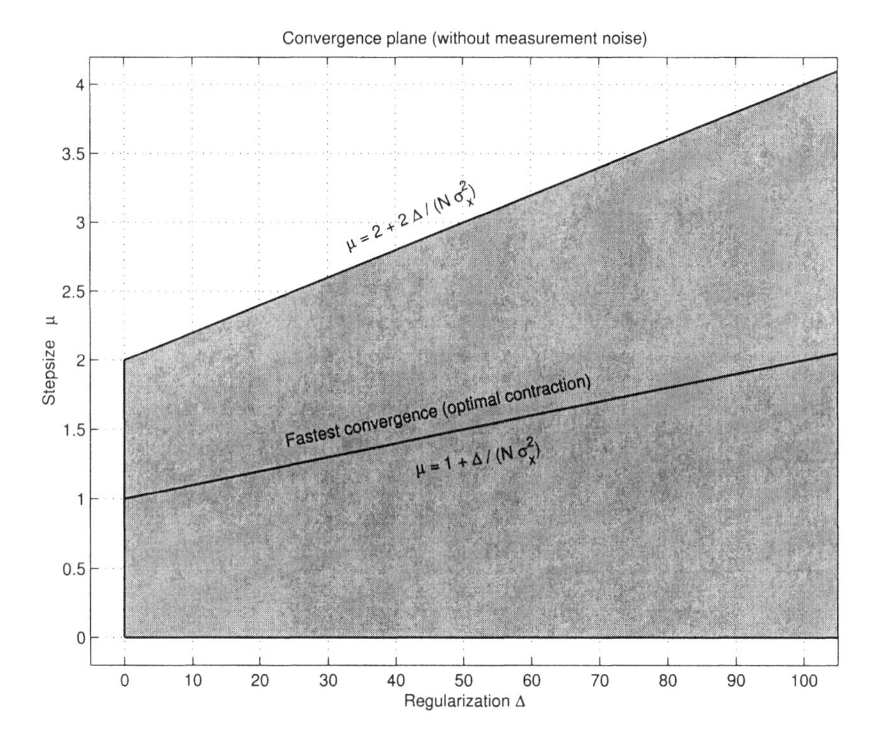

Fig. 7.8 Convergence plane for $\sigma_n^2 = 0$, $\sigma_x^2 = 1$, and $N = 100$. The gray plane depicts the control parameter area, where the system distance is contracting (with permission from [102]).

Comparing the factors in front of the expected system distances and taking the natural logarithm leads to

$$-\ln 10 \;\approx\; n_{10\,\mathrm{dB}} \ln\left(1 - \frac{1}{N}\right). \tag{7.48}$$

For a large filter length N the second logarithm in Eq. 7.48 can be approximated according to

$$\ln(1+x) \;=\; \sum_{k=0}^{\infty} \frac{(-1)^k}{k+1}\, x^{k+1} \;\approx\; x \quad \text{for} \quad |x| < 1. \tag{7.49}$$

Using this approximation as well as $\ln 10 \approx 2$, we finally get

$$n_{10\,\mathrm{dB}} \;\approx\; 2N. \tag{7.50}$$

This means that after a number of iterations that is equal to twice the filter length a decrease of the system distance of about 10 dB can be achieved. We can also learn from approximation 7.48 that short filters converge much faster than do long ones. In order to elucidate this relationship, two convergence examples with different filter orders are presented in Fig. 7.9.

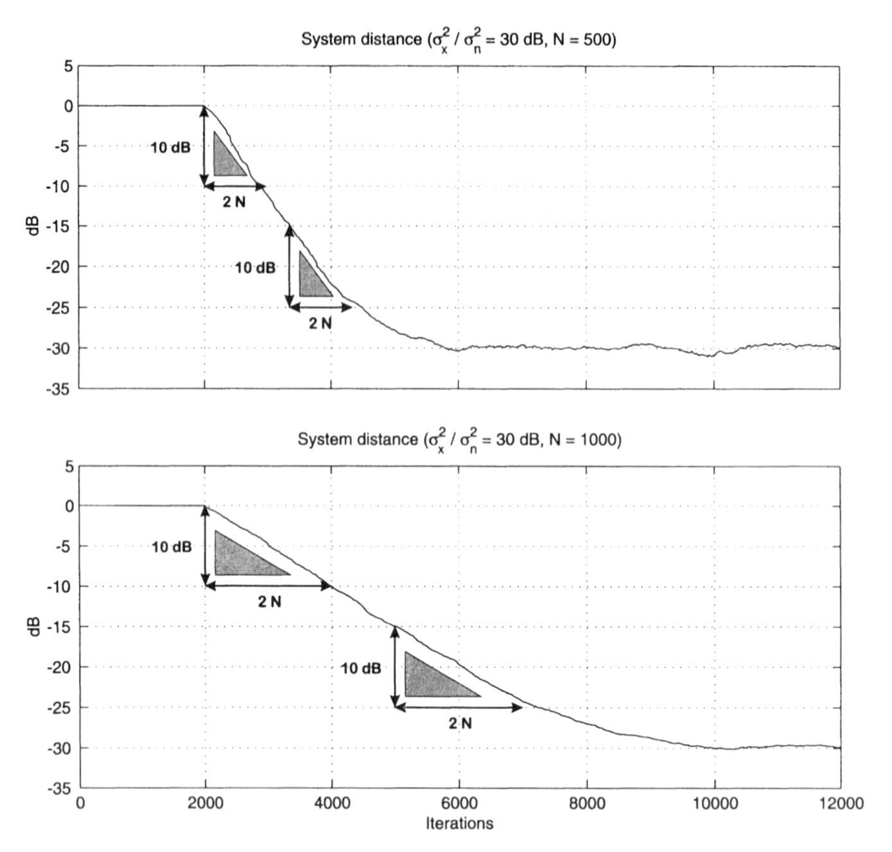

Fig. 7.9 Two convergence examples with different filter orders are depicted (upper diagram $N = 500$, lower diagram $N = 1000$). White noise was used for the excitation signal as well as for the local noise with a signal-to-noise ratio of about 30 dB (with permission from [102]).

In the first simulation a filter length of $N = 500$ for both filters [for the adaptive filter $\widehat{\boldsymbol{h}}(n)$ and for the unknown impulse response $\boldsymbol{h}(n)$] was implemented. White noise was used for the excitation signal as well as for the local noise with a signal-to-noise ratio of about 30 dB. The adaptation of the filter $\widehat{\boldsymbol{h}}(n)$ was enabled after iteration 2000. The control parameter were set to $\mu = 1$ and $\Delta = 0$.

The second convergence example, depicted in the lower part of Fig. 7.9, has the same boundary conditions except that the filter length was doubled to $N = 1000$. Even if both system distance curves are reaching the same theoretical minimum of about -30 dB (see Eq. 7.38), the speed of convergence is rather different. The larger the filter order is chosen, the slower is the speed of convergence.

To show the validity of approximation 7.50 triangles with edge lengths 10 dB and $2N$ are added to the convergence plots in Fig. 7.9. Especially where the convergence starts [here the assumption $\sigma_e^2(n) \gg \sigma_n^2(n)$ is certainly fulfilled] the theoretical declines of the system distances fit the measured ones very well.

Test or pilot runs are often performed with real excitation signals. If the speed of convergence drops distinctly below the theoretical threshold, decorrelation filters according to Fig. 7.5 should be applied. Nevertheless fixed decorrelation filters have also a drawback. They can only compensate the mean power spectral density of the excitation signal. The coefficients are therefore calculated using a representative set of input sequences. This, of course, can never be an optimal match to the actual excitation signal. The speed of convergence therefore is slower as in case of an adaptive renewal of the decorrelation filter coefficients. The great advantage of fixed decorrelation filters, however, is their low computational complexity.

Improved performance is achieved by the use of adaptive decorrelation filters (see Fig. 7.10). The filter coefficients are matched periodically to the short-term characteristics of the excitation signal. The coefficients for the decorrelation filter are calculated using the Levinson–Durbin algorithm [111]. It is important, however, to stop the recursion when the order of the correlation filter outnumbers the rank of the input autocorrelation matrix. In this case the autocorrelation matrix becomes singular. Due to the nonstationarity of most real excitation signals, this length can vary with time. However, since the Levinson–Durbin algorithm is rank-recursive, one can always fall back to the preceding order.

One should note that the increase of computational complexity is not negligible. In addition to the computation of the new decorrelation coefficients, both the excitation vector and its squared vector norm have to be updated. If necessary, one can compensate for this additional computational load by skipping updates of the echo canceler coefficients.

In a single-filter structure the update of the excitation vector $\boldsymbol{x}(n)$ requires a large computational effort. Whenever the coefficients of the prewhitening filter are exchanged (1) the excitation vector has to be recalculated by a convolution with the *old* inverse decorrelation filter and (2) a convolution with the *new* prewhitening filter has to be performed. Since the decorrelation filter has FIR structure, its inverse is of IIR structure. Especially the inverse decorrelation counts for a large computational load.

In order to reduce the amount of processing power, a two-filter structure according to Fig. 7.10 can be applied. The adaptation is performed only in the prewhitened signal domain. The coefficients of the filter vector can be copied to the filter in the original signal domain. After updating the coefficients of the decorrelation filters, the original excitation $x(n)$ is still available and only the new decorrelation has to be computed. Further details can be found in Section 9.1.1.1.

If the speed of convergence is still much lower than the theoretical limit or if the update periods of the decorrelation filter need to be too short, the application of more sophisticated algorithms like the AP algorithm (see Section 7.2) or the RLS algorithm (see Section 7.3) should be considered.

7.1.2.3 Convergence without Regularization

In some applications only stepsize control is implemented. In these cases the recursive computation of the expected system distance according to approximation

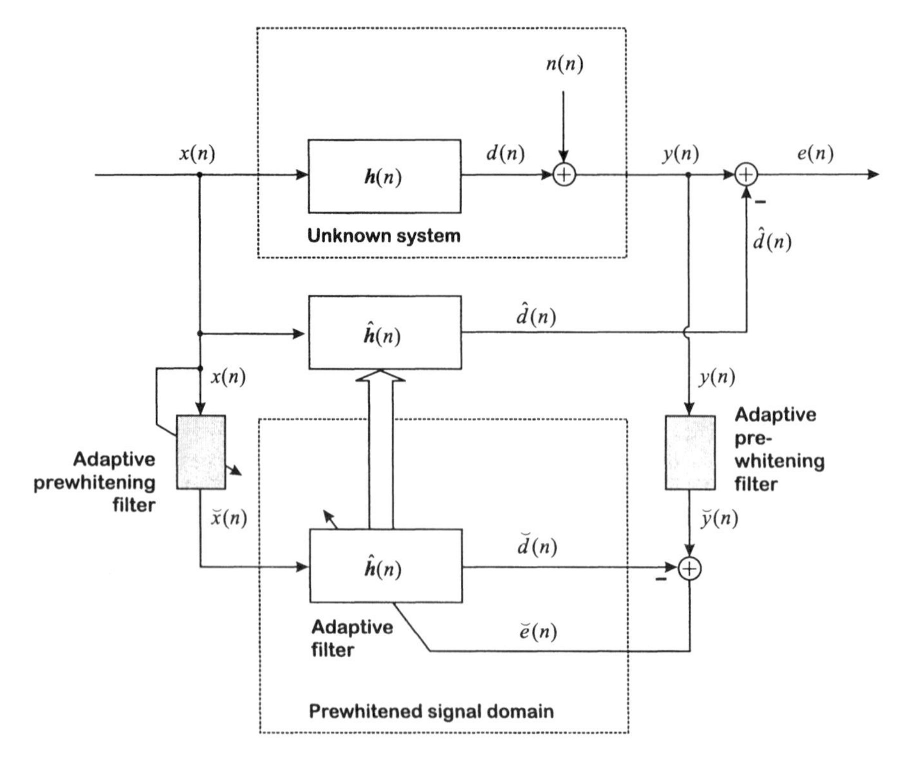

Fig. 7.10 Two-filter structure with time-variant decorrelation filters.

7.30 can be simplified to

$$\mathrm{E}\Big\{\|\boldsymbol{h}_\Delta(n+1)\|^2\Big\}\Big|_{\Delta=0} \approx \left(1-\frac{\mu\,(2-\mu)}{N}\right)\mathrm{E}\Big\{\|\boldsymbol{h}_\Delta(n)\|^2\Big\}\Big|_{\Delta=0} + \frac{\mu^2}{N}\,\frac{\sigma_n^2}{\sigma_x^2}\,. \tag{7.51}$$

For $0 \le \mu \le 2$ the system distance converges for $n \to \infty$ toward

$$\lim_{n\to\infty} \mathrm{E}\Big\{\|\boldsymbol{h}_\Delta(n)\|^2\Big\}\Big|_{\Delta=0} \approx \frac{\mu}{2-\mu}\,\frac{\sigma_n^2}{\sigma_x^2}\,. \tag{7.52}$$

In Fig. 7.11 three examples of convergences with different choices for the stepsize ($\mu = 1$, $\mu = 0.5$, and $\mu = 0.25$) are presented. The same boundary conditions (white noise for the excitation and the local noise, 30 dB signal-to-noise ratio, $N = 1000$) are used as in the first simulation series of Section 7.1.2.1.

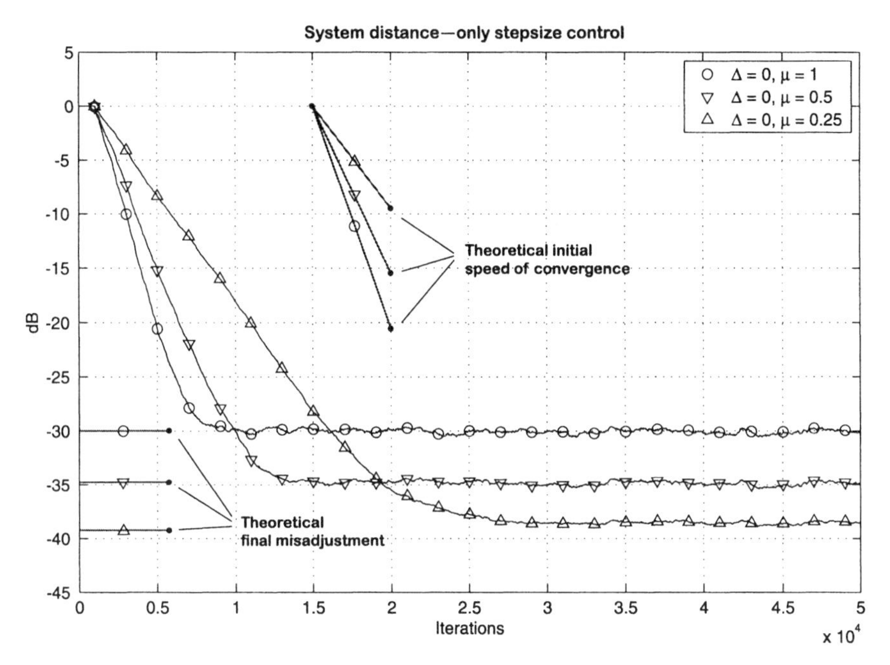

Fig. 7.11 Three examples of the convergence without regularization but with different choices for the stepsize ($\mu = 1$, $\mu = 0.5$, and $\mu = 0.25$) are depicted. The same boundary conditions (white noise for the excitation and the local noise, 30 dB signal-to-noise ratio, $N = 1000$) are used as in the first simulation series of Section 7.1.2.1 (with permission from [102]).

According to approximation 7.52 final filter misadjustments of

- -30 dB, for the choice $\mu = 1$
- -34.8 dB, for the choice $\mu = \frac{1}{2}$
- -38.5 dB, for the choice $\mu = \frac{1}{4}$

should be achieved. The speed of the initial convergence can be computed as

$$10 \log_{10} \left(1 - \frac{\mu (2 - \mu)}{N} \right) \frac{\text{dB}}{\text{iteration}} . \tag{7.53}$$

In Fig. 7.11 the theoretical final misadjustments as well as the approximated convergence speeds are inserted. A final misadjustment of the adaptive filter smaller than the signal-to-noise ratio (here 30 dB) can be achieved only if the stepsize is smaller than 1. This leads, however, to a decreased speed of convergence.

7.1.2.4 Convergence without Stepsize Control (only with Regularization)

By analogy with the last section, the recursive approximation of the mean squared norm of the system mismatch vector can be simplified only if regularization control ($\mu = 1$) is applied:

$$\mathrm{E}\Big\{\|\boldsymbol{h}_\Delta(n+1)\|^2\Big\}\Big|_{\mu=1} \approx \left(1 - \frac{N\sigma_x^4 + 2\sigma_x^2\Delta}{\left(N\sigma_x^2 + \Delta\right)^2}\right) \mathrm{E}\Big\{\|\boldsymbol{h}_\Delta(n)\|^2\Big\}\Big|_{\mu=1} + \frac{N\sigma_x^4}{\left(N\sigma_x^2 + \Delta\right)^2} \frac{\sigma_n^2}{\sigma_x^2}. \tag{7.54}$$

For $\Delta \geq 0$ the system distance converges for $n \to \infty$ toward

$$\lim_{n\to\infty} \mathrm{E}\left\{\|\boldsymbol{h}_\Delta(n)\|^2\right\}\Big|_{\mu=1} \approx \frac{N\sigma_x^2}{N\sigma_x^2 + 2\Delta} \frac{\sigma_n^2}{\sigma_x^2}. \tag{7.55}$$

Like in the previous section three simulation examples with regularization control only are presented. In Fig. 7.12 convergence curves for the same boundary conditions—except for the choice of the control parameters—as in the last example are given.

The simulation runs were performed with $\Delta = 0$, $\Delta = 1000$, and $\Delta = 4000$. According to these values and approximation 7.55 final misadjustments of the adaptive filter of -30 dB, -34.8 dB, and -39.5 dB should be achievable. The initial convergence speed (see recursion 7.54) can be stated as

$$10 \log_{10}\left(1 - \frac{N\sigma_x^4 + 2\sigma_x^2\Delta}{\left(N\sigma_x^2 + \Delta\right)^2}\right) \frac{\text{dB}}{\text{iteration}}. \tag{7.56}$$

The expected final misadjustments and the convergence speeds shown in Fig. 7.11 are also shown in Fig. 7.12.

7.1.2.5 Concluding Remarks

In this section the convergence properties of the NLMS algorithm in the presence of local noise were investigated. Approximations for the speed of convergence as well as the final misadjustment of the adaptive filter in dependence of the control parameters were derived. It was shown that using parameter sets leading to an optimal initial speed of convergence (e.g., $\mu = 1$, $\Delta = 0$) the final misadjustment can not be smaller than the inverse of the signal-to-noise ratio.

Especially for small signal-to-noise ratios this might not be adequate for a system identification. Reducing the stepsize or increasing the regularization parameter leads to smaller final misadjustment, but at the same time to a reduced speed of convergence.

A high speed of convergence as well as a good steady-state behavior can be achieved only with time-variant control parameters. Several control strategies are

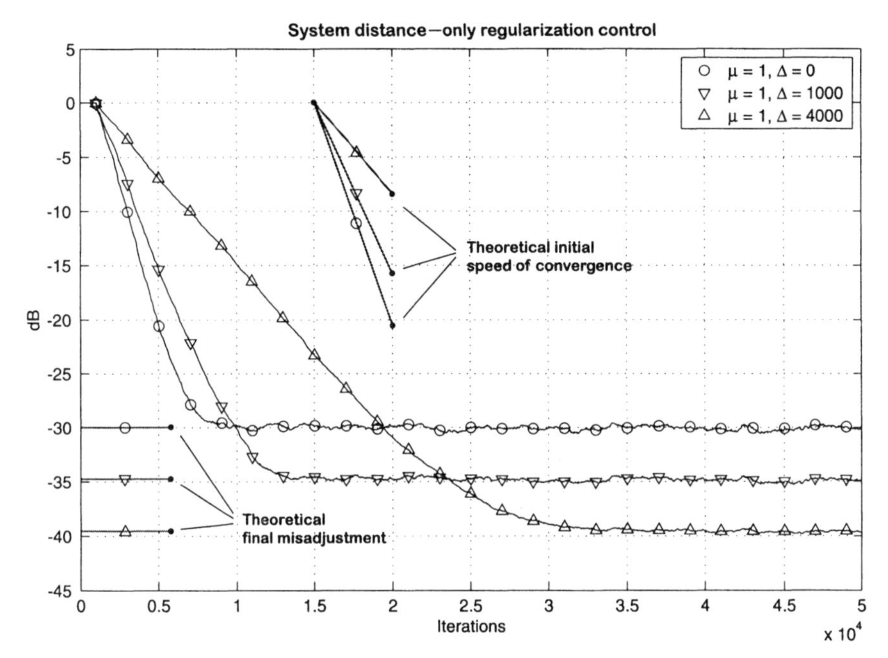

Fig. 7.12 Three examples of convergence without stepsize control but with different choices for the regularization parameter ($\Delta = 0$, $\Delta = 1000$, and $\Delta = 4000$) are depicted. The same boundary conditions (white noise for the excitation and the local noise, 30 dB signal-to-noise ratio, $N = 1000$) were used as in Fig. 7.11 (with permission from [102]).

depicted in Fig. 7.13. If only stepsize control ($\Delta = 0$) is implemented, one should start with a large stepsize ($\mu = 1$). The more the system distance reduces, the more the stepsize should be lessened.

Similar strategies apply if only regularization control or hybrid control are implemented. In Chapter 13 optimal choices for the stepsize and the regularization parameter in dependence of the convergence state and the signal-to-noise ratio will be derived.

7.2 THE AFFINE PROJECTION ALGORITHM

In contrary to the NLMS algorithm, the affine projection (AP) algorithm does not only consider the error at time n. It also takes into account "hypothetical" errors resulting from old data vectors filtered by the adaptive filter with current coefficient settings. This means that the computational complexity of the AP algorithm as compared to the NLMS algorithm is increased. The algorithm has memory. Further, we can expect faster convergence.

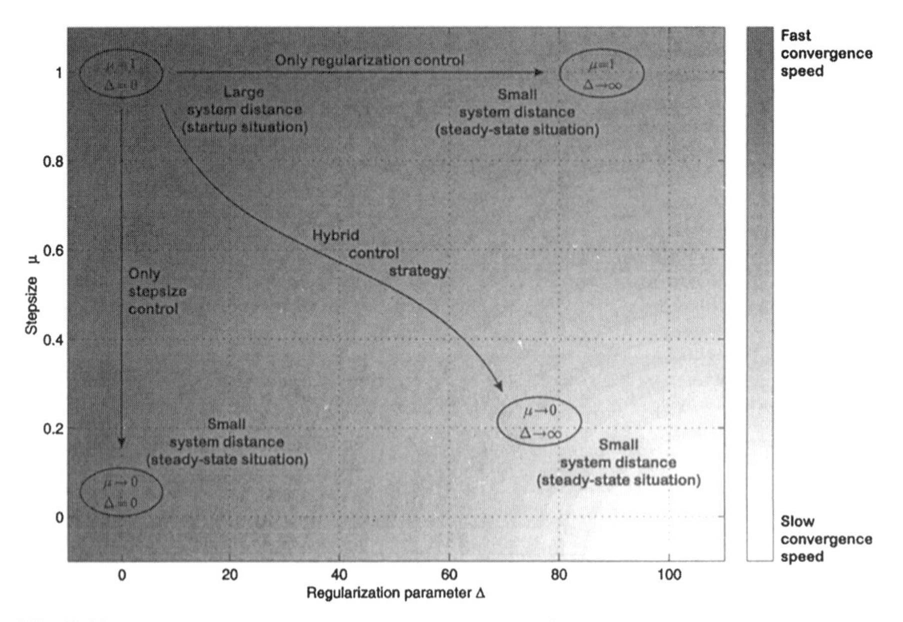

Fig. 7.13 Dependence of the speed of convergence and the final misadjustment on the control strategy.

7.2.1 Derivation of the Algorithm

To accomplish this, an *input signal matrix* $\boldsymbol{X}(n)$ is formed out of the current and the $M-1$ preceding input signal vectors

$$\boldsymbol{X}(n) = \left[\,\boldsymbol{x}(n),\, \boldsymbol{x}(n-1),\, \ldots,\, \boldsymbol{x}(n-M+1)\,\right], \tag{7.57}$$

where M marks the *order* of the AP algorithm. We use the notation

$$\mathbf{e}(n_1|n_2) = \boldsymbol{y}(n_1) - \boldsymbol{X}^{\mathrm{T}}(n_1)\,\widehat{\boldsymbol{h}}^*(n_2)\,, \tag{7.58}$$

and calculate an *a priori error vector* $\mathbf{e}(n|n)$

$$\mathbf{e}(n|n) = \boldsymbol{y}(n) - \widehat{\boldsymbol{d}}(n) = \boldsymbol{y}(n) - \boldsymbol{X}^{\mathrm{T}}(n)\,\widehat{\boldsymbol{h}}^*(n) \tag{7.59}$$

containing the errors for the current and the $M-1$ past input signal vectors applied to the echo cancellation filter with the coefficient settings at time n. The vector $\boldsymbol{y}(n)$ is defined by

$$\boldsymbol{y}(n) = \left[\,y(n),\, y(n-1),\, \ldots,\, y(n-M+1)\,\right]^{\mathrm{T}}. \tag{7.60}$$

The derivation of the update equation can follow the same lines as that for the NLMS algorithm. Using Eq. 7.9, we also define an *a posteriori error vector* $\mathbf{e}(n|n+1)$:

$$\begin{aligned}
\mathbf{e}(n|n+1) &= \boldsymbol{y}(n) - \boldsymbol{X}^{\mathrm{T}}(n)\,\widehat{\boldsymbol{h}}^*(n+1) \\
&= \boldsymbol{y}(n) - \boldsymbol{X}^{\mathrm{T}}(n)\,\widehat{\boldsymbol{h}}^*(n) - \boldsymbol{X}^{\mathrm{T}}(n)\,\boldsymbol{\Delta}\widehat{\boldsymbol{h}}^*(n) \\
&= \mathbf{e}(n|n) - \boldsymbol{X}^{\mathrm{T}}(n)\,\boldsymbol{\Delta}\widehat{\boldsymbol{h}}^*(n)\,.
\end{aligned} \tag{7.61}$$

A successful filter update requires that the Euclidean norm of the a posteriori error $\|\boldsymbol{e}(n|n+1)\|$ be at least not larger than that of the a priori error $\|\boldsymbol{e}(n|n)\|$:

$$\|\boldsymbol{e}(n|n+1)\| \leq \|\boldsymbol{e}(n|n)\| \tag{7.62}$$

or equivalently

$$\|\boldsymbol{e}(n|n+1)\|^2 \leq \|\boldsymbol{e}(n|n)\|^2 . \tag{7.63}$$

Evaluating $\|\boldsymbol{e}(n|n+1)\|^2$

$$\begin{aligned}\|\boldsymbol{e}(n|n+1)\|^2 &= \left[\boldsymbol{e}^{\mathrm{T}}(n|n) - \Delta\widehat{\boldsymbol{h}}^{\mathrm{H}}(n)\,\boldsymbol{X}(n)\right]\left[\boldsymbol{e}^*(n|n) - \boldsymbol{X}^{\mathrm{H}}(n)\Delta\widehat{\boldsymbol{h}}(n)\right] \\ &= \|\boldsymbol{e}(n|n)\|^2 - \boldsymbol{e}^{\mathrm{T}}(n|n)\,\boldsymbol{X}^{\mathrm{H}}(n)\Delta\widehat{\boldsymbol{h}}(n) \\ &\quad - \Delta\widehat{\boldsymbol{h}}^{\mathrm{H}}(n)\,\boldsymbol{X}(n)\,\boldsymbol{e}^*(n|n) \\ &\quad + \Delta\widehat{\boldsymbol{h}}^{\mathrm{H}}(n)\,\boldsymbol{X}(n)\,\boldsymbol{X}^{\mathrm{H}}(n)\,\Delta\widehat{\boldsymbol{h}}(n) ,\end{aligned} \tag{7.64}$$

condition 7.63 reads as follows:

$$\begin{aligned}\boldsymbol{e}^{\mathrm{T}}(n|n)\,\boldsymbol{X}^{\mathrm{H}}(n)\Delta\widehat{\boldsymbol{h}}(n) \quad &+ \quad \Delta\widehat{\boldsymbol{h}}^{\mathrm{H}}(n)\,\boldsymbol{X}(n)\,\boldsymbol{e}^*(n|n) \\ &- \quad \Delta\widehat{\boldsymbol{h}}^{\mathrm{H}}(n)\,\boldsymbol{X}(n)\,\boldsymbol{X}^{\mathrm{H}}(n)\,\Delta\widehat{\boldsymbol{h}}(n) \geq 0 .\end{aligned} \tag{7.65}$$

To find a rule for $\Delta\widehat{\boldsymbol{h}}(n)$ that satisfies this condition, we assume an equation very similar to Eq. 7.16:

$$\Delta\widehat{\boldsymbol{h}}(n) = \mu\,\boldsymbol{M}(n)\,\boldsymbol{X}(n)\,\left[\boldsymbol{X}^{\mathrm{H}}(n)\,\boldsymbol{M}(n)\,\boldsymbol{X}(n)\right]^{-1}\,\boldsymbol{e}^*(n|n) . \tag{7.66}$$

Inserting Eq. 7.66 into Eq. 7.65 again leads to the condition 7.17 on the range of the stepsize factor μ.

Assuming a unit matrix $\boldsymbol{I}$ for the matrix $\boldsymbol{M}(n)$ leads to the equation for the filter update:

$$\widehat{\boldsymbol{h}}(n+1) = \widehat{\boldsymbol{h}}(n) + \mu\,\boldsymbol{X}(n)\,\left[\boldsymbol{X}^{\mathrm{H}}(n)\,\boldsymbol{X}(n)\right]^{-1}\,\boldsymbol{e}^*(n|n) . \tag{7.67}$$

Comparing the update according to Eq. 7.19 of the NLMS algorithm with Eq. 7.67, it becomes clear that the NLMS algorithm can be considered as an AP algorithm of order $M = 1$. The update of the filter coefficients according to the AP algorithm requires the inversion of an $M \times M$ matrix. Numerical problems arising during the inversion of the matrix $\boldsymbol{X}^{\mathrm{H}}(n)\,\boldsymbol{X}(n)$ can be overcome by introducing a *regularization term*

$$\widehat{\boldsymbol{h}}(n+1) = \widehat{\boldsymbol{h}}(n) + \mu\,\boldsymbol{X}(n)\,\left[\boldsymbol{X}^{\mathrm{H}}(n)\,\boldsymbol{X}(n) + \Delta\,\boldsymbol{I}\right]^{-1}\,\boldsymbol{e}^*(n|n) , \tag{7.68}$$

where Δ is a small real positive constant. To arrive at a more compact notation, we define a normalized error vector $\boldsymbol{\varepsilon}(n|n)$:

$$\boldsymbol{\varepsilon}(n|n) = \left[\boldsymbol{X}^{\mathrm{T}}(n)\,\boldsymbol{X}^*(n) + \Delta\,\boldsymbol{I}\right]^{-1}\,\boldsymbol{e}(n|n) . \tag{7.69}$$

Then the update of the AP algorithm reads as follows:

$$\widehat{\boldsymbol{h}}(n+1) = \widehat{\boldsymbol{h}}(n) + \mu\, \boldsymbol{X}(n)\, \boldsymbol{\varepsilon}^*(n|n)\,. \tag{7.70}$$

As examples will show, for correlated input signals already the step from $M = 1$ to $M = 2$ speeds up the convergence of the filter coefficients considerably. In case of acoustic echo cancellation an increase of the order higher than $M = 4$ only slightly improves the performance of the adaptive filter. The AP algorithm is summarized in Table 7.2.

Table 7.2 Summary of the AP algorithm

Algorithmic part	Corresponding equations
Filtering	$\widehat{\boldsymbol{d}}(n) = \boldsymbol{X}^{\mathrm{T}}(n)\, \widehat{\boldsymbol{h}}^*(n)$
Error signal	$\boldsymbol{e}(n\|n) = \boldsymbol{y}(n) - \widehat{\boldsymbol{d}}(n)$
Normalization matrix update	$\tilde{\boldsymbol{x}}(n) = [x(n),\, x(n-1),\, \ldots,\, x(n-M+1)]$ $\boldsymbol{X}^{\mathrm{H}}(n)\, \boldsymbol{X}(n) = \boldsymbol{X}^{\mathrm{H}}(n-1)\, \boldsymbol{X}(n-1) - \tilde{\boldsymbol{x}}^{\mathrm{H}}(n-N)\, \tilde{\boldsymbol{x}}(n-N) + \tilde{\boldsymbol{x}}^{\mathrm{H}}(n)\, \tilde{\boldsymbol{x}}(n)$
Filter update	$\widehat{\boldsymbol{h}}(n+1) = \widehat{\boldsymbol{h}}(n) + \mu\, \boldsymbol{X}(n)\, \left[\boldsymbol{X}^{\mathrm{H}}(n)\, \boldsymbol{X}(n)\right]^{-1} \boldsymbol{e}^*(n\|n)$

Looking on the update equations 7.19 and 7.67 of the two algorithms from a geometric point of view, one learns that the NLMS algorithm updates the vector of the filter coefficients along a line whereas the update of the AP algorithm lies within an M-dimensional hyperplane (see Fig. 7.14). This explains the faster convergence to an optimal setting.

For $M = 1$, the AP algorithm is equal to the NLMS procedure. For speech input signals, even $M = 2$ leads to a considerably faster convergence of the filter coefficients (see Fig. 7.15). Suggested values for M are between 2 and 5 for the update of the echo cancellation filter.

It should be noted, however, that faster convergence of the coefficients of an adaptive filter also means faster divergence in case of strong noise like local signals in echo canceling applications. Therefore, faster control of the algorithm is required as well.

Optimal control values are based on estimated quantities (see Chapter. 13). Since the reliability of estimates depends on the lengths of the data records usable for the estimation, excessively high speed of convergence leading to fast changes of

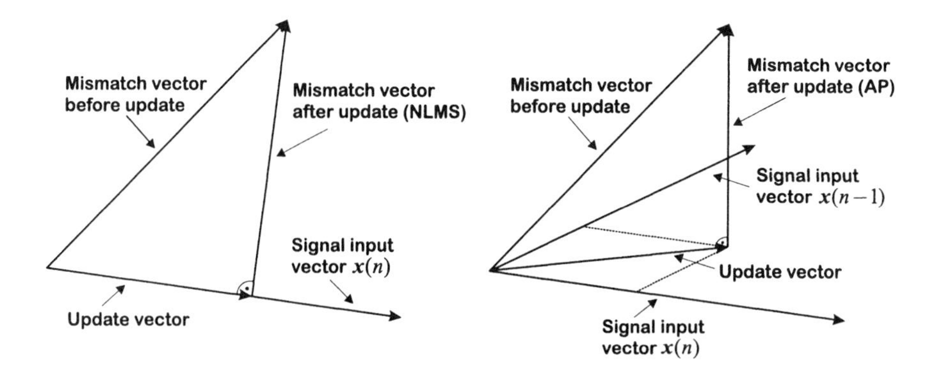

Fig. 7.14 Updates of the system mismatch vector according to the NLMS algorithm (left) and an AP algorithm of order $M = 2$ (right) (with permission from [101]).

the adaptive filter and its input/output relations may not be desirable. Nevertheless, the AP algorithm seems to be a good candidate to replace the NLMS algorithm in acoustic echo canceling applications.

7.2.2 Fast Versions of the AP Algorithm

The adaptation of the filter coefficients according to Eq. 7.68 requires vector and matrix operations including a matrix inversion. Even if the order M of the algorithm is low, the increased numerical complexity compared to the NLMS algorithm can prohibit its application. Therefore a number of methods have been suggested to reduce this complexity to a linear dependence of M [75, 76, 77, 78, 168, 173, 223]. The resulting algorithms are called *fast affine projection (FAP) algorithms.*

Basically, there are three steps that reduce the complexity of the algorithm:

- *A simplification of the update of the error vector:* To show this, we rewrite Eq. 7.59:

$$\begin{aligned} \boldsymbol{e}(n|n) &= \boldsymbol{y}(n) - \widehat{\boldsymbol{d}}(n) \qquad (7.71) \\ &= \begin{bmatrix} y(n) - \boldsymbol{x}^{\mathrm{T}}(n)\,\widehat{\boldsymbol{h}}^*(n) \\ \bar{\boldsymbol{y}}(n-1) - \overline{\boldsymbol{X}}^{\mathrm{T}}(n-1)\,\widehat{\boldsymbol{h}}^*(n) \end{bmatrix} \\ &= \begin{bmatrix} e(n|n) \\ \bar{\boldsymbol{e}}(n-1|n) \end{bmatrix}, \end{aligned}$$

where the bar marks that the related quantities contain only their $M-1$ uppermost elements. According to Eq. 7.61 and Eq. 7.68, the a posteriori error vector $\boldsymbol{e}(n-1|n)$ is given by

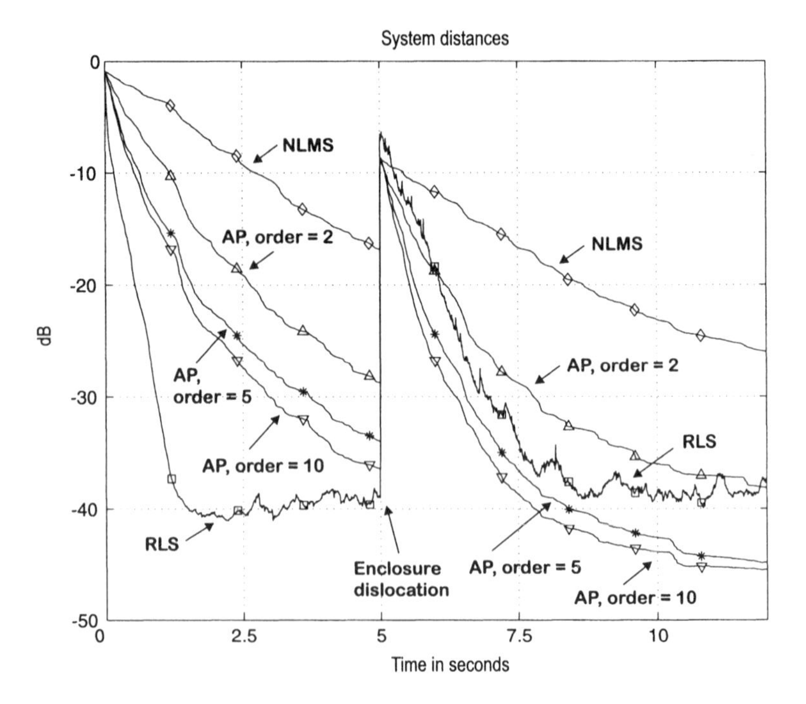

Fig. 7.15 Convergence of the system distance for different adaptation algorithms (filter length = 1024, sampling frequency = 8 kHz): $\diamond$: NLMS ($\mu = 1$), $\triangle$: AP of order 2, $*$: AP of order 5, ∇: AP of order 10 (all AP algorithms with $\mu = 1$), $\Box$: RLS ($\lambda = 0.9999$) (see Section 7.3). The impulse response of the LEM system is changed at $t = 5$ seconds. We face the fact that the algorithms do not converge to the same steady state (with permission from [25]).

$$\begin{aligned} \boldsymbol{e}(n-1|n) &= \boldsymbol{y}(n-1) - \boldsymbol{X}^{\mathrm{T}}(n-1)\,\widehat{\boldsymbol{h}}^{*}(n) \\ &= \boldsymbol{e}(n-1|n-1) - \boldsymbol{X}^{\mathrm{T}}(n-1)\,\boldsymbol{\Delta}\widehat{\boldsymbol{h}}^{*}(n-1) \\ &= \boldsymbol{e}(n-1|n-1) - \mu\,\boldsymbol{X}^{\mathrm{T}}(n-1)\,\boldsymbol{X}^{*}(n-1) \\ &\qquad \left[\boldsymbol{X}^{\mathrm{T}}(n-1)\,\boldsymbol{X}^{*}(n-1) + \Delta^{*}\,\boldsymbol{I}\right]^{-1}\,\boldsymbol{e}(n-1|n-1) \\ &\approx (1-\mu)\,\boldsymbol{e}(n-1|n-1)\,. \end{aligned} \tag{7.72}$$

In case of no regularization ($\Delta = 0$) approximation 7.72 holds as an equality. Now approximation 7.72 can be written as an update of $\boldsymbol{e}(n|n)$:

$$\boldsymbol{e}(n|n) \approx \begin{bmatrix} e(n|n) \\ (1-\mu)\,\bar{\boldsymbol{e}}(n-1|n-1) \end{bmatrix}. \tag{7.73}$$

For $\mu = 1$, this equation further simplifies to

$$\boldsymbol{e}(n|n) \approx \begin{bmatrix} e(n|n) \\ \bar{\boldsymbol{0}} \end{bmatrix}. \tag{7.74}$$

- *The introduction of a modified filter coefficient vector $\widehat{\boldsymbol{\eta}}(n)$ allowing simpler adaptation:* It is important that only the uppermost element $e(n|n)$ of the error vector $\boldsymbol{e}(n|n)$ constitutes the output of the echo canceling system. Therefore any modification of the filter coefficient vector $\widehat{\boldsymbol{h}}(n)$ is allowed as long as $e(n|n)$ remains unchanged.

  Regarding Eq. 7.70, the filter coefficient vector $\widehat{\boldsymbol{h}}(n+1)$ can be expressed by its starting value $\widehat{\boldsymbol{h}}(0)$ and the consecutive updates:

$$\begin{aligned}\widehat{\boldsymbol{h}}(n+1) &= \widehat{\boldsymbol{h}}(0) + \mu \sum_{i=0}^{n} \boldsymbol{X}(n-i)\,\boldsymbol{\varepsilon}^*(n-i|n-i) \\ &= \widehat{\boldsymbol{h}}(0) + \mu \sum_{i=0}^{n} \sum_{j=0}^{M-1} \boldsymbol{x}(n-i-j)\,\varepsilon_j^*(n-i|n-i)\,, \end{aligned} \tag{7.75}$$

  where $\varepsilon_j^*(n-i|n-i)$ is the jth element of the vector $\boldsymbol{\varepsilon}^*(n-i|n-i)$. To reduce the number of multiplications, we change the sequence in which we sum the terms in the double sum. To visualize this step, Eq. 7.75 is written out for $M=3$ and $n=6$ and split into three parts: an upper triangle, a central part, and a lower triangle:

$$\begin{array}{llllllll} \multicolumn{8}{l}{\widehat{\boldsymbol{h}}(7) = \widehat{\boldsymbol{h}}(0)} \\ +\mu & \big[\boldsymbol{x}(6) & \varepsilon_0^*(6|6) & + & \boldsymbol{x}(5) & \varepsilon_1^*(6|6) & + \ \boldsymbol{x}(4) & \varepsilon_2^*(6|6) \\ + & \boldsymbol{x}(5) & \varepsilon_0^*(5|5) & + & \boldsymbol{x}(4) & \varepsilon_1^*(5|5) & & \\ + & \boldsymbol{x}(4) & \varepsilon_0^*(4|4) & & & & & \\ & & & & & & + \ \boldsymbol{x}(3) & \varepsilon_2^*(5|5) \\ & & & + & \boldsymbol{x}(3) & \varepsilon_1^*(4|4) & + \ \boldsymbol{x}(2) & \varepsilon_2^*(4|4) \\ + & \boldsymbol{x}(3) & \varepsilon_0^*(3|3) & + & \boldsymbol{x}(2) & \varepsilon_1^*(3|3) & + \ \boldsymbol{x}(1) & \varepsilon_2^*(3|3) \\ + & \boldsymbol{x}(2) & \varepsilon_0^*(2|2) & + & \boldsymbol{x}(1) & \varepsilon_1^*(2|2) & + \ \boldsymbol{x}(0) & \varepsilon_2^*(2|2) \\ + & \boldsymbol{x}(1) & \varepsilon_0^*(1|1) & + & \boldsymbol{x}(0) & \varepsilon_1^*(1|1) & & \\ + & \boldsymbol{x}(0) & \varepsilon_0^*(0|0) & & & & & \\ & & & & & & + \ \boldsymbol{x}(-1) & \varepsilon_2^*(1|1) \\ & & & + & \boldsymbol{x}(-1) & \varepsilon_1^*(0|0) & + \ \boldsymbol{x}(-2) & \varepsilon_2^*(0|0)\,\big]\,. \end{array} \tag{7.76}$$

  For the signal vectors $\boldsymbol{x}(i)$, $i < 0$, we assume that they are zero. Then all elements in the lower triangle are zero and Eq. 7.75 can be expressed by two double sums over the upper triangle and the central part, respectively:

$$\begin{aligned}\widehat{\boldsymbol{h}}(n+1) &= \widehat{\boldsymbol{h}}(0) \\ &+ \mu \sum_{k=0}^{M-1} \boldsymbol{x}(n-k) \sum_{j=0}^{k} \varepsilon_j^*(n-k+j|n-k+j) \\ &+ \mu \sum_{k=M}^{n} \boldsymbol{x}(n-k) \sum_{j=0}^{M-1} \varepsilon_j^*(n-k+j|n-k+j)\,. \end{aligned} \tag{7.77}$$

After defining a vector $\phi(n)$

$$\phi(n) = \begin{bmatrix} \varepsilon_0(n|n) \\ \varepsilon_1(n|n) + \varepsilon_0(n-1|n-1) \\ \varepsilon_2(n|n) + \varepsilon_1(n-1|n-1) + \varepsilon_0(n-2|n-2) \\ \vdots \quad \vdots \\ \varepsilon_{M-1}(n|n) + \varepsilon_{M-2}(n-1|n-1) + \cdots \\ \cdots + \varepsilon_0(n-M+1|n-M+1) \end{bmatrix}, \quad (7.78)$$

we can express the first double sum in Eq. 7.77 by a matrix–vector multiplication:

$$\boldsymbol{X}(n)\,\boldsymbol{\phi}^*(n) = \sum_{k=0}^{M-1} \boldsymbol{x}(n-k) \sum_{j=0}^{k} \varepsilon_j^*(n-k+j|n-k+j)\,. \quad (7.79)$$

The remaining terms in Eq. 7.77 we associate with a modified filter coefficient vector $\widehat{\boldsymbol{\eta}}(n)$:

$$\widehat{\boldsymbol{\eta}}(n+1) = \widehat{\boldsymbol{h}}(0) + \mu \sum_{k=M}^{n} \boldsymbol{x}(n-k) \sum_{j=0}^{M-1} \varepsilon_j^*(n-k+j|n-k+j)\,. \quad (7.80)$$

Finally, we write the filter coefficient vector $\widehat{\boldsymbol{h}}(n+1)$ as follows:

$$\widehat{\boldsymbol{h}}(n+1) = \widehat{\boldsymbol{\eta}}(n+1) + \mu\,\boldsymbol{X}(n)\,\boldsymbol{\phi}^*(n)\,. \quad (7.81)$$

The vector $\boldsymbol{\phi}(n)$ can be updated recursively

$$\boldsymbol{\phi}(n+1) = \boldsymbol{\varepsilon}(n+1|n+1) + \begin{bmatrix} 0 \\ \bar{\boldsymbol{\phi}}(n) \end{bmatrix}, \quad (7.82)$$

where $\bar{\boldsymbol{\phi}}(n)$ represents the $M-1$ uppermost elements of the vector $\boldsymbol{\phi}(n)$.

It remains to give an efficient update for the error vector $\boldsymbol{e}(n|n)$ (see Eq. 7.59). Due to approximation 7.73, it is necessary only to find the new element $e(n|n)$ in this vector:

$$\begin{aligned} e(n|n) &= y(n) - \boldsymbol{x}^{\mathrm{T}}(n)\,\widehat{\boldsymbol{h}}^*(n) \\ &= y(n) - \boldsymbol{x}^{\mathrm{T}}(n)\,\widehat{\boldsymbol{\eta}}^*(n) - \mu\,\boldsymbol{x}^{\mathrm{T}}(n)\,\boldsymbol{X}^*(n-1)\,\boldsymbol{\phi}(n-1)\,. \end{aligned} \quad (7.83)$$

The product $\boldsymbol{x}^{\mathrm{T}}(n)\,\boldsymbol{X}^*(n)$ results in a row vector $\boldsymbol{z}(n)$, which can be updated recursively:

$$\boldsymbol{z}(n) = \boldsymbol{z}(n-1) - x(n-N)\,\tilde{\boldsymbol{x}}^*(n-N) + x(n)\,\tilde{\boldsymbol{x}}^*(n)\,, \quad (7.84)$$

where the row vector $\tilde{\boldsymbol{x}}(n)$ is defined as

$$\tilde{\boldsymbol{x}}(n) = [x(n),\, x(n-1),\, \ldots,\, x(n-M+1)]^{\mathrm{T}}\,. \quad (7.85)$$

Using Eqs. 7.84 and 7.85, the computation of the error signal $e(n|n)$ can be written as

$$e(n|n) = y(n) - \boldsymbol{x}^{\mathrm{T}}(n)\,\widehat{\boldsymbol{\eta}}^{*}(n) - \mu\,\boldsymbol{z}^{\mathrm{T}}(n)\,\boldsymbol{\phi}(n-1)\,. \tag{7.86}$$

Note that the computation of the vector product $\boldsymbol{z}^{\mathrm{T}}(n)\,\boldsymbol{\phi}(n-1)$ requires only M multiplication and $M-1$ additions (usually we have $M \ll N$).

Finally, the modified filter coefficient vector $\widehat{\boldsymbol{\eta}}(n)$ needs to be computed recursively in order to avoid the direct (and costly) computation according to Eq. 7.81. For this reason, we extract the first element from the outer sum of Eq. 7.80 and obtain

$$\begin{aligned}\widehat{\boldsymbol{\eta}}(n+1) \;&=\; \widehat{\boldsymbol{h}}(0) + \mu\,\boldsymbol{x}(n-M)\sum_{j=0}^{M-1}\varepsilon_j^{*}(n-M+j|n-M+j)) \\ &+\; \underbrace{\mu\sum_{k=M+1}^{n}\boldsymbol{x}(n-k)\sum_{j=0}^{M-1}\varepsilon_j^{*}(n-k+j|n-k+j)}_{(1)}\,. \end{aligned} \tag{7.87}$$

When exchanging the index n with $n-1$ in the last expression [denoted by (1)] and comparing the result with Eq. 7.80, we easily get

$$\mu\sum_{k=M+1}^{n}\boldsymbol{x}(n-k)\sum_{j=0}^{M-1}\varepsilon_j^{*}(n-k+j|n-k+j) = \widehat{\boldsymbol{\eta}}(n) - \widehat{\boldsymbol{h}}(0)\,. \tag{7.88}$$

By inserting this result in Eq. 7.87, we obtain

$$\widehat{\boldsymbol{\eta}}(n+1) = \widehat{\boldsymbol{\eta}}(n) + \mu\,\boldsymbol{x}(n-M)\sum_{j=0}^{M-1}\varepsilon_j^{*}(n-M+j|n-M+j)\,. \tag{7.89}$$

The last sum in Eq. 7.89

$$\begin{aligned}\sum_{j=0}^{M-1}\varepsilon_j^{*}(n-M+j|n-M+j) \;&=\; \\ &\varepsilon_0^{*}(n-M|n-M) + \ldots + \varepsilon_{M-1}^{*}(n-1|n-1)\end{aligned} \tag{7.90}$$

is equal to the last element of the vector $\boldsymbol{\phi}^{*}(n-1)$ (see Eq. 7.78). By denoting this element with $\phi_{M-1}^{*}(n-1)$, we end up with the recursive computation of the modified filter vector

$$\widehat{\boldsymbol{\eta}}(n+1) = \widehat{\boldsymbol{\eta}}(n) + \mu\,\boldsymbol{x}(n-M)\,\phi_{M-1}^{*}(n-1)\,. \tag{7.91}$$

- *Efficient inversion of the normalization matrix:* Finally an efficient method has to be found to update the normalized error vector $\boldsymbol{\varepsilon}(n|n)$ (see Eq. 7.70). If M is small (like $2 \cdots 4$), a direct inversion of the matrix can be advantageous since no stability problems arise. For larger M, techniques such as the *Levinson–Durbin recursion* (see Section 6.2) can be applied.

The most important parts of the fast affine projection algorithm are summarized in Table 7.3.

Table 7.3 Summary of the fast affine projection algorithm

Algorithmic part	Corresponding equations
Filtering	$\boldsymbol{z}(n) = \boldsymbol{z}(n-1) - x(n-N)\,\tilde{\boldsymbol{x}}^*(n-N) + x(n)\,\tilde{\boldsymbol{x}}^*(n)$
	with $\tilde{\boldsymbol{x}}(n) = [x(n),\, x(n-1),\, \ldots,\, x(n-M+1)]^{\mathrm{T}}$
	$\widehat{d}(n) = \boldsymbol{x}^{\mathrm{T}}(n)\,\widehat{\boldsymbol{\eta}}^*(n) + \mu\,\boldsymbol{z}^{\mathrm{T}}(n)\,\boldsymbol{\phi}(n-1)$
Error signal	$e(n\vert n) = y(n) - \widehat{d}(n)$
	$\mathbf{e}(n\vert n) = \begin{bmatrix} e(n\vert n) \\ (1-\mu)\,\bar{\mathbf{e}}(n-1\vert n-1) \end{bmatrix}$
	with $\bar{\mathbf{e}}(n\vert n)$ = uppermost $M-1$ elements of $\mathbf{e}(n\vert n)$
Normalization	$\boldsymbol{X}^{\mathrm{H}}(n)\,\boldsymbol{X}(n) = \boldsymbol{X}^{\mathrm{H}}(n-1)\,\boldsymbol{X}(n-1) - \tilde{\boldsymbol{x}}(n-N)\,\tilde{\boldsymbol{x}}^{\mathrm{H}}(n-N) + \tilde{\boldsymbol{x}}(n)\,\tilde{\boldsymbol{x}}^{\mathrm{H}}(n)$
	$\boldsymbol{\varepsilon}(n\vert n) = \left[\boldsymbol{X}^{\mathrm{H}}(n)\,\boldsymbol{X}(n) + \Delta\,\boldsymbol{I}\right]^{-1}\,\mathbf{e}(n\vert n)$
	$\boldsymbol{\phi}(n) = \boldsymbol{\varepsilon}(n\vert n) + \begin{bmatrix} 0 \\ \bar{\boldsymbol{\phi}}(n-1) \end{bmatrix}$
	where $\bar{\boldsymbol{\phi}}(n)$ is the uppermost $M-1$ element $\boldsymbol{\phi}(n)$
Filter update	$\widehat{\boldsymbol{\eta}}(n+1) = \widehat{\boldsymbol{\eta}}(n) + \mu\,\boldsymbol{x}(n-M)\,\phi_{M-1}^*(n-1)$
	where $\phi_{M-1}(n)$ is the last element of the vector $\boldsymbol{\phi}(n)$

7.3 THE RECURSIVE LEAST SQUARES ALGORITHM

The recursive least squares (RLS) algorithm belongs to another class of adaptive algorithms. Instead of minimizing the expectation of squared error signals, this algorithm minimizes the sum of squared errors.

7.3.1 Derivation of the Algorithm

The RLS algorithm is designed such that the adaptive filter minimizes the (windowed) sum of the absolute values of the squared errors

$$\mathcal{E}(n) = \sum_{l=0}^{n} \lambda^{(n-l)}\, e(l|n)\, e^*(l|n)\,, \tag{7.92}$$

where $e(l|n)$ represents the error arising when the data at time index l is filtered with $\widehat{\boldsymbol{h}}(n)$:

$$e(l|n) = y(l) - \widehat{\boldsymbol{h}}^{\mathrm{H}}(n)\, \boldsymbol{x}(l) = y(l) - \boldsymbol{x}^{\mathrm{T}}(l)\, \widehat{\boldsymbol{h}}^{*}(n)\,. \tag{7.93}$$

The *forgetting factor* λ – with $|\lambda| \leq 1$ – defines a window with exponential weighting; thus contributions of past errors are reduced exponentially according to their distance from the time index n.

Inserting the error as defined in Eq. 7.93 into Eq. 7.92 leads to

$$\mathcal{E}(n) = \sum_{l=0}^{n} \lambda^{(n-l)} \left[y(l) - \widehat{\boldsymbol{h}}^{\mathrm{H}}(n)\, \boldsymbol{x}(l)\right] \left[y^*(l) - \boldsymbol{x}^{\mathrm{H}}(l)\, \widehat{\boldsymbol{h}}(n)\right]. \tag{7.94}$$

Taking the derivatives with respect to the complex coefficients $h_i(n)$ of the adaptive filter (see Section 5.1) and equating the result to zero leads to

$$\sum_{l=0}^{n} \lambda^{(n-l)}\, \boldsymbol{x}(l)\, \boldsymbol{x}^{\mathrm{H}}(l)\, \widehat{\boldsymbol{h}}(n) = \sum_{l=0}^{n} \lambda^{(n-l)}\, \boldsymbol{x}(l)\, y^*(l)\,. \tag{7.95}$$

Defining

$$\widehat{\boldsymbol{R}}_{xx}(n) = \sum_{l=0}^{n} \lambda^{(n-l)}\, \boldsymbol{x}(l)\, \boldsymbol{x}^{\mathrm{H}}(l) \tag{7.96}$$

and

$$\widehat{\boldsymbol{r}}_{xy}(n) = \sum_{l=0}^{n} \lambda^{(n-l)}\, \boldsymbol{x}(l)\, y^*(l) \tag{7.97}$$

that may be considered as estimates of the autocorrelation matrix $\boldsymbol{R}_{xx}$ and the cross-correlation vector $\boldsymbol{r}_{xy}$. Then 7.95 reads as follows:

$$\widehat{\boldsymbol{R}}_{xx}(n)\, \widehat{\boldsymbol{h}}(n) = \widehat{\boldsymbol{r}}_{xy}(n)\,. \tag{7.98}$$

This equation represents a deterministic equivalent to the normal equation 5.21. To come to a recursive solution leading to an update equation for $\widehat{\boldsymbol{h}}(n+1)$ assuming that $\widehat{\boldsymbol{h}}(n)$ is already known, one can use the update properties of $\widehat{\boldsymbol{R}}_{xx}(n+1)$ and $\widehat{\boldsymbol{r}}_{xy}(n+1)$:

$$\begin{aligned}\widehat{\boldsymbol{R}}_{xx}(n+1) &= \lambda \sum_{l=0}^{n} \lambda^{(n-l)}\, \boldsymbol{x}(l)\, \boldsymbol{x}^{\mathrm{H}}(l) + \boldsymbol{x}(n+1)\, \boldsymbol{x}^{\mathrm{H}}(n+1) \\ &= \lambda\, \widehat{\boldsymbol{R}}_{xx}(n) + \boldsymbol{x}(n+1)\, \boldsymbol{x}^{\mathrm{H}}(n+1)\,, \qquad (7.99) \\ \widehat{\boldsymbol{r}}_{xy}(n+1) &= \lambda \sum_{l=0}^{n} \lambda^{(n-l)}\, \boldsymbol{x}(l)\, y^{*}(l) + \boldsymbol{x}(n+1)\, y^{*}(n+1) \\ &= \lambda\, \widehat{\boldsymbol{r}}_{xy}(n) + \boldsymbol{x}(n+1)\, y^{*}(n+1)\,. \qquad (7.100)\end{aligned}$$

To find the inverse of $\widehat{\boldsymbol{R}}_{xx}(n+1)$, the *matrix inversion lemma* [109]—also called *Woodbury's law*—can be applied. In general notation, this rule reads

$$\left(\boldsymbol{A} + \boldsymbol{u}\,\boldsymbol{v}^{\mathrm{H}}\right)^{-1} = \boldsymbol{A}^{-1} - \frac{\boldsymbol{A}^{-1}\,\boldsymbol{u}\,\boldsymbol{v}^{\mathrm{H}}\,\boldsymbol{A}^{-1}}{1 + \boldsymbol{v}^{\mathrm{H}}\,\boldsymbol{A}^{-1}\,\boldsymbol{u}}\,. \qquad (7.101)$$

Applied to Eq. 7.99, this leads to

$$\begin{aligned}\widehat{\boldsymbol{R}}_{xx}^{-1}(n+1) &= \lambda^{-1}\, \widehat{\boldsymbol{R}}_{xx}^{-1}(n) \\ &\quad - \frac{\lambda^{-2}\, \widehat{\boldsymbol{R}}_{xx}^{-1}(n)\, \boldsymbol{x}(n+1)\, \boldsymbol{x}^{\mathrm{H}}(n+1)\, \widehat{\boldsymbol{R}}_{xx}^{-1}(n)}{1 + \lambda^{-1}\, \boldsymbol{x}^{\mathrm{H}}(n+1)\, \widehat{\boldsymbol{R}}_{xx}^{-1}(n)\, \boldsymbol{x}(n+1)}\,. \qquad (7.102)\end{aligned}$$

To achieve a more concise notation, we define a *gain vector* $\boldsymbol{\gamma}(n+1)$:

$$\boldsymbol{\gamma}(n+1) = \frac{\lambda^{-1}\, \widehat{\boldsymbol{R}}_{xx}^{-1}(n)\, \boldsymbol{x}(n+1)}{1 + \lambda^{-1}\, \boldsymbol{x}^{\mathrm{H}}(n+1)\, \widehat{\boldsymbol{R}}_{xx}^{-1}(n)\, \boldsymbol{x}(n+1)}\,. \qquad (7.103)$$

Then, Eq. 7.102 reduces to

$$\widehat{\boldsymbol{R}}_{xx}^{-1}(n+1) = \lambda^{-1}\, \widehat{\boldsymbol{R}}_{xx}^{-1}(n) - \lambda^{-1}\, \boldsymbol{\gamma}(n+1)\, \boldsymbol{x}^{\mathrm{H}}(n+1)\, \widehat{\boldsymbol{R}}_{xx}^{-1}(n)\,. \qquad (7.104)$$

Multiplying both sides of Eq. 7.103 by the denominator of the right side and inserting Eq. 7.104 leads to a very simple expression for the gain vector $\boldsymbol{\gamma}(n+1)$:

$$\boldsymbol{\gamma}(n+1) = \widehat{\boldsymbol{R}}_{xx}^{-1}(n+1)\, \boldsymbol{x}(n+1)\,. \qquad (7.105)$$

Now we are prepared to formulate a recursion for the filter vector $\widehat{\boldsymbol{h}}(n+1)$. We start from Eq. 7.98 and first insert Eq. 7.100:

$$\begin{aligned}\widehat{\boldsymbol{h}}(n+1) &= \widehat{\boldsymbol{R}}_{xx}^{-1}(n+1)\, \widehat{\boldsymbol{r}}_{xy}(n+1) \\ &= \lambda\, \widehat{\boldsymbol{R}}_{xx}^{-1}(n+1)\, \widehat{\boldsymbol{r}}_{xy}(n) + \widehat{\boldsymbol{R}}_{xx}^{-1}(n+1)\, \boldsymbol{x}(n+1)\, y^{*}(n+1)\,. \qquad (7.106)\end{aligned}$$

Next, $\widehat{\boldsymbol{R}}_{xx}^{-1}(n+1)$ is replaced by $\widehat{\boldsymbol{R}}_{xx}^{-1}(n)$ (see Eq. 7.104) in the first term of the right side:

$$\begin{aligned} \widehat{\boldsymbol{h}}(n+1) &= \widehat{\boldsymbol{R}}_{xx}^{-1}(n)\,\widehat{\boldsymbol{r}}_{xy}(n) - \boldsymbol{\gamma}(n+1)\,\boldsymbol{x}^{\mathrm{H}}(n+1)\,\widehat{\boldsymbol{R}}_{xx}^{-1}(n)\,\widehat{\boldsymbol{r}}_{xy}(n) \\ &\quad + \widehat{\boldsymbol{R}}_{xx}^{-1}(n+1)\,\boldsymbol{x}(n+1)\,y^*(n+1) \\ &= \widehat{\boldsymbol{h}}(n) - \boldsymbol{\gamma}(n+1)\,\boldsymbol{x}^{\mathrm{H}}(n+1)\,\widehat{\boldsymbol{h}}(n) \\ &\quad + \widehat{\boldsymbol{R}}_{xx}^{-1}(n+1)\,\boldsymbol{x}(n+1)\,y^*(n+1)\,. \end{aligned} \tag{7.107}$$

To simplify this expression further, we use Eq. 7.105:

$$\widehat{\boldsymbol{h}}(n+1) = \widehat{\boldsymbol{h}}(n) + \boldsymbol{\gamma}(n+1)\left[y^*(n+1) - \boldsymbol{x}^{\mathrm{H}}(n+1)\,\widehat{\boldsymbol{h}}(n)\right]. \tag{7.108}$$

The term in brackets represents the complex conjugate of an error $e(n+1|n)$ that would arise if the new data would be applied to the not yet updated filter:

$$\begin{aligned} e(n+1|n) &= y(n+1) - \widehat{d}(n+1|n) \qquad (7.109) \\ &= y(n+1) - \widehat{\boldsymbol{h}}^{\mathrm{H}}(n)\,\boldsymbol{x}(n+1)\,. \end{aligned}$$

Also the term "gain vector" for $\boldsymbol{\gamma}(n+1)$ becomes clear in this equation.

The expression 7.108 provides a good example to understand the performance of a time-recursive algorithm; there exists a filter $\widehat{\boldsymbol{h}}(n)$ optimized for data up to and including time index n and there is a new pair of data samples $y(n+1)$ and $x(n+1)$ becoming available. First, the new data are processed using the "old" filter. Only if $e(n+1|n)$ is different from zero is the filter modified proportional to this error. One can say that $e(n+1|n)$ contains the new information made available by the new pair of data.

Finally, the update of the filter coefficients for the RLS algorithm is given by[1]

$$\begin{aligned} \widehat{\boldsymbol{h}}(n) &= \widehat{\boldsymbol{h}}(n-1) + \Delta\widehat{\boldsymbol{h}}(n-1) \qquad (7.110) \\ &= \widehat{\boldsymbol{h}}(n-1) + \widehat{\boldsymbol{R}}_{xx}^{-1}(n)\,\boldsymbol{x}(n)\,e^*(n|n-1)\,. \qquad (7.111) \end{aligned}$$

In order to obtain a more in-depth understanding of the meaning of the update vector $\Delta\widehat{\boldsymbol{h}}(n-1)$, we rewrite this term using Eq. 7.110:

$$\begin{aligned} \Delta\widehat{\boldsymbol{h}}(n-1) &= \widehat{\boldsymbol{R}}_{xx}^{-1}(n)\,\boldsymbol{x}(n)\,e^*(n|n-1) \\ &= \widehat{\boldsymbol{R}}_{xx}^{-1}(n)\left[\boldsymbol{x}(n)\,y^*(n) - \boldsymbol{x}(n)\,\boldsymbol{x}^{\mathrm{H}}(n)\,\widehat{\boldsymbol{h}}(n-1)\right]. \end{aligned} \tag{7.112}$$

Assuming that the past updates were such that Eq. 7.95 is valid, we extend Eq. 7.112 by adding Eq. 7.95 for $\widehat{\boldsymbol{h}}(n-1)$ multiplied by λ:

[1] For notational convenience, the time index is changed from $n+1$ to n.

$$\begin{aligned}\boldsymbol{\Delta}\widehat{\boldsymbol{h}}(n-1) &= \widehat{\boldsymbol{R}}_{xx}^{-1}(n)\,\boldsymbol{x}(n)\,e^*(n|n-1) \qquad (7.113)\\ &= \widehat{\boldsymbol{R}}_{xx}^{-1}(n)\left[\boldsymbol{x}(n)\,y^*(n) + \lambda\sum_{l=0}^{n-1}\lambda^{(n-1-l)}\,\boldsymbol{x}(l)\,y^*(l)\right.\\ &\left.-\left(\boldsymbol{x}(n)\,\boldsymbol{x}^{\mathrm{H}}(n)\,\widehat{\boldsymbol{h}}(n-1) + \lambda\sum_{l=0}^{n-1}\lambda^{(n-1-l)}\,\boldsymbol{x}(l)\,\boldsymbol{x}^{\mathrm{H}}(l)\,\widehat{\boldsymbol{h}}(n-1)\right)\right].\end{aligned}$$

Combining terms leads to

$$\begin{aligned}\boldsymbol{\Delta}\widehat{\boldsymbol{h}}(n-1) &= \widehat{\boldsymbol{R}}_{xx}^{-1}(n)\left[\sum_{l=0}^{n}\lambda^{(n-l)}\,\boldsymbol{x}(l)\,y^*(l)\right. \qquad (7.114)\\ &\qquad\left.-\sum_{l=0}^{n}\lambda^{(n-l)}\,\boldsymbol{x}(l)\,\boldsymbol{x}^{\mathrm{H}}(l)\,\widehat{\boldsymbol{h}}(n-1)\right]\\ &= \widehat{\boldsymbol{R}}_{xx}^{-1}(n)\sum_{l=0}^{n}\lambda^{(n-l)}\,\boldsymbol{x}(l)\left[y^*(l) - \boldsymbol{x}^{\mathrm{H}}(l)\,\widehat{\boldsymbol{h}}(n-1)\right]\\ &= \widehat{\boldsymbol{R}}_{xx}^{-1}(n)\sum_{l=0}^{n}\lambda^{(n-l)}\,\boldsymbol{x}(l)\,e^*(l|n-1)\,.\end{aligned}$$

Equivalent to Eq. 7.97, we introduce

$$\widehat{\boldsymbol{r}}_{xe}(n|n-1) = \sum_{l=0}^{n}\lambda^{(n-l)}\,\boldsymbol{x}(l)\,e^*(l|n-1)\,. \qquad (7.115)$$

Then the update term of the RLS algorithm is finally written as

$$\boldsymbol{\Delta}\widehat{\boldsymbol{h}}(n-1) = \widehat{\boldsymbol{R}}_{xx}^{-1}(n)\,\widehat{\boldsymbol{r}}_{xe}(n|n-1)\,. \qquad (7.116)$$

If we consider $\widehat{\boldsymbol{r}}_{xe}(n|n-1)$ as an estimate of the cross-correlation between the input data and the resulting error based on the filter at time instant $n-1$ the update given by Eq. 7.116 clearly shows that in the RLS algorithm this correlation controls the adaptation of the filter coefficients.

For the numerical complexity of this algorithm, it is important to note that the matrix $\widehat{\boldsymbol{R}}_{xx}(n)$ that has to be inverted is of size $N \times N$, where N is the length of the adaptive filter. N may be 256 for an echo canceller in a car and up to several thousands for a canceler in an office. Compared with the AP algorithm where $M = 2 \cdots 5$, this means a dramatic increase in numerical complexity.

The multiplication of the data vector $\boldsymbol{x}(n)$ by the inverse $\widehat{\boldsymbol{R}}_{xx}^{-1}(n)$ of the estimate of the autocorrelation matrix means a rotation of this vector into the optimal update direction. Therefore, the algorithm achieves a fast convergence during an initialization phase. On the other hand, $\widehat{\boldsymbol{R}}_{xx}^{-1}(n)$ is based on all the past data weighted according to the exponential window. If the forgetting factor λ is too close to 1, the reaction of the filter to a changing system $\boldsymbol{h}(n)$ is very slow. In general, the choice of the numerical value of the forgetting factor has to consider two aspects—on one hand, a factor very close to 1 stabilizes the algorithm against temporarily insufficient data (e.g., speech pauses); on the other hand, this forgetting factor leads to a low tracking performance. Table 7.4 summarizes the RLS algorithm.

Table 7.4 Summary of the RLS algorithm

Algorithmic part	Corresponding equations
Filtering	$\widehat{d}(n\|n-1) = \widehat{\boldsymbol{h}}^{\mathrm{H}}(n-1)\,\boldsymbol{x}(n)$
Error signal	$e(n\|n-1) = y(n) - \widehat{d}(n\|n-1)$
Gain vector	$\boldsymbol{\gamma}(n) = \dfrac{\lambda^{-1}\,\widehat{\boldsymbol{R}}_{xx}^{-1}(n-1)\,\boldsymbol{x}(n)}{1+\lambda^{-1}\,\boldsymbol{x}^{\mathrm{H}}(n)\,\widehat{\boldsymbol{R}}_{xx}^{-1}(n-1)\,\boldsymbol{x}(n)}$
	$\widehat{\boldsymbol{R}}_{xx}^{-1}(n) = \lambda^{-1}\,\widehat{\boldsymbol{R}}_{xx}^{-1}(n-1) - \lambda^{-1}\,\boldsymbol{\gamma}(n)\,\boldsymbol{x}^{\mathrm{H}}(n)\,\widehat{\boldsymbol{R}}_{xx}^{-1}(n-1)$
Filter update	$\widehat{\boldsymbol{h}}(n) = \widehat{\boldsymbol{h}}(n-1) + \boldsymbol{\gamma}(n)\,e^{*}(n\|n-1)$

An overview of the three algorithms discussed until now is given in Table 7.5, which lists their cost functions and their numerical complexities.

Table 7.5 Adaptive algorithms, their cost functions, and their complexities

Algorithm	Cost function	Complexity
NLMS	$\mathrm{E}\left\{\|e(n)\|^2\right\}$	$O_{\mathrm{NLMS}}(N) \sim 2\,N$
AP	$\mathrm{E}\left\{\sum_{i=0}^{L-1} \|e(n-i\|n)\|^2\right\}$	$O_{\mathrm{AP}}(N) \sim 2\,L\,N$, but fast versions exist
RLS	$\sum_{l=0}^{n} \lambda^{n-l}\,\|e(l\|n)\|^2$, with $0 < \lambda \leq 1$	$O_{\mathrm{RLS}}(N) \sim N^2$, but fast versions exist

7.3.2 Fast Versions of the RLS Algorithm

In the case of colored input signals, the RLS algorithm converges much faster than does the NLMS algorithm. This is due to an implicit decorrelation achieved by the multiplication with the inverse $\widehat{\boldsymbol{R}}_{xx}^{-1}(n)$ of the estimate of the autocorrelation matrix (see Eq. 7.111). The price to be paid is an increase in computational complexity from $O(N)$ to $O(N^2)$.

Adaptive algorithms are called *fast* if their complexity is $O(N)$ or, in other words, if their computational demand grows only linearly with the number of filter coefficients to be updated. For the RLS algorithm the step from $O(N^2)$ to $O(N)$ is achieved by taking advantage of the shift properties of the excitation vector $\boldsymbol{x}(n)$ (see Eq. 7.8), the properties of the estimates of the autocorrelation matrix $\widehat{\boldsymbol{R}}_{xx}(n)$ (see Eq. 7.96), and the cross-correlation vector $\widehat{\boldsymbol{r}}_{xy}(n)$ (see Eq. 7.97) resulting from it. To show the recursions over the order, we write $\boldsymbol{x}^{(N)}(n)$, $\widehat{\boldsymbol{R}}_{xx}^{(N)}(n)$, and $\widehat{\boldsymbol{r}}_{xy}^{(N)}(n)$, where (N) marks the order of the vector or matrix.

We start with the excitation vector

$$\begin{aligned} \boldsymbol{x}^{(N)}(n) &= \left[x(n),\, x(n-1),\, \ldots,\, x(n-N+1)\right]^{\mathrm{T}} \\ &= \begin{bmatrix} \boldsymbol{x}^{(N-1)}(n) \\ x(n-N+1) \end{bmatrix} \\ &= \begin{bmatrix} x(n) \\ \boldsymbol{x}^{(N-1)}(n-1) \end{bmatrix}. \end{aligned} \tag{7.117}$$

Inserting this into Eq. 7.96 leads to

$$\begin{aligned} \widehat{\boldsymbol{R}}_{xx}^{(N)}(n) &= \sum_{l=0}^{n} \lambda^{(n-l)}\, \boldsymbol{x}^{(N)}(l)\, \boldsymbol{x}^{(N)^{\mathrm{H}}}(l) \\ &= \sum_{l=0}^{n} \lambda^{(n-l)} \begin{bmatrix} \boldsymbol{x}^{(N-1)}(l) \\ x(l-N+1) \end{bmatrix} \left[\boldsymbol{x}^{(N-1)^{\mathrm{H}}}(l),\; x^*(l-N+1)\right] \\ &= \begin{bmatrix} \widehat{\boldsymbol{R}}_{xx}^{(N-1)}(n) & \star \\ \star & \star \end{bmatrix}, \end{aligned} \tag{7.118}$$

where $\star$ marks terms that are not of interest in this context. Similar manipulations result in

$$\widehat{\boldsymbol{R}}_{xx}^{(N)}(n) = \begin{bmatrix} \star & \star \\ \star & \widehat{\boldsymbol{R}}_{xx}^{(N-1)}(n-1) \end{bmatrix}. \tag{7.119}$$

For the cross-correlation vector the order recursion reads

$$\begin{aligned}
\widehat{\boldsymbol{r}}_{xy}^{(N)}(n) &= \sum_{l=0}^{n} \lambda^{(n-l)}\, \boldsymbol{x}^{(N)}(l)\, y^*(l) \\
&= \sum_{l=0}^{n} \lambda^{(n-l)} \begin{bmatrix} \boldsymbol{x}^{(N-1)}(l) \\ x(l-N+1) \end{bmatrix} y^*(l) \\
&= \begin{bmatrix} \widehat{\boldsymbol{r}}_{xy}^{(N-1)}(n) \\ \star \end{bmatrix} = \begin{bmatrix} \star \\ \widehat{\boldsymbol{r}}_{xy}^{(N-1)}(n-1) \end{bmatrix}. \qquad (7.120)
\end{aligned}$$

By interleaving time and order recursions, a number of fast algorithms have been developed [34, 133, 146]. One of the most efficient ones is the *fast transversal filter* (FTF) *algorithm* [39]. For real data, the FTF algorithm achieves complexity $7N$ as compared to $2N$ for the NLMS algorithm. It combines four transversal filters: a forward predictor, a backward predictor, a filter for the gain vector calculation, and the filter for the system estimation. A new problem, however, arises with fast RLS (or Kalman) algorithms. When implemented with finite-precision fixed-point arithmetic, numerical instabilities occur. A number of stabilized versions have been proposed (e.g., [147, 19, 214, 9, 194]). They introduce redundant calculations and, thus, increase the numerical complexities of the related algorithms. However, they preserve the complexity $O(N)$. A conceptually simple stabilization procedure consists of monitoring certain variables. Their approach to the borders of their allowed regions is taken as an indication for an impending instability. In this case the algorithm is restarted.

7.4 THE KALMAN ALGORITHM

Kalman in 1960 [131] and Kalman and Bucy one year later [132] published an algorithm that estimates the state vector of a system of first-order differential equations. Their publications ended the efforts to overcome one of the main disadvantages of the Wiener filter: the necessary assumption of stationary inputs. The Kalman filter algorithm generates an estimate that is

- Linear
- Unbiased
- With minimum error variance
- Recursive

It does not need the assumption of stationary inputs or time-invariant systems. In acoustic echo and noise control the algorithm may be applied for two purposes: noise reduction [188] and echo cancellation [143].

In noise reduction the algorithm performs as a filter. If it is possible to find realistic models for the local speech, the noise, and the echo signal, a high-quality echo and noise-reduced output is achieved.

Because of its recursiveness, the algorithm may also be considered for the update of the coefficients of the echo canceling filter and, thus, playing the role of the adaptation algorithm.

General descriptions of the Kalman filter can be found, for example, in [21, 38, 110, 219]. The description given here is especially tailored to the *filter coefficient update*.

We start by formulating a system of linear differential equations for the filter coefficient vector $\boldsymbol{h}(n+1)$:

$$\boldsymbol{h}(n+1) = \boldsymbol{A_h}(n+1|n)\,\boldsymbol{h}(n) + \boldsymbol{B_h}(n)\,\boldsymbol{w}(n)\,. \tag{7.121}$$

In state space terminology this is the *system equation* with the *transition matrix* $\boldsymbol{A_h}(n+1|n)$ and the *control matrix* $\boldsymbol{B_h}(n)$. The input or *system noise vector* $\boldsymbol{w}(n)$ represents the driving force for the system (see Fig. 7.16). If the system is time-independent, the arguments $n+1|n$ and n of the matrices can be omitted.

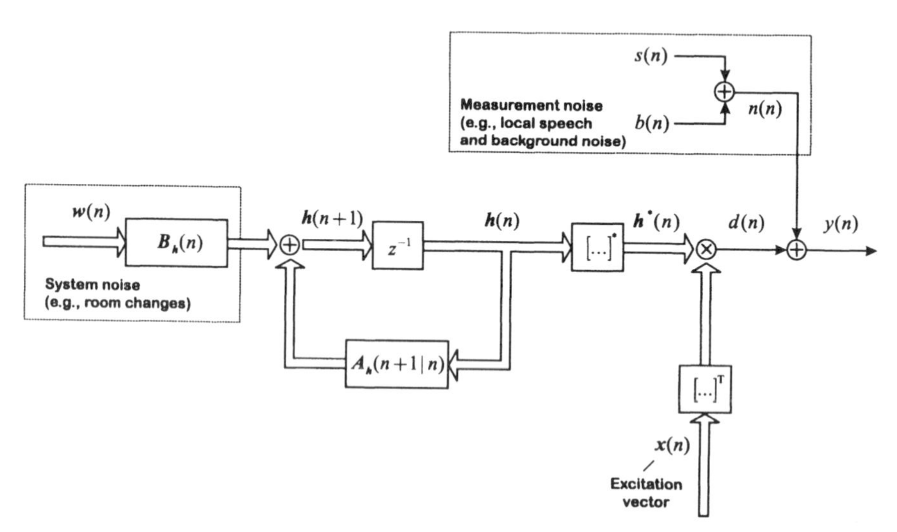

Fig. 7.16 Kalman filter: state space description of coefficient update.

If the algorithm is applied for the update of the filter coefficients the transition matrix $\boldsymbol{A_h}(n+1|n)$ may be assumed as $\alpha\,\boldsymbol{I}$, $0 < \alpha \leq 1$, and $\boldsymbol{B_h}(n)$ as unit matrix $\boldsymbol{I}$. We will, however, keep the general form in order to be prepared for later modifications of the state vector.

The filter output $y(n)$ is described by the *measurement equation*:

$$y(n) = \boldsymbol{c}_{\boldsymbol{h}}^{\mathrm{T}}(n)\,\boldsymbol{h}^*(n) + n(n)\,. \tag{7.122}$$

As long as the state vector $\boldsymbol{h}(n)$ is not extended, the *measurement vector* $\boldsymbol{c_h}(n)$ is equal to the signal vector $\boldsymbol{x}(n)$ of the echo canceling filter:

$$\boldsymbol{c_h}(n) = \boldsymbol{x}(n)\,. \tag{7.123}$$

In order to be consistent with the definition of the convolution in Eq. 5.2, we have to use the conjugate complex state vector $\boldsymbol{h}^*(n)$ in Eq. 7.122. The locally generated speech signal $s(n)$ and the local noise $b(n)$ are combined in $n(n)$ (see Fig. 3.13).

According to Eq. 5.3, the filter coefficient vector exhibits N coefficients. Consequently, the size of the vectors $\boldsymbol{c_h}(n)$ and $\boldsymbol{x}(n)$ is $(N \times 1)$ and the size of the matrix $\boldsymbol{A_h}(n+1|n)$ is $(N \times N)$. We assume that $\boldsymbol{w}(n)$ contains one element per filter coefficient, that is, its size is $(N \times 1)$. Then the size of the matrix $\boldsymbol{B_h}(n)$ is $(N \times N)$.

Before deriving the algorithm, we have to formulate a number of assumptions. They concern the statistical properties of the initial value of the filter vector $\boldsymbol{h}(0)$, the system noise vector $\boldsymbol{w}(n)$, and the measurement noise $n(n)$ and the relations between them:

$$\mathrm{E}\{\, \boldsymbol{h}(0) \,\} = \boldsymbol{0}\,, \tag{7.124}$$

$$\mathrm{E}\{\, \boldsymbol{w}(n) \,\} = \boldsymbol{0} \quad \text{for all } n\,, \tag{7.125}$$

$$\mathrm{E}\{\, n(n) \,\} = 0 \quad \text{for all } n\,. \tag{7.126}$$

All three quantities are assumed to be pairwise uncorrelated:

$$\mathrm{E}\{\, \boldsymbol{h}(0)\, n(n) \,\} = \boldsymbol{0} \quad \text{for all } n\,, \tag{7.127}$$

$$\mathrm{E}\left\{\, \boldsymbol{w}(n)\, \boldsymbol{h}^{\mathrm{H}}(0) \,\right\} = \boldsymbol{0} \quad \text{for all } n\,, \tag{7.128}$$

$$\mathrm{E}\{\, n(n_1)\, \boldsymbol{w}(n_2) \,\} = \boldsymbol{0} \quad \text{for all } n_1 \text{ and } n_2\,. \tag{7.129}$$

Due to the definition of the convolution (see Eq. 5.2) and the measurement equation 7.122 that relates $n(n)$ and $\boldsymbol{h}^*(n)$, we have to use $n(n)$ and not $n^*(n)$ in Eq. 7.127 (see also Eq. 7.138 and its derivation) and $\boldsymbol{w}(n)$ in Eq. 7.129, accordingly.

We also assume that we know the covariance matrices

$$\mathrm{E}\left\{\, \boldsymbol{h}(0)\, \boldsymbol{h}^{\mathrm{H}}(0) \,\right\} = \boldsymbol{P_h}(0|0)\,, \tag{7.130}$$

$$\mathrm{E}\left\{\, \boldsymbol{w}(n_1)\, \boldsymbol{w}^{\mathrm{H}}(n_2) \,\right\} = \begin{cases} \sigma_w^2(n_1)\, \boldsymbol{I} & \text{for } n_1 = n_2 \\ \boldsymbol{0} & \text{for } n_1 \neq n_2 \end{cases}, \tag{7.131}$$

and the covariance of the measurement noise

$$\mathrm{E}\{\, n(n_1)\, n^*(n_2) \,\} = \begin{cases} \sigma_n^2(n_1) & \text{for } n_1 = n_2 \\ 0 & \text{for } n_1 \neq n_2 \end{cases}. \tag{7.132}$$

This assumptions are realistic except Eqs. 7.132 and 7.131. Assumption 7.132 postulates that the local speech and the local noise are white; assumption 7.131

claims that system changes are also white. Both have to be modified in echo canceling applications.

The algorithm will be derived in three steps. In the first step we formulate the *initial conditions*. In the second step an estimate $\widehat{\boldsymbol{h}}(n|n-1)$ will be *predicted* on the basis of data available up to and including sampling instant $n-1$. The final step *corrects* this estimate to $\widehat{\boldsymbol{h}}(n|n)$ by using the data available up to and including sampling instant n. To show the optimality we always use the *orthogonality principle* (see Eq. 5.13).

7.4.1 Initialization

We assume for the initial estimate of the filter coefficient vector $\widehat{\boldsymbol{h}}(0|0)$:

$$\widehat{\boldsymbol{h}}(0|0) = \boldsymbol{K}(0)\, y^*(0)\,. \tag{7.133}$$

This estimate is linear and due to the assumptions, also unbiased. The vector $\boldsymbol{K}(0)$—in general $\boldsymbol{K}(n)$—is called the *Kalman gain*. If the estimate is optimal with respect to a minimal error variance the signal $e_{\mathrm{u}}(0)$, then

$$e_{\mathrm{u}}(0) = \boldsymbol{h}_{\Delta}^{\mathrm{H}}(0|0)\, \boldsymbol{c}_{\boldsymbol{h}}(0) = \boldsymbol{c}_{\boldsymbol{h}}^{\mathrm{T}}(0)\, \boldsymbol{h}_{\Delta}^{*}(0|0)\,, \tag{7.134}$$

caused by the estimation error

$$\boldsymbol{h}_{\Delta}(0|0) = \boldsymbol{h}(0) - \widehat{\boldsymbol{h}}(0|0) \tag{7.135}$$

has to be orthogonal to the measurement $y(0)$:

$$\mathrm{E}\left\{ e_{\mathrm{u}}(0)\, y^*(0) \right\} = \boldsymbol{c}_{\boldsymbol{h}}^{\mathrm{T}}(0)\, \mathrm{E}\left\{ \boldsymbol{h}_{\Delta}^{*}(0|0)\, y^*(0) \right\} = 0\,. \tag{7.136}$$

The input vector $\boldsymbol{c}_{\boldsymbol{h}}^{\mathrm{T}}(n)$ represents a *measured* signal and, consequently, has to be modeled by a deterministic quantity. Therefore, it can be taken out of the expectation. The condition for orthogonality then reads

$$\mathrm{E}\left\{ \boldsymbol{h}_{\Delta}^{*}(0|0)\, y^*(0) \right\} = \boldsymbol{0}\,, \tag{7.137}$$

or, taking the conjugate complex form,

$$\mathrm{E}\left\{ \boldsymbol{h}_{\Delta}(0|0)\, y(0) \right\} = \boldsymbol{0}\,. \tag{7.138}$$

Using Eq. 7.135 it follows:

$$\underbrace{\mathrm{E}\left\{ \boldsymbol{h}(0)\, y(0) \right\}}_{(1)} = \underbrace{\mathrm{E}\left\{ \widehat{\boldsymbol{h}}(0|0)\, y(0) \right\}}_{(2)}\,. \tag{7.139}$$

First, we extend (1) by inserting Eq. 7.122:

$$\begin{aligned}
\mathrm{E}\left\{ \boldsymbol{h}(0)\, y(0) \right\} &= \mathrm{E}\left\{ \boldsymbol{h}(0)\left[\boldsymbol{c}_{\boldsymbol{h}}^{\mathrm{T}}(0)\, \boldsymbol{h}^*(0) + n(0)\right] \right\} \\
&= \mathrm{E}\left\{ \boldsymbol{h}(0)\left[\boldsymbol{h}^{\mathrm{H}}(0)\, \boldsymbol{c}_{\boldsymbol{h}}(0) + n(0)\right] \right\} \\
&= \mathrm{E}\left\{ \boldsymbol{h}(0)\, \boldsymbol{h}^{\mathrm{H}}(0) \right\} \boldsymbol{c}_{\boldsymbol{h}}(0) \\
&= \boldsymbol{P}_{\boldsymbol{h}}(0|0)\, \boldsymbol{c}_{\boldsymbol{h}}(0)\,.
\end{aligned} \tag{7.140}$$

In the third line of the last equation we used Eq. 7.127. It should also be remarked again that $\boldsymbol{c_h}(n) = \boldsymbol{x}(n)$ represents the *measured* input signal vector. Therefore, we have to treat it as a *deterministic* quantity.

To evaluate (2) in Eq. 7.139 we use Eq. 7.133, the measurement equation 7.122, as well as Eqs. 7.127 and 7.132:

$$\begin{aligned} \mathrm{E}\left\{ \widehat{\boldsymbol{h}}(0|0)\, y(0) \right\} &= \boldsymbol{K}(0)\, \mathrm{E}\left\{ y^*(0)\, y(0) \right\} \\ &= \boldsymbol{K}(0)\, \mathrm{E}\left\{ \left[\boldsymbol{c}_{\boldsymbol{h}}^{\mathrm{H}}(0)\, \boldsymbol{h}(0) + n^*(0)\right] \left[\boldsymbol{h}^{\mathrm{H}}(0)\, \boldsymbol{c}_{\boldsymbol{h}}(0) + n(0)\right] \right\} \\ &= \boldsymbol{K}(0) \left[\boldsymbol{c}_{\boldsymbol{h}}^{\mathrm{H}}(0)\, \boldsymbol{P}_{\boldsymbol{h}}(0|0)\, \boldsymbol{c}_{\boldsymbol{h}}(0) + \sigma_n^2\right]. \end{aligned} \tag{7.141}$$

The term in brackets in the last line is scalar. Therefore, we can express the Kalman gain $\boldsymbol{K}(0)$ by a fraction:

$$\boldsymbol{K}(0) = \frac{\boldsymbol{P}_{\boldsymbol{h}}(0|0)\, \boldsymbol{c}_{\boldsymbol{h}}(0)}{\boldsymbol{c}_{\boldsymbol{h}}^{\mathrm{H}}(0)\, \boldsymbol{P}_{\boldsymbol{h}}(0|0)\, \boldsymbol{c}_{\boldsymbol{h}}(0) + \sigma_n^2}. \tag{7.142}$$

This result should be compared with the transfer function 5.33 of the Wiener filter.

As the last step of the initialization we have to calculate the covariance matrix $\boldsymbol{P}_{\boldsymbol{\Delta}}(0|0)$ of the estimation error vector $\boldsymbol{h}_{\boldsymbol{\Delta}}(0|0)$ (see Eq. 7.135):

$$\begin{aligned} \boldsymbol{P}_{\boldsymbol{\Delta}}(0|0) &= \mathrm{E}\left\{ \boldsymbol{h}_{\boldsymbol{\Delta}}(0|0)\, \boldsymbol{h}_{\boldsymbol{\Delta}}^{\mathrm{H}}(0|0) \right\} \\ &= \mathrm{E}\left\{ \left[\boldsymbol{h}(0) - \widehat{\boldsymbol{h}}(0|0)\right] \boldsymbol{h}_{\boldsymbol{\Delta}}^{\mathrm{H}}(0|0) \right\} \\ &= \mathrm{E}\left\{ \boldsymbol{h}(0)\, \boldsymbol{h}_{\boldsymbol{\Delta}}^{\mathrm{H}}(0|0) \right\} - \mathrm{E}\left\{ \widehat{\boldsymbol{h}}(0|0)\, \boldsymbol{h}_{\boldsymbol{\Delta}}^{\mathrm{H}}(0|0) \right\} \\ &= \mathrm{E}\left\{ \boldsymbol{h}(0)\, \boldsymbol{h}_{\boldsymbol{\Delta}}^{\mathrm{H}}(0|0) \right\} - \boldsymbol{K}(0)\, \mathrm{E}\left\{ y^*(0)\, \boldsymbol{h}_{\boldsymbol{\Delta}}^{\mathrm{H}}(0|0) \right\}. \end{aligned} \tag{7.143}$$

The second expectation is zero because of the orthogonality of the estimation error and the data (see Eq. 7.136). Therefore, we continue as follows:

$$\begin{aligned} \boldsymbol{P}_{\boldsymbol{\Delta}}(0|0) &= \mathrm{E}\left\{ \boldsymbol{h}(0) \left[\boldsymbol{h}^{\mathrm{H}}(0) - \widehat{\boldsymbol{h}}^{\mathrm{H}}(0|0)\right] \right\} \\ &= \boldsymbol{P}_{\boldsymbol{h}}(0|0) - \mathrm{E}\left\{ \boldsymbol{h}(0)\, \widehat{\boldsymbol{h}}^{\mathrm{H}}(0|0) \right\} \\ &= \boldsymbol{P}_{\boldsymbol{h}}(0|0) - \mathrm{E}\left\{ \boldsymbol{h}(0)\, y(0) \right\} \boldsymbol{K}^{\mathrm{H}}(0) \\ &= \boldsymbol{P}_{\boldsymbol{h}}(0|0) - \mathrm{E}\left\{ \boldsymbol{h}(0) \left[\boldsymbol{c}_{\boldsymbol{h}}^{\mathrm{T}}(0)\, \boldsymbol{h}^*(0) + n(0)\right] \right\} \boldsymbol{K}^{\mathrm{H}}(0) \\ &= \boldsymbol{P}_{\boldsymbol{h}}(0|0) - \boldsymbol{P}_{\boldsymbol{h}}(0|0)\, \boldsymbol{c}_{\boldsymbol{h}}(0)\, \boldsymbol{K}^{\mathrm{H}}(0), \end{aligned} \tag{7.144}$$

and finally

$$\boldsymbol{P}_{\boldsymbol{\Delta}}(0|0) = \boldsymbol{P}_{\boldsymbol{h}}(0|0) \left[\boldsymbol{I} - \boldsymbol{c}_{\boldsymbol{h}}(0)\, \boldsymbol{K}^{\mathrm{H}}(0)\right]. \tag{7.145}$$

This completes the initialization.

7.4.2 Prediction

We assume that the estimate $\widehat{\boldsymbol{h}}(n-1|n-1)$ and the estimation error covariance matrix $\boldsymbol{P}_{\boldsymbol{\Delta}}(n-1|n-1)$ are known. The estimate is optimal with respect to the data $y(k)$ for $k=0, \ldots, n-1$. On the basis of these data, one can predict an estimate $\widehat{\boldsymbol{h}}(n|n-1)$ for the sampling instant n:

$$\widehat{\boldsymbol{h}}(n|n-1) = \boldsymbol{A}_{\boldsymbol{h}}(n|n-1)\,\widehat{\boldsymbol{h}}(n-1|n-1)\,. \tag{7.146}$$

This assumption is based on the homogeneous part of the system equation. The system noise $\boldsymbol{w}(n-1)$ does not enter the prediction since it is assumed white (see Eq. 7.131) and zero mean (see Eq. 7.125). For optimality the estimation error $\boldsymbol{h}_{\boldsymbol{\Delta}}(n|n-1) = \boldsymbol{h}(n) - \widehat{\boldsymbol{h}}(n|n-1)$ has to be orthogonal to the data $y(k)$ for $k=0, \ldots, n-1$:

$$\mathrm{E}\Big\{ \Big[\boldsymbol{h}(n) - \widehat{\boldsymbol{h}}(n|n-1)\Big]\; y(k) \Big\} = \boldsymbol{0} \quad \text{for } k=0, \ldots, n-1\,. \tag{7.147}$$

Replacing $\boldsymbol{h}(n)$ by the system equation 7.121 and $\widehat{\boldsymbol{h}}(n|n-1)$ by Eq. 7.146, this reads as follows:

$$\begin{aligned} &\boldsymbol{A}_{\boldsymbol{h}}(n|n-1)\,\underbrace{\mathrm{E}\Big\{ \Big[\boldsymbol{h}(n-1) - \widehat{\boldsymbol{h}}(n-1|n-1)\Big]\; y(k) \Big\}}_{(1)} \\ &\quad + \boldsymbol{B}_{\boldsymbol{h}}(n-1)\,\underbrace{\mathrm{E}\{\, \boldsymbol{w}(n-1)\, y(k) \,\}}_{(2)} = \boldsymbol{0} \;\; \text{for } k=0, \ldots, n-1. \end{aligned} \tag{7.148}$$

Expectation (1) is zero since $\widehat{\boldsymbol{h}}(n-1|n-1)$ is assumed to be an optimal estimate and expectation (2) vanishes since $\boldsymbol{w}(n-1)$ is assumed to be white and, therefore, via the system equation 7.121 correlated with $y(n)$ but not with $y(k)$ for $k \leq n-1$. Thus, Eq. 7.146 is justified.

Next, we have to calculate the covariance matrix $\boldsymbol{P}_{\boldsymbol{\Delta}}(n|n-1)$ of the error

$$\boldsymbol{h}_{\boldsymbol{\Delta}}(n|n-1) = \boldsymbol{h}(n) - \widehat{\boldsymbol{h}}(n|n-1) \tag{7.149}$$

of the predicted estimate of the filter coefficient vector $\widehat{\boldsymbol{h}}(n|n-1)$:

$$\begin{aligned} \boldsymbol{P}_{\boldsymbol{\Delta}}(n|n-1) &= \mathrm{E}\{\, \boldsymbol{h}_{\boldsymbol{\Delta}}(n|n-1)\, \boldsymbol{h}_{\boldsymbol{\Delta}}^{\mathrm{H}}(n|n-1) \,\} \\ &= \mathrm{E}\Big\{ \Big[\boldsymbol{h}(n) - \boldsymbol{A}_{\boldsymbol{h}}(n|n-1)\,\widehat{\boldsymbol{h}}(n-1|n-1)\Big] \\ &\qquad \Big[\boldsymbol{h}^{\mathrm{H}}(n) - \widehat{\boldsymbol{h}}^{\mathrm{H}}(n-1|n-1)\,\boldsymbol{A}_{\boldsymbol{h}}^{\mathrm{H}}(n|n-1)\Big]\Big\}\,. \end{aligned} \tag{7.150}$$

Inserting the system equation further leads to

$$\begin{aligned} &\boldsymbol{P}_{\boldsymbol{\Delta}}(n|n-1) \\ &\quad = \boldsymbol{A}_{\boldsymbol{h}}(n|n-1)\,\mathrm{E}\{\, \boldsymbol{h}_{\boldsymbol{\Delta}}(n-1|n-1)\, \boldsymbol{h}_{\boldsymbol{\Delta}}^{\mathrm{H}}(n-1|n-1) \,\}\, \boldsymbol{A}_{\boldsymbol{h}}^{\mathrm{H}}(n|n-1) \\ &\qquad + \boldsymbol{B}_{\boldsymbol{h}}(n-1)\,\mathrm{E}\{\, \boldsymbol{w}(n-1)\, \boldsymbol{w}^{\mathrm{H}}(n-1) \,\}\, \boldsymbol{B}_{\boldsymbol{h}}^{\mathrm{H}}(n-1)\,, \end{aligned} \tag{7.151}$$

or finally

$$\begin{aligned}\boldsymbol{P}_{\Delta}(n|n-1) &= \boldsymbol{A}_{\boldsymbol{h}}(n|n-1)\,\boldsymbol{P}_{\Delta}(n-1|n-1)\,\boldsymbol{A}_{\boldsymbol{h}}^{\mathrm{H}}(n|n-1) \\ &\quad +\boldsymbol{B}_{\boldsymbol{h}}(n-1)\,\sigma_w^2(n-1)\,\boldsymbol{B}_{\boldsymbol{h}}^{\mathrm{H}}(n-1)\,. \end{aligned} \qquad (7.152)$$

Here we also used the orthogonality between $\boldsymbol{h}(n-1)$ and $\boldsymbol{w}(n-1)$.

7.4.3 Correction

For the final step we assume that the measurement data $y(n)$ are available and should be used to correct the predicted filter coefficient vector $\widehat{\boldsymbol{h}}(n|n-1)$ to $\widehat{\boldsymbol{h}}(n|n)$. Again we start with an assumption:

$$\begin{aligned}\widehat{\boldsymbol{h}}(n|n) &= \widehat{\boldsymbol{h}}(n|n-1) + \boldsymbol{K}(n)\left[y^*(n) - \boldsymbol{c}_{\boldsymbol{h}}^{\mathrm{H}}(n)\,\widehat{\boldsymbol{h}}(n|n-1)\right] \\ &= \widehat{\boldsymbol{h}}(n|n-1) + \boldsymbol{K}(n)\,e^*(n)\,. \end{aligned} \qquad (7.153)$$

This equation means that the predicted estimate is updated proportional to the conjugate complex of the difference between the measured value $y(n)$ and the output that would be generated by the predicted estimate $\widehat{\boldsymbol{h}}(n|n-1)$ in a noise-free measurement. This difference is weighted by the Kalman gain $\boldsymbol{K}(n)$ that has to be determined such that the orthogonality principle is satisfied:

$$\begin{aligned}&\mathrm{E}\left\{\left[\boldsymbol{h}(n) - \widehat{\boldsymbol{h}}(n|n)\right]\,y(k)\right\} \\ &\quad = \mathrm{E}\{\,\boldsymbol{h}_{\Delta}(n|n)\,y(k)\,\} = \boldsymbol{0} \quad \text{for } k = 0,\,\ldots\,,n\,. \end{aligned} \qquad (7.154)$$

Inserting Eq. 7.153 and the measurement equation 7.122 leads to

$$\begin{aligned}&\mathrm{E}\Big\{\Big[\boldsymbol{h}(n) - \widehat{\boldsymbol{h}}(n|n-1) - \boldsymbol{K}(n) \\ &\qquad \left(\boldsymbol{c}_{\boldsymbol{h}}^{\mathrm{H}}(n)\,\boldsymbol{h}(n) + n^*(n) - \boldsymbol{c}_{\boldsymbol{h}}^{\mathrm{H}}(n)\,\widehat{\boldsymbol{h}}(n|n-1)\right)\Big]\,y(k)\Big\} = \boldsymbol{0} \\ &\qquad\qquad \text{for } k = 0,\,\ldots\,,n\,. \end{aligned} \qquad (7.155)$$

After combining the terms, this reads

$$\begin{aligned}&\left(\boldsymbol{I} - \boldsymbol{K}(n)\,\boldsymbol{c}_{\boldsymbol{h}}^{\mathrm{H}}(n)\right)\,\mathrm{E}\left\{\left[\boldsymbol{h}(n) - \widehat{\boldsymbol{h}}(n|n-1)\right]y(k)\right\} \\ &\qquad -\boldsymbol{K}(n)\,\mathrm{E}\{\,n^*(n)\,y(k)\,\} = \boldsymbol{0} \quad \text{for } k = 0,\,\ldots\,,n\,. \end{aligned} \qquad (7.156)$$

According to Eq. 7.147, this condition is satisfied for $k = 0,\,\ldots\,,n-1$. Therefore, it remains only to show that Eq. 7.156 also can be fulfilled for $k = n$ by an optimal Kalman gain vector $\boldsymbol{K}(n)$. For $k = n$, the equation is given by

$$\begin{aligned}&\left(\boldsymbol{I} - \boldsymbol{K}(n)\,\boldsymbol{c}_{\boldsymbol{h}}^{\mathrm{H}}(n)\right)\,\underbrace{\mathrm{E}\left\{\left[\boldsymbol{h}(n) - \widehat{\boldsymbol{h}}(n|n-1)\right]y(n)\right\}}_{(1)} \\ &\qquad -\boldsymbol{K}(n)\,\underbrace{\mathrm{E}\{\,n^*(n)\,y(n)\,\}}_{(2)} = \boldsymbol{0}\,. \end{aligned} \qquad (7.157)$$

To evaluate the expectations (1) and (2), we use the measurement equation 7.122. Under the assumption that $n(n)$ is white (see Eq. 7.132), expectation (1) can be modified as follows:

$$\begin{aligned} &\mathrm{E}\Big\{ \Big[\boldsymbol{h}(n) - \widehat{\boldsymbol{h}}(n|n-1)\Big] \left[\boldsymbol{c}_{\boldsymbol{h}}^{\mathrm{T}}(n)\, \boldsymbol{h}^{*}(n) + n(n)\right] \Big\} \\ &\qquad = \underbrace{\mathrm{E}\Big\{ \Big[\boldsymbol{h}(n) - \widehat{\boldsymbol{h}}(n|n-1)\Big]\, \boldsymbol{h}^{\mathrm{H}}(n) \Big\}}_{(3)}\, \boldsymbol{c}_{\boldsymbol{h}}(n)\,, \end{aligned} \tag{7.158}$$

where

$$\mathrm{E}\Big\{ \Big[\boldsymbol{h}(n) - \widehat{\boldsymbol{h}}(n|n-1)\Big]\, n(n) \Big\} = \boldsymbol{0} \tag{7.159}$$

because $n(n)$ is white and uncorrelated with $\boldsymbol{h}(0)$ (see Eqs. 7.127 and 7.129) and consequently with all $\boldsymbol{h}(k)$, $k = 1, \ldots, n$. Furthermore, the predicted filter coefficient vector $\widehat{\boldsymbol{h}}(n|n-1)$ uses data up to $n-1$ only.

Expectation (3) is extended:

$$\begin{aligned} &\mathrm{E}\Big\{ \Big[\boldsymbol{h}(n) - \widehat{\boldsymbol{h}}(n|n-1)\Big]\, \boldsymbol{h}^{\mathrm{H}}(n) \Big\} \\ &\quad = \mathrm{E}\Big\{ \Big[\boldsymbol{h}(n) - \widehat{\boldsymbol{h}}(n|n-1)\Big] \Big[\boldsymbol{h}^{\mathrm{H}}(n) - \widehat{\boldsymbol{h}}^{\mathrm{H}}(n|n-1)\Big] \Big\} \\ &\qquad\qquad + \mathrm{E}\Big\{ \Big[\boldsymbol{h}(n) - \widehat{\boldsymbol{h}}(n|n-1)\Big]\, \widehat{\boldsymbol{h}}^{\mathrm{H}}(n|n-1) \Big\} \\ &\quad = \boldsymbol{P}_{\boldsymbol{\Delta}}(n|n-1) + \underbrace{\mathrm{E}\Big\{ \Big[\boldsymbol{h}(n) - \widehat{\boldsymbol{h}}(n|n-1)\Big]\, \widehat{\boldsymbol{h}}^{\mathrm{H}}(n|n-1) \Big\}}_{(4)}\,. \end{aligned} \tag{7.160}$$

Expectation (4) is zero. To show this, we insert the system equation 7.121 and the predicted estimate 7.146:

$$\begin{aligned} &\mathrm{E}\Big\{ \Big[\boldsymbol{h}(n) - \widehat{\boldsymbol{h}}(n|n-1)\Big]\, \widehat{\boldsymbol{h}}^{\mathrm{H}}(n|n-1) \Big\} \\ &\quad = \boldsymbol{A}_{\boldsymbol{h}}(n|n-1) \\ &\qquad\quad \mathrm{E}\Big\{ \Big[\boldsymbol{h}(n-1) - \widehat{\boldsymbol{h}}(n-1|n-1)\Big]\, \widehat{\boldsymbol{h}}^{\mathrm{H}}(n-1|n-1) \Big\}\, \boldsymbol{A}_{\boldsymbol{h}}^{\mathrm{H}}(n|n-1) \\ &\quad + \boldsymbol{B}_{\boldsymbol{h}}(n-1)\, \mathrm{E}\Big\{ \boldsymbol{w}(n-1)\, \widehat{\boldsymbol{h}}^{\mathrm{H}}(n-1|n-1) \Big\}\, \boldsymbol{A}_{\boldsymbol{h}}^{\mathrm{H}}(n|n-1)\,. \end{aligned} \tag{7.161}$$

Both expectations in this equation are zero: the first one since $\widehat{\boldsymbol{h}}(n-1|n-1)$ is an optimal estimate and has to obey the principle of orthogonality [111] and the second one due to the assumption that $\boldsymbol{w}(n-1)$ is white.

Again, under the assumption that $n(n)$ is white, it follows for the expectation (2) in Eq. 7.157 that

$$\mathrm{E}\left\{ n^{*}(n) \left[\boldsymbol{c}_{\boldsymbol{h}}^{\mathrm{T}}(n)\, \boldsymbol{h}^{*}(n) + n(n)\right] \right\} = \sigma_n^2(n)\,. \tag{7.162}$$

Inserting the results for expectations (1) and (2) leads to an equation that can be solved for the still unknown Kalman gain vector $\boldsymbol{K}(n)$

$$\left[\boldsymbol{I} - \boldsymbol{K}(n)\,\boldsymbol{c}_{\boldsymbol{h}}^{\mathrm{H}}(n)\right]\,\boldsymbol{P}_{\boldsymbol{\Delta}}(n|n-1)\,\boldsymbol{c}_{\boldsymbol{h}}(n) - \boldsymbol{K}(n)\,\sigma_n^2(n) = \boldsymbol{0}\,, \tag{7.163}$$

and finally

$$\boldsymbol{K}(n) = \frac{\boldsymbol{P}_{\boldsymbol{\Delta}}(n|n-1)\,\boldsymbol{c}_{\boldsymbol{h}}(n)}{\boldsymbol{c}_{\boldsymbol{h}}^{\mathrm{H}}(n)\,\boldsymbol{P}_{\boldsymbol{\Delta}}(n|n-1)\,\boldsymbol{c}_{\boldsymbol{h}}(n) + \sigma_n^2(n)}\,. \tag{7.164}$$

If we define $\boldsymbol{P}_{\boldsymbol{\Delta}}(0|-1) = \boldsymbol{P}_{\boldsymbol{h}}(0|0)$, this equation also covers the initialization (see Eq. 7.142).

In order to complete the recursion, the covariance matrix $\boldsymbol{P}_{\boldsymbol{\Delta}}(n|n)$

$$\begin{aligned}\boldsymbol{P}_{\boldsymbol{\Delta}}(n|n) &= \mathrm{E}\left\{\boldsymbol{h}_{\boldsymbol{\Delta}}(n|n)\,\boldsymbol{h}_{\boldsymbol{\Delta}}^{\mathrm{H}}(n|n)\right\}\\ &= \mathrm{E}\left\{\left[\boldsymbol{h}(n) - \widehat{\boldsymbol{h}}(n|n)\right]\left[\boldsymbol{h}^{\mathrm{H}}(n) - \widehat{\boldsymbol{h}}^{\mathrm{H}}(n|n)\right]\right\}\end{aligned} \tag{7.165}$$

of the estimation error at time instant n has to be determined as a function of $\boldsymbol{P}_{\boldsymbol{\Delta}}(n|n-1)$.

As a first step, we find a relation for the estimation error $\boldsymbol{h}_{\boldsymbol{\Delta}}(n|n)$ by applying the equation for the filter coefficient estimate 7.153 and inserting the system equation 7.121 and the measurement equation 7.122:

$$\begin{aligned}\boldsymbol{h}_{\boldsymbol{\Delta}}(n|n) &= \boldsymbol{h}(n) - \widehat{\boldsymbol{h}}(n|n-1) - \boldsymbol{K}(n)\left[y^*(n) - \boldsymbol{c}_{\boldsymbol{h}}^{\mathrm{H}}(n)\,\widehat{\boldsymbol{h}}(n|n-1)\right]\\
&= \boldsymbol{A}_{\boldsymbol{h}}(n|n-1)\,\boldsymbol{h}(n-1) + \boldsymbol{B}_{\boldsymbol{h}}(n-1)\,\boldsymbol{w}(n-1)\\
&\quad - \boldsymbol{A}_{\boldsymbol{h}}(n|n-1)\,\widehat{\boldsymbol{h}}(n-1|n-1)\\
&\quad - \boldsymbol{K}(n)\left[\boldsymbol{c}_{\boldsymbol{h}}^{\mathrm{H}}(n)\left(\boldsymbol{A}_{\boldsymbol{h}}(n|n-1)\,\boldsymbol{h}(n-1) + \boldsymbol{B}_{\boldsymbol{h}}(n-1)\,\boldsymbol{w}(n-1)\right)\right.\\
&\quad \left. + n^*(n) - \boldsymbol{c}_{\boldsymbol{h}}^{\mathrm{H}}(n)\,\boldsymbol{A}_{\boldsymbol{h}}(n|n-1)\,\widehat{\boldsymbol{h}}(n-1|n-1)\right]\\
&= \left[\boldsymbol{I} - \boldsymbol{K}(n)\,\boldsymbol{c}_{\boldsymbol{h}}^{\mathrm{H}}(n)\right]\\
&\qquad \left[\boldsymbol{A}_{\boldsymbol{h}}(n|n-1)\left(\boldsymbol{h}(n-1) - \widehat{\boldsymbol{h}}(n-1|n-1)\right)\right.\\
&\qquad \left. + \boldsymbol{B}_{\boldsymbol{h}}(n-1)\,\boldsymbol{w}(n-1)\right] - \boldsymbol{K}(n)\,n^*(n)\\
&= \left[\boldsymbol{I} - \boldsymbol{K}(n)\,\boldsymbol{c}_{\boldsymbol{h}}^{\mathrm{H}}(n)\right]\\
&\qquad \left[\boldsymbol{A}_{\boldsymbol{h}}(n|n-1)\,\boldsymbol{h}_{\boldsymbol{\Delta}}(n-1|n-1) + \boldsymbol{B}_{\boldsymbol{h}}(n-1)\,\boldsymbol{w}(n-1)\right]\\
&\qquad - \boldsymbol{K}(n)\,n^*(n)\,.\end{aligned} \tag{7.166}$$

For the final transformation we make use of the fact that $\boldsymbol{w}(n-1)$ and $n(n)$ are assumed to be pairwise orthogonal meaning that all cross-correlations disappear:

$$\begin{aligned}\boldsymbol{P}_{\boldsymbol{\Delta}}(n|n) &= \left[\boldsymbol{I} - \boldsymbol{K}(n)\,\boldsymbol{c}_{\boldsymbol{h}}^{\mathrm{H}}(n)\right]\,\boldsymbol{P}_{\boldsymbol{\Delta}}(n|n-1)\,\underbrace{\left[\boldsymbol{I} - \boldsymbol{c}_{\boldsymbol{h}}(n)\,\boldsymbol{K}^{\mathrm{H}}(n)\right]}_{(1)}\\
&\qquad + \boldsymbol{K}(n)\,\sigma_n^2(n)\,\boldsymbol{K}^{\mathrm{H}}(n)\,.\end{aligned} \tag{7.167}$$

To achieve further simplifications, the term (1) has to be broken up:

$$\begin{aligned}\boldsymbol{P_\Delta}(n|n) &= \left[\boldsymbol{I} - \boldsymbol{K}(n)\,\boldsymbol{c}_{\boldsymbol{h}}^{\mathrm{H}}(n)\right]\boldsymbol{P_\Delta}(n|n-1) \\ &- \left\{\left[\boldsymbol{I} - \boldsymbol{K}(n)\,\boldsymbol{c}_{\boldsymbol{h}}^{\mathrm{H}}(n)\right]\boldsymbol{P_\Delta}(n|n-1)\,\boldsymbol{c_h}(n) \right.\\ &\left. - \boldsymbol{K}(n)\,\sigma_n^2(n)\right\}\boldsymbol{K}^{\mathrm{H}}(n)\,. \end{aligned} \tag{7.168}$$

The expression in curly brackets complies with Eq. 7.163. Therefore the equation for the covariance matrix of the estimation error simplifies to

$$\boldsymbol{P_\Delta}(n|n) = \left[\boldsymbol{I} - \boldsymbol{K}(n)\,\boldsymbol{c}_{\boldsymbol{h}}^{\mathrm{H}}(n)\right]\boldsymbol{P_\Delta}(n|n-1)\,. \tag{7.169}$$

To show the equivalence to Eq. 7.145, we can also write the Hermitian—the conjugate complex transpose—of this equation:

$$\boldsymbol{P_\Delta}(n|n) = \boldsymbol{P_\Delta}(n|n-1)\left[\boldsymbol{I} - \boldsymbol{c_h}(n)\,\boldsymbol{K}^{\mathrm{H}}(n)\right]. \tag{7.170}$$

This completes the recursive algorithm.

The matrices $\boldsymbol{A_h}(n|n-1)$, $\boldsymbol{B_h}(n)$, and $\boldsymbol{c_h}(n)$ have been introduced to enable an extension of the state vector. If white inputs $\boldsymbol{w}(n)$ and $n(n)$ are allowed, in an echo cancellation application $\boldsymbol{A_h}(n|n-1) = \boldsymbol{I}$ or (to avoid stability problems) $\boldsymbol{A_h}(n|n-1) = \alpha\,\boldsymbol{I}$ with $0 < \alpha < 1$, $\boldsymbol{B_h}(n) = \boldsymbol{I}$, and $\boldsymbol{c_h}(n) = \boldsymbol{x}(n)$ (see Eq. 7.123). For $\boldsymbol{A_h}(n|n-1) = \alpha\,\boldsymbol{I}$ and $\boldsymbol{B_h}(n) = \boldsymbol{I}$, the equations for the Kalman algorithm are summarized in Table 7.6.

Fig. 7.17 shows a diagram of the Kalman filter as given in Table 7.6. The filter algorithm copies the system and measurement model. It is updated by the difference between the measurement $y(n)$ (the output of the real system) and the output of the model. The difference is weighted by the Kalman gain.

The sequence in which the quantities of the Kalman filter algorithm are calculated is depicted in Fig. 7.18. It becomes visible that there are two loops: one for the calculation of the error covariance matrices and the Kalman gain and a second one for the determination of the quantity to be estimated—here the filter coefficient vector.

7.4.4 Colored Noise

The filter algorithm derived above can be extended for colored system noise $\boldsymbol{w}(n)$ and/or colored measurement noise $n(n)$. In this case the quantities are assumed to be the outputs of linear systems driven by white input signals. The systems are described by linear differential equations with —to shorten the notation—fixed coefficients. To explain the procedure, we use the measurement noise $n(n)$ as an example:

$$n(n) = \sum_{k=1}^{q} \alpha_k\, n(n-k) + n_{\mathrm{w}}(n)\,. \tag{7.171}$$

Table 7.6 Summary of the Kalman algorithm

Algorithmic part	Corresponding equations
System equation	$\boldsymbol{h}(n) = \alpha\, \boldsymbol{h}(n-1) + \boldsymbol{w}(n-1)$
Measurement equation	$y(n) = \boldsymbol{x}^{\mathrm{T}}(n)\, \boldsymbol{h}^{*}(n) + n(n)$
Initialization	$\boldsymbol{K}(0) = \dfrac{\boldsymbol{P}_{\boldsymbol{h}}(0\|0)\, \boldsymbol{x}(0)}{\boldsymbol{x}^{\mathrm{H}}(0)\, \boldsymbol{P}_{\boldsymbol{h}}(0\|0)\, \boldsymbol{x}(0) + \sigma_n^2(0)}$
	$\widehat{\boldsymbol{h}}(0\|0) = \boldsymbol{K}(0)\, y^{*}(0)$
	$\boldsymbol{P}_{\boldsymbol{\Delta}}(0\|0) = \boldsymbol{P}_{\boldsymbol{h}}(0\|0)\, \left[\boldsymbol{I} - \boldsymbol{x}(0)\, \boldsymbol{K}^{\mathrm{H}}(0)\right]$
Prediction	$\widehat{\boldsymbol{h}}(n\|n-1) = \alpha\, \widehat{\boldsymbol{h}}(n-1\|n-1)$
	$\boldsymbol{P}_{\boldsymbol{\Delta}}(n\|n-1) = \alpha^2\, \boldsymbol{P}_{\boldsymbol{\Delta}}(n-1\|n-1) + \sigma_w^2(n-1)$
Correction	$\boldsymbol{K}(n) = \dfrac{\boldsymbol{P}_{\boldsymbol{\Delta}}(n\|n-1)\, \boldsymbol{x}(n)}{\boldsymbol{x}^{\mathrm{H}}(n)\, \boldsymbol{P}_{\boldsymbol{\Delta}}(n\|n-1)\, \boldsymbol{x}(n) + \sigma_n^2(n)}$
	$\widehat{\boldsymbol{h}}(n\|n) = \widehat{\boldsymbol{h}}(n\|n-1) + \boldsymbol{K}(n)\left[y^{*}(n) - \boldsymbol{x}^{\mathrm{H}}(n)\, \widehat{\boldsymbol{h}}(n\|n-1)\right]$
	$\boldsymbol{P}_{\boldsymbol{\Delta}}(n\|n) = \boldsymbol{P}_{\boldsymbol{\Delta}}(n\|n-1)\, \left[\boldsymbol{I} - \boldsymbol{x}(n)\, \boldsymbol{K}^{\mathrm{H}}(n)\right]$

Here q is the order of the system and $n_{\mathrm{w}}(n)$ the white input signal. In state space notation Eq. 7.171 reads as

$$\boldsymbol{n}(n+1) = \boldsymbol{A}_{\boldsymbol{n}}(n+1|n)\, \boldsymbol{n}(n) + \begin{bmatrix} 0 \\ 0 \\ \vdots \\ 0 \\ 1 \end{bmatrix} n_{\mathrm{w}}(n+1)\,, \tag{7.172}$$

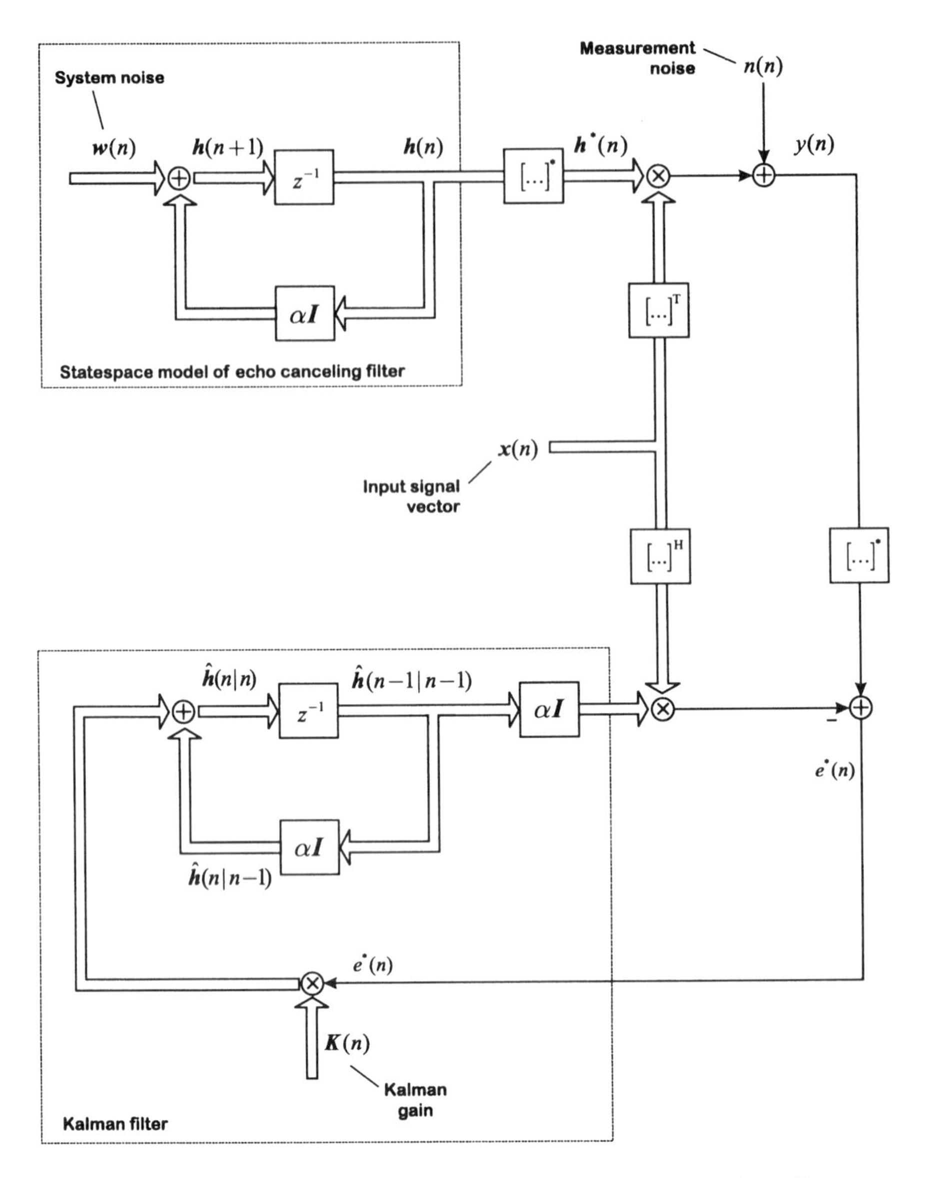

Fig. 7.17 Statespace model of the echo canceling filter and the Kalman filter.

with the following $(q \times 1)$ vector and $(q \times q)$ matrix notation:

$$\boldsymbol{n}(n) = \left[\begin{array}{ccccc} n(n-q+1) & n(n-q+2) & \cdots & n(n-1) & n(n) \end{array} \right]^{\mathrm{T}}, \quad (7.173)$$

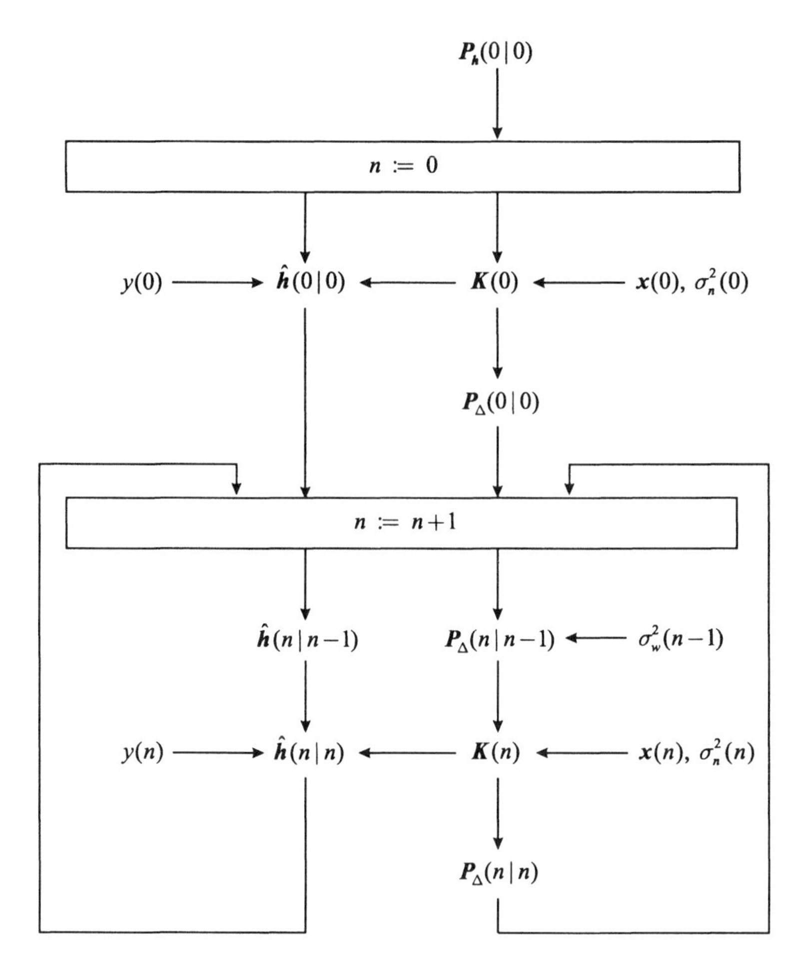

Fig. 7.18 Sequence of steps of the Kalman filter algorithm applied to echo cancellation filter update (with permission from [99]).

$$\boldsymbol{A_n}(n+1|n) = \begin{bmatrix} 0 & 1 & 0 & \dots & 0 \\ 0 & 0 & 1 & \dots & 0 \\ \vdots & \vdots & \vdots & \ddots & \vdots \\ 0 & 0 & 0 & \dots & 1 \\ \alpha_q & \alpha_{q-1} & \alpha_{q-2} & \dots & \alpha_1 \end{bmatrix}. \tag{7.174}$$

Combining the measurement equation 7.122 and the state space equation 7.172 for the measurement noise leads to

$$\begin{aligned} y(n) &= \boldsymbol{c}_{\boldsymbol{h}}^{\mathrm{T}}(n)\, \boldsymbol{h}^*(n) + [0\ 0\ \cdots\ 0\ 1]\ \boldsymbol{n}(n) \\ &= \boldsymbol{c}_{\boldsymbol{h}}^{\mathrm{T}}(n)\, \boldsymbol{h}^*(n) \end{aligned}$$

$$+ [0\ 0\ \cdots\ 0\ 1] \left[\boldsymbol{A_n}(n|n-1)\, \boldsymbol{n}(n-1) + \begin{bmatrix} 0 \\ 0 \\ \vdots \\ 0 \\ 1 \end{bmatrix} n_\mathrm{w}(n) \right]. \tag{7.175}$$

This expression can be simplified:

$$\begin{aligned} y(n) &= \boldsymbol{c}_{\boldsymbol{h}}^\mathrm{T}(n)\, \boldsymbol{h}^*(n) + [0\ 0\ \cdots\ 0\ 1]\ \boldsymbol{A_n}(n|n-1)\, \boldsymbol{n}(n-1) + n_\mathrm{w}(n) \\ &= \left[\boldsymbol{c}_{\boldsymbol{h}}^\mathrm{T}(n), [0\ 0\ \cdots\ 0\ 1]\right] \begin{bmatrix} \boldsymbol{I} & \boldsymbol{0} \\ \boldsymbol{0} & \boldsymbol{A_n}(n|n-1) \end{bmatrix} \begin{bmatrix} \boldsymbol{h}^*(n) \\ \boldsymbol{n}(n-1) \end{bmatrix} \\ &\quad + n_\mathrm{w}(n)\,, \end{aligned} \tag{7.176}$$

where $\boldsymbol{0}$ denotes an all zero matrix of appropriate size. Now one can extend the state vector and the system matrix:

$$\boldsymbol{h}_\mathrm{ext}(n) = \begin{bmatrix} \boldsymbol{h}(n) \\ \boldsymbol{n}^*(n-1) \end{bmatrix}, \tag{7.177}$$

$$\begin{aligned} \boldsymbol{c}_\mathrm{ext}(n) &= \begin{bmatrix} \boldsymbol{I} & \boldsymbol{0} \\ \boldsymbol{0} & \boldsymbol{A}_{\boldsymbol{n}}^\mathrm{T}(n|n-1) \end{bmatrix} \begin{bmatrix} \boldsymbol{c}_{\boldsymbol{h}}(n) \\ \begin{bmatrix} 0 \\ 0 \\ \vdots \\ 0 \\ 1 \end{bmatrix}_{(q\times 1)} \end{bmatrix} \\ &= \begin{bmatrix} \boldsymbol{c}_{\boldsymbol{h}}(n) \\ \alpha_q \\ \alpha_{q-1} \\ \vdots \\ \alpha_2 \\ \alpha_1 \end{bmatrix}, \end{aligned} \tag{7.178}$$

where the sizes of $\boldsymbol{h}_\mathrm{ext}(n)$ and of $\boldsymbol{c}_\mathrm{ext}(n)$ are equal to $((N+q) \times 1)$.

Using those quantities we arrive at a measurement equation that has the same form as Eq. 7.122:

$$y(n) = \boldsymbol{c}_\mathrm{ext}^\mathrm{T}(n)\, \boldsymbol{h}_\mathrm{ext}^*(n) + n_\mathrm{w}(n)\,. \tag{7.179}$$

To come to a state equation for the extended state vector $\boldsymbol{h_{ext}}(n)$, we have to extend the system matrix

$$\boldsymbol{A}_\mathrm{ext}(n+1|n) = \begin{bmatrix} \boldsymbol{A_h}(n+1|n) & \boldsymbol{0} \\ \boldsymbol{0} & \boldsymbol{A}_{\boldsymbol{n}}^*(n|n-1) \end{bmatrix}, \tag{7.180}$$

as well as the control matrix

$$\boldsymbol{B}_{\text{ext}}(n) = \begin{bmatrix} \boldsymbol{B}_{\boldsymbol{h}}(n) & \mathbf{0} \\ \mathbf{0} & \boldsymbol{B}_{\boldsymbol{n}} \end{bmatrix}, \tag{7.181}$$

and the system noise vector

$$\boldsymbol{w}_{\text{ext}}(n) = \begin{bmatrix} \boldsymbol{w}(n) \\ \begin{bmatrix} 0 \\ 0 \\ \vdots \\ 0 \\ n_{\text{w}}^{*}(n) \end{bmatrix}_{(q\times 1)} \end{bmatrix}, \tag{7.182}$$

where $\boldsymbol{B}_{\boldsymbol{n}}$ is given by

$$\boldsymbol{B}_{\boldsymbol{n}} = \begin{bmatrix} 0 & 0 & \dots & 0 & 0 \\ 0 & 0 & \dots & 0 & 0 \\ \vdots & \vdots & \dots & \vdots & \vdots \\ 0 & 0 & \dots & 0 & 0 \\ 0 & 0 & \dots & 0 & 1 \end{bmatrix}_{(q\times q)}. \tag{7.183}$$

Using those notations, one finally arrives at the following system equation (see also Eq. 7.121):

$$\boldsymbol{h}_{\text{ext}}(n+1) = \boldsymbol{A}_{\text{ext}}(n+1|n)\,\boldsymbol{h}_{\text{ext}}(n) + \boldsymbol{B}_{\text{ext}}(n)\,\boldsymbol{w}_{\text{ext}}(n)\,. \tag{7.184}$$

Since the measurement noise $n(n)$ is a sum of the local speech signal $s(n)$ and the local noise $b(n)$ (see Fig. 7.1), it may be advisable to have individual models for both components [188].

An equivalent technique can be applied for the system noise $\boldsymbol{w}(n)$ that models the variations of the LEM system. White noise is certainly not a realistic description of these changes. At the time this text was written, however, no suitable model was known to the authors.

Part III

Acoustic Echo and Noise Control

8

Traditional Methods for Stabilization of Electroacoustic Loops

Before we start describing algorithms for echo cancellation, noise reduction, and beamforming in the next chapters, we will mention here some traditional methods. These methods were invented, when processing power and memory were highly limited. In the meantime the algorithms have been advanced and have proved their robustness in various systems and products.

In today's systems, algorithms like echo cancellation or beamforming have not replaced traditional schemes completely. Both have been combined. An example might be the interaction of loss control and echo cancellation. While the first one (see Section 8.3) is only able to allow half-duplex communication the latter can achieve a full-duplex one. In terms of user comfort, only echo cancellation should be applied. Nevertheless, in some situations, such as at the beginning of phone calls, adaptive filters have not yet converged and cannot guarantee a sufficiently high amount of echo attenuation. In these situations the application of loss control is required. With increasing convergence of the echo cancellation filter the attenuation of the loss control can be reduced, bringing the system from half-duplex to full-duplex.

In this chapter we will mention three traditional methods:

- Adaptive line enhancers
- Frequency shifts
- Loss controls

Adaptive line enhancement was designed at first for enhancing sinusoid signals that were corrupted by wideband noise. Areas of application are any kind of amplitude modulated transmission schemes. By exchanging the role of noise and desired signal, line enhancement algorithms can also be used to suppress sinusoidal distortions (of unknown frequency) in speech. In Section 8.1 we will focus on this type of application.

Frequency shift approaches are utilized for stabilizing closed acoustic loops as they appear in public address systems or in-car communication systems. As the term indicates, the recorded signals are shifted in frequency by a few cycles before they are played via the loudspeakers. When increasing the amplification of the loudspeakers, such systems would start howling at a certain frequency. The frequency shift, however, moves the howling frequency by a few cycles at each loop through the system. As soon as a spectral valley in the frequency responses of loudspeaker–enclosure–microphone (LEM) systems is reached, the howling is attenuated.

Finally we will describe loss controls or nonlinear[1] processors in Section 8.3. The latter term is often used in ITU or ETSI recommendations. As already explained, these systems place multiplication units in the sending and receiving paths of electroacoustic systems. A first task is to attenuate the direction that is not active, respectively to distribute a certain amount of "artificial" attenuation in both channels. Furthermore, loss controls are often combined with automatic gain controls. In this case, the incoming and the outgoing signal should have certain levels (during periods of speech activity). Signals of speakers with loud voices or speakers close to the microphone will be attenuated; soft signals or signals from sources far away from the microphone will be amplified.

8.1 ADAPTIVE LINE ENHANCEMENT

Adaptive line enhancement belongs to the group of adaptive noise cancellation algorithms as described by Widrow [238]. Figure 8.1 illustrates the basic concept of this type of application. A signal $s(n)$ is transmitted to a sensor. Besides this desired signal, the sensor also picks up background noise $n_1(n)$. The noise $n_1(n)$ is assumed to be uncorrelated with the desired signal $s(n)$. A second sensor receives a noise $n_2(n)$ that is also uncorrelated with $s(n)$ but correlated in some (linear) way with the background noise $n_1(n)$ at the primary sensor. This sensor provides the reference input to the noise cancellation filter.

If the correlation between $n_1(n)$ and $n_2(n)$ has also noncausal parts[2] a delay of N_C samples is introduced in the primary path. The filter $\boldsymbol{w}(n)$ is adapted in order to produce an output $\hat{n}_1(n-N_C)$ that is an as close as possible replica to $n_1(n-N_C)$. By subtracting the replica from the delayed signal of the primary sensor, noise reduction or signal enhancement, respectively, is achieved.

[1] To be precise, the adjective "time-variant" should also be used here.

[2] This would be the case if the reference sensor were located closer to the noise source than the primary sensor.

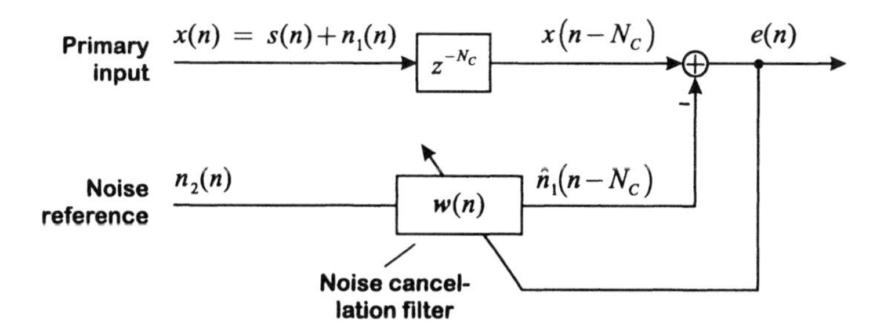

Fig. 8.1 Basic scheme of adaptive noise cancellation.

A variety of applications exist for adaptive noise cancellation. A few examples are the following:

- *Communication headsets for pilots:* Besides a microphone located close to the pilot's mouth, these devices are equipped with a second microphone for environmental noise. This second microphone blocks the direction to the speaker's mouth and records only the environmental noise. With this mechanism the signal-to-noise ratio within the sending path of the communication system can be improved considerably.

- *Enhancement of fetal electrocardiography:* For monitoring the heartbeat of the fetus, a primary sensor is placed on the mother's stomach. The heartbeats of both the mother and the fetus are received with this sensor. In order to cancel that "noisy component," one or more reference sensors are placed close to the mother's heart. After a short adaptation period, the mother's signal can be reduced significantly.

A problem occurs when it is not possible to get a noise reference. In this case one has two possibilities: (1) the reference signal can be generated artificially or (2) one can try to extract the reference out of the primary signal. An example of the first case might be any kind of humbling noise (e.g., interferences caused by electrical ground loops) suppression. Depending on the country, a 50- or 60-Hz sinusoidal signal is generated and used as a noise reference. The task of the adaptive filter then consists of adjusting amplitude and phase of the periodic reference.

If there is no a priori knowledge except periodicity about the interference, a delayed version of the primary signal can be used as a reference signal. The delay must be chosen sufficiently long in order to decorrelate the signal components in the reference channel from those in the primary channel. Due to the periodicity of the interference, this component is still correlated in the reference and the primary path. Applications for this scheme are public address systems, hearing aids, and other types of closed acoustic loops. In case of high loudspeaker amplification these systems can produce howling which is a periodic interference. A delay of 1–2 ms (e.g., 8–16 samples at 8 kHz sampling rate) is sufficient to reduce the short-term correlation of speech significantly. Due to periodic components of speech signals, the memory of the adaptive filter should not exceed a time interval equivalent to the pitch period.[3] For this reason the filter should not contain more than 40–60 coefficients (again at 8 kHz sampling rate). Otherwise also the periodic components of the speech signal would be suppressed.

Several structures for adaptive line enhancement have been established—two of these structures will be discussed and compared in more detail in the next three sections.

8.1.1 FIR Structure

The basic scheme of an adaptive line enhancer with FIR structure is depicted in Fig. 8.2. Because of its structure, the stability of the filter can be guaranteed as long as the stepsize μ of the adaptation process is chosen from the intervals defined in Chapter 7. In most applications the NLMS algorithm with a fixed, but small stepsize

$$\mu \in \{0.01,\, 0.00001\} \tag{8.1}$$

is utilized. It is needless to say that also other adaptation algorithms can be used. Besides the ones mentioned in Chapter 7, algorithms that use higher-order statistics have also been applied to this problem [45]. Nevertheless, we will use the NLMS algorithm because of its robustness and simplicity.

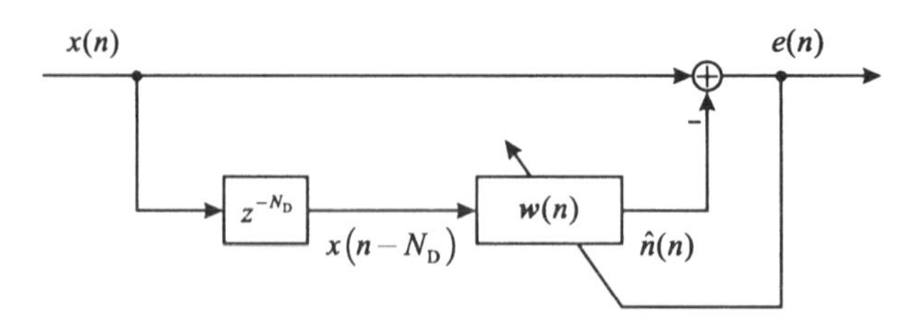

Fig. 8.2 Structure of an FIR filter for adaptive line enhancement.

When defining a delayed excitation vector

$$\boldsymbol{x}(n-N_\mathrm{D}) = \left[x(n-N_\mathrm{D}),\, x(n-N_\mathrm{D}-1),\, \dots,\, x(n-N_\mathrm{D}-N+1)\right]^\mathrm{T} \tag{8.2}$$

and a vector containing the filter coefficients at time index n

$$\boldsymbol{w}(n) = \left[w_0(n),\, w_1(n),\, \dots,\, w_{N-1}(n)\right]^\mathrm{T}, \tag{8.3}$$

the output of the adaptive filter[4] can be described as

$$\hat{n}(n) = \boldsymbol{w}^\mathrm{T}(n)\,\boldsymbol{x}(n-N_\mathrm{D}) = \boldsymbol{x}^\mathrm{T}(n-N_\mathrm{D})\,\boldsymbol{w}(n)\,. \tag{8.4}$$

[3]The specification is only true for FIR structures. In case of IIR schemes the group delay should not exceed the specified range.

[4]Adaptive line enhancers are applied usually in fullband. For this reason, we will mention here only real input signals and real coefficient vectors.

By subtracting the noise estimate $\hat{n}(n)$ from the primary input signal $x(n)$, an enhanced signal

$$e(n) = x(n) - \hat{n}(n) \tag{8.5}$$

is generated. The enhanced signal $e(n)$ serves as an error signal for the NLMS adaptation process:

$$\boldsymbol{w}(n+1) = \boldsymbol{w}(n) + \mu \frac{\boldsymbol{x}(n - N_{\mathrm{D}})\, e(n)}{\|\boldsymbol{x}(n - N_{\mathrm{D}})\|^2} . \tag{8.6}$$

Due to short-term periodic components within the speech signal, the filter tries to suppress also parts of the speech signal. By using a small stepsize μ this behavior can be avoided and only periodic distortions that are present for a longer time interval are canceled. A small stepsize, on the other hand, leads to a slow convergence. Suddenly appearing (periodic) distortions would be attenuated only after a nonnegligible period of time. For this reason a compromise for the stepsize has to be found.

Figure 8.3 shows an example for adaptive line enhancement. The primary sensor picks up a speech signal that is corrupted by two strong sinusoidal signals at 600 and at 2700 Hz. A delay of $N_{\mathrm{D}} = 12$ samples and an adaptive filter with $N = 48$ coefficients have been used. The stepsize was chosen to $\mu = 0.007$. With this setup the sinusoidal signals were sufficiently suppressed after 1 second (see center diagram of Fig. 8.3). The lowest diagram shows the frequency response of the entire structure at the end of the simulation. According to Fig. 8.2, the frequency response $G_{\mathrm{FIR}}(e^{j\Omega}, n)$ of the FIR adaptive line enhancer at time index n can be computed as

$$G_{\mathrm{FIR}}\left(e^{j\Omega}, n\right) = 1 - e^{-j\Omega N_D} W\left(e^{j\Omega}, n\right) . \tag{8.7}$$

In this case $W(e^{j\Omega}, n)$ denotes the Fourier transform of the impulse response $w_i(n)$ at time index n. The two notches at 600 and at 2700 Hz are clearly visible. Nevertheless, also the area around 200 Hz is attenuated by about 10 dB. This effect can be reduced by using an even smaller stepsize. The price to be paid for this reduction is a slower convergence. Thus, the time where the periodic distortions are audible would be extended.

If the width of the notches should be reduced, the number of coefficients of the adaptive filter $\boldsymbol{w}(n)$ needs to be increased. This demand is contradictory to the limitation caused by periodic speech components. A possibility to reduce the notch widths without increasing the filter order is to use IIR structures, discussed in the next section.

8.1.2 IIR Structure

The basic structure of an adaptive line enhancer (see Fig. 8.2) can be extended by a weighted feedback path [36] as depicted in Fig. 8.4. By varying the feedback gain α, it is possible to modify the filter from an FIR structure ($\alpha = 0$) to an adaptive oscillator ($\alpha = 1$) [88].

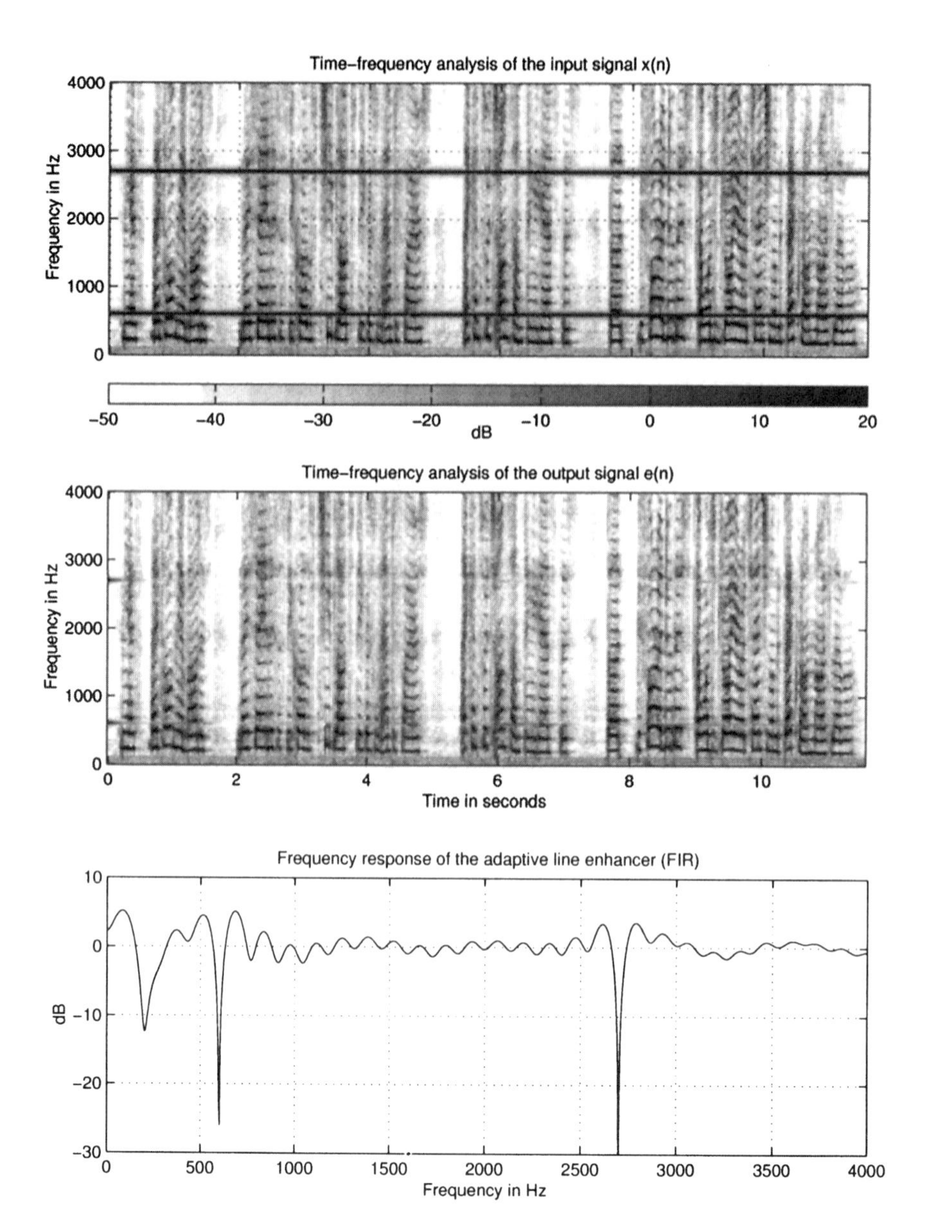

Fig. 8.3 Simulation example for FIR adaptive line enhancement. The upper diagram shows a time–frequency analysis of the primary input signal $x(n)$. The two sinusoidal distortions at 600 and 2700 Hz are clearly visible. By applying an adaptive (FIR) line enhancer with the parameters $N_{\mathrm{D}} = 12$, $N = 48$, and $\mu = 0.007$ the distortions are attenuated sufficiently after one second (see diagram in the middle, showing a time–frequency analysis of the output signal $e(n)$). In the lowest diagram the frequency response of the filter at the end of the simulation is depicted.

The motivation behind this IIR configuration is to achieve some of the benefits of a noise canceller with a separate pure periodic reference. With the extended structure it is possible to achieve very narrow notches. Nevertheless, due to the IIR structure, the filter might become instable if the adaptation process forces the poles to move out of the unit circle within the z domain.

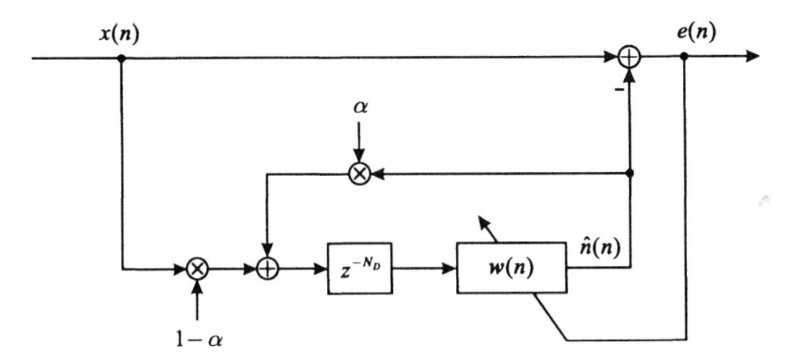

Fig. 8.4 Structure of an IIR filter for adaptive line enhancement.

In order to demonstrate the benefits of the new structure, the simulation presented in the Section 8.1.1 was repeated, with 70% ($\alpha = 0.7$) of the estimated distortion $\hat{n}(n)$ fed back to the input of the adaptive filter. As in the FIR structure, the coefficients $w_i(n)$ were adapted using the NLMS algorithm. Instead of monitoring the poles of the filter, a simple leakage control was implemented:

$$\boldsymbol{w}(n+1) = \lambda(n)\,\boldsymbol{w}(n) + \mu\,\frac{\tilde{\boldsymbol{x}}(n-N_{\mathrm{D}})\,e(n)}{\|\tilde{\boldsymbol{x}}(n-N_{\mathrm{D}})\|^2}\,. \tag{8.8}$$

The excitation vector $\tilde{\boldsymbol{x}}(n-N_{\mathrm{D}})$ now consists of mixtures of the primary signal $x(n)$ and the estimated noise $\hat{n}(n)$:

$$\begin{aligned}\tilde{\boldsymbol{x}}(n-N_{\mathrm{D}}) &= \left[\tilde{x}(n-N_{\mathrm{D}}),\, \tilde{x}(n-N_{\mathrm{D}}-1),\, \ldots,\, \tilde{x}(n-N_{\mathrm{D}}-N+1)\right]^{\mathrm{T}} && (8.9)\\ \text{with } \tilde{x}(n) &= (1-\alpha)\,x(n) + \alpha\,\hat{n}(n)\,. && (8.10)\end{aligned}$$

Even if control by leakage cannot guarantee stability of an IIR filter, we have used it here for simplicity. Short-term powers of the excitation signal and the error signal were computed. If the power of the error signal $\overline{|e(n)|^2}$ exceeds the one of the input signal $\overline{|x(n)|^2}$ by 3 dB, the leakage constant $\lambda(n)$ was set to 0.95. Otherwise it was set to one:

$$\lambda(n) = \begin{cases} 0.95, & \text{if } \overline{|e(n)|^2} > 2\,\overline{|x(n)|^2}, \\ 1, & \text{else.} \end{cases} \tag{8.11}$$

The short-term powers of both signals were estimated using a first-order IIR filter with the squared magnitude of the signal as an input:

$$\begin{aligned}\overline{|x(n)|^2} &= 0.999\,\overline{|x(n-1)|^2} + 0.001\,|x(n)|^2\,, && (8.12)\\ \overline{|e(n)|^2} &= 0.999\,\overline{|e(n-1)|^2} + 0.001\,|e(n)|^2\,. && (8.13)\end{aligned}$$

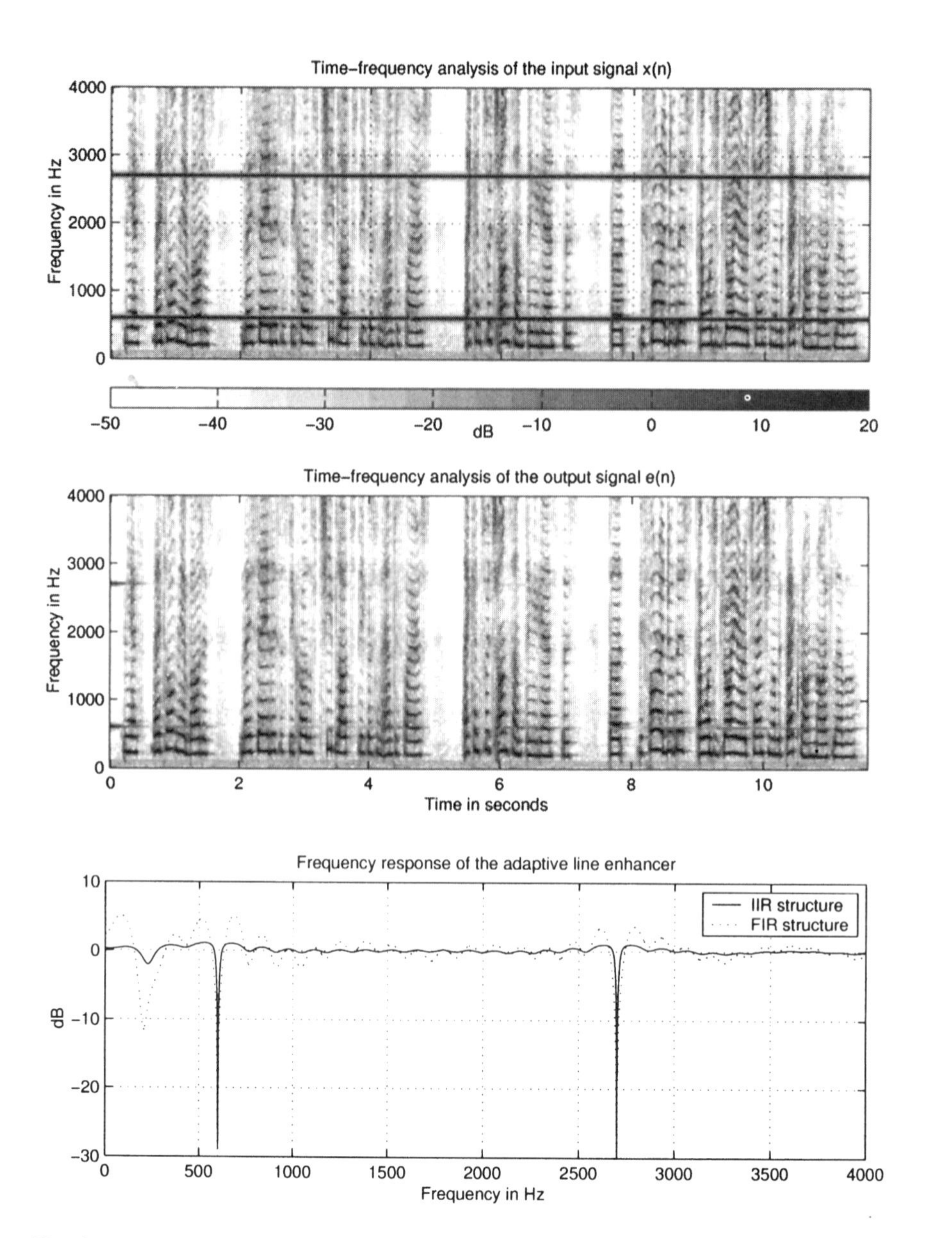

Fig. 8.5 Simulation example for IIR adaptive line enhancement. The upper diagram shows a time–frequency analysis of the primary input signal $x(n)$. By applying an adaptive (IIR) line enhancer with the parameters $N_{\mathrm{D}} = 12$, $N = 48$, $\alpha = 0.7$, and $\mu = 0.007$ the distortions can be attenuated [see diagram in the middle, showing a time–frequency analysis of the output signal $e(n)$]. In the lowest diagram the frequency response of the filter (solid line) at the end of the simulation is depicted. In order to allow a comparison with the FIR structure, the frequency response of the FIR structure is repeated (dotted line).

The results of the simulation are presented in Fig. 8.5. As in the previous example, the frequency response of the line enhancement filter at the end of the simulation run is depicted in the lowest diagram. Due to the feedback path, the frequency response is given by

$$G_{\mathrm{IIR}}\left(e^{j\Omega}, n\right) = 1 - (1-\alpha)\,\frac{e^{-j\Omega N_{\mathrm{D}}}\,W\left(e^{j\Omega}, n\right)}{1 - \alpha\, e^{-j\Omega N_{\mathrm{D}}}\,W\left(e^{j\Omega}, n\right)}\,. \tag{8.14}$$

When comparing the final frequency responses of the FIR (dotted line) and the IIR (solid line) structures, the sharper and deeper notches of the IIR filter are clearly visible. As in the FIR example the filter was adapted with a fixed stepsize $\mu = 0.007$.

8.1.3 Comparison of Both Structures

The simulation results presented in Fig. 8.5 suggest that the IIR structure is superior. Indeed, when comparing the estimated distortions of both filters, the advantages of the IIR structure become even more obvious. Time–frequency analyses of the signal $\hat{n}(n)$ are depicted in Fig. 8.6. The speech components in the lower diagram (IIR structure) are much smaller than in the upper diagram (FIR structure).

Nevertheless, the IIR structure has also disadvantages. In the last two examples the acoustic loop was not closed. This means that the outputs of the line enhancement filters were not fed back in any kind to the input of the primary sensor. If this is the case (e.g., in hearing aids or public address systems), the power of the sinusoidal components of the excitation signal would be reduced according to the depth of the notches. Consequently, the periodic distortions are not as dominant as before in the cost function of the adaptive filter, leading to a reduction of the notch depth. After moving the zeros away from the unit circle within the z domain, the periodic distortions dominate again in the cost function. For this reason the notch depths will be oscillating in case of closed-loop operation.

If the stepsize is chosen small (as done in our example), these oscillations can be tolerated and a significant reduction of periodic distortions is possible, especially with the FIR structure. The IIR structure tends to short oscillations—especially when the primary signal consists of short noise bursts, such as the sound of clapping hands or smashing doors. If these oscillations within the IIR structure should be avoided, the simple leakage control based on comparison of short-term powers has to be replaced by expensive stability tests (e.g., modified Schur–Cohn tests [109, 111]). Another possibility is to use a cascade of second-order IIR filters. Kwan and Martin have described these approaches in detail [137].

8.2 FREQUENCY SHIFT

Another possibility for increasing the stability margin of closed acoustic loops is to use frequency shift methods. The most common application of these algorithms are

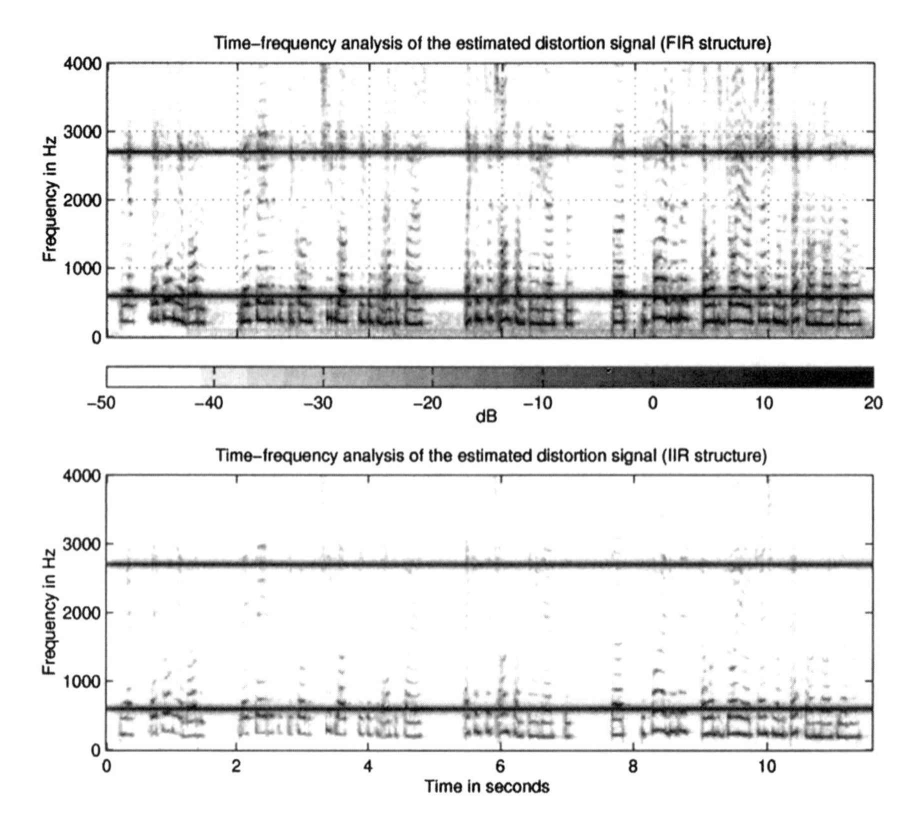

Fig. 8.6 Time–frequency analysis of the estimated distortion signals $\hat{n}(n)$. The result of the FIR structure is depicted in the upper diagram. The lower one shows the analysis of $\hat{n}(n)$ of the IIR structure.

public address systems. We will use this application in order to explain the basic ideas and give a few results concerning achievable gain increase.

8.2.1 Basic Idea

Public address systems that use frequency shifts were first presented in the early 1960s [205]. In those systems parts of the amplified signal are coupled back into the speaker's microphone (see Fig. 8.7). As a consequence, the audience hears an echoic speech signal. Furthermore, the entire system may become unstable if the closed-loop gain exceeds the stability margin.

By using a frequency shift, the maximum gain of a public address system can be increased by a few decibels. The approach exploits the transmission characteristics of large rooms. When measuring the transfer function from the loudspeaker to the microphone, spectral peaks and valleys, as depicted in Fig. 8.8, will appear. The

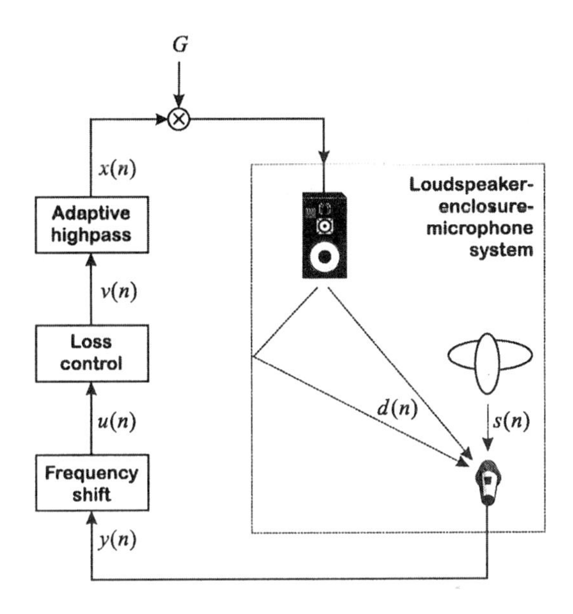

Fig. 8.7 Structure of a public address system. The system consists of a frequency shift, a speech-activated attenuation (loss control), and an adaptive highpass [197].

maxima and minima have an average distance of about 10 Hz [205]. They differ in amplitude typically by 10 dB, but also differences up to 30 dB can be observed.

Feedback systems are stable only if the closed-loop gain is smaller than one at all frequencies. We will assume here that the amplifier of the loudspeaker is linear and that its gain does not depend on frequency. For this reason, howling will first appear at the frequency where the response of the LEM system has its maximum. Usually, the howling needs a few loops through the system until it has an unpleasant loudness. The basic idea is now to shift the entire signal spectrum by a few cycles at each loop through the system. The spectral component that is responsible for the howling will be shifted into a spectral valley after a few turnarounds. The stability condition is now fulfilled.

However, a large amplification is required only in speech activity passages. For this reason a loss control is also implemented for evaluating the public address system. Whenever a speech pause is detected, a moderate amount of attenuation is introduced within the system; otherwise the attenuation factor is set to one. Details about the attenuation unit can be found in Section 8.3.

Furthermore, Schertler and Heitkämper [197] suggested the use of adaptive highpass filtering (in combination with shifting toward lower frequencies) in order to improve the system quality. Feedback howling occurs usually at medium or high frequencies. If an oscillation appears, such as at 500 Hz, and a shift of 10 Hz toward lower frequencies is implemented, in the worst case it could last up to 50 loops until the oscillation is attenuated. By applying highpass filtering the process can be

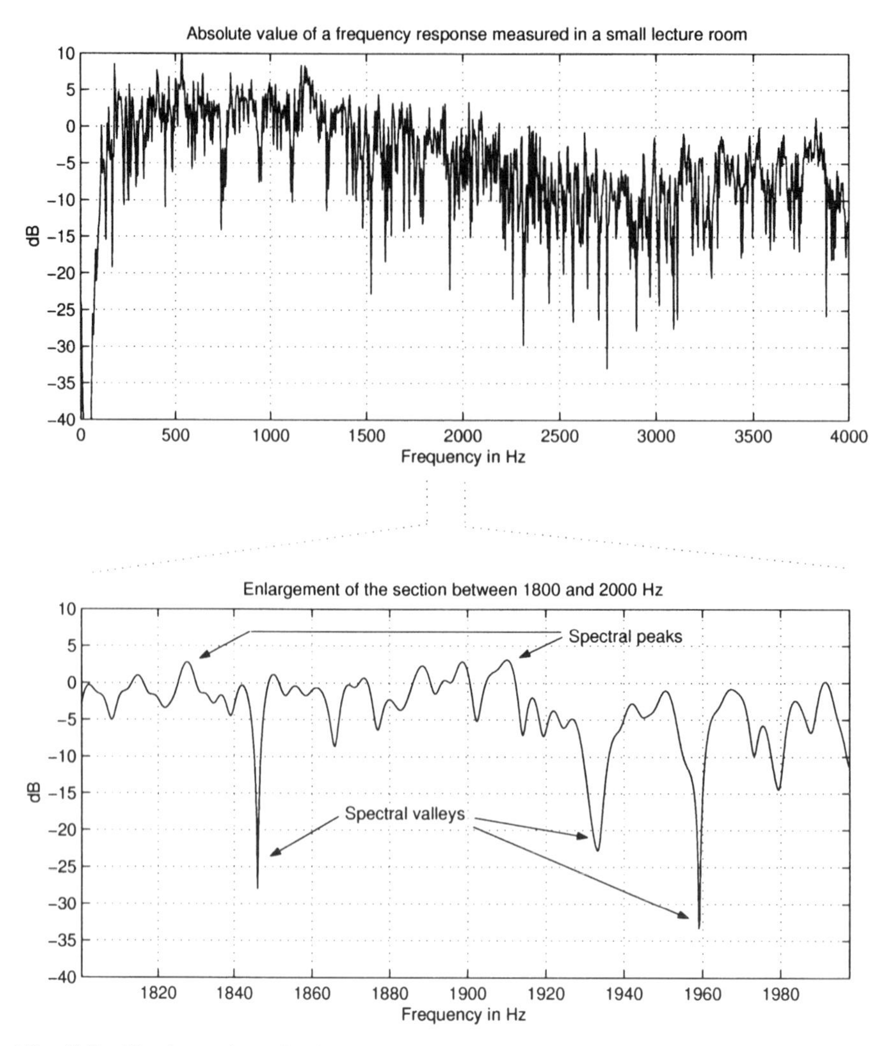

Fig. 8.8 Absolute value of a frequency response measured in a small lecture room. The spectral peaks and valleys are clearly visible, especially in the lower diagram, where the range from 1800 to 2000 Hz has been enlarged.

accelerated. Schertler and Heitkämper [197] suggested the use of a first-order IIR filter

$$H_{\mathrm{HP}}\big(e^{j\Omega}, n\big) = \frac{1 + a(n)}{2} \, \frac{1 - e^{j\Omega}}{1 - a(n)\, e^{j\Omega}} \,. \tag{8.15}$$

Depending on a speech activity decision two different coefficient sets are used:

$$a(n) = \begin{cases} 0.9244, & \text{during speech activity,} \\ 0.7673, & \text{otherwise.} \end{cases} \tag{8.16}$$

At a sampling rate of 24 kHz this results in a filter with a 3-dB cutoff frequency of about 300 Hz during speech intervals. In speech pauses the 3-dB cutoff frequency is moved to about 1000 Hz.

8.2.2 Hilbert Filter

In this section we will describe how a frequency shift can be implemented. A band-limited spectrum according to Fig. 8.9 is given. We call the spectrum at negative frequencies *lower sideband* and at positive frequencies *upper sideband*:

$$S\left(e^{j\Omega}\right) = S_{\mathrm{L}}\left(e^{j\Omega}\right) + S_{\mathrm{U}}\left(e^{j\Omega}\right) . \tag{8.17}$$

In case of spectral movements toward higher frequencies, it is desired to move the upper sideband toward $+\pi$ and the lower one toward $-\pi$. The shift frequency will be denoted by Ω_0.

The desired spectrum can be generated (in a first step) by multiplying the input signal with a cosine function:

$$s_{\cos}(n) = s(n)\, \cos\left(n\Omega_0\right) . \tag{8.18}$$

In the frequency domain, the multiplication results in two shifted versions of the original spectrum:

$$S_{\cos}\left(e^{j\Omega}\right) = \frac{1}{2}\, S\left(e^{j\,(\Omega+\Omega_0)}\right) + \frac{1}{2}\, S\left(e^{j\,(\Omega-\Omega_0)}\right) . \tag{8.19}$$

By replacing the entire spectrum by its upper and lower sidebands according to Eqn. 8.17, we obtain

$$\begin{aligned} S_{\cos}\left(e^{j\Omega}\right) = \frac{1}{2} \Bigg(& \underbrace{S_{\mathrm{L}}\left(e^{j\,(\Omega-\Omega_0)}\right) + S_{\mathrm{U}}\left(e^{j\,(\Omega+\Omega_0)}\right)}_{\text{desired spectral components}} \\ & + \underbrace{S_{\mathrm{L}}\left(e^{j\,(\Omega+\Omega_0)}\right) + S_{\mathrm{U}}\left(e^{j\,(\Omega-\Omega_0)}\right)}_{\text{undesired spectral components}} \Bigg) . \end{aligned} \tag{8.20}$$

The resulting spectrum contains the two desired components. Unfortunately, two spectral components that were shifted in the opposite direction are also part of the spectrum. In order to cancel them, a second carrier signal is generated. This time a sine signal is used as a modulation function. Before modulating the signal, a Hilbert filter is applied:

$$s_{\mathrm{Hil}}(n) = s(n) * h_{\mathrm{Hil},\,n} . \tag{8.21}$$

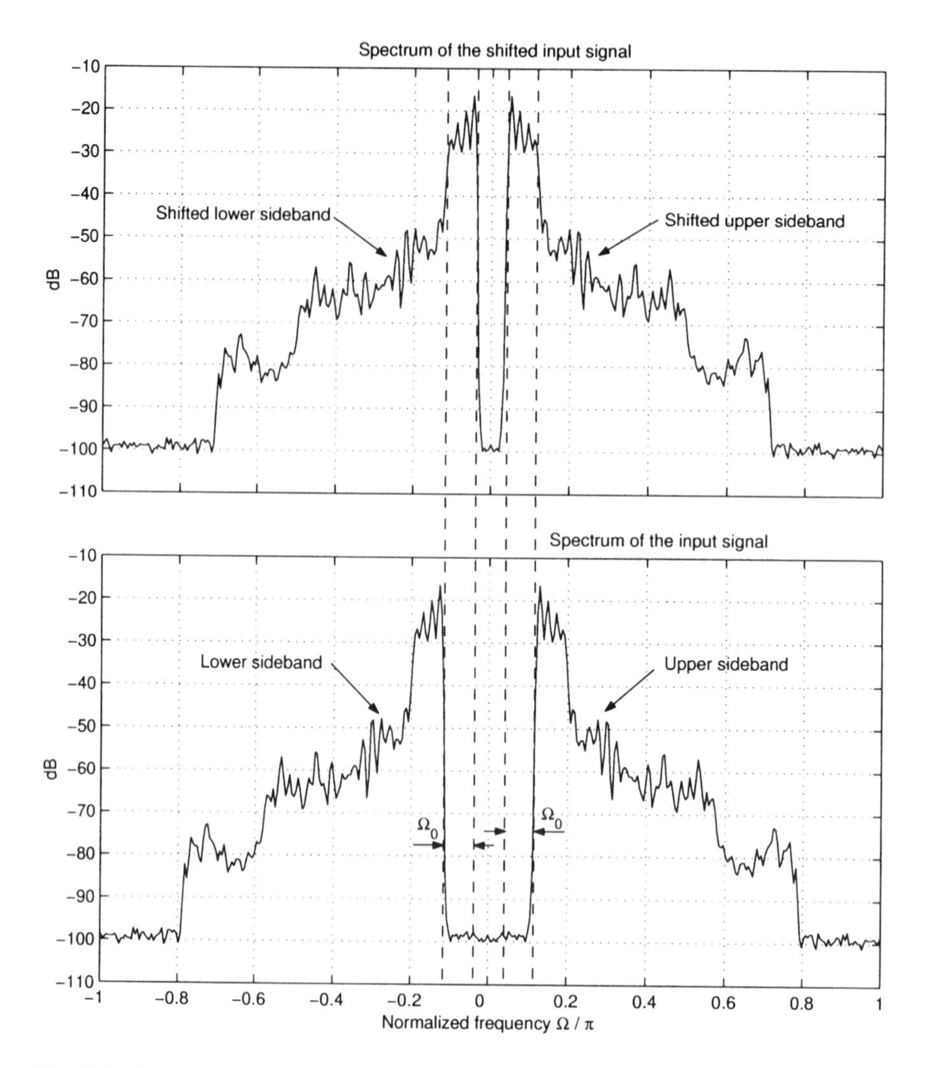

Fig. 8.9 Spectra of original and shifted signals. The shift frequency Ω_0 would be much smaller in reality. We have chosen a large Ω_0 only for clarification.

The task of a Hilbert filter is to shift all negative components of the input spectrum by $\pi/2$ and all positive frequencies by $-\pi/2$:

$$S_{\text{Hil}}\left(e^{j\Omega}\right) = -j\,\text{sgn}(\Omega)\,S\left(e^{j\,(\Omega)}\right). \tag{8.22}$$

Hence the frequency response of $h_{\text{Hil},\,i}$ is

$$H_{\text{Hil}}\left(e^{j\Omega}\right) = -j\,\text{sgn}(\Omega)\,. \tag{8.23}$$

When multiplying the Hilbert filtered signal with the sine function

$$s_{\text{sin}}(n) = s_{\text{Hil}}(n) \, \sin(n\Omega_0) \ , \tag{8.24}$$

we obtain in the spectral domain:

$$S_{\text{sin}}\left(e^{j\Omega}\right) = j\frac{1}{2}\, S_{\text{Hil}}\left(e^{j\,(\Omega+\Omega_0)}\right) - j\frac{1}{2}\, S_{\text{Hil}}\left(e^{j\,(\Omega-\Omega_0)}\right) . \tag{8.25}$$

After inserting Eqs. 8.22 and 8.17, the cancellation process becomes obvious:

$$\begin{aligned} S_{\text{sin}}\left(e^{j\Omega}\right) &= \frac{1}{2}\Big(S_{\text{L}}\left(e^{j\,(\Omega-\Omega_0)}\right) + S_{\text{U}}\left(e^{j\,(\Omega+\Omega_0)}\right) \\ &\quad - S_{\text{L}}\left(e^{j\,(\Omega+\Omega_0)}\right) - S_{\text{U}}\left(e^{j\,(\Omega-\Omega_0)}\right)\Big) . \end{aligned} \tag{8.26}$$

By adding the two modulated signals, only the upper sideband, which is shifted toward $+\pi$ and the lower sideband, which is shifted toward $-\pi$, remain:

$$\begin{aligned} U\left(e^{j\Omega}\right) &= S_{\text{cos}}\left(e^{j\Omega}\right) + S_{\text{sin}}\left(e^{j\Omega}\right) \\ &= S_{\text{L}}\left(e^{j\,(\Omega-\Omega_0)}\right) + S_{\text{U}}\left(e^{j\,(\Omega+\Omega_0)}\right) . \end{aligned} \tag{8.27}$$

Figure 8.10 shows the structure of the shifting device.

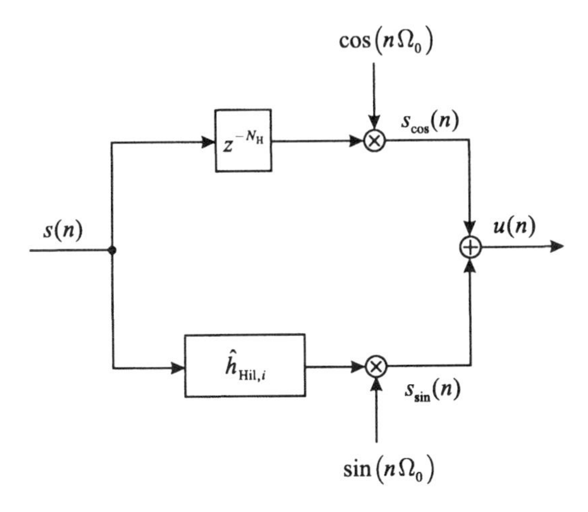

Fig. 8.10 Structure of the frequency shift.

It remains to compute the impulse response of the Hilbert filter. Until now we have only specified the frequency response. Applying a Fourier transform for discrete signals on Eqn. 8.23 results in

$$
\begin{aligned}
h_{\text{Hil},\,i} &= \frac{1}{2\pi}\int_{-\pi}^{\pi} H_{\text{Hil}}\left(e^{j\Omega}\right) e^{j\Omega i}\,d\Omega \\
&= -\frac{j}{2\pi}\int_{-\pi}^{\pi} \text{sgn}(\Omega)\, e^{j\Omega i}\,d\Omega \\
&= \frac{j}{2\pi}\left(\int_{-\pi}^{0} e^{j\Omega i}\,d\Omega - \int_{0}^{\pi} e^{j\Omega i}\,d\Omega\right) \\
&= \frac{1}{\pi}\int_{0}^{\pi} \sin(\Omega i)\,d\Omega\,.
\end{aligned}
\tag{8.28}
$$

Solving the integral leads to

$$
h_{\text{Hil},\,i} = \begin{cases} 0, & \text{if } i \text{ is even,} \\ \dfrac{2}{i\pi}, & \text{else.} \end{cases}
\tag{8.29}
$$

This direct design exhibits two main problems. The resulting impulse response is infinitely long as well as noncausal. For this reason, the impulse response $h_{\text{Hil},\,i}$ is first truncated to a range from $i = -N_{\text{H}}$ to $i = N_{\text{H}}$. Direct truncating (applying a rectangle window) would lead to the Gibbs phenomenon. Therefore, typically lowpass filters like a Hamming window centered at $i = 0$ are applied. Figure 8.11 shows the frequency responses of a windowed and a direct design. The oscillations around integer multiples of π are clearly visible. In order to get a causal impulse response, the windowed function is finally shifted by N_{H} coefficients:

$$
\hat{h}_{\text{Hil},\,i} = h_{\text{Hil},\,i-N_{\text{H}}}\, h_{\text{Win},\,i-N_{\text{H}}}\,.
\tag{8.30}
$$

Due to this shift, the cosine modulated branch in Fig. 8.10 also has to be delayed by N_{H} samples.

8.2.3 Results

Finally we present some results which based on investigations described by Schertler and Heitkämper [197]. Increases of the stability margin caused by the frequency shift were determined for three different shift frequencies (6, 9, and 12 Hz). In contrast to Schroeder's method [205],[5] speech signals with different powers were used as excitation. The public address system was tested in three different locations. The first one was a lecture room at the University of Darmstadt. In this room special arrangements for avoidance of acoustical reflections are installed. The second room was the entrance hall of the lecture rooms. This room was even larger than the lecture

[5] Schroeder [205] determined the gain increase with white noise.

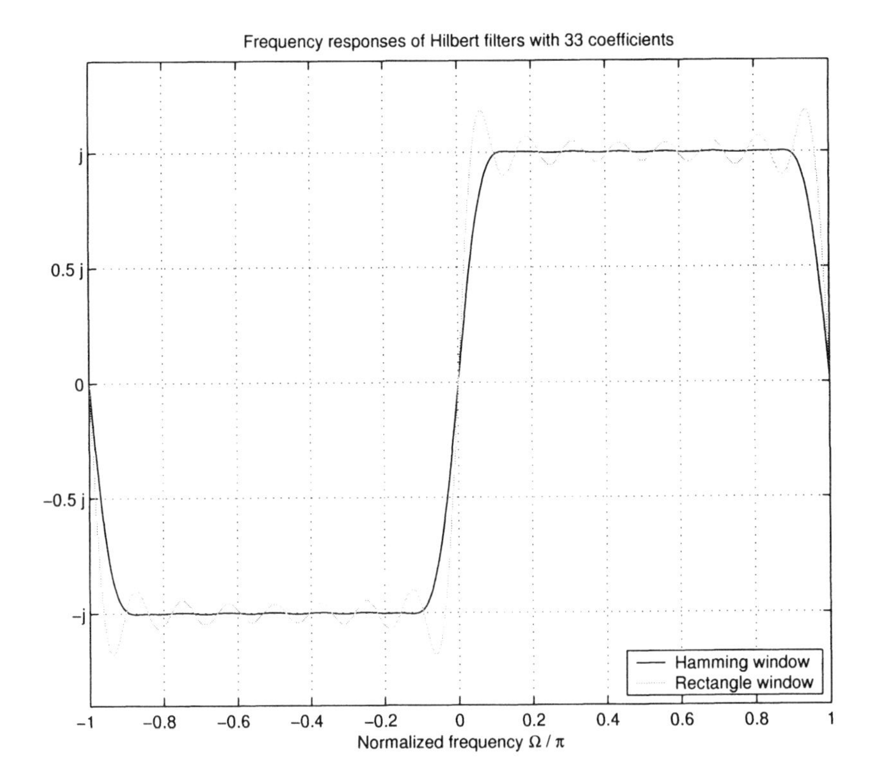

Fig. 8.11 Frequency responses of Hilbert filters. For the design 33 coefficients (16 noncausal) were used. Depicted are the direct design without any window function and a windowed version using a Hamming lowpass.

hall, resulting in a long reverberation time. The last room was the echoic chamber of the acoustical research department. This enclosure has a reverberation time more than a second.

First, all measurements were performed with white noise. In this case the additional gain allowed by the frequency shift can be determined precisely. Large gains up to 11 dB (in the echoic chamber) as reported by Schroeder [205] were measured. However, this setup does not include any information about speech quality degradation. For this reason, the measurements were repeated with speech signals at different amplitude levels. In a first step, the gain of the system without the frequency shift was increased until the first quality degradations were audible. This took place a few decibels below the stability margin. Afterward the frequency shift was activated and the gain was adjusted a second time. Again, when the first quality degradations were audible the process was terminated.

Using this measurement setup smaller additional gains were determined. The achieved gains (averaged over all loudness levels) are depicted in Fig. 8.12. In the lecture room gains of only 1–2 dB (that are hardly audible) were measured. If the

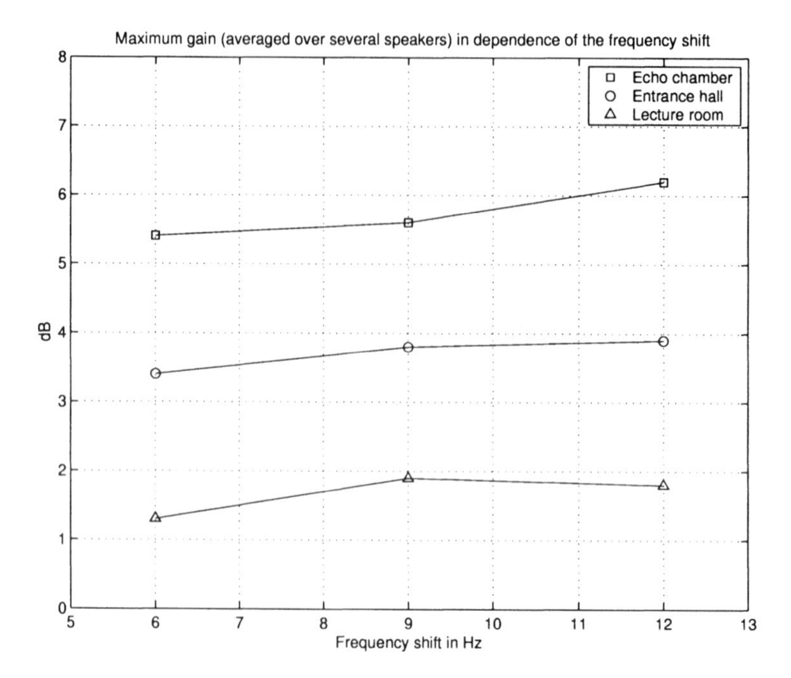

Fig. 8.12 Additional gains due to the frequency shift. The public address system was installed at three different locations: a lecture room, an entrance hall, and an echoic chamber. Depicted are the average gain enhancements at 6, 9, and 12 Hz shift frequency.

frequency shift was set to values larger than 12 Hz artifacts were audible. For this reason, only shift frequencies up to 12 Hz were investigated. Larger gains were determined in the entrance hall. Here the frequency shift caused additional enhancements up to 4 dB. The best results were achieved in the echoic chamber with 5–6 dB.

Even if the enhancements are not that large, the frequency shift reduces the correlation of the microphone signal and the signals of the loudspeakers. This eases the implementation and increases the effectiveness of additional techniques for closed-loop stabilization (e.g., feedback cancellation).

8.3 CONTROLLED ATTENUATION

One of the oldest attempts to solve the acoustic feedback problem is the use of speech-activated switching or loss controls. Attenuation units are placed into the receiving path $a_{\rm r}(n)$ and into the transmitting path $a_{\rm t}(n)$ of a hands-free telephone as shown in Fig. 8.16. In case of single talk of the remote speaker only the transmitting path is attenuated. If only the local speaker is active, all attenuation is inserted into the receiving path. During double talk (both partners speaking simultaneously) the control circuit decides which direction (transmitting or receiving) will be opened.

Beside this disadvantage only loss controls can guarantee the amount of attenuation required by the ITU-T or ETSI recommendations. Loss controls will be described in more detail in Section 8.3.2. Loss control circuits have the longest history in hands-free communication systems; a device for hands-free telephone conversation using voice switching was presented in 1957 [41].

This basic principle has been enhanced by adding several features. Before introducing the attenuation of the loss control, the incoming signals are weighted with a slowly varying gain or attenuation in order to achieve a determined signal level during speech activity. Low voices will be amplified while loud speech is attenuated (up to a certain level). This mechanism is called *automatic gain control.* In case of hands-free telephones, automatic gain controls are also used to adjust the signal level according to the distance of the local speaker from the microphone. In Section 8.3.1 a basic scheme for computing the desired gain factor is presented.

Before the attenuated or amplified signals are converted from digital to analog, it must be ensured that the amplitudes do not exceed the range of the converters. If this is the case, clipping would occur, which reduces the speech quality significantly. For this reason *dynamic limiters* are introduced. Such a device monitors the short-term envelope of a signal. Only large peaks will be attenuated. All other parts of the signal are not attenuated in order to keep the dynamics of the signal. A basic scheme for this component is presented in Section 8.3.3.

Figure 8.13 shows an overview of the three components as they would be implemented in a hands-free telephone. Further algorithms that belong to the group of dynamic processing can be found, for example, in [249].

The algorithms presented in the next three sections should be regarded only as "rough frameworks." When applying them, the special boundary conditions of each

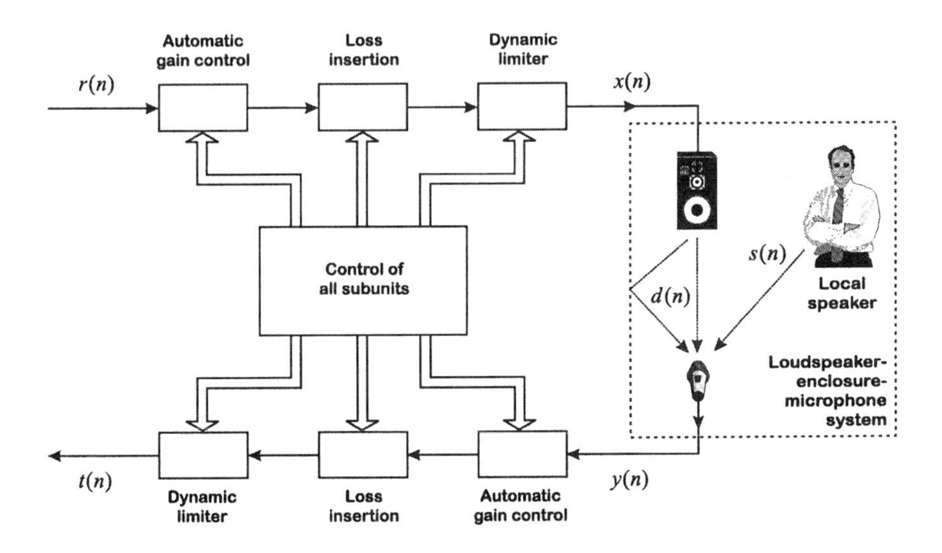

Fig. 8.13 Dynamic processing in hands-free telephone systems.

application have to be taken into account and the algorithms have to be modified in order to cope with these new conditions. If a loss control should be used in a car environment, for example, the background noise level would be much higher than in a calm environment (which was assumed here). In this case it might be necessary to replace fixed thresholds with adaptive ones, controlled by estimation of the background noise level. Furthermore, in such an environment the signal-to-noise ratio is quite small, especially at low (below the pitch frequency) and at high frequencies (e.g., above 3–4 kHz). For this reason, the signals that were utilized to compute the gain factors should be replaced by their bandpass-filtered counterparts. Nevertheless, the algorithms presented here give insight into the principles of gain control, voice-activated switching, and dynamic signal limiting, and it should be straightforward for the reader to do necessary modifications.

8.3.1 Automatic Gain Control

Gain control circuits exhibit very slow changes of the gain factor only. The basic structure of an automatic gain control (AGC) is shown in Fig. 8.14. During speech intervals the short-term power of the input signal is estimated and compared to a desired value. If the speech level exceeds an upper bound, the gain factor $g(n)$ is decreased—in the other case the gain is increased. In order to avoid fast gain variations during short speech pauses $g(n)$ is incremented or decremented in very small steps.

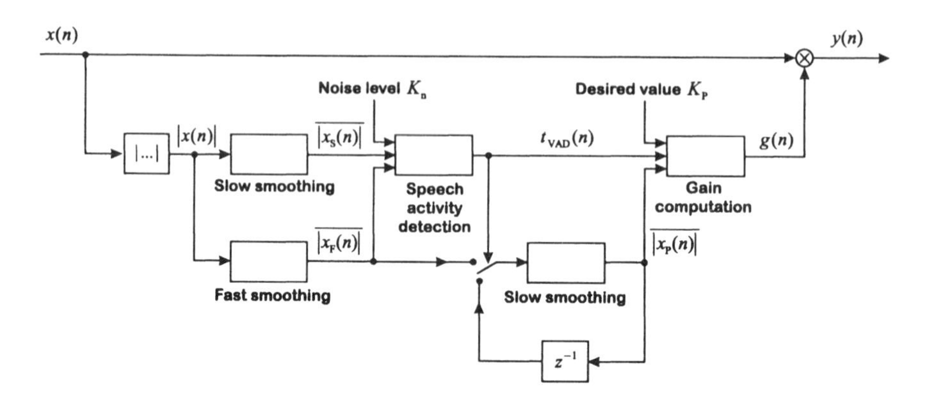

Fig. 8.14 Basic structure of an automatic gain control circuit.

8.3.1.1 Implementation Example

In order to distinguish between periods of speech activity and speech pauses, a voice activity detection is required. For this reason both slow and fast envelope trackers are used. For the slow envelope tracker, first-order IIR smoothing of the absolute value of the input signal is performed:

$$\overline{|x_{\rm S}(n)|} = (1 - \gamma_{\rm S}(n))\,|x(n)| + \gamma_{\rm S}(n)\,\overline{|x_{\rm S}(n-1)|}\,. \tag{8.31}$$

The (time-variant) time constant $\gamma_{\rm S}(n)$ of the IIR filter is chosen differently for rising and falling signal edges:

$$\gamma_{\rm S}(n) = \begin{cases} \gamma_{\rm S,r}\,, & \text{if } |x(n)| > \overline{|x_{\rm S}(n-1)|}\,, \\ \gamma_{\rm S,f}\,, & \text{else.} \end{cases} \tag{8.32}$$

A rising signal edge is followed faster than a falling one:

$$0 \ll \gamma_{\rm S,r} < \gamma_{\rm S,f} < 1\,. \tag{8.33}$$

The fast estimation $\overline{|x_{\rm F}(n)|}$ is computed in the same manner as the slow one

$$\overline{|x_{\rm F}(n)|} = (1 - \gamma_{\rm F}(n))\,|x(n)| + \gamma_{\rm F}(n)\,\overline{|x_{\rm F}(n-1)|} \tag{8.34}$$

$$\text{with } \gamma_{\rm F}(n) = \begin{cases} \gamma_{\rm F,r}\,, & \text{if } |x(n)| > \overline{|x_{\rm F}(n-1)|} \\ \gamma_{\rm F,f}\,, & \text{else,} \end{cases} \tag{8.35}$$

except that different time constants are used:

$$\gamma_{\rm S,r} > \gamma_{\rm F,r}\,, \tag{8.36}$$

$$\gamma_{\rm S,f} > \gamma_{\rm F,f}\,. \tag{8.37}$$

At the end of the section we will present a simulation example. The involved signals have been sampled at 8 kHz. For this sampling rate the following time constants have been used:

$$\gamma_{\rm S,r} = 0.999\,, \qquad \gamma_{\rm S,f} = 0.9997\,, \tag{8.38}$$

$$\gamma_{\rm F,r} = 0.992\,, \qquad \gamma_{\rm F,f} = 0.9990\,. \tag{8.39}$$

Comparing the outputs of the fast and the slow estimator is the first stage of a simple voice activity detector. At the beginning of a speech interval, the output of the fast estimator rises faster than that of the slow one. Due to short-term power variations during speech activity on one hand and the choices of the time constants on the other hand, the output signal of the fast envelope tracker is always larger than its slow counterpart. At the end of the voice activity period, the output of the slow estimator does not decrease as rapidly as does that of the fast IIR smoothing filter, and now the output of the slow estimator exceeds that of the fast one.

In order to avoid wrong decisions during speech pauses, the simple comparison is enhanced by bounding the output of the slow estimator to a constant $K_{\rm n}$. This constant should be chosen a few decibels above the background noise level. If this information is not known a priori, a background noise-level estimation (see Section 13.4.1.2) is necessary. Voice activity ($t_{\rm VAD}(n) = 1$) is detected as follows:

$$t_{\rm VAD}(n) = \begin{cases} 1\,, & \text{if } \overline{|x_{\rm F}(n)|} > \max\left\{\overline{|x_{\rm S}(n)|},\, K_{\rm n}\right\}\,, \\ 0\,, & \text{else.} \end{cases} \tag{8.40}$$

The voice activity detection is used twofold. During voiced intervals the amplitude level is tracked. This is performed again by first-order IIR smoothing:

$$\overline{|x_{\rm P}(n)|} = \begin{cases} (1-\gamma_{\rm P}(n))\,\overline{|x_{\rm F}(n)|} + \gamma_{\rm P}(n)\,\overline{|x_{\rm P}(n-1)|}\,, & \text{if } t_{\rm VAD}(n) = 1\,, \\ \overline{|x_{\rm P}(n-1)|}\,, & \text{else}\,. \end{cases} \tag{8.41}$$

The filter is updated only if speech activity was detected, otherwise the filter keeps its old output value. The fast envelope tracker is used as an input. In order to avoid an excessively rapid decrease $\overline{|x_{\rm P}(n)|}$ during wrong decisions of voice activity detection, a very long time constant for falling signal edges was chosen:

$$\gamma_{\rm P}(n) = \begin{cases} \gamma_{\rm P,r}\,, & \text{if } \overline{|x_{\rm F}(n)|} > \overline{|x_{\rm P}(n-1)|}\,, \\ \gamma_{\rm P,f}\,, & \text{else.} \end{cases} \tag{8.42}$$

For a sampling rate of 8 kHz, the following time constants have been implemented:

$$\gamma_{\rm P,r} = 0.995\,, \tag{8.43}$$

$$\gamma_{\rm P,f} = 0.9997\,. \tag{8.44}$$

Also the gain factor $g(n)$ is updated only during (detected) periods of speech activity:

$$g(n) = \begin{cases} g_{\rm mod}(n)\,g(n-1)\,, & \text{if } t_{\rm VAD}(n) = 1\,, \\ g(n-1)\,, & \text{else.} \end{cases} \tag{8.45}$$

The multiplicative modification $g_{\rm mod}(n)$ is computed by comparing the amplified or attenuated [depending on $g(n-1)$] speech-level estimator $\overline{|x_{\rm P}(n)|}$ with a desired value $K_{\rm P}$. If the speech level after weighting with $g(n-1)$ is still to small, the gain factor is increased:

$$g_{\rm mod}(n) = \begin{cases} g_{\rm inc}\,, & \text{if } \overline{|x_{\rm p}(n)|}\,g(n-1) < K_{\rm P}\,, \\ g_{\rm dec}\,, & \text{else.} \end{cases} \tag{8.46}$$

Consequently, if an excessive value is achieved with the old gain factor, the new one will be decreased. We have used the following modification factors:

$$g_{\rm inc} = 1.0001\,, \tag{8.47}$$

$$g_{\rm dec} = \frac{1}{g_{\rm inc}} = 0.9999\,. \tag{8.48}$$

The desired speech level is dependent on the time constants of the fast envelope tracker and the level estimator.[6] For the values given in Eqs. 8.39 and 8.44, a desired value of

$$K_{\rm P} = \frac{4000}{V_0} \tag{8.49}$$

[6]Caused by different time constants for rising and falling signal edges, the smoothed magnitude estimation has a multiplicative bias that depends on the value of the time constants.

is suitable. V_0 is a normalization constant, representing, for instance, the voltage of signal which will be converted to the least significant bit of the AD converter. The output signal $y(n)$ is computed by multiplying the input signal $x(n)$ with the gain factor $g(n)$:

$$y(n) = x(n)\ g(n)\,. \tag{8.50}$$

In order to demonstrate this basic algorithm for automatic gain control a simulation example is presented in Fig. 8.15. The input signal consists of a speech signal spoken by two different speakers. The first one is a very loud speaker (see first 12 seconds in the upper diagram of Fig. 8.15, while the second one has a soft voice (13th second until end).

In the second diagram the outputs of the three envelope trackers (fast, slow, and peak-level estimator) are depicted. Due to the slow decrease of the smoothing filter $\overline{|x_S(n)|}$, the peak-level estimation $\overline{|x_P(n)|}$ is stopped directly after a voice activity interval. In the third diagram the resulting gain factor is shown. While the signal of the speaker with the loud voice is attenuated by 5–10 dB, the signal of the speaker with the soft voice is amplified by 7–12 dB. The output signal (depicted in the lowest part of Fig. 8.15) has a nearly constant speech level—except for the time when the speakers change.

8.3.2 Loss Control

The task of loss control circuits or nonlinear processors[7] within hands-free telephone systems is very simple; the system designer specifies an overall attenuation (e.g., 50 dB) and the loss control unit decides in which path (sending or transmitting) the attenuation should be placed, specifically, what fraction of the total attenuation should be introduced in which path. The basic structure of a loss control is depicted in Fig. 8.16. Loss control circuits can also be used in one-way communication systems, such as in public address systems (see Section 8.2.1). In these cases only those parts of the algorithm concerning the receiving channel should be used.

8.3.2.1 Implementation Example

In order to decide which path of the system should be attenuated, the microphone signal $y(n)$ as well as the remote excitation signal $r(n)$ are connected to voice activity detection units. In both detectors short-term magnitude estimations are computed:

$$\overline{|r(n)|} = (1-\gamma_\mathrm{r}(n))\,|r(n)| + \gamma_\mathrm{r}(n)\,\overline{|r(n-1)|}\,, \tag{8.51}$$

$$\overline{|y(n)|} = (1-\gamma_\mathrm{y}(n))\,|y(n)| + \gamma_\mathrm{y}(n)\,\overline{|y(n-1)|}\,. \tag{8.52}$$

[7] This term is often used in ITU-T and ETSI recommendations.

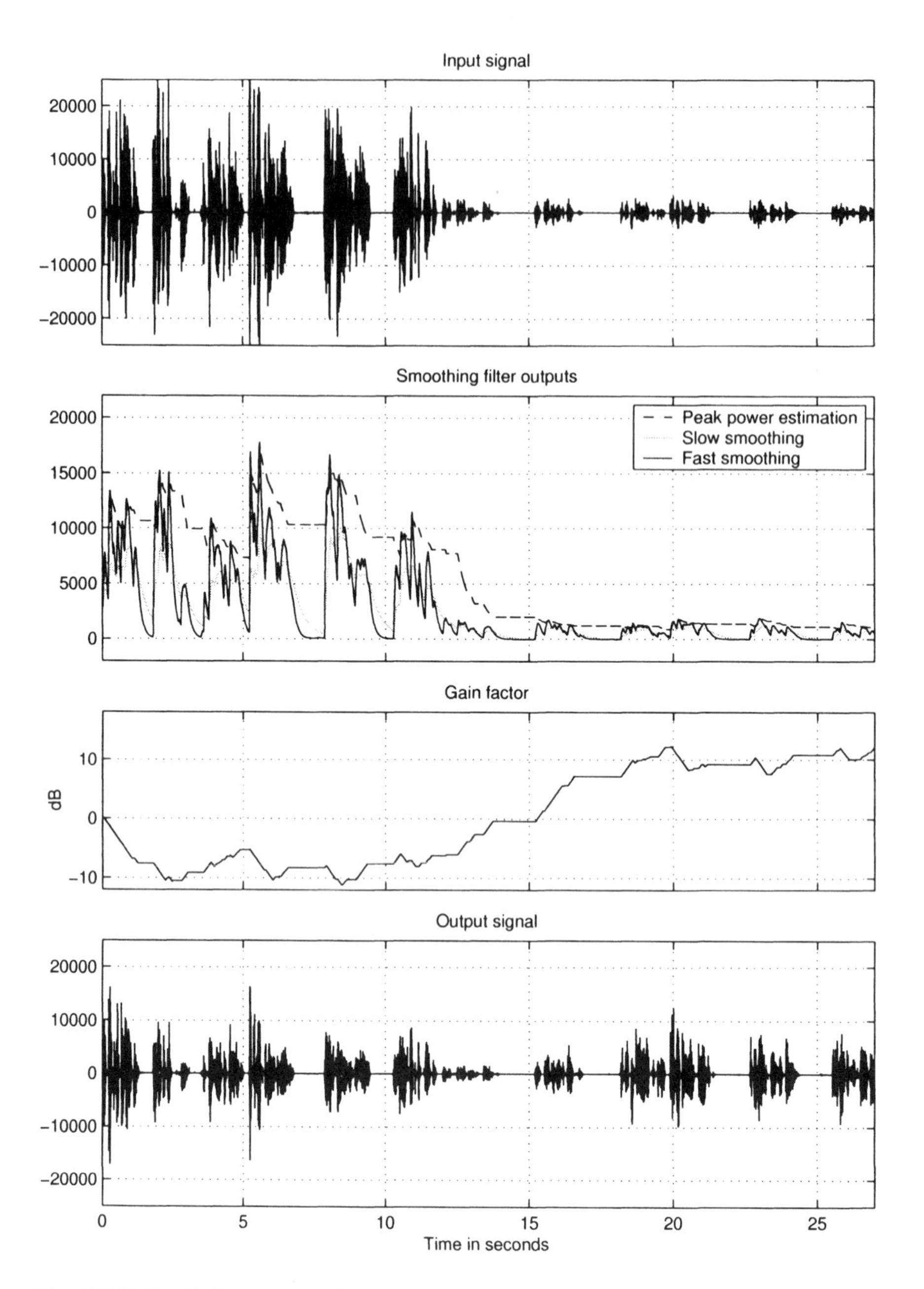

Fig. 8.15 Simulation example for automatic gain control. The input signal (depicted in the upper diagram) consists of a speech signal spoken by two different speakers. The first one is a very loud speaker (first 12 seconds), while the second one has a soft voice (13th second until end). In the second diagram the output of the three smoothing filters are shown. The third and the last parts show the resulting gain factor and the output signal, respectively.

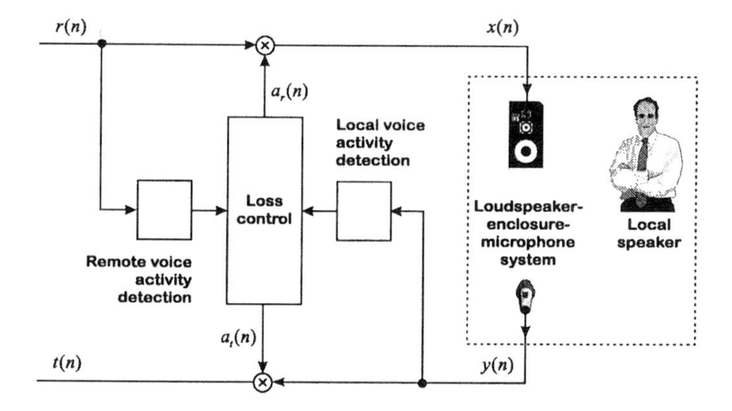

Fig. 8.16 Structure of a loss control. Depending on the remote and the local speech activity, the transmitting and/or the receiving path are attenuated.

As in the previous section, different time constants for rising and falling signal edges are used:

$$\gamma_{\text{r}}(n) = \begin{cases} \gamma_{\text{r}}, & \text{if } |r(n)| > \overline{|r(n-1)|}, \\ \gamma_{\text{f}}, & \text{else}, \end{cases} \tag{8.53}$$

$$\gamma_{\text{y}}(n) = \begin{cases} \gamma_{\text{r}}, & \text{if } |y(n)| > \overline{|y(n-1)|}, \\ \gamma_{\text{f}}, & \text{else}. \end{cases} \tag{8.54}$$

For a sampling rate of 8 kHz we have chosen the following constants:

$$\gamma_{\text{r}} = 0.995, \tag{8.55}$$

$$\gamma_{\text{f}} = 0.997. \tag{8.56}$$

In contrast to the voice activity detection within automatic gain controls (presented in the last section), it is not necessary to detect the end of a speech interval quickly. In loss controls it is rather important to detect the beginning of voice activity immediately (in order to remove the attenuation) and to hold the detection result over short periods of time. The latter is required to avoid gain variations during short pauses (e.g., between or even within words).

In order to fulfill these demands, a simple counting mechanism is used. Voice activity of the far-end partner is detected if the short-term magnitude $\overline{|r(n)|}$ exceeds a predefined threshold K_{far}. In this case a counter $t_{\text{act,far}}(n)$ is set to a value N_{far}. Otherwise the counter is decreased by one. For avoiding an underflow of the counter bounding by zero is applied:

$$t_{\text{act,far}}(n) = \begin{cases} N_{\text{far}}, & \text{if } \overline{|r(n)|} > K_{\text{far}}, \\ \max\{t_{\text{act,far}}(n-1) - 1, \, 0\}, & \text{else}. \end{cases} \tag{8.57}$$

Voice activity is detected as long as the counter $t_{\text{act,far}}(n) > 0$. A simulation example is presented at the end of this section. The involved signals have been sampled at

8 kHz. In the example we have implemented

$$N_{\text{far}} = 2000 \tag{8.58}$$

for the remote activity detection. This means that a once detected speech activity is kept for at least 2000 samples, which is a quarter of a second. The threshold K_{far} should be chosen to be a few decibels above the far-end background noise level. We have implemented

$$K_{\text{far}} = \frac{500}{V_0}\,. \tag{8.59}$$

Detecting local voice activity is more complicated. The signal of the far-end partner couples (over the loudspeaker through the local room) back into the microphone. This means that the microphone signal consists of a (reverberant) far-end speech component, a local speech component, and local background noise. For avoiding speech activity detections caused by the remote echo component an estimated echo level is computed first. This is performed by multiplying the product $\overline{|r(n)|}\ a_{\text{r}}(n-1)$ with a coupling factor K_{c}. The coupling from the loudspeaker to the microphone depends on the (analog and digital) gains of the hands-free telephone system and varies typically in the range from -10 to 10 dB. The constant K_{c} should be chosen to be a few decibels below this coupling. If the room coupling is not known a priori, it has to be estimated, using, for instance, the method presented in Section 13.4.3.1. The incoming signal of the far-end speaker $r(n)$ will be weighted with the gain factor $g_{\text{r}}(n)$. The value of $g_{\text{r}}(n)$ depends on the result of the voice activity detection. Nevertheless, $g_{\text{r}}(n)$ can vary only slowly from sample to sample. For this reason, $\overline{|r(n)|}\ g_{\text{r}}(n-1)$ is a suitable estimation for the short-term amplitude of the signal emitted by the loudspeaker.

In order to detect local voice activity the short-term amplitude estimation of the microphone signal should be larger than the expected echo level on one hand. On the other hand, $\overline{|y(n)|}$ should also be larger than the local background noise level. Thus, $\overline{|y(n)|}$ is compared to the maximum of $K_{\text{c}}\,\overline{|r(n)|}\,a_{\text{r}}(n-1)$ and the constant K_{loc}, which was chosen to be a few decibels above the local background noise level:

$$t_{\text{act,loc}}(n) = \begin{cases} N_{\text{loc}}\,, & \text{if } \overline{|y(n)|} > \ldots \\ & \ldots \max\left\{K_{\text{loc}}\,, K_{\text{c}}\,\overline{|r(n)|}\,a_{\text{r}}(n-1)\right\}, \\ \max\left\{t_{\text{act,loc}}(n-1)-1,\,0\right\}\,, & \text{else}\,. \end{cases} \tag{8.60}$$

In the simulation presented in Figs. 8.17 and 8.18 we have chosen

$$K_{\text{loc}} = \frac{500}{V_0} \quad \text{and} \quad K_{\text{c}} = 2\,. \tag{8.61}$$

A once detected local speech activity is kept for a short period of time using a counting mechanism such as that used in far-end detection. Due to the more difficult

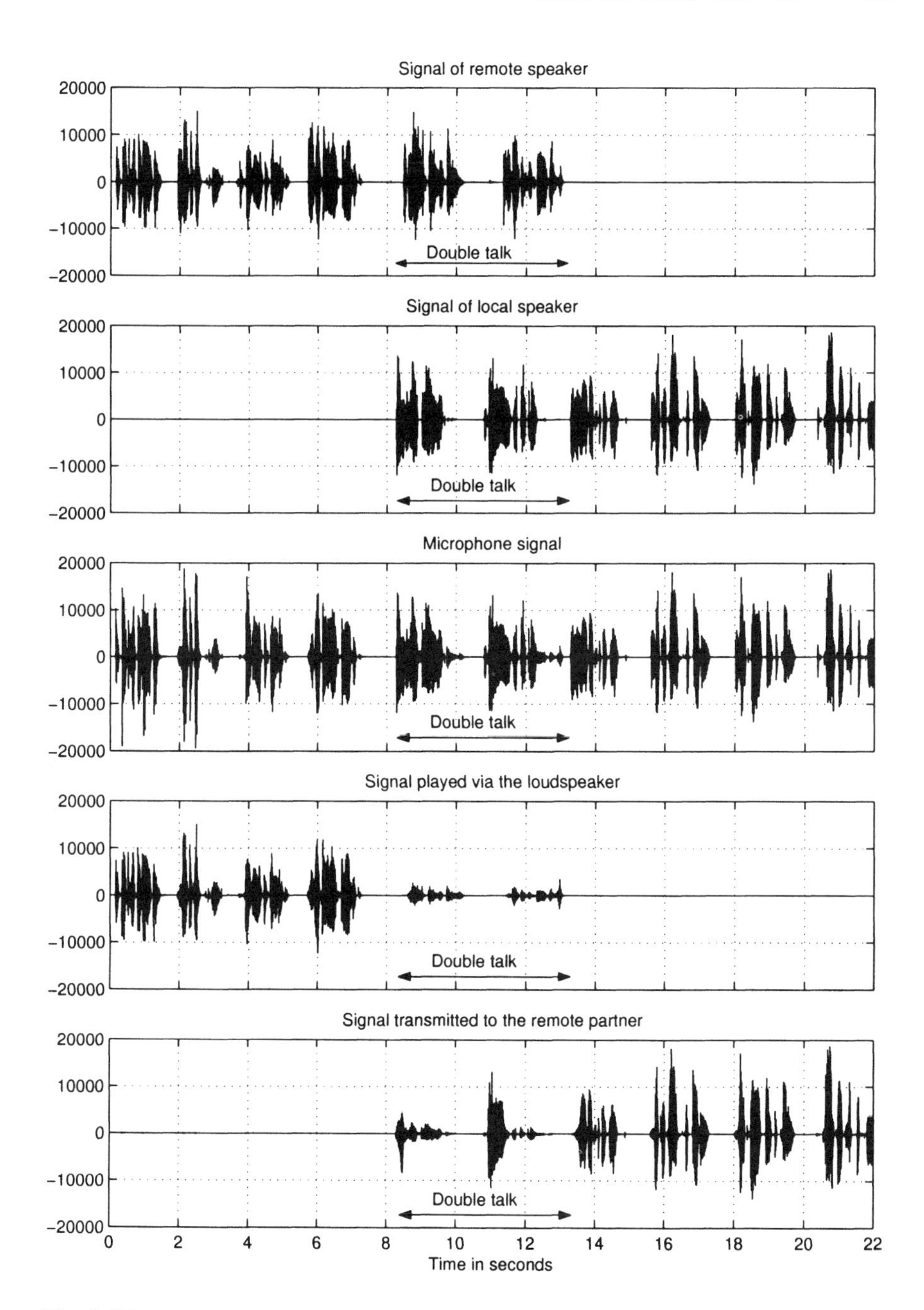

Fig. 8.17 Simulation example for a loss control circuit (part I). In the upper two diagrams the signals of the remote and the local communication partners are depicted. The third diagram shows the microphone signal consisting of an echo component and local speech. The last two graphs show the signal of the loudspeaker and the transmitted signal. These are the incoming signals weighted with the gain factors $a_{\mathrm{r}}(n)$ and $a_{\mathrm{t}}(n)$, respectively.

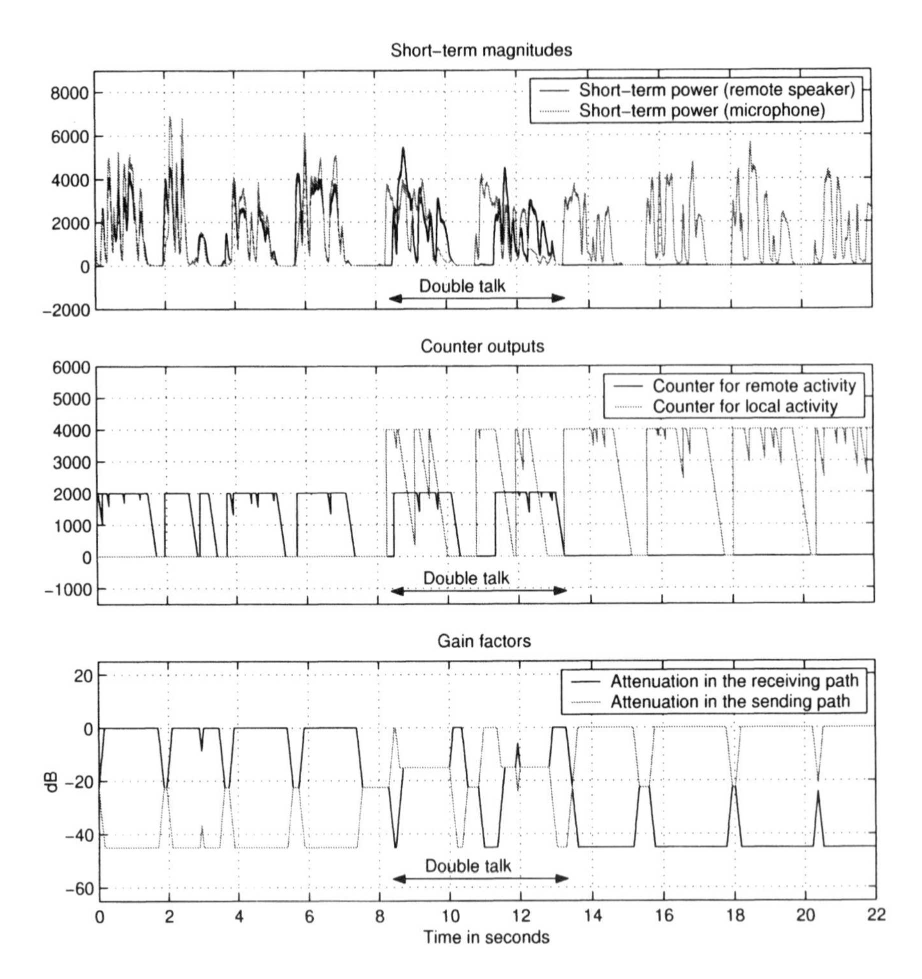

Fig. 8.18 Simulation example for a loss control circuit (part II). In the upper diagram the short-term magnitudes of the incoming signals $\overline{|r(n)|}$ and $\overline{|y(n)|}$ are depicted. The second graph shows the remote and the local voice activity detection. The lowest diagram presents the gain factors in the receiving and transmitting paths, respectively.

detection conditions, we hold the result for half a second which means that

$$N_{\text{loc}} = 4000\,. \tag{8.62}$$

This is twice as long as in the remote voice activity detector. In the upper two diagrams of Fig. 8.17 the speech signals of a remote and a local speaker are depicted. After 8 seconds of far-end single talk both communication partners speak simultaneously for about 5 seconds. Then only the local partner continues. The echo signal is generated by convolution of the remote signal weighted by $a_{\text{r}}(n)$ and an impulse response measured in an office with a reverberation time of about 300 ms. Adding the echo component and the local speech signal leads to the microphone signal presented in the third diagram of Fig. 8.17.

The resulting short-term amplitude estimation as well as the voice activity detections are presented in Fig. 8.18. Finally, both detections [$t_{\text{act,far}}(n) > 0$ and $t_{\text{act,loc}}(n) > 0$] need to be "converted" into the attenuation factors $a_{\text{r}}(n)$ and $a_{\text{t}}(n)$.

In case of single talk of the remote speaker, $a_{\text{r}}(n)$ should be increased until the final value $a_{\text{r}}(n) = 1$ is reached. If both partners are silent, half of the attenuation should be inserted into both channels, and if only the local speaker is "active," $a_{\text{r}}(n)$ should be decreased until $a_{\text{r}}(n) = \alpha_{\text{min}}$ is achieved. During double talk both channels should insert half of the attenuation that is required for this state. Depending on the present value of $a_{\text{r}}(n)$, it should be increased or decreased until the final attenuation $a_{\text{r}}(n) = \sqrt{\alpha_{\text{dt}}}$ is reached. The six states can be described by the following set of conditions:

$$\Big(t_{\text{act,far}}(n) \le 0\Big) \wedge \Big(t_{\text{act,loc}}(n) > 0\Big)\,, \tag{8.63}$$

$$\Big(t_{\text{act,far}}(n) \le 0\Big) \wedge \Big(t_{\text{act,loc}}(n) \le 0\Big) \wedge \Big(\alpha_{\text{r}}(n) > \sqrt{\alpha_{\text{min}}}\Big)\,, \tag{8.64}$$

$$\Big(t_{\text{act,far}}(n) \le 0\Big) \wedge \Big(t_{\text{act,loc}}(n) \le 0\Big) \wedge \Big(\alpha_{\text{r}}(n) \le \sqrt{\alpha_{\text{min}}}\Big)\,, \tag{8.65}$$

$$\Big(t_{\text{act,far}}(n) > 0\Big) \wedge \Big(t_{\text{act,loc}}(n) \le 0\Big)\,, \tag{8.66}$$

$$\Big(t_{\text{act,far}}(n) > 0\Big) \wedge \Big(t_{\text{act,loc}}(n) > 0\Big) \wedge \Big(\alpha_{\text{r}}(n) > \sqrt{\alpha_{\text{dt}}}\Big)\,, \tag{8.67}$$

$$\Big(t_{\text{act,far}}(n) > 0\Big) \wedge \Big(t_{\text{act,loc}}(n) > 0\Big) \wedge \Big(\alpha_{\text{r}}(n) \le \sqrt{\alpha_{\text{dt}}}\Big)\,. \tag{8.68}$$

Conditions 8.63–8.68 build a closed and orthogonal set. Depending on these conditions, the attenuation factor in the receiving path of the loss control system is set as follows:

$$\alpha_{\text{r}}(n) = \begin{cases} \max\left\{\alpha_{\text{r}}(n-1)\,\alpha_{\text{dec}},\, \alpha_{\text{min}}\right\}\,, & \text{if condition 8.63 is true,} \\ \max\left\{\alpha_{\text{r}}(n-1)\,\alpha_{\text{dec}},\, \sqrt{\alpha_{\text{min}}}\right\}\,, & \text{if condition 8.64 is true,} \\ \min\left\{\alpha_{\text{r}}(n-1)\,\alpha_{\text{inc}},\, \sqrt{\alpha_{\text{min}}}\right\}\,, & \text{if condition 8.65 is true,} \\ \min\left\{\alpha_{\text{r}}(n-1)\,\alpha_{\text{inc}},\, \alpha_{\text{max}}\right\}\,, & \text{if condition 8.66 is true,} \\ \max\left\{\alpha_{\text{r}}(n-1)\,\alpha_{\text{dec}},\, \sqrt{\alpha_{\text{dt}}}\right\}\,, & \text{if condition 8.67 is true,} \\ \min\left\{\alpha_{\text{r}}(n-1)\,\alpha_{\text{inc}},\, \sqrt{\alpha_{\text{dt}}}\right\}\,, & \text{if condition 8.68 is true.} \end{cases} \tag{8.69}$$

According to the ITU-T recommendation G.167 [120], we have chosen

$$\alpha_{\text{max}} = 1 = 0\,\text{dB}\,, \tag{8.70}$$

$$\alpha_{\text{min}} = 0.0056 \approx -45\,\text{dB}\,, \tag{8.71}$$

$$\alpha_{\text{dt}} = 0.0316 \approx -30\,\text{dB}\,. \tag{8.72}$$

The attenuation factor in the transmitting path $\alpha_{\text{t}}(n)$ is computed in the same way—only all remote quantities in conditions 8.63–8.68 and in Eq. 8.69 have to be replaced by their local counterparts and vice versa.

The lowest diagram of Fig. 8.18 shows the computed gain factor in our example. During single talk only the "nonactive" channel is attenuated. In speech pauses the attenuation is distributed half into the sending path and half into the receiving path. In the double-talk situation both channels are attenuated by 15 dB each. The lowest two diagrams of Fig. 8.17 show the corresponding output signals of the loss control. During the double-talk situation the disadvantage of loss control circuits is clearly noticeable. Both signals are attenuated, leading to a reduction in speech intelligibility and communication comfort.

This drawback can be compensated if an echo cancellation filter is implemented in addition. If the filter has converged, the maximal attenuation (α_{max} and α_{dt}, respectively) can be reduced according to the amount of echo attenuation achieved by the echo cancellation filter. Details about those combinations can be found in the literature (e.g., see [186, 198]).

8.3.3 Limiter

Processing of the incoming signal may result in an amplification. An example of this type of signal processing is automatic gain control (see Section 8.3.1). Signals of speakers with soft voices are amplified by a few decibels. After the amplification, short signal bursts (e.g., noise caused by hand clapping) might exceed the range of the DA converters for outgoing signals. The simplest way to avoid any malfunctioning of the converters is to limit the signal. A hard limitation according to

$$u(n) = \begin{cases} K_{\text{H}}\,, & \text{if } y(n) > K_{\text{H}} \\ y(n)\,, & \text{if } |y(n)| \leq K_{\text{H}} \\ -K_{\text{H}}\,, & \text{if } y(n) < -K_{\text{H}} \end{cases} \tag{8.73}$$

avoids any overflow but adds audible distortions to a speech signal. The constant K_{H} in Eq. 8.73 is often set to the maximum amplitude allowed by the AD and DA converters. For 16-bit converters, K_{H} would be

$$K_{\text{H}} = \frac{2^{15}-1}{V_0} = \frac{32767}{V_0}\,, \tag{8.74}$$

where V_0 is a normalization constant, representing, for instance, the voltage of the signal, which will be converted to the most significant bit of the AD converter.

A better method to attenuate large amplitudes is the usage of soft or dynamic limiters. These units monitor the maximum of the short-term amplitude. If this quantity exceeds a predefined threshold an attenuation is introduced in order to avoid overflow of the DA converters. A combination of a hard and a soft limiter is presented in Fig. 8.19.

8.3.3.1 Implementation Example

For tracking the amplitude peaks, an IIR filter that smooths the modulus of the input signal is used:

$$\overline{|x(n)|} = (1-\gamma(n))\,|x(n)| + \gamma(n)\,\overline{|x(n-1)|}\,. \tag{8.75}$$

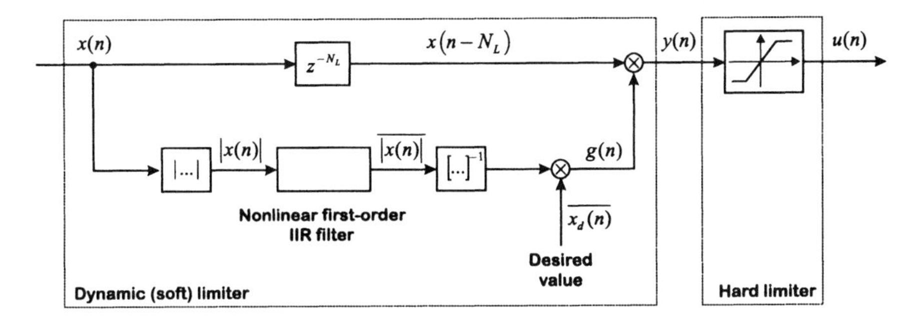

Fig. 8.19 Hard–soft limiter combination.

An increase of the magnitudes is followed quickly while decreasing magnitudes are tracked slowly:

$$\gamma(n) \quad = \quad \begin{cases} \gamma_{\rm r}, & \text{if } |x(n)| > \overline{|x(n-1)|}, \\ \gamma_{\rm f}, & \text{else}. \end{cases} \tag{8.76}$$

For this purpose the time constant for falling signal edges is chosen much larger than for rising edges. A simulation example is presented at the end of this section. All involved signals have been sampled at 8 kHz. For this sampling rate, we suggest the following time constants:

$$\gamma_{\rm r} \quad = \quad 0.9\,, \tag{8.77}$$

$$\gamma_{\rm f} \quad = \quad 0.995\,. \tag{8.78}$$

In order to bound $\overline{|x(n)|}$ to a maximum, a gain factor $g(n)$ according to

$$g(n) = \frac{\overline{x_{\rm d}(n)}}{\overline{|x(n)|}} \tag{8.79}$$

is computed. The desired value $\overline{x_{\rm d}(n)}$ is chosen to be equal to the smoothed magnitude as long as the latter does not exceed a maximum value $K_{\rm d}$:

$$\overline{x_{\rm d}(n)} = \min\left\{\overline{|x(n)|},\, K_{\rm d}\right\}\,. \tag{8.80}$$

Only for large signals, the desired value is chosen to be equal to the constant $K_{\rm d}$, leading to an attenuation of signal. In the simulation example $K_{\rm d}$ was set to

$$K_{\rm d} = \frac{16{,}000}{V_0}\,. \tag{8.81}$$

The output of the dynamic limiter is computed by weighting the delayed input signal $x(n - N_{\rm L})$ with the gain factor $g(n)$:

$$y(n) = x(n - N_{\rm L})\, g(n)\,. \tag{8.82}$$

The short-term magnitude follows a sudden increase of the signal only with a few samples (depending on the choice of time constants) of delay. In some implementations (e.g., see [249]) the input signal is also delayed by a few samples. We have omitted this delay to avoid introducing any delay in the signal paths. For avoiding overflows at the beginning of signal edges, we utilized a series connection of a soft limiter and a hard limiter (see Fig. 8.19).

A simulation example is presented in Fig. 8.20. The input signal $x(n)$ is depicted in the upper diagram. After 5 seconds an internal gain was increased by 6 dB. Therefore, the amplitudes of the output signal would exceed the desired maximum a few times (without any processing). In order to avoid this the soft–hard limiter combination described above has been implemented. The quantities $\overline{|x(n)|}$ and $g(n)$ are depicted in the second and the third diagrams, respectively. The output signal stays within the desired amplitude range and has nearly no audible distortions.

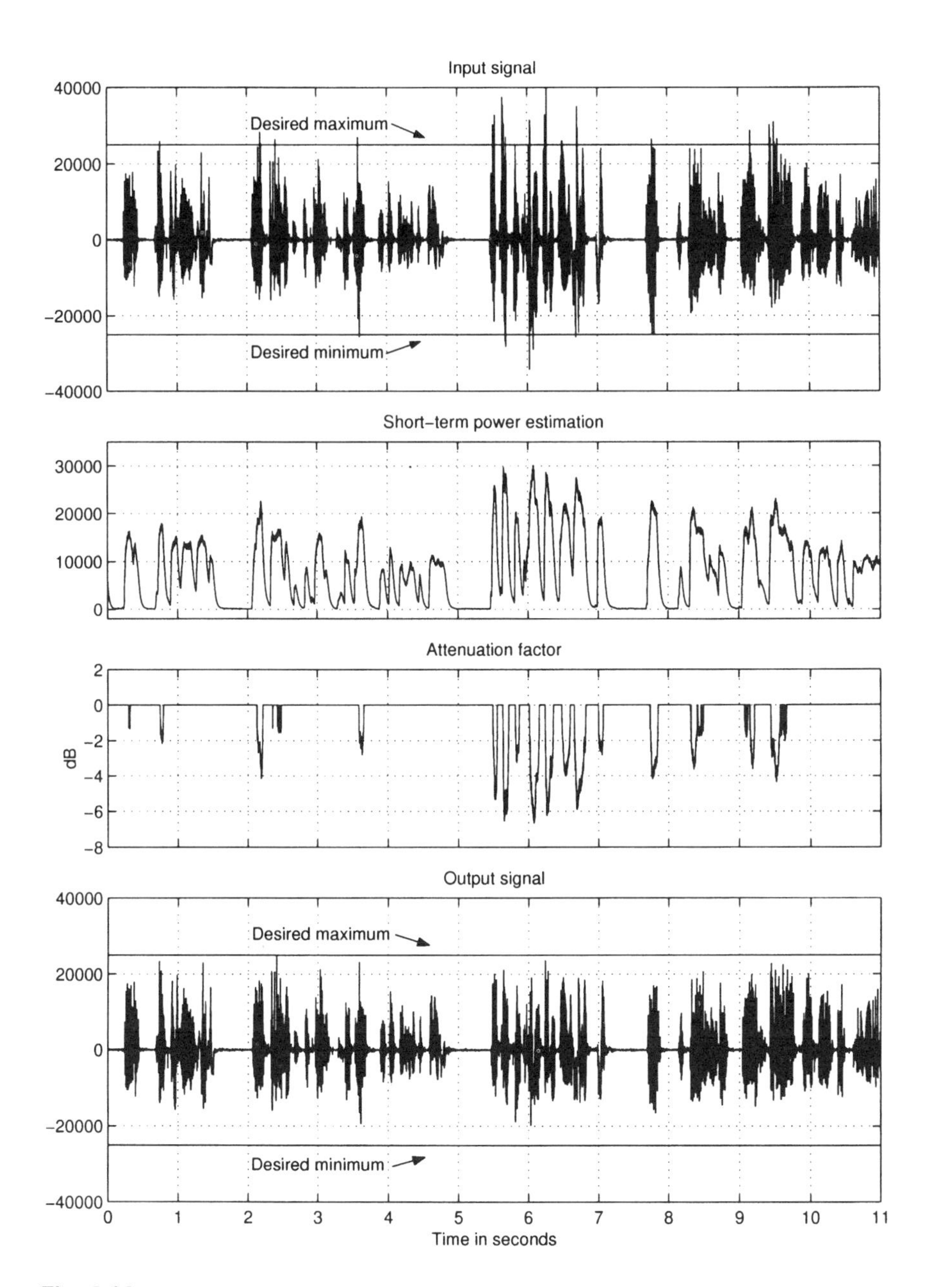

Fig. 8.20 Simulation example for dynamic limiting. The input signal is depicted in the upper diagram. Without any processing, the signal would reach the overload range of the DA converter. The second and the third diagrams show the short-term magnitude and the resulting attenuation factor, respectively. The output signal is depicted in the lowest part of the figure.

9

Echo Cancellation

In contrast to loss controls, which have been described in the last section, echo cancellation by means of an adaptive filter offers the possibility of full-duplex communication. As already explained in Section 2.2.1, an adaptive filter is used as a digital replica of the loudspeaker–enclosure–microphone (LEM) system (see Fig. 9.1). By subtracting the estimated echo $\widehat{d}(n)$ from the microphone signal $y(n)$, a nearly perfect decoupling of the system is possible (assuming that the echo cancellation filter has converged).

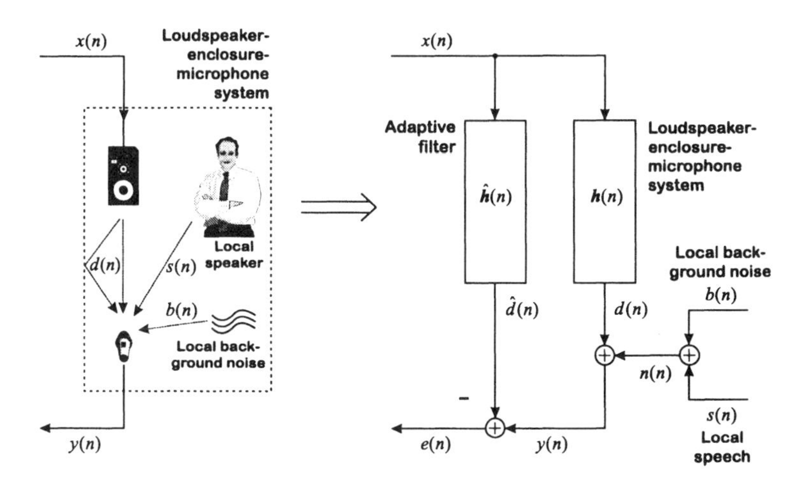

Fig. 9.1 Structure of a single-channel acoustic echo control system.

In most applications the adaptive filter requires a few hundred or even more than thousand coefficients (see Chapter 3) in order to model the LEM system with sufficient accuracy. Even for powerful modern signal processors, this means a large computational load, especially if more than one microphone is used and echo cancellation is performed independently for each microphone. For this reason, other structures than fullband processing are often applied.

When changing from monophonic excitation signals to stereophonic or even multichannel signals, not only the complexity increases but also a nonuniqueness problem [11, 12] within the system identification arises.

In this chapter we will discuss processing structures as well as multichannel systems in more detail. The control of those structures and systems will be presented in Part IV of this book.

9.1 PROCESSING STRUCTURES

Beside selecting an adaptive algorithm such as LMS and RLS, the system designer has also the possibility to choose between different processing structures. The most popular ones are

- Fullband processing
- Block processing[1]
- Subband processing

The application of high-order adaptive filters in the time domain (fullband processing) results in an immense amount of computational complexity. Frequency-domain adaptive filtering (block processing) and subband structures reduce the complexity problem. Furthermore, both structures allow a frequency-dependent normalization of the input signal. This method improves the convergence rate of simple adaptive algorithms like NLMS in case of colored excitation signals.

Before the processing structures listed above are described in more detail (in Sections 9.1.1–9.1.3), Table 9.1 summarizes their advantages and disadvantages. In fullband structures it is possible to adjust the control parameters of an algorithm differently for each sample interval. For this reason we have a very good time resolution for control purposes. On the other hand, it is not possible to chose frequency-dependent control parameters differently. Subband processing and block processing in the frequency domain offer this possibility. Due to the subsampling and the collection of a block of input signals, respectively, the time-resolution decreases.

Block processing with large block sizes offers the highest reduction in computational complexity—but also introduces the largest delay. Subband structures can be seen as a compromise. If the analysis and synthesis filterbanks are well designed, it

[1]By block processing we mean here performing the convolution and/or the adaptation in the frequency domain and using overlap-add or overlap-save techniques.

Table 9.1 Advantages and disadvantages of different processing structures (where t_{resol} and ν_{resol} = time and frequency resolution, respectively)

<table>
<tr><th></th><th colspan="6">Processing structure</th></tr>
<tr><th></th><th colspan="2">Fullband</th><th colspan="2">Block</th><th colspan="2">Subband</th></tr>
<tr><th></th><th>t_{resol}</th><th>ν_{resol}</th><th>t_{resol}</th><th>ν_{resol}</th><th>t_{resol}</th><th>ν_{resol}</th></tr>
<tr><td>Control alternatives</td><td>++</td><td>− −</td><td>− −</td><td>++</td><td>+</td><td>+</td></tr>
<tr><td>Computational complexity</td><td colspan="2">− −</td><td colspan="2">++</td><td colspan="2">+</td></tr>
<tr><td>Introduced delay</td><td colspan="2">++</td><td colspan="2">− −</td><td colspan="2">−</td></tr>
</table>

is possible to achieve acceptable time and frequency resolutions, to introduce only a moderate delay, and to keep the computational load at a moderate level. Nevertheless, as we will see later, block and subband processing are closely related to each other.

In the following sections we will give several examples of processing structures, possibilities to enhance the convergence speed, and methods to reduce the computational complexity. In order to keep all schemes comparable, in all examples we use the NLMS algorithm to adapt the filter coefficients.

9.1.1 Fullband Cancellation

Fullband processing exhibits the best time resolution among all processing structures. Especially for impulse responses that concentrate their energy on only a few coefficients and if a matrix stepsize according to Section 7.1.1 is used, this is an important advantage for control purposes.

Fullband processing structures do not introduce any delay into the signal path. This is the very basic advantage of fullband processing. For some applications this is an essential feature. For nonmobile hands-free telephones, for instance, the International Telecommunication Union (ITU) recommends that the delay caused by front-end signal processing should not exceed 2 ms [120]. Due to the long reverberation time of rooms (several hundred milliseconds), a huge number of coefficients is required to achieve a sufficient amount of echo reduction. Even when using the (in terms of complexity) simple NLMS algorithms for updating the coefficients, the computational load might still be prohibitive. Therefore, several algorithms for reducing the complexity by updating only a subset of the coefficients within each iteration have been proposed [200]. In contrast to block processing and subband systems, these schemes slow down the convergence, but keep the advantage of being delayless structures.

To increase the convergence speed in case of speech excitation, decorrelation filters can be applied. Details about these filters as well as a brief description of selective coefficient update schemes will follow in the next two sections.

9.1.1.1 Decorrelation Filters

Even though various adaptive algorithms are applicable for acoustic echo cancellation filters, simple and robust algorithms may outperform more sophisticated solutions. Therefore, in most applications with limited precision and processing power, the NLMS algorithm is applied. However, the adaptation performance of the NLMS algorithm, in the case of speech excitation, is rather poor because of the strong correlation of neighboring samples of the speech signal. One way to overcome this problem is to prewhiten or decorrelate the incoming excitation signal according to

$$x_{\mathrm{d}}(n) = x(n) - \sum_{i=1}^{N_{\mathrm{d}}-1} a_i(n)\, x(n-i) \tag{9.1}$$

before passing it to the adaptive algorithm [111]. Especially in the field of acoustic echo cancellation, decorrelation filters are widely applied [70, 245, 247]. Decorrelation filters are predictor error filters (see Chapter 6) and their coefficients $a_i(n)$ are matched with the correlation properties of the speech signal. Since speech signals are nonstationary, one has to decide whether the prediction coefficients should be adapted periodically to be matched with the current section of the speech signal, or simply to the long-term statistics of speech

$$a_i(n) = a_i\,. \tag{9.2}$$

This, as we will see later, again means a tradeoff between processing power and performance.

There are various possibilities for placing the decorrelation filters within the system. For identification purposes, it would be sufficient to simply filter the excitation signal. But since this signal also serves as loudspeaker signal and is therefore transmitted into the room, preprocessing is not possible. One can move this linear system into the filter branch and also across the LEM system (see Fig. 9.2). Since the filter is still in the signal path to the far-end listener, one has to apply an inverse filter after calculation of the error value. To invert the linear prediction, one would have to use a recursive filter and apply the same coefficients as in the prediction filter:

$$e(n) = e_{\mathrm{d}}(n) + \sum_{i=1}^{N_{\mathrm{d}}-1} a_i(n)\, e(n-i)\,. \tag{9.3}$$

Since a linear prediction error filter is minimum phase, its inverse is always causal and therefore stable [111].

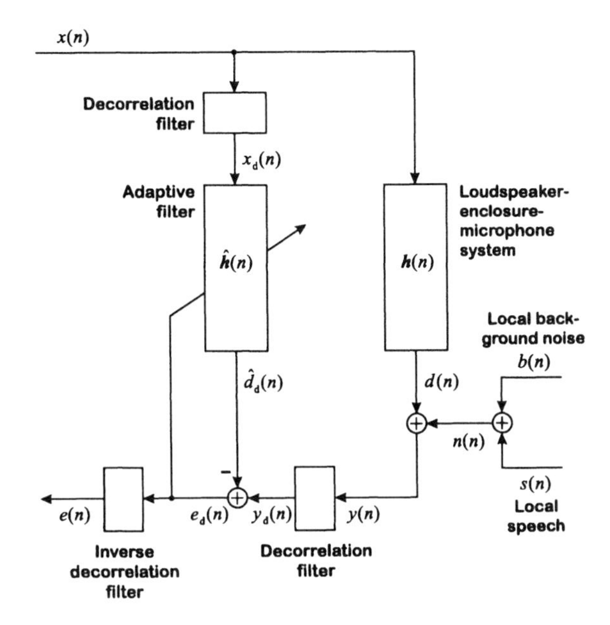

Fig. 9.2 Placement of decorrelation and inverse decorrelation filters.

One can omit the decorrelation filter (and therefore also its inverse) in the signal path, in which case the adaptive filter also has to model the inverse decorrelation filter [247].

Especially for low-order decorrelation filters, the language of the remote partner does not affect the adjustment of the filter. In Fig. 9.3, power spectral densities (averaged over a few minutes of speech sequences of each language) as well as the frequency responses of first- and eight-order decorrelation filters are depicted for six languages: American English, British English, French, German, Japanese, and Spanish. When comparing the power spectral densities and the filter transfer functions only marginal differences appear for the different languages.

A different approach for the application of decorrelation filters, especially if adaptive decorrelation filters are used, is the implementation of an auxiliary loop for adaptation as shown in Fig. 9.4 [70]. Since the coefficients of the echo cancellation filter are now copied to the echo cancellation filter located in the signal path, there is no more need for a decorrelation filter or its inverse in the signal path. In the case of fixed-point processing, where recursive filters of a high order may cause stability problems, this is of considerable advantage. Moreover, for adaptive decorrelation filters it is preferable not to have constantly changing decorrelation filters in the path of the estimated echo signal. Otherwise, the data vectors within the echo cancellation filter would have to be newly calculated every time the decorrelation filter is updated, which could lead to distortions in the near-end signal.

Implementing this second path, however, results in an increased processor load in terms of both memory and computation. The filtering operation of the echo canceller

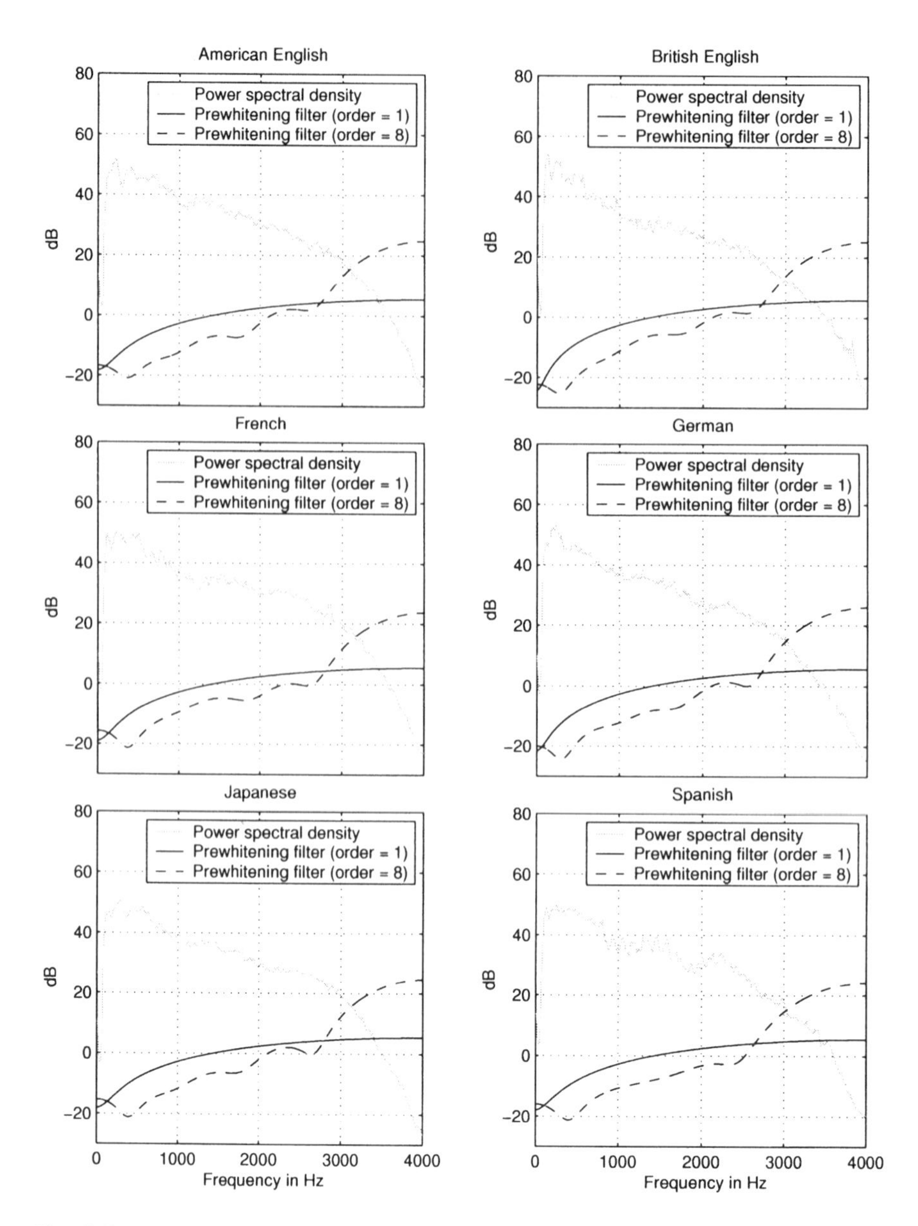

Fig. 9.3 Power spectral densities and decorrelation filters for different languages. When comparing the power spectral densities and the prewhitening filters, only marginal differences appear for the different languages.

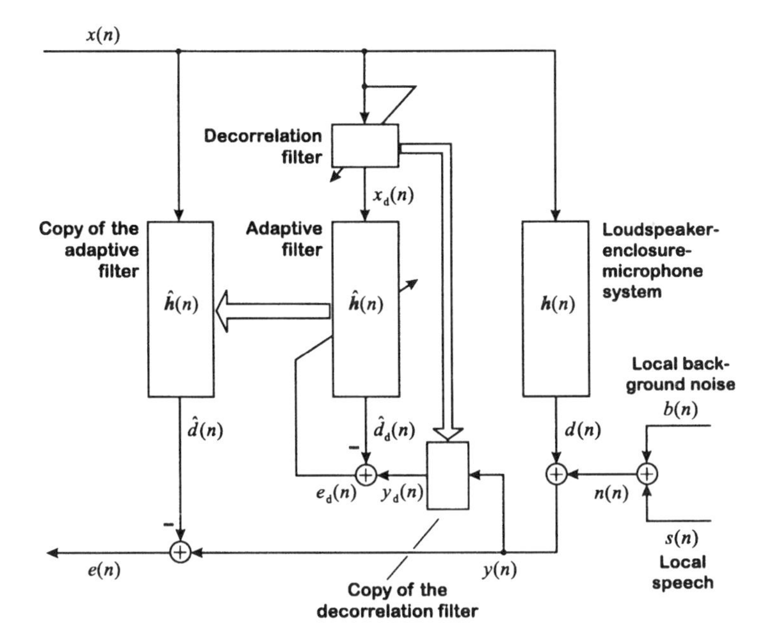

Fig. 9.4 Auxiliary loop for the adaptation algorithm—the inverse decorrelation is no longer necessary.

has to be performed twice, once with the original signal of the far-end speaker and once with the decorrelated signal. This also means a second buffer for the decorrelated excitation signal.

Fixed decorrelation filters can account only for the mean power spectral density of speech. The coefficients are therefore calculated using a representative set of speech sequences. This, of course, is in general not an optimal match to the actual excitation signal. The speed of convergence of the adaptive filters is therefore not as fast as with adaptively renewed coefficients. The great advantage of fixed decorrelation filters, however, is their very low computational complexity.

The coefficients of adaptive decorrelation filters are updated periodically to the short-term characteristics of the excitation signal [70, 245]. They are calculated using the Levinson–Durbin algorithm (see Section 6.2). It is important, however, to stop the recursion when they outnumber the rank of the input autocorrelation matrix, since in this case the autocorrelation matrix becomes singular. Due to the instationarity of speech signals, this length can vary with time. However, since the Levinson–Durbin algorithm is order recursive, one can always fall back to the lower order.

One should note that the increase of computational complexity is not negligible. In addition to the computation of the new decorrelation coefficients, both the excitation vector and its vector norm have to be updated. If necessary, one can compensate for this additional computational load by skipping updates of the echo canceller coefficients.

When discussing the order N_d of the decorrelation filter, one has to distinguish again between fixed and adaptive decorrelation filters. For fixed decorrelation filters, an order higher than 2 is not advisable, since only the long-term characteristic can be modeled. For adaptive decorrelation filters, however, additional gain can be achieved by further enlarging the number of decorrelation filter coefficients. Simulation results have been published by several authors [25, 247].

Figure 9.6 shows the improvement in the speed of convergence gained by decorrelation filters of various orders. The effect of first-order filters is remarkable. Already this very simple form of decorrelation achieves a significant decrease in the residual system error norm in our simulation. In contrast, the additional gain of a second order but still fixed decorrelation filter is negligible. When faster convergence is required, adaptive filters have to be used. However, as depicted in Fig. 9.5, at least $N_d = 8$ coefficients should be implemented.

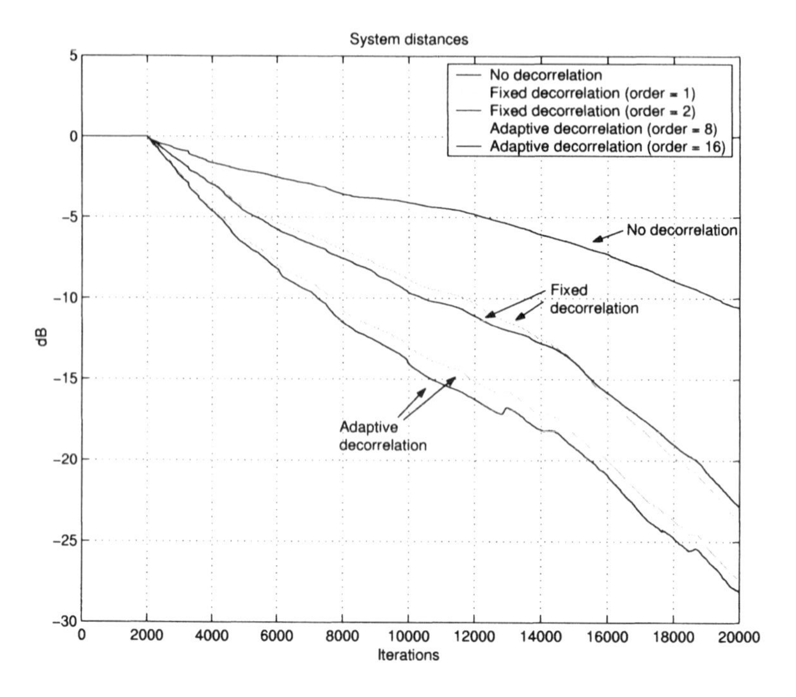

Fig. 9.5 Convergence of the NLMS algorithm using decorrelation filters with speech excitation.

Also the number of data points needed for proper estimation of the correlation coefficients increases with the order of the decorrelation filter. Hence, one has to make sure that the block size required for estimation of the correlation coefficients does not exceed the time interval in which speech signals can be assumed to be stationary. Simulations have shown that, for this application, one should consider the order N_d of the decorrelation filters in the range of 8–16 (depending on the sampling rate).

As was shown in Fig. 9.5, the gain yielded by only a first-order decorrelation filter is considerable. We now want to examine the influence of this prediction coefficient on the speed of convergence. Since speech signals have a lowpass characteristic, the first-order decorrelation filter will be a simple highpass filter. If an analysis is performed for a representative set of speech signals,[2] including both male and female voices, a mean value of about

$$a_1 = \frac{s_{xx}(1)}{s_{xx}(0)} = 0.95 \tag{9.4}$$

for 8 kHz sampling rate is calculated for the decorrelation coefficient. Finally we remark that the final misalignment is not affected by prewhitening. All convergence curves lie within the range of a few dB of final system error norm (not depicted in Figs. 9.5 and 9.6).

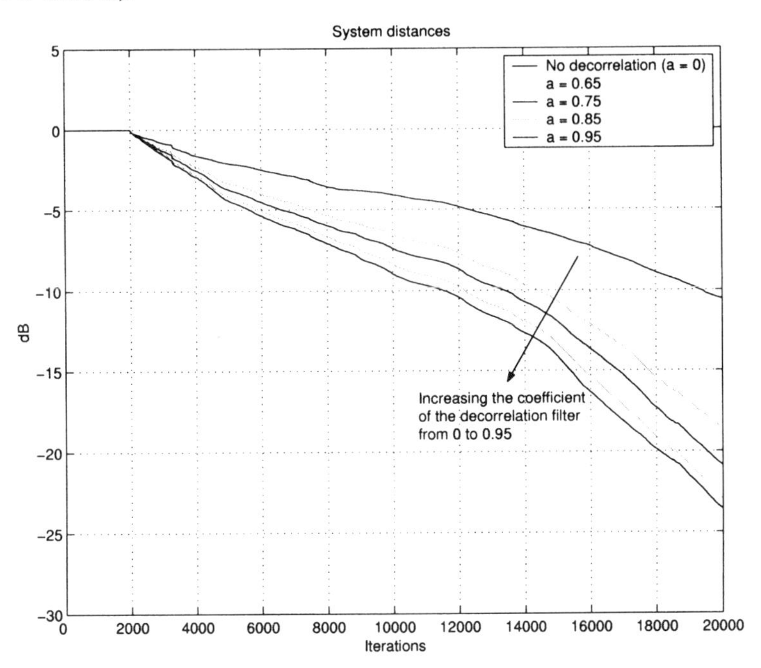

Fig. 9.6 First-order decorrelation, excited with speech sequences from both male and female speakers.

9.1.1.2 Selective Coefficient Update

To decrease the computational load, it is possible to update only a portion of the adaptive filter coefficients at a time. The subset of coefficients that should be updated can be selected in several ways.

[2]The speech database consisted of both American and British English.

9.1.1.2.1 Static Schemes

Static schemes, like the periodic NLMS algorithm [226] or the sequential NLMS algorithm [54], use predefined subsets to update the filter coefficients. The latter algorithm can be described as follows. Let N be the filter length and M the number of coefficients to be updated. For simplicity, we will assume $N/M = B$ to be an integer. In this case the filter vector $\widehat{\boldsymbol{h}}(n)$ and the coefficient vector $\boldsymbol{x}(n)$ can be partitioned into B subfilters and B subsignal vectors, respectively, of length M:

$$\begin{aligned} \widehat{\boldsymbol{h}}(n) &= \left[\hat{h}_0(n),\, \hat{h}_1(n),\, \ldots,\, \hat{h}_{N-1}(n)\right]^{\mathrm{T}} \\ &= \left[\widehat{\boldsymbol{h}}_0^{\mathrm{T}}(n),\, \widehat{\boldsymbol{h}}_1^{\mathrm{T}}(n),\, \ldots,\, \widehat{\boldsymbol{h}}_{B-1}^{\mathrm{T}}(n)\right]^{\mathrm{T}}, \end{aligned} \tag{9.5}$$

$$\begin{aligned} \boldsymbol{x}(n) &= \left[x(n),\, x(n-1),\, \ldots,\, x(n-N+1)\right]^{\mathrm{T}} \\ &= \left[\boldsymbol{x}_0^{\mathrm{T}}(n),\, \boldsymbol{x}_1^{\mathrm{T}}(n),\, \ldots,\, \boldsymbol{x}_{B-1}^{\mathrm{T}}(n)\right]^{\mathrm{T}}. \end{aligned} \tag{9.6}$$

The subfilters $\widehat{\boldsymbol{h}}_i(n)$ and the corresponding signal vectors $\boldsymbol{x}_i(n)$ are defined as

$$\widehat{\boldsymbol{h}}_{iB}(n) = \left[\hat{h}_{iB}(n),\, \hat{h}_{iB+1}(n),\, \ldots,\, \hat{h}_{iB+B-1}(n)\right]^{\mathrm{T}}, \tag{9.7}$$

$$\boldsymbol{x}_{iB}(n) = \left[x(n-iB),\, x(n-iB-1),\, \ldots,\, x(n-iB-B+1)\right]^{\mathrm{T}}. \tag{9.8}$$

Using these definitions the sequential block NLMS algorithms can be denoted as

$$\widehat{\boldsymbol{h}}_{iB}(n+1) = \begin{cases} \widehat{\boldsymbol{h}}_{iB}(n) + \mu \dfrac{e^*(n)\, \boldsymbol{x}_{iB}(n)}{\|\boldsymbol{x}(n)\|^2}, & \text{if } n + i \bmod B \equiv 0\,, \\ \widehat{\boldsymbol{h}}_{iB}(n), & \text{else}\,. \end{cases} \tag{9.9}$$

The error signal is computed in the same manner as for the conventional NLMS algorithm:

$$e(n) = y(n) - \widehat{\boldsymbol{h}}^{\mathrm{H}}(n)\, \boldsymbol{x}(n)\,. \tag{9.10}$$

According to Eq. 9.9 only one block of coefficients is updated within each iteration. The other blocks remain unchanged. Therefore, the computational load is decreased, but also the speed of convergence is reduced. Since the update scheme is determined in a static manner, the overhead for testing the modulo condition in Eq. 9.9 is only marginal.

At a first glance the reduction concerning computational complexity by updating, for example, only a quarter of the coefficients at each iteration, is not very large. When counting the multiplications required for adaptation and convolution a reduction of only about 25% appears. This is mainly because the complexity of the convolution remains unchanged.

At a second glance, especially when we have a DSP implementation with fixedpoint arithmetic in mind, the advantages become more evident. A convolution can usually be implemented by a loop of N MAC (multiply and accumulate) instructions and one round instruction at the end. This means that the processor requires approximately N cycles to perform a convolution. Implementing the coefficient update turns out to be more complex. One needs typically one load instruction, one MAC command, one round, and one store instruction. For several DSPs it is not possible perform all of these instructions in parallel. If four operations are required to perform an update of one coefficient, the computational complexity in terms of DSP cycles is reduced by 60% (if again only one-quarter of the coefficients is updated at each iteration). Nevertheless, the speed of convergence is reduced to a quarter of the standard NLMS (see Fig. 9.7).

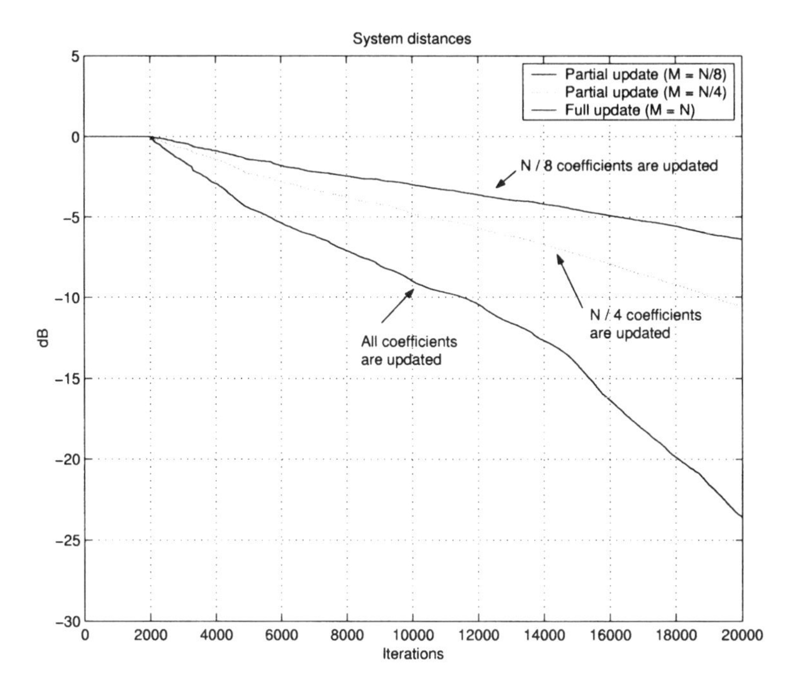

Fig. 9.7 System distance for different static update schemes. All convergence curves were averaged over 8 male and 8 female speech sequences. Furthermore, fixed, first-order decorrelation filtering was applied.

9.1.1.2.2 Adaptive Schemes

Another method to reduce the computational complexity without decreasing the convergence speed as much as in the static schemes is to update the most important coefficients only. The term "important" means here to exchange only these coefficients that exhibit a large gradient component on the error surface [2]. These coefficients can be selected by sorting the elements of the input vector $x(n)$ with respect to their magnitude. Only those coefficients associated with the M largest amplitudes are updated:

$$\widehat{h}_i(n+1) = \begin{cases} \widehat{h}_i(n) + \mu \, \dfrac{e^*(n)\, x(n-i)}{\|\boldsymbol{x}(n)\|^2}, & \text{if } x(n-i) \text{ belongs to the } M \\ & \text{largest amplitudes of } \boldsymbol{x}(n), \\ \widehat{h}_i(n), & \text{else}. \end{cases} \tag{9.11}$$

In order to find the M largest magnitude values of the excitation vector $\boldsymbol{x}(n)$, a running ordering algorithm called *sortline* [135, 184] is used. Since only one new value is entering the observation window and only one value is leaving it at each iteration, the sorting procedure requires at most $2\log_2 N + 2$ comparisons per iteration.

Even if the number of multiplications or additions is not affected by the sorting scheme, the overall complexity in terms of processor cycles exceeds the one of the full update NLMS algorithms on most signal processors. Nevertheless, the convergence speed, which can be achieved for speech excitation, is close to the full update NLMS algorithm (see Fig. 9.8).

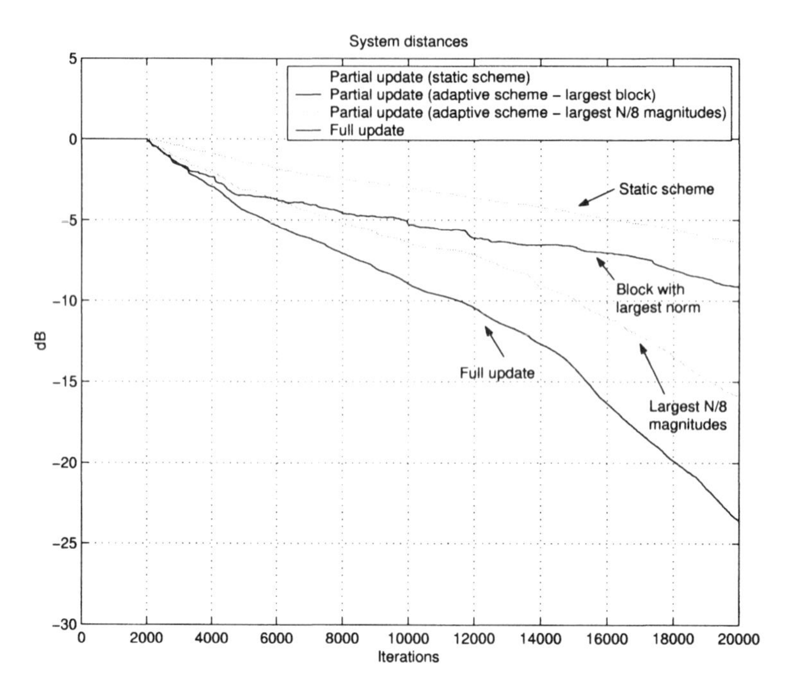

Fig. 9.8 System distance for different adaptive update schemes. All simulation results were averaged over 8 male and 8 female speech sequences. As in the previous simulation, fixed, first-order decorrelation filters were applied to increase the speed of convergence.

To overcome the complexity problem, it is possible to apply adaptive block update schemes. Instead of looking for the M largest magnitudes one selects a subblock of the excitation vector with the largest vector norm [198]. If the excitation and the filter vector are decomposed according to Eqs. 9.5–9.8, the filter update can be denoted as as follows:

$$\widehat{\boldsymbol{h}}_{ib}(n+1) = \begin{cases} \widehat{\boldsymbol{h}}_{iB}(n) + \mu \, \dfrac{e^*(n)\, \boldsymbol{x}_{iB}(n)}{\|\boldsymbol{x}(n)\|^2}\,, & \text{if } \boldsymbol{x}_{iB}(n) \text{ has the largest norm among all subblocks}\,, \\ \widehat{\boldsymbol{h}}_{iB}(n), & \text{else}\,. \end{cases} \tag{9.12}$$

The number of blocks is much lower than the number of coefficients. Thus, the complexity of the sorting procedure is not as large as in the previous case. Furthermore, the norms of the subblocks can be computed recursively. Even if this block-selective scheme does not reach the convergence speed of the coefficient selective algorithm, the convergence is still better than in the static scheme (see Fig. 9.8).

9.1.2 Block Processing Algorithms

Selective coefficient update schemes as presented in the previous section are able to reduce the computational complexity. Nevertheless, for many applications, such as acoustic echo or noise control, algorithms with even lower numerical complexity are necessary. To remedy the complexity problem, adaptive filters based on block processing [165, 171, 211] can be used.

In general, most block processing algorithms collect B input signal samples before they calculate a block of B output signal samples. Consequently, the filter is adapted only once every B sampling instants. To reduce the computational complexity the convolution and the adaptation can be performed in the frequency domain.

Besides the advantage of reduced computational complexity, block processing has also disadvantages. Due to the computation of only one adaptation every B samples, the time resolution for control purposes is reduced. If the signal-to-noise ratio changes in the middle of the signal block, for example, the control parameters can be adjusted only to the mean signal-to-noise ratio (averaged over the block length). Especially for a large block length B and therefore for a large reduction of computational complexity, the impacts of the reduced time resolution clearly turn out as a disadvantage.

If the filter update is performed in the frequency domain, a new degree of freedom arises. Each frequency bin of the update of the transformed filter vector can be weighted individually. Besides the advantages of block processing, another inherent disadvantage of this processing structure should also be mentioned—because of the collection of B samples, a delay of respective length is introduced into the signal paths.

Before describing frequency-domain solutions in Section 9.1.2.2, we will start with a brief description of time-domain block processing structures.

9.1.2.1 Time Domain

The complexity of the convolution and the adaptation of a block-processed FIR filter can be reduced by rearranging the additions and multiplications [166]. We will show the basic principle by performing a convolution [3]

$$\begin{aligned} y(n) &= \boldsymbol{x}^{\mathrm{T}}(n)\,\boldsymbol{h}^* \\ &= \big[x(n),\, x(n-1),\, \ldots,\, x(n-N+1)\big] \begin{bmatrix} h_0^* \\ h_1^* \\ \vdots \\ h_{N-1}^* \end{bmatrix} \end{aligned} \tag{9.13}$$

of an input signal $x(n)$ with a fixed filter h_i. For a block size $B = 2$, two convolutions according to Eq. 9.13 have to be performed within one block:

$$\begin{bmatrix} y(n) \\ y(n-1) \end{bmatrix} = \begin{bmatrix} x(n), & \ldots, & x(n-N+1) \\ x(n-1), & \ldots, & x(n-N) \end{bmatrix} \begin{bmatrix} h_0^* \\ \vdots \\ h_{N-1}^* \end{bmatrix}. \tag{9.14}$$

In the following B samples of the output signal $y(n)$ will be combined as

$$\boldsymbol{y}(n) = \big[y(n),\, y(n-1)\big]^{\mathrm{T}}. \tag{9.15}$$

The computational complexity can be reduced by nearly 25% by reusing interim results. Therefore, the entire filter vector $\boldsymbol{h}$ is split into one containing the coefficients with even index and one with the coefficients having an odd index:

$$\boldsymbol{h}_{\text{even}} = \big[h_0,\, h_2,\, \ldots,\, h_{N-2}\big]^{\mathrm{T}}, \tag{9.16}$$

$$\boldsymbol{h}_{\text{odd}} = \big[h_1,\, h_3,\, \ldots,\, h_{N-1}\big]^{\mathrm{T}}. \tag{9.17}$$

Note that the vectors $\boldsymbol{h}_{\text{even}}$ and $\boldsymbol{h}_{\text{odd}}$ contain only half of the coefficients of the original filter vector $\boldsymbol{h}$. Accordingly, also the signal vector is partitioned:

$$\boldsymbol{x}_{\text{even}}(n) = \big[x(n),\, x(n-2),\, \ldots,\, x(n-N+2)\big]^{\mathrm{T}}, \tag{9.18}$$

$$\boldsymbol{x}_{\text{odd}}(n) = \big[x(n-1),\, x(n-3),\, \ldots,\, x(n-N+1)\big]^{\mathrm{T}}, \tag{9.19}$$

where we assume that N is even. Inserting definitions 9.16–9.19 into Eq. 9.14 leads to

$$\boldsymbol{y}(n) = \begin{bmatrix} \boldsymbol{x}_{\text{even}}^{\mathrm{T}}(n) & \boldsymbol{x}_{\text{odd}}^{\mathrm{T}}(n) \\ \boldsymbol{x}_{\text{even}}^{\mathrm{T}}(n-1) & \boldsymbol{x}_{\text{odd}}^{\mathrm{T}}(n-1) \end{bmatrix} \begin{bmatrix} \boldsymbol{h}_{\text{even}}^* \\ \boldsymbol{h}_{\text{odd}}^* \end{bmatrix}. \tag{9.20}$$

[3]Block processing can be also combined with subband structures. In general, subband signals and impulse responses are complex. For this reason, we use the definition of a convolution for complex signals and systems, respectively (see Eq. 5.2).

The complexity can be reduced by using the following properties of the signal vectors:

$$\boldsymbol{x}_{\text{odd}}(n) = \boldsymbol{x}_{\text{even}}(n-1)\,. \tag{9.21}$$

Inserting Eq. 9.21 in Eq. 9.20 and rearranging the summands leads to

$$y(n) = \begin{bmatrix} \left(\boldsymbol{x}_{\text{even}}(n) - \boldsymbol{x}_{\text{odd}}(n)\right)^{\text{T}} \boldsymbol{h}^{*}_{\text{even}} + \boldsymbol{x}^{\text{T}}_{\text{odd}}(n)\left(\boldsymbol{h}^{*}_{\text{odd}} - \boldsymbol{h}^{*}_{\text{even}}\right) \\ -\boldsymbol{x}^{\text{T}}_{\text{odd}}(n)\left(\boldsymbol{h}^{*}_{\text{odd}} - \boldsymbol{h}^{*}_{\text{even}}\right) + \left(\boldsymbol{x}_{\text{even}}(n-2) + \boldsymbol{x}_{\text{odd}}(n)\right)^{\text{T}} \boldsymbol{h}^{*}_{\text{odd}} \end{bmatrix}. \tag{9.22}$$

When comparing the four convolutions of the subsampled signals and filter vectors, it becomes obvious that two of the vector products are equal. The addition and the subtraction of the signal vectors can be done recursively. In this case the amount of memory for storing the signals is increased by 50% while the computational complexity is decreased by 25 percent.

Figure 9.9 shows a possible structure for computing a block version of an FIR filter according to Eq. 9.22. The downsampling and upsampling units will be discussed in more detail in Section 9.1.3.2.1.

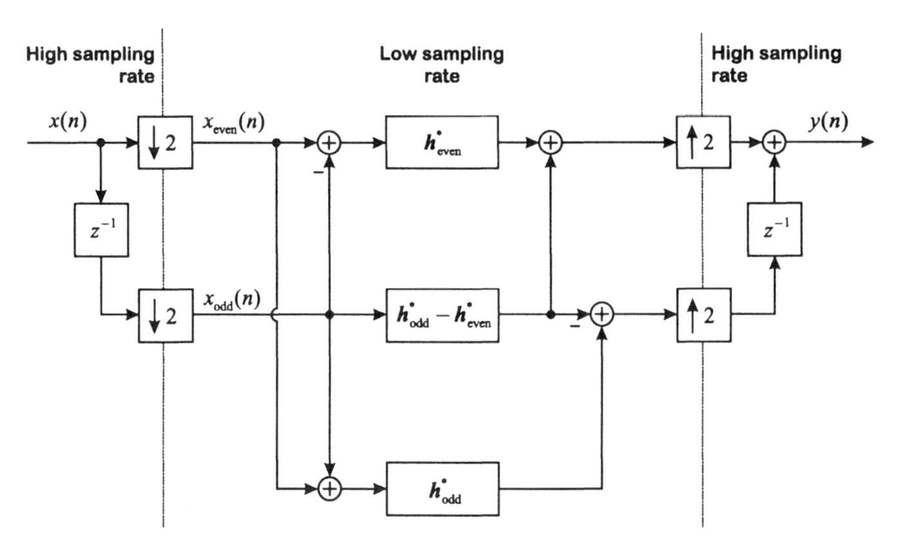

Fig. 9.9 Time-domain block processing.

Rearranging the coefficients according to Eqs. 9.16 and 9.17 is also possible for higher-order radix algorithms leading to further reduction of computational complexity. Furthermore, the basic principle can be extended to block sizes other than $B = 2$, and mixed radix structures are also possible.

9.1.2.2 Frequency Domain

If the computational complexity should be reduced further block processing within the frequency domain can be applied. If the block length B is chosen equal to the filter length N, this processing structure offers the lowest complexity among all methods presented in Section 9.1.

As we will see in Section 9.1.2.2.3, frequency-domain structures offer the possibility to normalize the excitation signal frequency selective. This accelerates the adaptation process. On the other hand, the filter is updated only once within each block. This reduces the speed of convergence. Long story short, both effects nearly compensate. Figure 9.10 compares the convergence of a fullband and a block processing structure—one with white noise and the other with speech excitation. In both cases the power of the residual echo signal decreases with nearly the same speed.

Even if both structures converge with nearly the same speed the block processing scheme requires only around 10% of the processing power of the fullband scheme. The price to be paid for this reduction is also visible in Fig. 9.10—the error signal of the block processing structure is delayed by $B = 1024$ samples. The block processing structure will be discussed in detail in the next three sections.

9.1.2.2.1 Convolution

For the derivation of frequency-domain block processing structures, we start—as in the last section—with a block-oriented notation of the convolution resulting in the microphone signal. The vector of the last B samples of the microphone signal

$$\boldsymbol{y}_{\mathrm{B}}(n) = \left[y(n-B+1),\, \ldots,\, y(n-1),\, y(n)\right]^{\mathrm{T}} \tag{9.23}$$

can be computed as

$$\begin{aligned} \boldsymbol{y}_{\mathrm{B}}(n) &= \begin{bmatrix} x(n-B+1) & \ldots & x(n-N-B+2) \\ \vdots & \ddots & \vdots \\ x(n-1) & \ldots & x(n-N) \\ x(n) & \ldots & x(n-N+1) \end{bmatrix} \begin{bmatrix} h_0^* \\ h_1^* \\ \vdots \\ h_{N-1}^* \end{bmatrix} \\ &= \boldsymbol{X}_{\mathrm{B}}(n)\,\boldsymbol{h}^*\,. \end{aligned} \tag{9.24}$$

If efficient FFT structures should be applied, it is necessary to extend the equations until the signal matrix $\boldsymbol{X}_{\mathrm{B}}(n)$ has circular characteristics. For this reason we append $B-1$ zeros at the end of the coefficient vector

$$\begin{aligned} \tilde{\boldsymbol{h}} &= \left[h_0,\, h_1,,\, \ldots,\, h_{N-1},\, 0,\, \ldots,\, 0\right]^{\mathrm{T}} \\ &= \left[\boldsymbol{h}^{\mathrm{T}},\, 0,\, \ldots,\, 0\right]^{\mathrm{T}}. \end{aligned} \tag{9.25}$$

If the vector containing the microphone signals is also extended by $N-1$ samples

$$\begin{aligned} \tilde{\boldsymbol{y}}_{\mathrm{B}}(n) &= \left[\ldots,\, y(n-B+1),\, \ldots,\, y(n-1),\, y(n)\right]^{\mathrm{T}} \\ &= \left[\ldots,\, \boldsymbol{y}^{\mathrm{T}}(n)\right]^{\mathrm{T}}, \end{aligned} \tag{9.26}$$

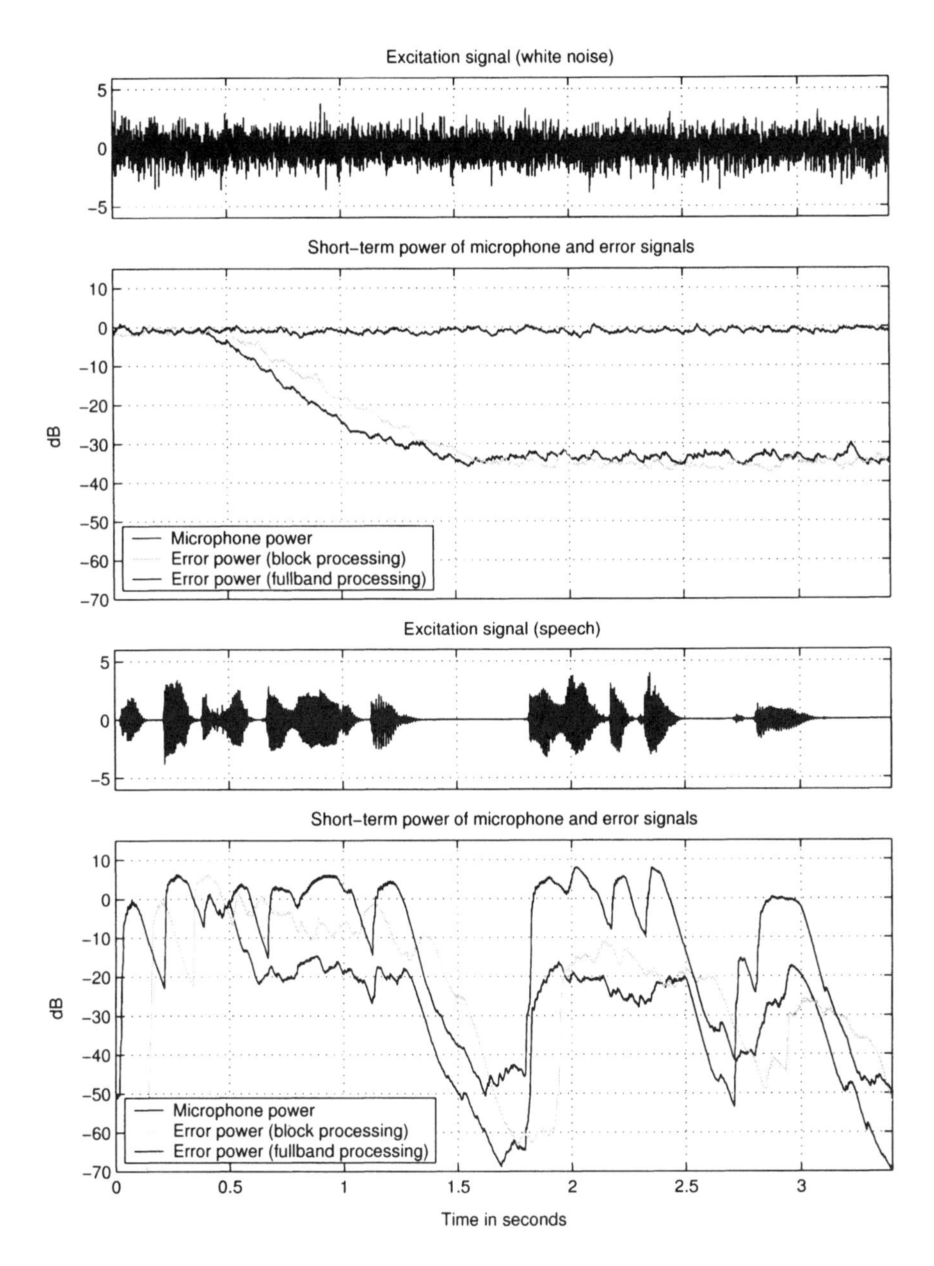

Fig. 9.10 Convergences using fullband NLMS and normalized block LMS. The upper two diagrams show the excitation signal (white noise) as well as the power of the microphone signal and the short-term power of the error signal of the block processing structure and of the fullband structure. The lower two diagrams show the same quantities, but this time speech was used as excitation. The adaptation started after 0.5 second. In both cases the block processing structure introduces a delay of $B = 1024$ samples (approximately 125 ms).

the system of equations can be rewritten with a circular excitation matrix

$$\tilde{\boldsymbol{y}}_{\mathrm{B}}(n) = \tilde{\boldsymbol{X}}_{\mathrm{B}}(n)\, \tilde{\boldsymbol{h}}^{*}\,. \tag{9.27}$$

The first $N-1$ elements of the extended microphone signal vector $\tilde{\boldsymbol{y}}_{\mathrm{B}}(n)$ will be discarded after applying efficient FFT structures. Thus, we do not name these coefficients explicitly. The extended excitation matrix now has circular characteristics:

$$\tilde{\boldsymbol{X}}_{\mathrm{B}}(n) = \begin{bmatrix} x(n-N-B+1) & \dots & x(n-N-B+3) & x(n-N-B+2) \\ \vdots & \ddots & \vdots & \vdots \\ x(n-2) & \dots & x(n) & x(n-1) \\ x(n-1) & \dots & x(n-B-B+1) & x(n) \\ x(n) & \dots & x(n-N-B+2) & x(n-N-B+1) \end{bmatrix}. \tag{9.28}$$

The term "circular characteristics" means that the matrix $\tilde{\boldsymbol{X}}_{\mathrm{B}}(n)$ is completely defined by, for instance, the last row

$$\tilde{\boldsymbol{x}}_{\mathrm{B}}^{\mathrm{T}}(n) = \big[x(n),\, x(n-1),\, \dots,\, x(n-N-B+1)\big]\,. \tag{9.29}$$

All rows above can be generated by rotating the row below by one element to the left. For $B=2$ and $N=3$, Eq. 9.27 can be written as

$$\begin{bmatrix} - \\ - \\ y(n-1) \\ y(n) \end{bmatrix} = \begin{bmatrix} x(n-3) & x(n) & x(n-1) & x(n-2) \\ x(n-2) & x(n-3) & x(n) & x(n-1) \\ x(n-1) & x(n-2) & x(n-3) & x(n) \\ x(n) & x(n-1) & x(n-2) & x(n-3) \end{bmatrix} \begin{bmatrix} h_0^* \\ h_1^* \\ h_2^* \\ 0 \end{bmatrix}. \tag{9.30}$$

In order to extract the last B samples out of the vector $\tilde{\boldsymbol{y}}_{\mathrm{B}}(n)$ without destroying the circular characteristics of the matrix $\tilde{\boldsymbol{X}}_{\mathrm{B}}(n)$ a so-called projection matrix

$$\boldsymbol{P}_y = \big[\boldsymbol{0}_{B,N-1},\, \boldsymbol{I}_{B,B}\big] \tag{9.31}$$

can be applied. By premultiplication with the extended microphone signal vector, we obtain

$$\boldsymbol{y}_{\mathrm{B}}(n) = \boldsymbol{P}_y\, \tilde{\boldsymbol{y}}_{\mathrm{B}}(n)\,. \tag{9.32}$$

Inserting Eq. 9.27 leads to

$$\boldsymbol{y}_{\mathrm{B}}(n) = \boldsymbol{P}_y\, \tilde{\boldsymbol{X}}_{\mathrm{B}}(n)\, \tilde{\boldsymbol{h}}^{*}\,. \tag{9.33}$$

Equation. 9.33 can be solved very efficiently in the frequency domain. Applying a DFT to a signal or impulse response vector of length $N+B-1$ according to

$$\tilde{H}\big(e^{j\frac{2\pi\, m}{N+B-1}}\big) = \sum_{i=0}^{N+B-2} \tilde{h}_i\, e^{-j\frac{2\pi\, im}{N+B-1}}\,, \quad \text{for } m \in \{0,\, \dots,\, N+B-2\} \tag{9.34}$$

can also be written as a matrix–vector multiplication

$$\tilde{\boldsymbol{H}} = \boldsymbol{F}\,\tilde{\boldsymbol{h}}. \tag{9.35}$$

The matrix

$$\boldsymbol{F} = \begin{bmatrix} 1 & 1 & \cdots & 1 \\ 1 & e^{-j\frac{2\pi}{N+B-1}} & \cdots & e^{-j\frac{2\pi\,(N+B-2)}{N+B-1}} \\ \vdots & \vdots & \ddots & \vdots \\ 1 & e^{-j\frac{2\pi\,(N+B-2)}{N+B-1}} & \cdots & e^{-j\frac{2\pi\,(N+B-2)^2}{N+B-1}} \end{bmatrix} \tag{9.36}$$

is called a *Fourier matrix*. In the same way an inverse DFT can be written as

$$\tilde{\boldsymbol{h}} = \boldsymbol{F}^{-1}\,\tilde{\boldsymbol{H}}. \tag{9.37}$$

The inverse Fourier matrix is defined as

$$\boldsymbol{F}^{-1} = \frac{1}{N+B-1}\begin{bmatrix} 1 & 1 & \cdots & 1 \\ 1 & e^{j\frac{2\pi}{N+B-1}} & \cdots & e^{j\frac{2\pi\,(N+B-2)}{N+B-1}} \\ \vdots & \vdots & \ddots & \vdots \\ 1 & e^{j\frac{2\pi\,(N+B-2)}{N+B-1}} & \cdots & e^{j\frac{2\pi\,(N+B-2)^2}{N+B-1}} \end{bmatrix}. \tag{9.38}$$

Consequently, we have

$$\boldsymbol{F}^{-1}\,\boldsymbol{F} = \boldsymbol{I}. \tag{9.39}$$

Inserting the unity matrix as $\boldsymbol{F}^{-1}\boldsymbol{F}$ 2 times in Eq. 9.33 leads to

$$\begin{aligned} \boldsymbol{y}_{\mathrm{B}}(n) &= \boldsymbol{P}_y\left(\boldsymbol{F}^{-1}\boldsymbol{F}\right)\tilde{\boldsymbol{X}}_{\mathrm{B}}(n)\left(\boldsymbol{F}^{-1}\boldsymbol{F}\right)\tilde{\boldsymbol{h}}^{*} \\ &= \boldsymbol{P}_y\,\boldsymbol{F}^{-1}\left(\boldsymbol{F}\,\tilde{\boldsymbol{X}}_{\mathrm{B}}(n)\,\boldsymbol{F}^{-1}\right)\left(\boldsymbol{F}\,\tilde{\boldsymbol{h}}^{*}\right). \end{aligned} \tag{9.40}$$

Now an important property of Fourier matrices can be exploited. If a circular matrix is premultiplied with a Fourier matrix and postmultiplied with an inverse Fourier matrix the resulting matrix is diagonal. Furthermore, the diagonal elements are equal to the Fourier transform of the last row of the circular matrix. Thus, we have

$$\boldsymbol{F}\,\tilde{\boldsymbol{X}}_{\mathrm{B}}(n)\,\boldsymbol{F}^{-1} = \mathrm{diag}\Big\{\mathrm{DFT}\Big\{\tilde{\boldsymbol{x}}_{\mathrm{B}}(n)\Big\}\Big\}. \tag{9.41}$$

Inserting this property into Eq. 9.40, we obtain a very efficient procedure to compute the vector containing the last B microphone signals:

$$\boldsymbol{y}_{\mathrm{B}}(n) = \text{last } B \text{ elements of IDFT}\Big\{\mathrm{DFT}\Big\{\tilde{\boldsymbol{x}}_{\mathrm{B}}(n)\Big\} \otimes \mathrm{DFT}\Big\{\tilde{\boldsymbol{h}}^{*}\Big\}\Big\}. \tag{9.42}$$

The sign $\otimes$ denotes elementwise multiplication. If $N + B - 1$ is a power of 2, efficient FFT structures can be used. Before we transfer this result to the adaptation process (see Section 9.1.2.2.3), we will first analyze the computational complexity of this type of block processing structures.

9.1.2.2.2 Computational Complexity and Memory Requirements

Computing the matrix–vector product $\boldsymbol{y}_{\mathrm{B}}(n) = \boldsymbol{X}_{\mathrm{B}}(n)\,\boldsymbol{h}^*$ directly would require approximately NB additions and NB multiplications. If real input signals and a real impulse response are assumed, all addition and multiplications would be real ones, too. Normalizing this result to a single sample period leads to the following fullband complexity and memory requirements:

$$C_{\mathrm{FB,\,conv}}(N) \sim N\,, \tag{9.43}$$
$$M_{\mathrm{FB,\,conv}}(N) \sim N\,. \tag{9.44}$$

For the memory requirements only the memory for storing the input signal $x(n)$ is counted.[4] If the microphone vector $\boldsymbol{y}_{\mathrm{B}}(n)$ is computed according to Eq. 9.42, only one FFT, one IFFT, and a complex vector product of length $N + B + 1$ have to be computed every B samples. We assumed the filter vector $\boldsymbol{h}$ to be fixed. Thus, the Fourier transform of this vector has to be computed just once and will not be counted here. If $N + B - 1$ is not equal to a power of 2, zero padding can be applied. With

$$L_{\mathrm{B}}(N, B) = 2^{\lceil \log_2(N+B-1) \rceil} \tag{9.45}$$

we denote the power of 2 that is closest from above to $N + B - 1$. The signs $\lceil \cdots \rceil$ denote rounding toward infinity. If L_{Op} describes the average ratio between real and complex instructions, we can express the computational complexity (normalized to a single sample period) and the memory requirements of block processing structures as

$$C_{\mathrm{BP,conv}}(N, B) \sim 2L_{\mathrm{B}}(N, B)\,\log_2\big(L_{\mathrm{B}}(N, B)\big) + L_{\mathrm{Op}}\,L_{\mathrm{B}}(N, B)\,, \tag{9.46}$$
$$M_{\mathrm{BP,conv}}(N, B) \sim 4L_{\mathrm{B}}(N, B)\,. \tag{9.47}$$

Radix 2 algorithms were assumed for the FFTs. Thus, their complexity was estimated with $M\log_2(M)$, where M denotes the length of the input vector. In order to show the advantages here, the memory requirements and the complexities of fullband and block processing structures, are plotted in Fig. 9.11 for the special case $B = N$. Furthermore, an average complexity ratio for instructions operating real and complex data of $L_{\mathrm{Op}} = 3$ was assumed.

The immense decrease in complexity for large filter orders is clearly visible. For a filter length $N = 1000$, for instance, the computational complexity can be reduced by more than 90%. Nevertheless, the memory requirements are increased by more than 300% and also the delay introduced by block processing cannot be neglected.

[4]The impulse response is assumed to be fixed. Thus, the coefficients can be stored in nonversatile memory.

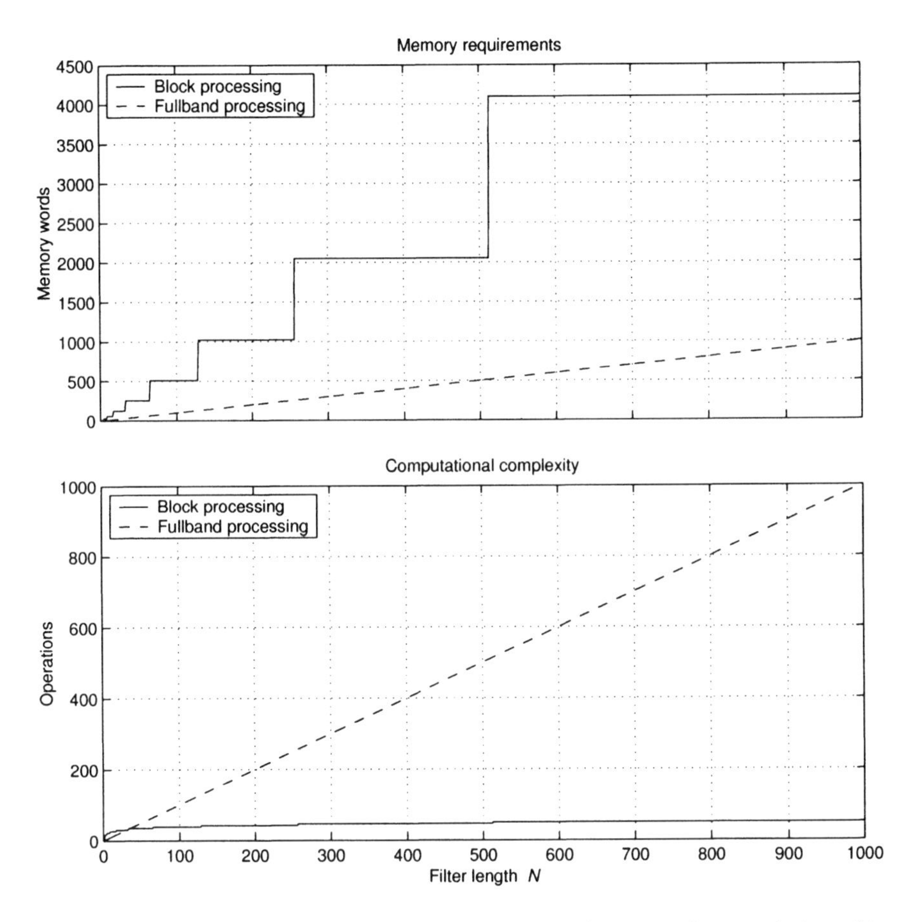

Fig. 9.11 Computational complexities and memory requirements of a convolution with a fixed coefficient vector by means of fullband and block processing in the frequency domain. It is assumed that the filter length N is equal to the block size B, resulting in minimal complexity.

9.1.2.2.3 Adaptation and Constraint

The same techniques that have been applied to the block convolution can also be used to reduce the complexity of the adaptation process. In general, efficient block structures can be derived for each algorithm presented in Chapter 7. We will describe here only the block version of the LMS algorithm. As in the derivation of efficient structures for performing a convolution, we start with the definition of a vector containing the last B samples of the error signal

$$\boldsymbol{e}_{\mathrm{B}}(n) = \left[e(n-B+1),\, \ldots,\, e(n-1),\, e(n)\right]^{\mathrm{T}}. \tag{9.48}$$

For the derivation of the block LMS algorithm, the expected squared L_2 norm of this error vector is minimized:

$$\mathrm{E}\left\{ \boldsymbol{e}_{\mathrm{B}}^{\mathrm{H}}(n)\, \boldsymbol{e}_{\mathrm{B}}(n) \right\} \longrightarrow \min\,. \tag{9.49}$$

Applying the same derivation as in the sample-by-sample LMS version (see Chapter 7) and assuming that the filter order N is equal to the block size B leads to the following update of the filter coefficients every B samples:

$$\widehat{\boldsymbol{h}}(n+B) = \widehat{\boldsymbol{h}}(n) + \tilde{\mu}\, \boldsymbol{X}_{\mathrm{B}}(n)\, \boldsymbol{e}_{\mathrm{B}}^{*}(n)\,. \tag{9.50}$$

The vector containing the coefficients of the adaptive filter is defined as

$$\widehat{\boldsymbol{h}}(n) = \left[\hat{h}_0(n),\, \hat{h}_1(n),\, \ldots,\, \hat{h}_{N-1}(n)\right]^{\mathrm{T}} \tag{9.51}$$

where $\tilde{\mu}$ describes the stepsize parameter. If the same techniques for reducing the complexity as in the convolution part should be applied, an extended error vector according to

$$\begin{aligned} \tilde{\boldsymbol{e}}_{\mathrm{B}}(n) &= \left[0,\, \ldots,\, 0,\, e(n-B+1),\, \ldots,\, e(n-1),\, e(n)\right]^{\mathrm{T}} \\ &= \left[0,\, \ldots,\, 0,\, \boldsymbol{e}^{\mathrm{T}}(n)\right]^{\mathrm{T}} \end{aligned} \tag{9.52}$$

is introduced. The new error vector $\tilde{\boldsymbol{e}}_{\mathrm{B}}(n)$ consists of $N+B-1$ elements. By inserting the new error vector and the extended excitation matrix according to Eq. 9.28 and by using a projection matrix

$$\boldsymbol{P}_{\hat{h}} = \left[\boldsymbol{I}_{N,N},\, \boldsymbol{0}_{B-1,N}\right], \tag{9.53}$$

we can replace the filter update in Eq. 9.50 by

$$\widehat{\boldsymbol{h}}(n+B) = \widehat{\boldsymbol{h}}(n) + \tilde{\mu}\, \boldsymbol{P}_{\hat{h}}\, \tilde{\boldsymbol{X}}_{\mathrm{B}}(n)\, \tilde{\boldsymbol{e}}_{\mathrm{B}}^{*}(n)\,. \tag{9.54}$$

Introducing twice a unity matrix in terms of $\boldsymbol{F}^{-1}\boldsymbol{F}$

$$\begin{aligned} \widehat{\boldsymbol{h}}(n+B) &= \widehat{\boldsymbol{h}}(n) + \tilde{\mu}\, \boldsymbol{P}_{\hat{h}} \left(\boldsymbol{F}^{-1}\boldsymbol{F}\right) \tilde{\boldsymbol{X}}_{\mathrm{B}}(n) \left(\boldsymbol{F}^{-1}\boldsymbol{F}\right) \tilde{\boldsymbol{e}}_{\mathrm{B}}^{*}(n) \\ &= \widehat{\boldsymbol{h}}(n) + \tilde{\mu}\, \boldsymbol{P}_{\hat{h}}\, \boldsymbol{F}^{-1} \left(\boldsymbol{F}\, \tilde{\boldsymbol{X}}_{\mathrm{B}}(n)\, \boldsymbol{F}^{-1}\right) \left(\boldsymbol{F}\, \tilde{\boldsymbol{e}}_{\mathrm{B}}^{*}(n)\right) \end{aligned} \tag{9.55}$$

and exploiting the properties of Fourier matrices leads to a very efficient implementation of the filter update:

$$\widehat{\boldsymbol{h}}(n+B) = \widehat{\boldsymbol{h}}(n) + \tilde{\mu}\, \boldsymbol{P}_{\hat{h}}\, \mathrm{IDFT}\Big\{ \mathrm{DFT}\big\{\tilde{\boldsymbol{x}}_{\mathrm{B}}(n)\big\} \otimes \mathrm{DFT}\big\{\tilde{\boldsymbol{e}}_{\mathrm{B}}^{*}(n)\big\} \Big\}. \tag{9.56}$$

If the adaptation process is excited by speech, very small values for the stepsize parameter $\tilde{\mu}$ have to be chosen. In analogy to the normalized time-domain LMS algorithm, frequency-selective power normalization is usually applied. This is achieved by smoothing the squared excitation spectrum according to

$$\boldsymbol{P}_{x}(n) = \gamma\, \boldsymbol{P}_{x}(n-B) + (1-\gamma) \left|\mathrm{DFT}\big\{\tilde{\boldsymbol{x}}_{\mathrm{B}}(n)\big\}\right|^2. \tag{9.57}$$

In the normalized version of the block LMS the original stepsize $\tilde{\mu}$ is replaced by a matrix control

$$\tilde{\mu} \longrightarrow \mu \, \mathrm{diag}\{\boldsymbol{P}_x^{-1}(n)\}\,. \tag{9.58}$$

The new stepsize parameter μ can be chosen from the interval

$$0 \leq \mu \leq 1\,. \tag{9.59}$$

Note that the matrix $\boldsymbol{P}_x(n)$ is a diagonal matrix. Thus, inverting the matrix requires only $N + B - 1$ divisions (every B samples). Finally, the normalized adaptation of the filter coefficients can be written as

$$\widehat{\boldsymbol{h}}(n+B) = \widehat{\boldsymbol{h}}(n) + \boldsymbol{P}_{\hat{h}} \; \mu \; \mathrm{IDFT}\Big\{\mathrm{DFT}\big\{\tilde{\boldsymbol{x}}_{\mathrm{B}}(n)\big\} \otimes \boldsymbol{P}_x^{-1}(n) \otimes \mathrm{DFT}\big\{\tilde{\boldsymbol{e}}_{\mathrm{B}}^{*}(n)\big\}\Big\}\,. \tag{9.60}$$

Figure 9.12 shows the signal flow of the entire algorithm. Note that the fast convolution is applied to the adaptive filter vector $\widehat{\boldsymbol{h}}(n)$ and not—as in Section 9.1.2.2.1—to the vector $\boldsymbol{h}$. The adaptation process can be performed in the time domain—as described above—or in the frequency domain. The latter can be derived by multiplying Eq. 9.33 with a Fourier matrix of appropriate size. Figure 9.12 shows a convenient way to combine the convolution and the adaptation process by keeping the filter vector $\widehat{\boldsymbol{h}}(n)$ in the frequency domain.

Applying an inverse DFT to the filter spectrum, multiplying with the projection matrix $\boldsymbol{P}_{\hat{h}}$, appending zeros, and applying a DFT finally is often called *constraint*. If these operations are omitted, the final misalignment is slightly worsened. On the other hand, two transforms can be omitted and the computational complexity can be reduced.

Finally, the normalized block LMS algorithm is summarized in Table 9.2.

9.1.3 Subband Cancellation

Another possibility to reduce the computational complexity is to apply subband processing. By using filterbanks [47, 229, 230], the excitation signal $x(n)$ and the microphone signal $y(n)$ are split up into several subbands (see Fig. 9.13). Depending on the properties of the lowpass, bandpass, and highpass filters, the sampling rate in the subbands can be reduced. According to this reduction, the adaptive filters can be shortened. Instead of filtering the fullband signal and adapting one filter at the high sampling rate, M (number of subbands) convolutions and adaptations with subsampled signals are performed in parallel at reduced sampling rates.

The subscript $\mu \in \{0, \ldots, M-1\}$ is used to address the subband channels. The final output signal $\hat{s}(n) = e(n)$ is obtained by recombining and upsampling the subband signals $e_\mu(n)$.

In contrast to block (frequency-domain) processing, the subband structure offers the system designer a second additional degree of freedom. Besides the possibility

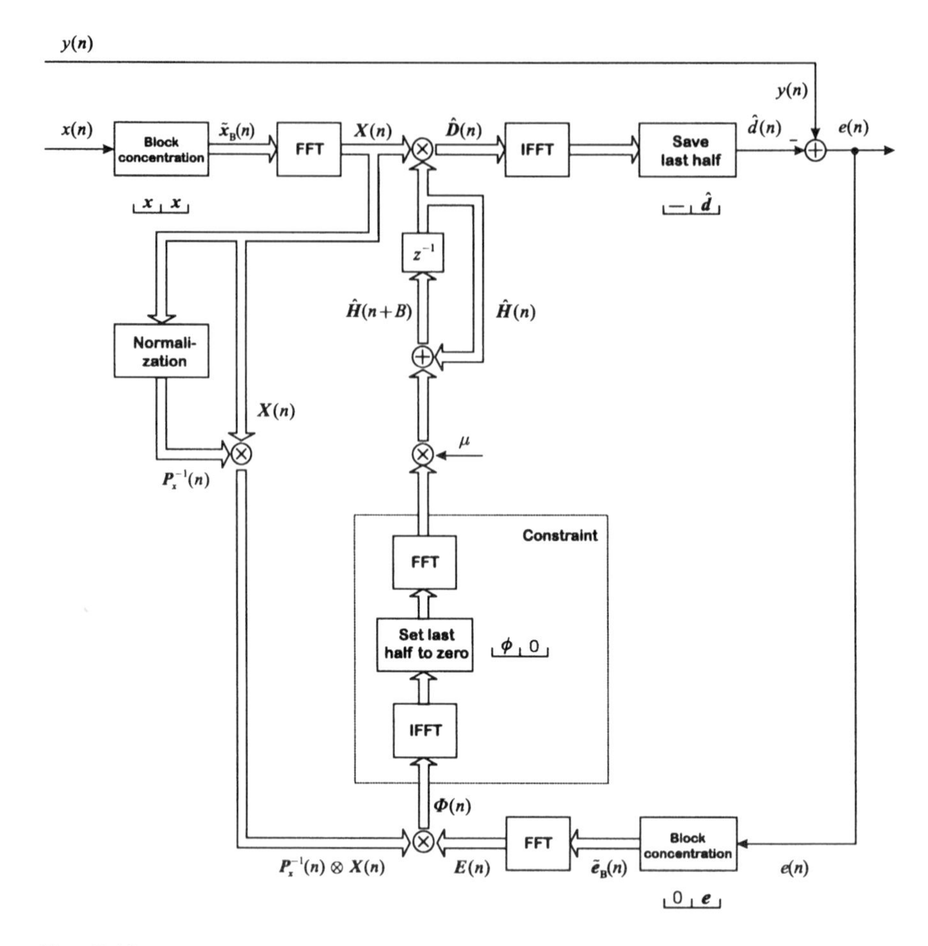

Fig. 9.12 Normalized fast block LMS algorithm for real input data. The following abbreviations have been used: $\boldsymbol{X}(n) = \text{DFT}\{\boldsymbol{x}_\text{B}(n)\}$, $\boldsymbol{E}(n) = \text{DFT}\{\tilde{\boldsymbol{e}}_\text{B}(n)\}$, and $\hat{\boldsymbol{H}}(n) = \text{DFT}\{[\hat{\boldsymbol{h}}(n), \mathbf{0}_{B-1,1}]\}$. The subsampling units have been omitted. Thus, delaying the filter vector $\hat{\boldsymbol{H}}(n)$ by B fullband samples requires only one (vector) delay element.

of having a frequency dependent control, dimensioning of the subsystem can also be frequency-selective. As an example, the orders of the adaptive filters can be adjusted individually in each channel according to the statistical properties of the excitation signal, the measurement noise, and the impulse response of the system to be identified.

Using subsampled signals leads to a reduction of computational complexity. All necessary forms of control and detection can operate independently in each channel. The price to be paid for these advantages is—as in block processing structures—a delay introduced into the signal path by the analysis and synthesis filterbanks.

Table 9.2 Normalized fast block LMS algorithm

Algorithmic part	Corresponding equations
Filtering	$\tilde{\boldsymbol{x}}_{\mathrm{B}}(n) = [x(n),\, x(n-1),\, \ldots,\, x(n-N-B+1)]^{\mathrm{T}}$ $\tilde{\boldsymbol{X}}_{\mathrm{B}}(n) = \mathrm{diag}\,[\mathrm{FFT}\,\{\tilde{\boldsymbol{x}}_{\mathrm{B}}(n)\}]$ $\hat{\boldsymbol{d}}(n) = \text{last } B \text{ elements of IFFT} \left\{\tilde{\boldsymbol{X}}_{\mathrm{B}}(n)\,\hat{\boldsymbol{H}}(n)\right\}$
Error signal	$\boldsymbol{y}(n) = [y(n),\, y(n-1),\, \ldots,\, y(n-B+1)]^{\mathrm{T}}$ $\boldsymbol{e}(n) = \boldsymbol{y}(n) - \hat{\boldsymbol{d}}(n)$ $\boldsymbol{E}(n) = \mathrm{IFFT}\,\{[\boldsymbol{0}_{1\times(N-1)},\, \boldsymbol{e}^{*}(n)]\}$
Normalization	$\boldsymbol{P}(n) = \gamma\,\boldsymbol{P}(n-1) + (1-\gamma)\,\boldsymbol{X}(n)\,\boldsymbol{X}^{\mathrm{H}}(n)$
Filter update	$\boldsymbol{\Phi}(n) = \boldsymbol{P}^{-1}(n)\,\boldsymbol{X}^{H}(n)\,\boldsymbol{E}(n)$ $\boldsymbol{\phi}(n) = \text{first } N \text{ elements of IFFT}\,\{\boldsymbol{\Phi}(n)\}$ $\hat{\boldsymbol{H}}(n+B) = \hat{\boldsymbol{H}}(n) + \mu\,\mathrm{FFT}\,\{[\boldsymbol{\phi}(n),\, \boldsymbol{0}_{1\times(B-1)}]\}$
All equations are computed only if $n \bmod B \equiv 0$	

9.1.3.1 Computational Complexity and Memory Requirements

The main advantage of applying subband processing (compared to fullband processing) is the reduction of computational complexity. The reduction is not as large as in block processing structures, but the memory requirements and the signal delay are smaller. To illustrate this we will use—as in Section 9.1.2.2.2—a convolution of a real input signal with a real fixed FIR filter of length N as an example. The computational complexity $C_{\mathrm{FB,conv}}(N)$ of such an operation is proportional to N for

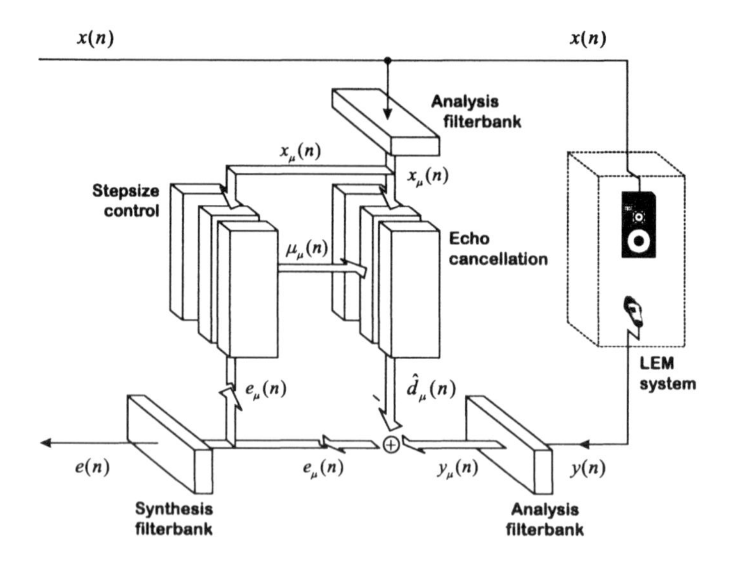

Fig. 9.13 Structure of an acoustic echo control system operating in subbands. By using two analysis filterbanks, the far-end speaker and the microphone signal are decomposed into M subbands. In each subband, a digital replica of the echo signal is generated by convolution of the subband excitation signal and the estimated subband impulse response. After subtraction of the estimated subband echo signals from the microphone signals the fullband output signal $e(n)$ is synthesized using a third filterbank.

time-domain processing. The required memory $M_{\text{FB, conv}}(N)$ is also proportional to N, if only volatile memory is counted:[5]

$$C_{\text{FB, conv}}(N) \sim N\,, \tag{9.61}$$

$$M_{\text{FB, conv}}(N) \sim N\,. \tag{9.62}$$

If the input signal is split up into M subbands, downsampling can be applied. The subsampling rate r can be chosen—depending on the analysis and synthesis filters—close to the number of channels. Later in this chapter we will present a filterbank with $M = 16$ channels (2 real ones and 14 complex ones) and a subsampling rate $r = 12$. Due to spectral symmetries of real input signals, only half of the complex subbands need to be processed.

Because of the subsampling, the memory requirements are reduced for real channels to N/r words and for complex channels to $2N/r$ words. When summing up all necessary channels, we obtain a slight degradation compared to the fullband memory requirements.

[5]The memory for storing the coefficients of the fixed filter is not counted here.

An improvement can be achieved concerning computational complexity. Subband convolutions have to be performed only every r fullband samples. Furthermore, only subsampled signals are used, leading to a complexity (per fullband sample period) of N/r^2 for real channels and $L_{\text{Op}}\, N/r^2$ for complex channels, respectively. As in Section 9.1.2.2.2, L_{Op} corresponds to the complexity ratio of complex and real operations. For additions, for instance, one gets $L_{\text{Add}} = 2$, for multiplications $L_{\text{Mul}} = 6$ (4 real multiplications and 2 additions), and for loading and storing of data $L_{\text{Mem}} = 2$.

When summing over all channels, we obtain for the computational complexity and for the memory requirements of subband processing

$$C_{\text{SB,conv}}(N) \sim \left(2 + L_{\text{Op}}\left(\frac{M}{2} - 1\right)\right)\frac{N}{r^2} + C_{\text{An,Syn}}\,, \tag{9.63}$$

$$M_{\text{SB,conv}}(N) \sim 2\frac{N}{r} + \left(\frac{M}{2} - 1\right)\frac{2N}{r} + M_{\text{An,Syn}}\,, \tag{9.64}$$

where $C_{\text{An,Syn}}$ and $M_{\text{An,Syn}}$ denote the computational complexity and the memory requirements of the analysis and the synthesis filterbanks. For the DFT-modulated polyphase filterbanks with $M = 16$ channels, a subsampling rate $r = 12$, and a prototype lowpass filter length $N_{\text{TP}} = 128$, we get

$$C_{\text{An,Syn}} = 32\,, \tag{9.65}$$

$$M_{\text{An,Syn}} = 256\,. \tag{9.66}$$

Details about the polyphase filterbank will be given in Section 9.1.3.2. In Fig. 9.14, the complexity and the memory demands of FIR filtering in fullband and in subband structures are compared with respect to the filter length N. An average complexity ratio between complex and real operations of $L_{\text{Op}} = 3$ has been used for the comparison.

In order to clarify the details of the comparison ,we will use a convolution with a fixed filter of length $N = 1024$ as an example. In subband structures (with $M = 16$, $r = 12$, and $N_{\text{LP}} = 128$ the memory requirements increase by about 40%. On the other hand the complexity is reduced by 80% (compared to fullband structures). The price to be paid for this reduction is a delay of 128 samples, which is 16 ms at 8 kHz sampling rate.

9.1.3.2 Filterbanks

Before describing special features of subband systems for echo cancellation, we will briefly introduce some aspects of filterbank theory. First, upsampling and downsampling units will be explained, because they are basic building blocks of subband systems. Furthermore, two types of filterbanks will be introduced: quadrature mirror filterbanks (QMF) and oversampled, DFT-modulated filterbanks. Whereas the first

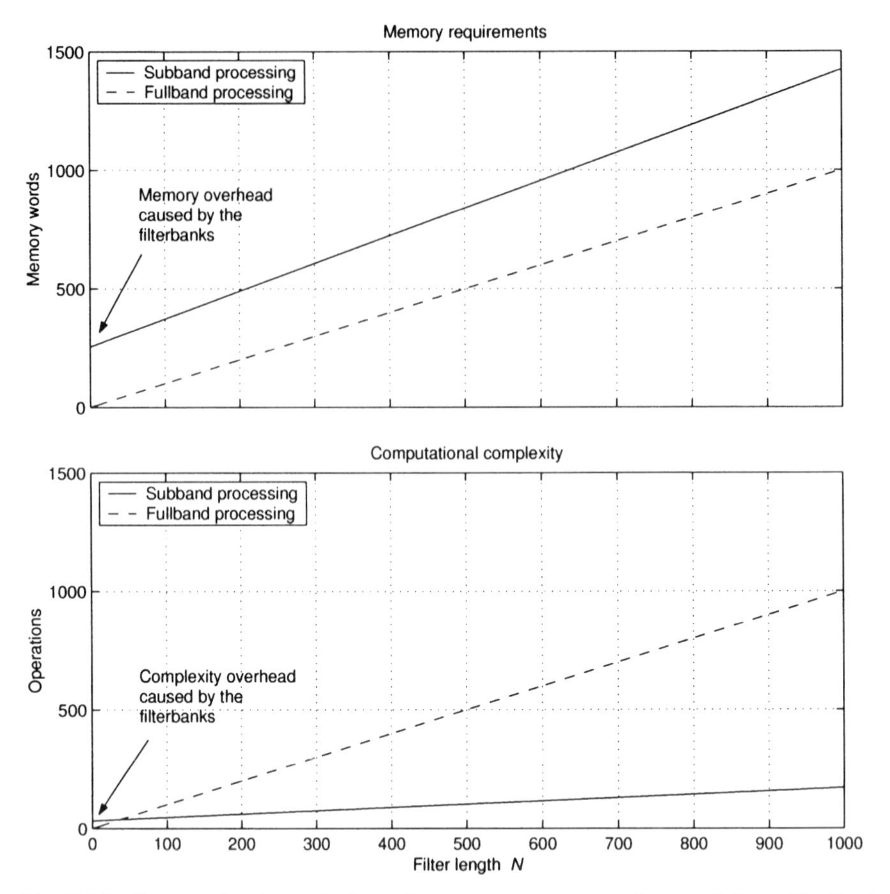

Fig. 9.14 Computational complexity and memory requirements of a convolution with a fixed coefficient vector by means of fullband and subband processing. The analysis and synthesis filterbanks were designed with $M = 16$ channels, a subsampling rate of $r = 12$, and a prototype lowpass filter length of $N_{\text{LP}} = 128$.

one is a special type of critically subsampled[6] filterbank, the latter type is most often used for the application of acoustic echo cancellation. For this reason, a detailed design procedure for this filterbank type is presented in Appendix B.

9.1.3.2.1 Basics

Besides additions and multiplication units as well as delay elements, two further building blocks are used in multirate signal processing: upsampling and downsampling

[6]The term *critically subsampled* means that the subsampling rate r is equal to the number of subbands M.

pling units. Both will be explained briefly in this section. Further details can be found, for example, in [47, 230].

Downsampling by a factor r is performed by simply removing $r-1$ neighboring samples out of r samples of the input signal $x(n)$:

$$y(n) = x(n\,r)\,. \tag{9.67}$$

In the following figures we will use the symbol presented in Fig. 9.15 for downsampling units. When applying a Fourier transform to the signal $y(n)$ we obtain for the

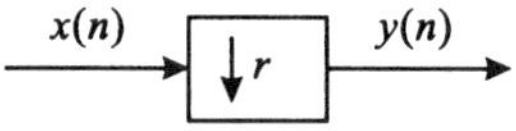

Fig. 9.15 Downsampling by a factor r.

corresponding spectrum

$$\begin{aligned} Y\big(e^{j\Omega}\big) &= \sum_{n=-\infty}^{\infty} y(n)\, e^{j\Omega n} \\ &= \sum_{n=-\infty}^{\infty} x(n\,r)\, e^{j\Omega n}. \end{aligned} \tag{9.68}$$

In the second line Eq. 9.67 has been inserted. By using the finite geometric series

$$\frac{1}{r}\sum_{m=0}^{r-1} e^{j\frac{2\pi}{r}nm} = \begin{cases} 1\,, & \text{if}\, n \bmod r \equiv 0\,, \\ 0\,, & \text{else}\,, \end{cases} \tag{9.69}$$

it is possible to compute the spectrum of the downsampled signal in dependence of the input spectrum:

$$\begin{aligned} Y\big(e^{j\Omega}\big) &= \sum_{n=-\infty}^{\infty} x(n) \left(\frac{1}{r}\sum_{m=0}^{r-1} e^{j\frac{2\pi}{r}nm}\right) e^{j\Omega\frac{n}{r}} \\ &= \frac{1}{r}\sum_{m=0}^{r-1}\sum_{n=-\infty}^{\infty} x(n)\; e^{j\left(\frac{\Omega}{r}-\frac{2\pi}{r}m\right)n} \\ &= \frac{1}{r}\sum_{m=0}^{r-1} X\Big(e^{j\left(\frac{\Omega}{r}-\frac{2\pi}{r}m\right)}\Big)\,. \end{aligned} \tag{9.70}$$

This equation shows that the output spectrum $Y(e^{j\Omega})$ is the sum of r stretched and shifted images of the input spectrum $X(e^{j\Omega})$. Thus, any overlap of the stretched input spectra results in aliasing. If this should be avoided it is necessary that the input spectrum $X(e^{j\Omega})$ is zero within the interval

$$\frac{r}{\pi} \le |\Omega| \le \pi\,. \tag{9.71}$$

Condition 9.71 is usually fulfilled by applying appropriate lowpass filtering before downsampling the signal $x(n)$.

Upsampling of a signal by a factor r is performed by inserting $r-1$ zeros between two adjacent samples of the input signal

$$y(n) = \begin{cases} x\left(\frac{n}{r}\right), & \text{if } n \bmod r \equiv 0, \\ 0, & \text{else}. \end{cases} \tag{9.72}$$

The symbol presented in Fig. 9.16 will be used whenever we draw upsampling units in the following figures. The spectrum of the upsampled signal $y(n)$ can be computed

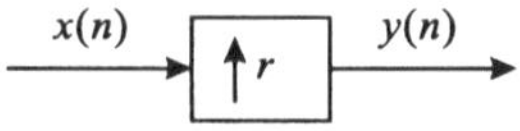

x(n) ↑r y(n)

Fig. 9.16 Upsampling by a factor r.

by applying a Fourier transform:

$$Y\left(e^{j\Omega}\right) = \sum_{n=-\infty}^{\infty} y(n)\, e^{j\Omega n}. \tag{9.73}$$

By replacing the index $n = m\,r$ the dependence of the spectrum of the upsampled signal on the input spectrum can be shown to be

$$\begin{aligned} Y\left(e^{j\Omega}\right) &= \sum_{m=-\infty}^{\infty} x(m)\, e^{j\Omega r m} \\ &= X\left(e^{j\Omega r}\right). \end{aligned} \tag{9.74}$$

Due to the 2π periodicity of $X(e^{j\Omega})$, the spectrum of the upsampled signal $Y(e^{j\Omega})$ consists of the compressed input spectrum as well as $r-1$ compressed replica.

9.1.3.2.2 Quadrature Mirror Filterbanks

In this section we describe a first type of filterbanks called *quadrature mirror filterbanks* (QMF). This filterbank type is based on tree structures in which the incoming signal is successively divided into subbands with smaller bandwidth at each state of the tree. Fig. 9.17 illustrates a simple example of a two-stage tree structure for a four-band quadrature mirror (analysis) filterbank.

The input signal $y(n)$ is filtered in the first stage of the tree by a pair of complementary lowpass and highpass filters, $g_{\text{LP},i}$ and $g_{\text{HP},i}$, respectively. These filters divide the input spectrum $S_{yy}(\Omega)$ into two parts of equal bandwidth as depicted in the lower part of Fig. 9.17. The sample rate of both signals is reduced by a factor of 2.

For the spectra of the downsampled signals $\tilde{y}_0(n)$ and $\tilde{y}_1(n)$, we obtain the following by applying Eq. 9.70 with $r = 2$:

$$\tilde{Y}_0\left(e^{j\Omega}\right) = \frac{1}{2}\left[Y\left(e^{j\frac{\Omega}{2}}\right) G_{\text{LP}}\left(e^{j\frac{\Omega}{2}}\right) + Y\left(e^{j(\frac{\Omega}{2}-\pi)}\right) G_{\text{LP}}\left(e^{j(\frac{\Omega}{2}-\pi)}\right)\right], \tag{9.75}$$

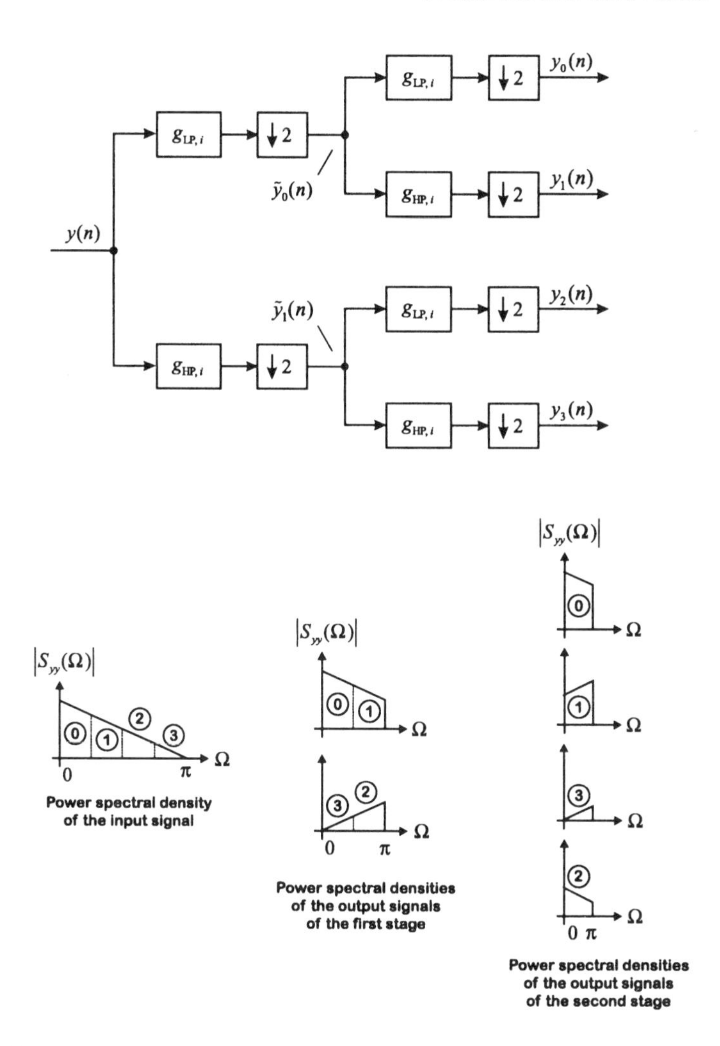

Fig. 9.17 Example of a two-stage tree structure resulting in a four-band uniform quadrature mirror filterbank.

$$\tilde{Y}_1\left(e^{j\Omega}\right) = \frac{1}{2}\left[Y\left(e^{j\frac{\Omega}{2}}\right) G_{\text{HP}}\left(e^{j\frac{\Omega}{2}}\right) + Y\left(e^{j\left(\frac{\Omega}{2}-\pi\right)}\right) G_{\text{HP}}\left(e^{j\left(\frac{\Omega}{2}-\pi\right)}\right)\right]. \quad (9.76)$$

If we assume that the lowpass filter has sufficient attenuation for $\Omega \in [\frac{\pi}{2},\ \pi]$ and the highpass filter for $\Omega \in [0,\ \frac{\pi}{2}]$, respectively, the spectra can be approximated by

$$\tilde{Y}_0\left(e^{j\Omega}\right) \approx \frac{1}{2} Y\left(e^{j\frac{\Omega}{2}}\right) G_{\text{LP}}\left(e^{j\frac{\Omega}{2}}\right), \quad (9.77)$$

$$\tilde{Y}_1\left(e^{j\Omega}\right) \approx \frac{1}{2} Y\left(e^{j\left(\frac{\Omega}{2}-\pi\right)}\right) G_{\text{HP}}\left(e^{j\left(\frac{\Omega}{2}-\pi\right)}\right). \quad (9.78)$$

This equation shows that the highpass branch of the first stage produces an inverted part of the original spectrum. In the second stage of the tree the signals $\tilde{y}_0(n)$ and $\tilde{y}_1(n)$ are filtered again by complementary pairs of lowpass and highpass filters. The sampling rate of the resulting signals is again reduced by a factor of 2. Again both highpass spectra are inverted. The lower part of Fig. 9.17 shows the four final spectra (again all aliasing components were neglected). Since the four output signals, $y_0(n)$ to $y_3(n)$ are reduced in sample rate by a factor of 4, the resulting design is a so-called critically sampled filterbank.

Figure 9.18 shows another example of a tree-structured quadrature mirror filterbank. Also four subbands are computed and all bands are critically subsampled, but this time the bandwidth of the channels is octave-spaced. Only the upper branches (i.e., the lowpass channels) of each stage are subdivided, resulting in a nonuniform band arrangement. In this filterbank realization all channels—except for the upper most two subbands—are sampled at different sampling rates.

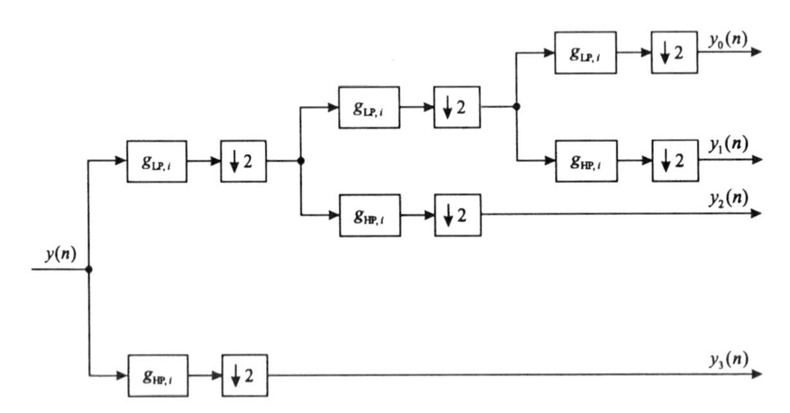

Fig. 9.18 Example of a three-stage tree structure for a four-band nonuniform quadrature mirror filterbank.

Fast realizations exist for most types of quadrature mirror filterbanks [47]. Nonuniform subband decomposition by means of this type of filterbank is often utilized in coding of speech and audio signals [177, 220].

When analyzing quadrature mirror filterbanks it is sufficient to discuss a two-band analysis–synthesis structure as depicted in Fig. 9.19. The filters $g_{\text{LP},i}$ and $g_{\text{HP},i}$ are antialiasing filters within the analysis stage and $\tilde{g}_{\text{LP},i}$ and $\tilde{g}_{\text{HP},i}$ are antiimaging filters of the synthesis stage, respectively. We assume a through-connected analysis synthesis system.

By using Eqs. 9.74, 9.75, and 9.76, we obtain the following for the spectra of the upsampled signals $\bar{s}_\mu(n)$:

$$\bar{S}_0\left(e^{j\Omega}\right) = \frac{1}{2}\left[Y\left(e^{j\Omega}\right) G_{\text{LP}}\left(e^{j\Omega}\right) + Y\left(e^{j(\Omega-\pi)}\right) G_{\text{LP}}\left(e^{j(\Omega-\pi)}\right)\right], \quad (9.79)$$

$$\bar{S}_1\left(e^{j\Omega}\right) = \frac{1}{2}\left[Y\left(e^{j\Omega}\right) G_{\text{HP}}\left(e^{j\Omega}\right) + Y\left(e^{j(\Omega-\pi)}\right) G_{\text{HP}}\left(e^{j(\Omega-\pi)}\right)\right]. \quad (9.80)$$

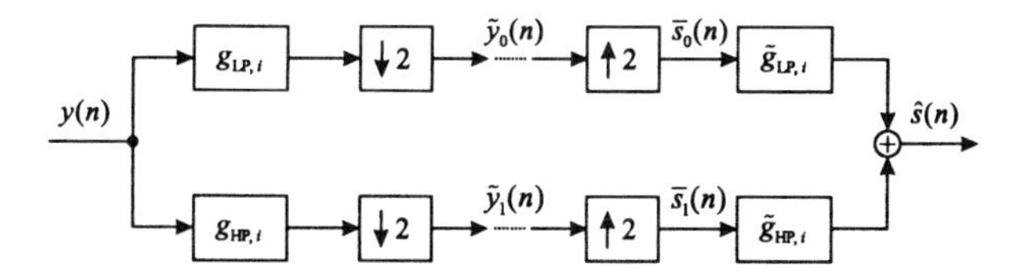

Fig. 9.19 Two-band analysis synthesis framework for a quadrature mirror filterbank.

After antiimaging filtering and combining both channels, we get the input–output frequency-domain relationship of the filterbank:

$$\begin{aligned}\hat{S}(e^{j\Omega}) &= \frac{1}{2}\left[G_{\text{LP}}(e^{j\Omega})\,\tilde{G}_{\text{LP}}(e^{j\Omega}) + G_{\text{HP}}(e^{j\Omega})\,\tilde{G}_{\text{HP}}(e^{j\Omega})\right] Y(e^{j\Omega}) \\ &\quad + \frac{1}{2}\left[G_{\text{LP}}\left(e^{j(\Omega-\pi)}\right)\tilde{G}_{\text{LP}}(e^{j\Omega}) + \right. \\ &\qquad \left. G_{\text{HP}}\left(e^{j(\Omega-\pi)}\right)\tilde{G}_{\text{HP}}(e^{j\Omega})\right] Y\left(e^{j(\Omega-\pi)}\right). \end{aligned} \tag{9.81}$$

The first term in this relation represents the dependence of the (desired) input spectrum $Y(e^{j\Omega})$. The second term represents the (undesired) aliasing components. If they should disappear, the analysis and synthesis filters have to fulfill the following condition:

$$G_{\text{LP}}\left(e^{j(\Omega-\pi)}\right)\tilde{G}_{\text{LP}}(e^{j\Omega}) + G_{\text{HP}}\left(e^{j(\Omega-\pi)}\right)\tilde{G}_{\text{HP}}(e^{j\Omega}) = 0\,. \tag{9.82}$$

This could be achieved by setting [47]

$$G_{\text{HP}}(e^{j\Omega}) = G_{\text{LP}}\left(e^{j(\Omega-\pi)}\right), \tag{9.83}$$

$$\tilde{G}_{\text{LP}}(e^{j\Omega}) = K\,G_{\text{LP}}(e^{j\Omega}), \tag{9.84}$$

$$\tilde{G}_{\text{HP}}(e^{j\Omega}) = -K\,G_{\text{LP}}\left(e^{j(\Omega-\pi)}\right), \tag{9.85}$$

where K is a real constant. Equivalently, in the time domain this implies that

$$g_{\text{HP},i} = (-1)^i\,g_{\text{LP},i}\,, \tag{9.86}$$

$$\tilde{g}_{\text{LP},i} = K\,g_{\text{LP},i}\,, \tag{9.87}$$

$$\tilde{g}_{\text{HP},i} = -K\,(-1)^i\,g_{\text{LP},i}\,. \tag{9.88}$$

Equation 9.83 states that the pair of lowpass and highpass filters, $G_{\text{LP}}(e^{j\Omega})$ and $G_{\text{HP}}(e^{j\Omega})$, have mirror-image symmetry with respect to the frequency $\Omega = \pi/2$. By introducing the conditions 9.83–9.85 the design of the analysis and synthesis filters has been translated to the common design of a lowpass filter with impulse response $g_{\text{LP},i}$. By inserting definitions 9.83–9.85 into Eq. 9.81, we obtain

$$\hat{S}(e^{j\Omega}) = \frac{K}{2}\left[G_{\text{LP}}^2(e^{j\Omega}) - G_{\text{LP}}^2\left(e^{j(\Omega-\pi)}\right)\right] Y(e^{j\Omega})\,. \tag{9.89}$$

The filterbank achieves unity gain for the choice $K = 2$ and appropriate lowpass filter design, which requires

$$\left| G_{\text{LP}}^2\left(e^{j\Omega}\right) - G_{\text{LP}}^2\left(e^{j(\Omega-\pi)}\right) \right| = 1 . \tag{9.90}$$

Aliasing components are canceled within the interpolation process of the synthesis filterbank. Additionally, it is desired that the magnitude of the filter $G_{\text{LP}}(e^{j\Omega})$ approximate that of an ideal half-band filter:

$$|G_{\text{LP}}(e^{j\Omega})| = \begin{cases} 1, & \text{for } 0 \leq |\Omega| \leq \dfrac{\pi}{2}, \\ 0, & \text{for } \dfrac{\pi}{2} < |\Omega| < \pi . \end{cases} \tag{9.91}$$

In principle, the demands 9.90 and 9.91 cannot be fulfilled exactly. However, they can be approximated for FIR and IIR filters of moderate order [47]. Figure 9.20 shows the frequency responses of a 32-tap FIR lowpass filter and a highpass filter as well as the analysis–synthesis reconstruction error.

Due to the relations between lowpass and highpass filters as well as analysis and synthesis filters, aliasing is canceled within the synthesis stage of the filterbank. For

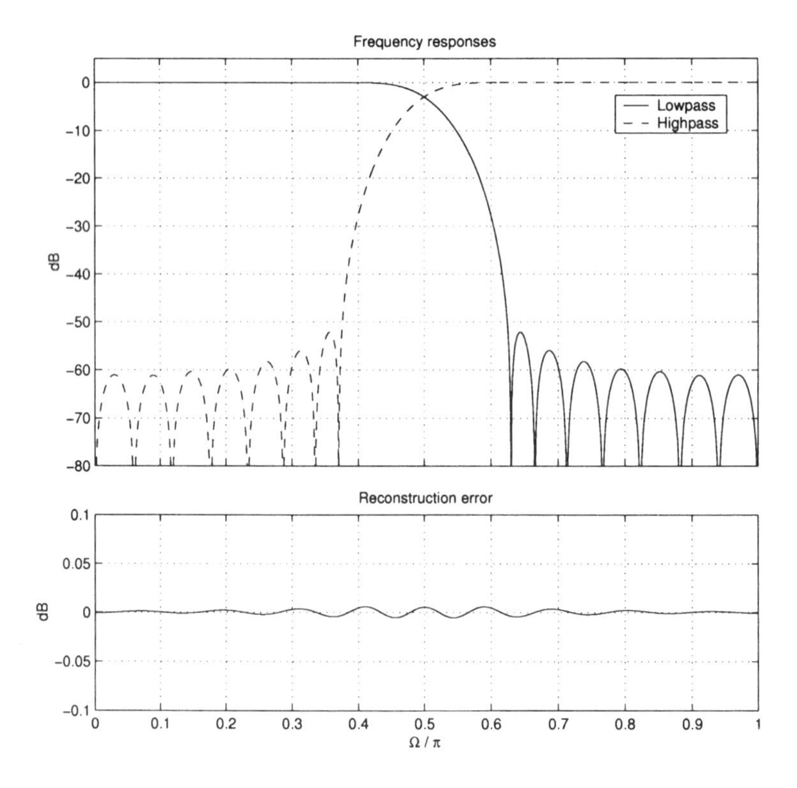

Fig. 9.20 Frequency responses of 32-tap lowpass and highpass filters (top) of a two-stage quadrature mirror filterbank and resulting analysis–synthesis reconstruction error (bottom).

noise reduction or speech coding purposes, this is sufficient. Subband echo cancellation, however, requires nearly aliasing free subband signals. For this reason, quadrature mirror filterbanks should be used only in case of very long prototype lowpass filters or when having spectral gaps around $\Omega = \pi/2$. In the latter case aliasing can be avoided, but signal transmission is also impossible around $\Omega = \pi/2$. A better choice are certain types of polyphase filterbanks, which are described in the next section.

9.1.3.2.3 DFT-Modulated Polyphase Filterbanks

DFT-modulated polyphase filterbanks[7] are another type of filterbank that allows very efficient realizations. Furthermore, with this type of filterbank it is possible to design nearly aliasing free analysis stages, making them well suited for the application of acoustic echo cancellation. In contrast to nonuniform quadrature mirror filterbanks, we will assume here to have equal subsampling rates in each channel. Figure 9.21 shows the basic structure of a polyphase analysis filterbank. The input signal $y(n)$ is decomposed by means of a set of lowpass, bandpass, and highpass filters.

The bandpass filters are generated by frequency shift of a *prototype lowpass filter* with impulse response g_i (see Fig. 9.22):

$$g_{\mu,\,i} = g_i \, e^{j \frac{2\pi}{M} \mu i} \, . \tag{9.92}$$

The bandwidths of the prototype lowpass filter B_{LP}, of the bandpass filters B_{BP}, and of the highpass filter B_{HP} are specified on one hand by the sampling rate f_{s} and on other hand by the number of subbands M. For the single-sided bandwidth of the lowpass and highpass filter, we obtain

$$B_{\mathrm{LP}} = B_{\mathrm{HP}} = \frac{1}{2} \frac{f_{\mathrm{s}}}{M} \, . \tag{9.93}$$

For the bandwidth of the bandpass filters, we obtain analogously

$$B_{\mathrm{BP}} = \frac{f_{\mathrm{s}}}{M} \, . \tag{9.94}$$

After convolution with the impulse responses of the frequency-shifted lowpass filters, the subband signals[8]

$$\begin{aligned} \bar{y}_\mu(n) &= y(n) * \left(g_n \, e^{j \frac{2\pi}{M} n\mu}\right) \\ &= \sum_{\kappa=-\infty}^{\infty} y(n-\kappa) \, g_\kappa \, e^{j \frac{2\pi}{M} \kappa\mu} \\ &= \sum_{\kappa=-\infty}^{\infty} y(\kappa) \, g_{n-\kappa} \, e^{j \frac{2\pi}{M} (n-\kappa)\, \mu} \end{aligned} \tag{9.95}$$

[7]Polyphase decompositions will be mentioned here only briefly. The interested reader is referred to Vaidyanathan [229, 230].

[8]For the reasons of simplicity, we assume real input signals.

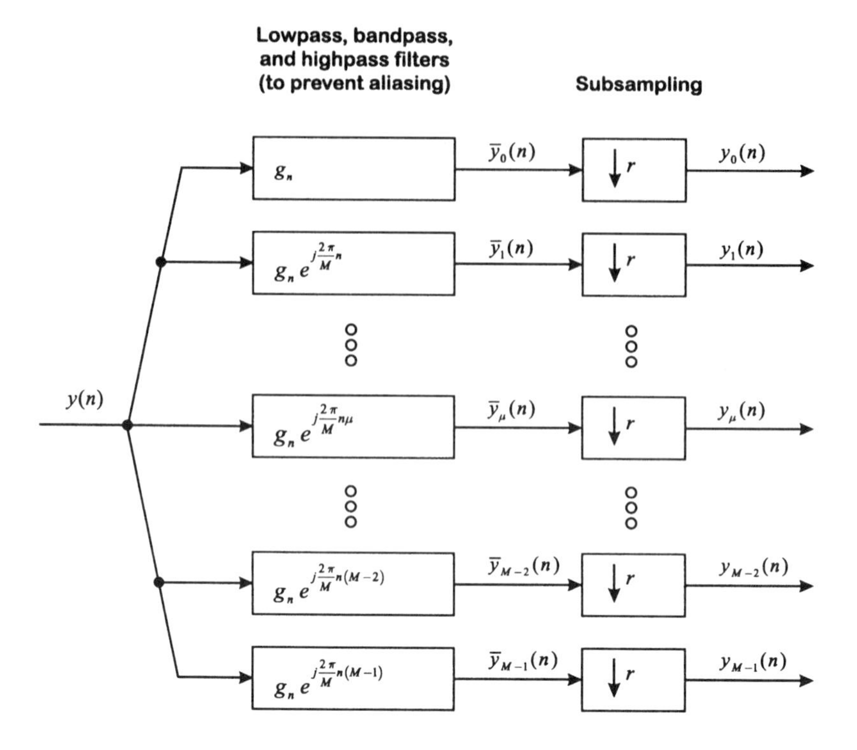

Fig. 9.21 Basic structure of a polyphase analysis filterbank.

occupy only a part of the frequency range. Thus, subsampling (reduction factor r) can be applied:

$$y_\mu(n) = \bar{y}_\mu(r\,n)\,. \tag{9.96}$$

If we assume the impulse response of the prototype lowpass filter g_n to be causal

$$g_i = 0 \qquad \text{for} \qquad i < 0\,, \tag{9.97}$$

we obtain for the subband signals $y_\mu(n)$:

$$y_\mu(n) = \sum_{\kappa=0}^{\infty} y(r\,n - \kappa)\, g_\kappa\, e^{j\frac{2\pi}{M}\kappa\mu}\,. \tag{9.98}$$

Computation of the signals $y_\mu(n)$, however, can be performed much more efficiently [231, 236]. In the following we will show that the computation of all subband signals can be done with one convolution and one weighted IFFT of order M. Furthermore, both have to be performed only every r (fullband) samples.

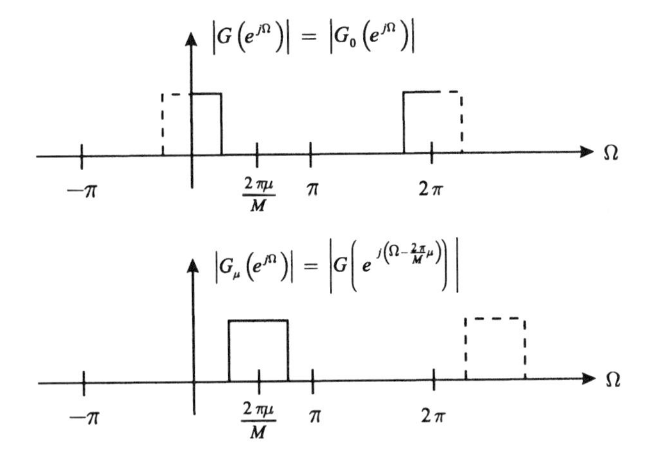

Fig. 9.22 Prototype lowpass filter and resulting bandpass filter.

When rearranging the summation index[9] in Eq. 9.98 according to

$$\kappa = lM + \nu \qquad \text{with} \qquad l \in \mathbf{N_0} \text{ and } \nu \in \{0, \ldots, M-1\} \tag{9.99}$$

we obtain the following for the subband signals:

$$\begin{aligned} y_\mu(n) &= \sum_{l=0}^{\infty} \sum_{\nu=0}^{M-1} y(rn - lM - \nu)\, g_{lM+\nu}\, e^{j\frac{2\pi}{M}(lM+\nu)\mu} \\ &= \sum_{\nu=0}^{M-1} \sum_{l=0}^{\infty} y(rn - lM - \nu)\, g_{lM+\nu}\, e^{j\frac{2\pi}{M}(lM+\nu)\mu} \\ &= \sum_{\nu=0}^{M-1} e^{j\frac{2\pi}{M}\nu\mu} \sum_{l=0}^{\infty} y(rn - lM - \nu)\, g_{lM+\nu}\, \underbrace{e^{j\frac{2\pi}{M}lM\mu}}_{=1} \\ &= \sum_{\nu=0}^{M-1} e^{j\frac{2\pi}{M}\nu\mu} \sum_{l=0}^{\infty} y(rn - lM - \nu)\, g_{lM+\nu}\,. \end{aligned} \tag{9.100}$$

By introducing the following abbreviations

$$y^{(M,p)}(m) = y(mM - p) \tag{9.101}$$

$$g_m^{(M,p)} = g_{mM+p} \tag{9.102}$$

with $m,\, p \in \mathbf{Z}$ (set of integers), we can rewrite the last sum in Eq. 9.100 to

[9] $\mathbf{N_0}$ describes the set of nonnegative integers.

$$\begin{aligned}
\breve{y}_\nu(n) &= \sum_{l=0}^{\infty} y(rn - lM - \nu)\, g_{lM+\nu} \\
&= \sum_{l=0}^{\infty} y(rn - (rn \bmod M) - lM - \nu + (rn \bmod M))\, g_{lM+\nu} \\
&= \sum_{l=0}^{\infty} y^{(M,\nu-(rn \bmod M))}\left(\frac{rn - (rn \bmod M)}{M} - l\right) g_l^{(M,\nu)} \\
&= y^{(M,\nu-(rn \bmod M))}\left(\frac{rn - (rn \bmod M)}{M}\right) * g^{(M,\nu)}_{(rn-(rn \bmod M))/M}\,.
\end{aligned} \tag{9.103}$$

On first glance Eq. 9.103 does not look very "simplified." However, implementing it leads directly to the upper part of the structure presented in Fig. 9.23. Now the advantages of the index rearrangements and modulo addressing become evident. Finally, when inserting Eq. 9.99 into Eq. 9.100, we obtain

$$y_\mu(n) = \sum_{\nu=0}^{M-1} e^{j\frac{2\pi}{M}\nu\mu}\, \breve{y}_\nu(n) \tag{9.104}$$

with $\mu \in \{0, \ldots, M-1\}$.

The inverse DFT, weighted with a factor M, needs to be computed every r fullband samples. Figure 9.23 shows the structure of the efficient polyphase filterbank realization. In comparison to the direct structure (see Fig. 9.21), the computational complexity has been reduced considerably.

After describing the decomposition of a fullband signal into subbands, we will now consider the synthesis of the subbands back into a fullband signal. In a hands-free telephone system the speech signal of the local speaker $s(n)$ should be estimated, which means that local background noise and echo signals should be suppressed. For this reason, we will denote the input signals of the synthesis by $\hat{s}_\mu(n)$ and the output signal $\hat{s}(n)$, respectively. Figure 9.24 shows the basic structure of a polyphase synthesis filterbank.

The subsampled (echo and noise reduced microphone) signals $\hat{s}_\mu$ will be upsampled by a factor r according to Eq. 9.72 in order to bring them on the fullband sampling rate:

$$\bar{s}_\mu(n) = \begin{cases} \hat{s}_\mu\left(\frac{n}{r}\right) & : \ n \bmod r \equiv 0\,, \\ 0 & : \ \text{else}\,. \end{cases} \tag{9.105}$$

Due to the upsampling, the resulting spectra of the signals $\bar{s}_\mu(n)$ consist of a sum of r shifted spectra (imaging spectra) of the input signals (see Eq. 9.74). If the subsampling rate r was chosen less than or equal to the number of channels M, it is possible to extract the original spectrum by appropriate lowpass ($\mu = 0$), bandpass, or highpass filtering ($\mu = M/2$):

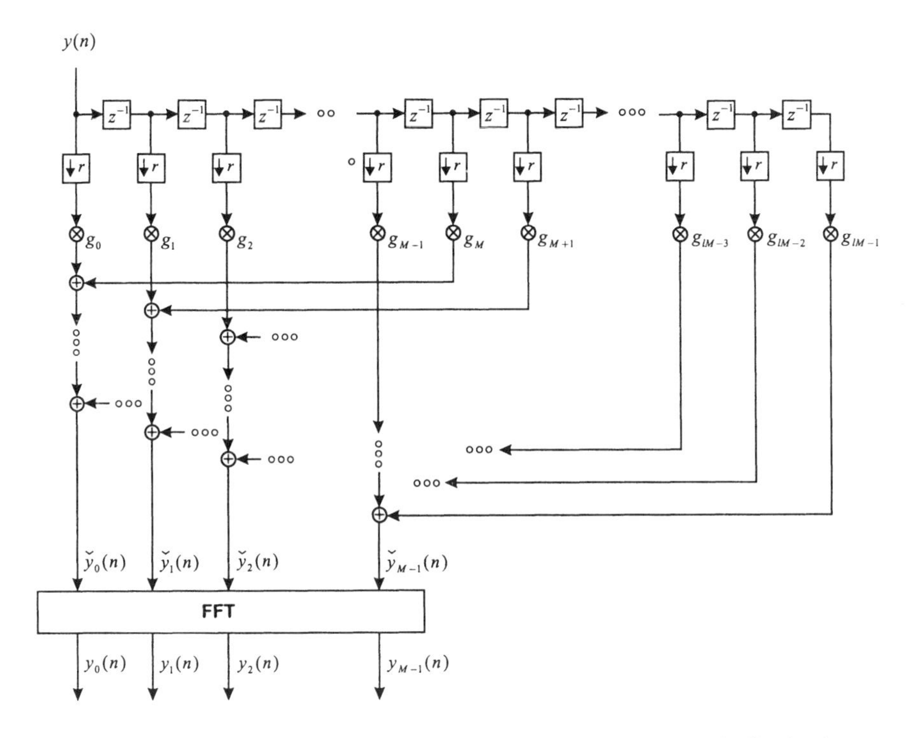

Fig. 9.23 Efficient structure of a DFT-modulated polyphase analysis filterbank.

$$\begin{aligned}\tilde{s}_\mu(n) &= \bar{s}_\mu(n) * \tilde{g}_{\mu,n} \\ &= \sum_{\kappa=-\infty}^{\infty} \bar{s}_\mu(n-\kappa)\,\tilde{g}_{\mu,\kappa}\,.\end{aligned} \tag{9.106}$$

As in the analysis filterbank, the bandpass and highpass filters are realized by frequency shifting of a lowpass filter with impulse response $\tilde{g}_i$:

$$\tilde{g}_{\mu,i} = \tilde{g}_i\, e^{j\frac{2\pi}{M}\mu i}. \tag{9.107}$$

By inserting Eq. 9.107 into Eq. 9.106, we obtain

$$\tilde{s}_\mu(n) = \sum_{\kappa=-\infty}^{\infty} \bar{s}_\mu(n-\kappa)\,\tilde{g}_\kappa\, e^{j\frac{2\pi}{M}\mu\kappa}\,. \tag{9.108}$$

When assuming the impulse response of the lowpass filter to be causal

$$\tilde{g}_i = 0 \qquad \text{for} \qquad i < 0, \tag{9.109}$$

we obtain

$$\tilde{s}_\mu(n) = \sum_{\kappa=0}^{\infty} \bar{s}_\mu(n-\kappa)\,\tilde{g}_\kappa\, e^{j\frac{2\pi}{M}\mu\kappa}\,. \tag{9.110}$$

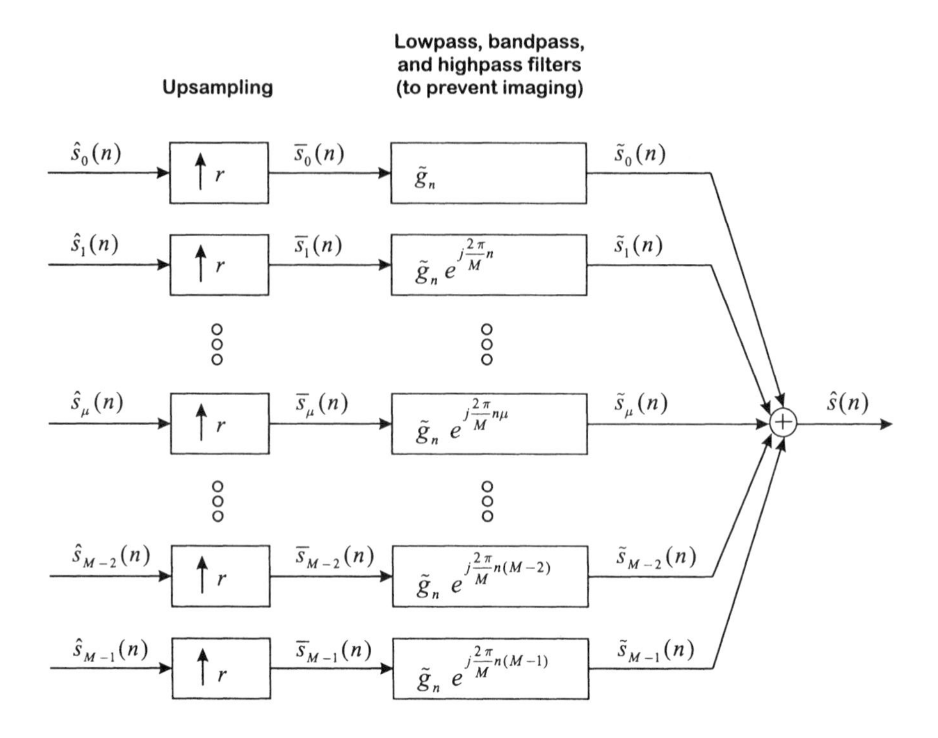

Fig. 9.24 Basic structure of a polyphase synthesis filterbank.

Finally the upsampled and filtered signals $\tilde{s}_\mu(n)$ can be combined to a fullband signal by summing up all channels:

$$\begin{aligned} \hat{s}(n) &= \sum_{\mu=0}^{M-1} \tilde{s}_\mu(n) \\ &= \sum_{\mu=0}^{M-1} \sum_{\kappa=0}^{\infty} \bar{s}_\mu(n-\kappa)\, \tilde{g}_\kappa\, e^{j\frac{2\pi}{M}\mu\kappa}\,. \end{aligned} \tag{9.111}$$

In the same way as for the analysis filterbank, it is possible to derive very efficient structures for the synthesis stage. Figure 9.25 shows such a structure. First, an inverse DFT of the subband signals $\hat{s}_\mu(n)$ is computed. The resulting (time-domain) vector is extended by simple duplication of its coefficients. The elements of the extended vector are multiplied with the impulse response of the lowpass filter $\tilde{g}_i$ and added to a tapped delay line. Details about the derivation can be found in the literature [231, 236].

As in the synthesis stage, the IDFT and the convolution have to be performed only at the lower sampling rate, leading to a considerable reduction in computational complexity (compared to the direct structure). When applying DFT-modulated polyphase

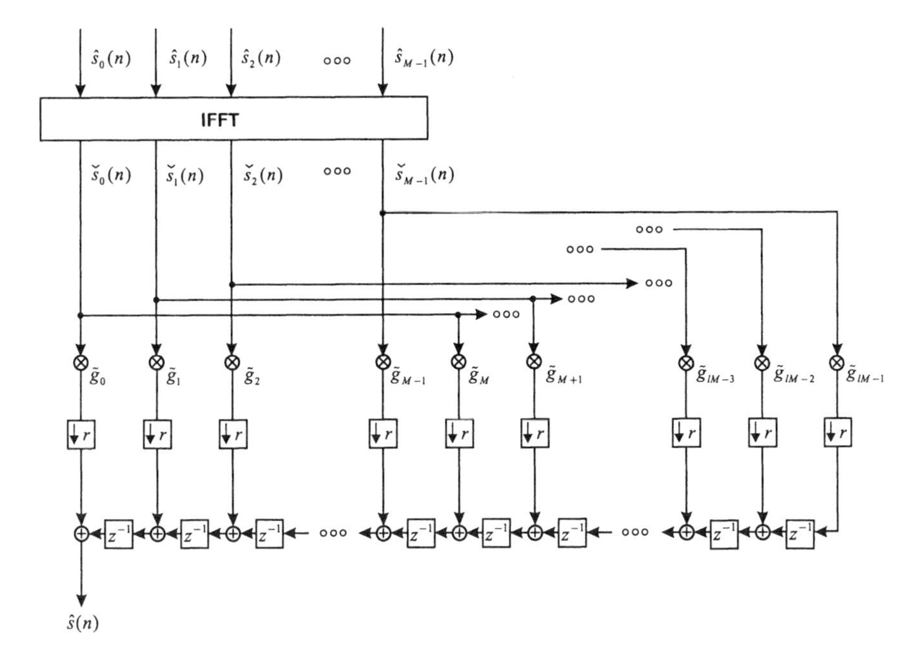

Fig. 9.25 Efficient structure of a DFT-modulated polyphase synthesis filterbank.

filterbanks, the symmetry of the subband signals

$$y_\mu(n) = y^*_{M-\mu}(n) \quad \text{for } \mu = 1, \ldots, \frac{M}{2} - 1, \tag{9.112}$$

can be exploited and only the first $\frac{M}{2} + 1$ subbands need to be computed.

Finally, the strong relation of subband structures and block processing in the frequency domain should be highlighted. When using a different addressing mode in Eq. 9.99, the IDFT within the analysis filterbank can be replaced by a DFT [in this case also the direction of the tapped delay line of the input signal $y(n)$ is reversed]. If, furthermore, the filter length of the analysis filterbank is set equal to the DFT length, one ends up with simple overlap-add structures. For this reason, DFT can also be interpreted as a (low-quality) filterbank. In other words, the presented filterbank type can be seen as an extended DFT.

9.1.3.3 Design of Prototype Filters

As described in the previous section, filterbanks require two lowpass filters: g_i (within the analysis) and $\tilde{g}_i$ (within the synthesis). Until now nothing has been said about the requirements these filters have to meet. This will be the topic of this section. The requirements on the prototype lowpass filters can be described as follows:

- Both prototype lowpass filters should exhibit sufficient stopband attenuation in order to keep the aliasing and the imaging components sufficiently small. For subband echo cancellation systems the avoidance of aliasing within the analysis stage is very important [76]. These terms disturb the convergence process—an attenuation of aliasing components within the synthesis stage would not have any effect on the echo cancellation.

- The analysis prototype lowpass filter g_i should have a steep slope. In this case the subsampling rate can be chosen close to the number of subbands leading to low computational complexity.

- A lossless or nearly lossless combination of the subband signals within the synthesis filterbank is possible only if neither the analysis nor the synthesis lowpass has zeros (on the unit circle) in its passband $\Omega \in [0 \ldots \frac{\pi}{M}]$.

- Special restrictions have to be fulfilled within the transition bands of both filters in order to achieve a lossless superposition of the subband signals within the synthesis filterbank.

A variety of methods that (more or less) fulfill the demands stated above have been published. Mentioning just a few of them in detail would go far beyond the scope of this chapter. For this reason, we have described a computational effective design procedure in Appendix B. Using this method a maximum phase prototype filter for the analysis stage and a minimum phase prototype filter for the synthesis stage have been designed. The results of this design for $M = 16$ channels, a filter length $N_{\text{LP}} = 128$, and a subsampling rate $r = 12$ are depicted in Fig. 9.26.

The advantages of the design method presented in Appendix B are wide-ranging. First, the prototype filter offers sufficient stopband attenuation (in the example 70 dB) in order to ensure nearly independent subband signals. Due to noncritical subsampling ($M > r$), the aliasing components can be kept small. Even if the filterbank is not a perfect reconstruction system, the group delay and magnitude variations are very small and do not provide any audible artifacts. Finally, the maximum and minimum phase properties, respectively, can be exploited to also model noncausal parts of the subband impulse responses (see Section 9.1.3.4.1).

9.1.3.4 Two Remarks on Subband Systems

After describing the basics of subband systems, like efficient polyphase structures and prototype lowpass filter design (see Appendix B), we highlight two further aspects:

- If adaptive filters are able to also model noncausal parts of the subband room impulse responses the performance, in terms of echo reduction, can be increased by a few decibels. Modeling noncausal systems can be achieved either by introducing artificial delay in the loudspeaker or the microphone path or by using different analysis filterbanks. Explanations for the noncausality

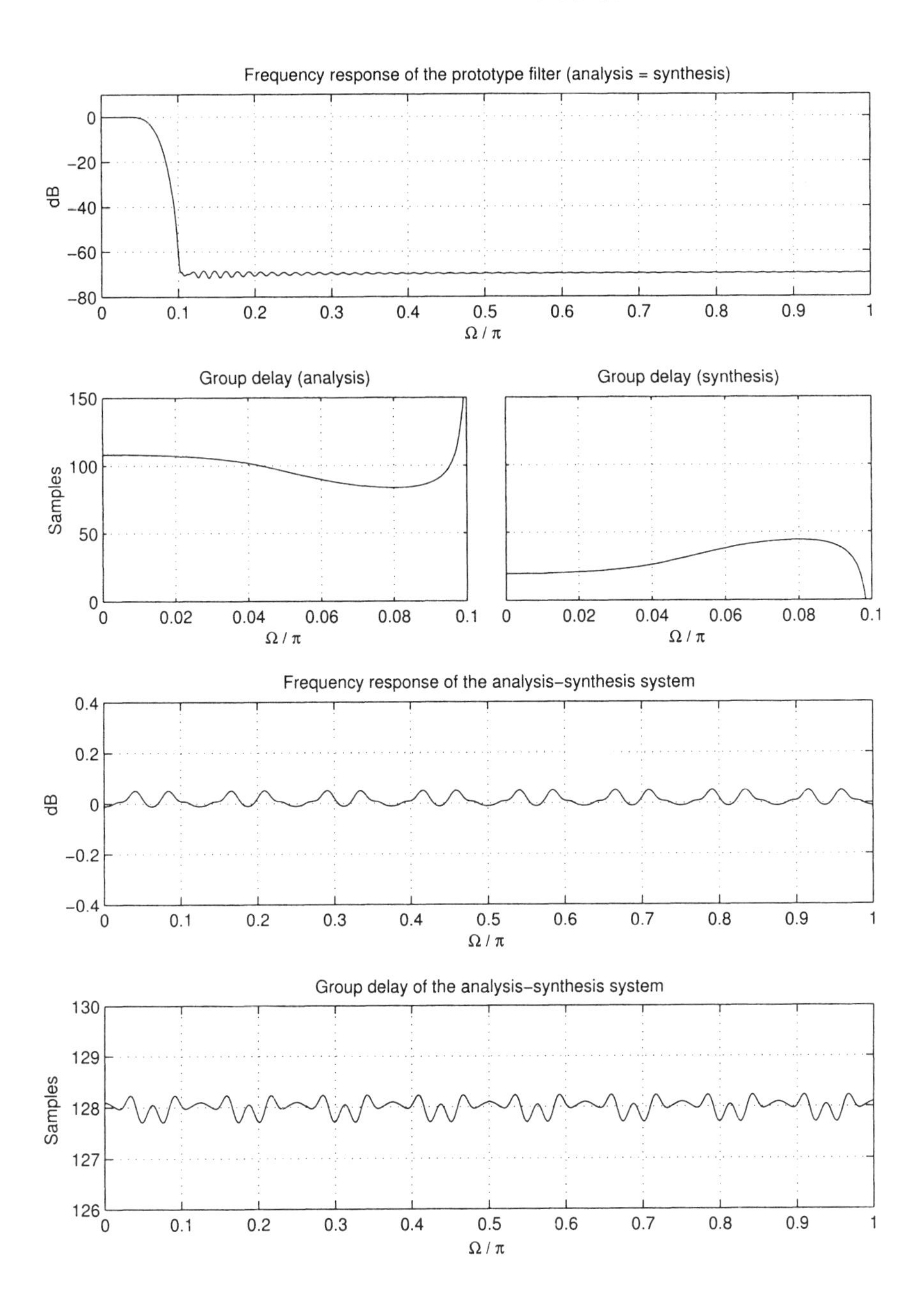

Fig. 9.26 Example of a prototype lowpass filter for DFT-modulated polyphase filterbanks. With $N_{\mathrm{LP}} = 128$ coefficients for each analysis and synthesis prototype lowpass filters, a stopband attenuation of about 70 dB can be achieved. The analysis filter is designed to have maximum phase; and the synthesis lowpass filter, minimum phase. If a subsampling rate $r = 12$ is applied, the 16-channel filterbank system will have group delay variations below 0.5 samples and gain variations below 0.1 dB.

as well as simulation results for both modeling methods will be presented in Appendix A and Section 9.1.3.4.1.

- Subband structures offer the possibility of distributing the number of filter coefficients nonuniformly over the adaptive filters. In Section 9.1.3.4.2, the advantages of this design will be explained and examples of filter order distributions will be given.

9.1.3.4.1 Modeling Noncausal Parts of Subband Impulse Responses

In Appendix A fullband impulse responses h_i have been transformed into equivalent sets of subband impulse responses $h_{\mu,\,i}$. Even if the fullband systems are assumed to be causal, the subband systems are not. In this section we will investigate the consequences that result from the noncausality.

The relationship between the broadband impulse response h_i and the impulse response in subband μ is given by (see Eq. A.9):

$$h_{\mu,\,i} = \sum_{m=-\infty}^{\infty} h_m \, \frac{\sin\left(\pi\left(i - \frac{m}{r}\right)\right)}{\pi\left(i - \frac{m}{r}\right)} \, e^{j\frac{2\pi\mu}{M}(ir-m)} \,. \tag{9.113}$$

Beside the multiplication with an exponential term e^{jx}, all reflections h_i are weighted with a $\sin(x)/x$ function. This leads to a temporal spreading of the subband impulse responses. The spreading has largest impact in portions where the broadband coefficients h_i are also large. In general this will be the coefficients belonging to the direct path—at the beginning of impulse response. As a result of the spreading, coefficients of the subband impulse response with negative index can also have significant amplitude.

If these coefficients should also be modeled by the subband adaptive filters $\hat{\boldsymbol{h}}_\mu(n)$, either the loudspeaker signal or the microphone signal has to be delayed by a few samples. To illustrate the enhancement of introducing artificial delay, two convergence examples are depicted in Fig. 9.27. The system was excited by white noise and the filter length was set to $N_\mu = 100$ in all subbands. The filterbank was designed for 16 channels and a subsampling rate $r = 12$ was used. No local signals were present. If the microphone signal is connected directly to the analysis filterbank, an echo reduction of about 25 dB can be achieved. By inserting a delay of 50 (fullband) samples, the performance can be enhanced by about 7–8 dB.

Besides the advantage of lower final misalignment the introduction of delay in the microphone or the loudspeaker channel has two major drawbacks:

- Delaying the microphone or the loudspeaker signal requires additional memory. If this system resource is limited the memory needs to be taken from the echo cancellation filters—resulting in shorter filters. Thus, the final misalignment is increased again.

- The delay is introduced in the signal path of the echo. Thus, the round-trip delay of the echo cancellation system is increased. If a sampling rate of 8 kHz

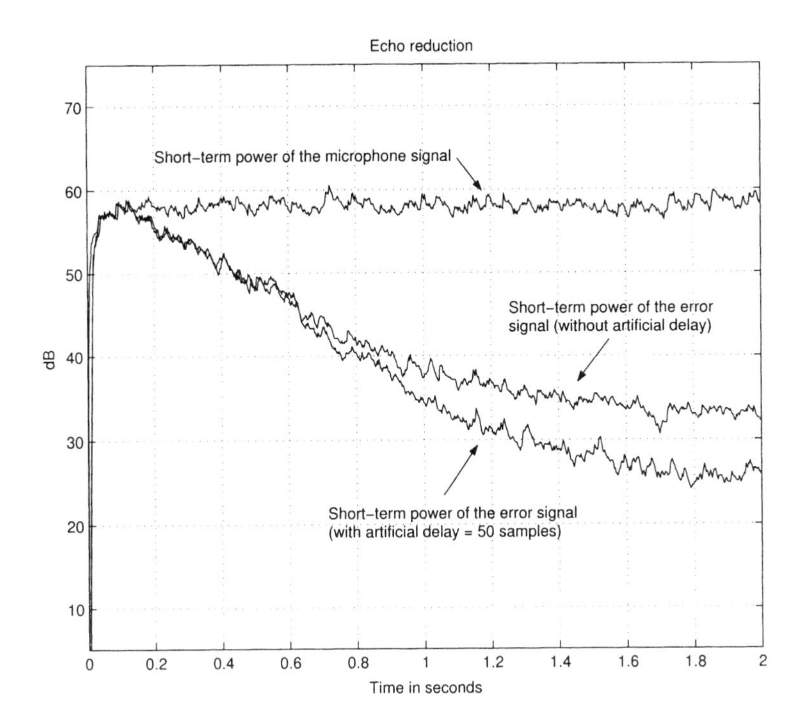

Fig. 9.27 Convergences of the echo reduction with and without artificially delayed microphone signal. White noise was used as excitation signal. The orders of the subband echo cancellation filter were chosen to be $N_\mu = 100$. By introducing a delay of 50 (fullband) samples, the final misalignment can be reduced by 7–8 dB.

is used, a delay of 50 samples increases the round-trip delay by more than 6 ms. According to this additional delay, the demands on the overall echo reduction are also increased (see Fig. 3.17).

These drawbacks can be avoided by using two different analysis filterbanks instead of the delay. Analysis of the excitation signal is performed with a maximum phase filterbank and analysis of the microphone signal, with a minimum phase filterbank. The minimum phase ability of the excitation analysis filterbank can be achieved by using the prototype lowpass filter of the synthesis stage. Thus, no additional memory is required and no additional delay is introduced into the signal path.

Figure 9.28 shows the group delays of both analysis filterbanks. Even in the transition band of the prototype filters ($\frac{\pi}{M} \cdots \frac{\pi}{r}$), the difference in the group delays exceeds 30 samples. In the passband ($0 \cdots \frac{\pi}{M}$) the difference varies between 40 and 85 samples.

When using different analysis filterbanks, the frequency responses of the echo cancellation filters do not converge toward the frequency responses of the subband impulse responses of the LEM system any more. Due to the different prototype lowpass filters, the new solution of the system identification will be (in case of negligible

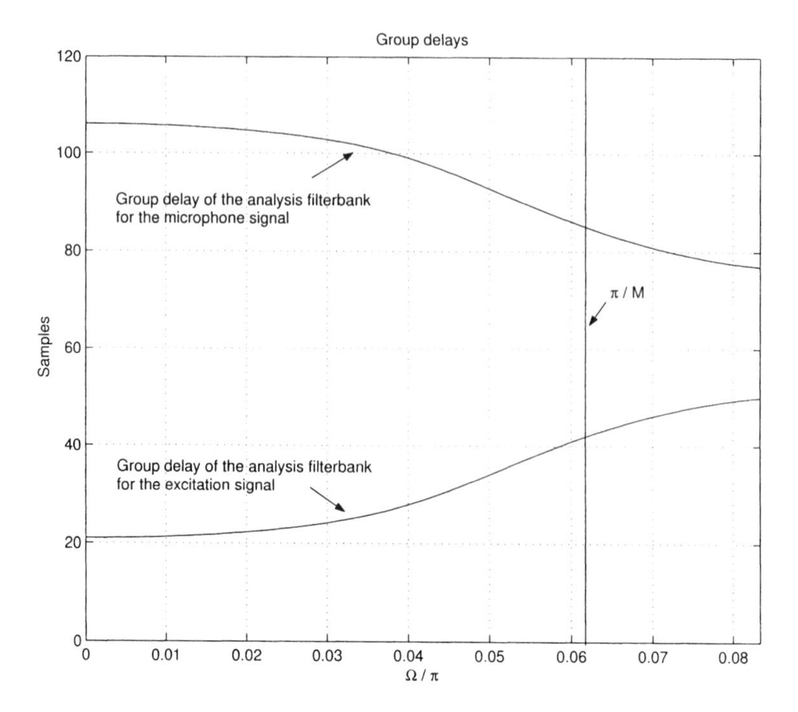

Fig. 9.28 Group delays of the two analysis filterbanks. The difference of both group delays is (within the interesting frequency range $0 \cdots \frac{\pi}{M}$) always larger than 40 samples. The difference is used by the adaptive filters to model noncausal parts of the subband impulse responses of LEM systems.

aliasing terms)

$$\hat{H}_\mu\big(e^{j\Omega}\big) = H_\mu\big(e^{j\Omega}\big)\,\frac{G_{\mu,\,y}\big(e^{j\Omega}\big)}{G_{\mu,\,x}\big(e^{j\Omega}\big)}\,. \tag{9.114}$$

The expression $G_{\mu,\,y}(e^{j\Omega})$ denotes the frequency response of the frequency-shifted prototype lowpass filter within the analysis of the microphone signal $y(n)$ and $G_{\mu,\,x}(e^{j\Omega})$ is the frequency response of the bandpass filter of the analysis of the excitation signal $x(n)$. If the prototype filters are designed using the method presented in Appendix B, the filters satisfy the condition

$$\left|\tilde{G}\big(e^{j\Omega}\big)\right| = \left|G_{0,\,x}\big(e^{j\Omega}\big)\right| > 0\,. \tag{9.115}$$

Furthermore, the filter $\tilde{g}_n$ was designed to be minimum phase, which means that all zeros within the z domain are located inside the unit circle. Thus, the frequency response $G_{\mu,\,x}(e^{j\Omega})$ can be inverted in principle.

However, inverting $G_{\mu,\,x}(e^{j\Omega})$ cannot be realized in full precision. The echo cancellation filters are realized in FIR structure. The inverse of the analysis filterbanks would require an all-pole filter. For this reason, the subband adaptive filter can only approximate the solution presented in Eq. 9.114. Nevertheless, the final misalignment can be improved by a few decibels as shown in Fig. 9.29. The previous simulation (see Fig. 9.27) has been repeated, but now, instead of an artificial delay, two different analysis filterbanks have been applied.

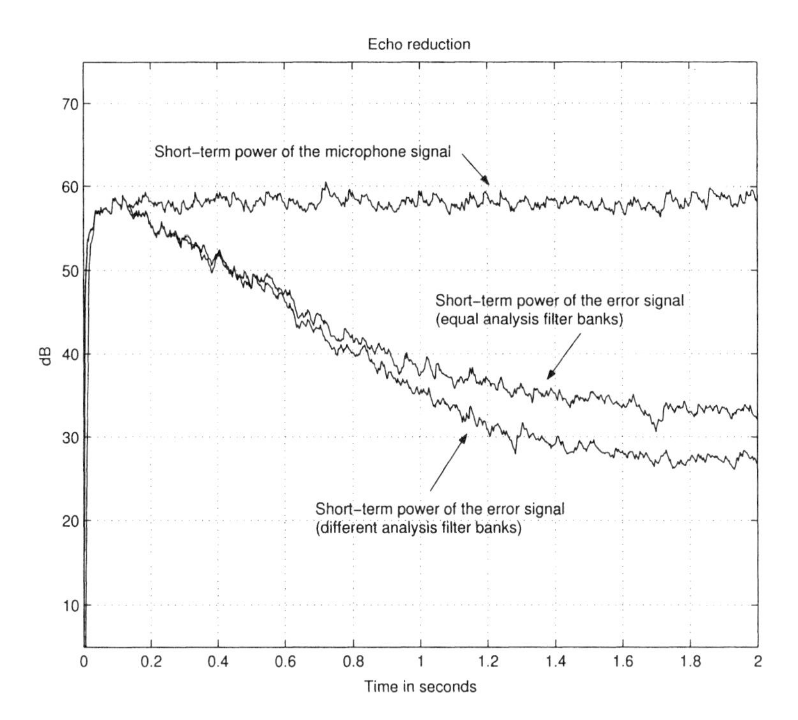

Fig. 9.29 Convergences with the same and with different analysis filterbanks. As in Fig. 9.27, white noise was used as excitation signal and the orders of the subband echo cancellation filters were chosen to be $N_\mu = 100$. By using different analysis filterbanks, the final misalignment can be reduced by 6–7 dB.

As presented in Fig. 9.29, the power of the residual error signal can be decreased by 6–7 dB if a minimum phase analysis filterbank is used for the excitation signal and a maximum phase filterbank for the microphone signal. Generally also a combination of both—different analysis filterbanks and artificial delay—is possible.

9.1.3.4.2 Choice of Filter Order

Subband adaptive filtering offers the possibility of allocating the filter orders in a nonuniform manner. Hence, subband impulse responses with longer reverberation time and higher excitation power can be modeled with more coefficients than others.

For visualization of this relationship the broadband echo reduction in dependence of the filter order was depicted in the lowest part of Fig. 3.16. Applying the same investigations on subband structures, we obtain for the minimum power of the residual echo signal in subband μ

$$\mathrm{E}\left\{|e_\mu(n)|^2\right\}\Big|_{\min} = \mathrm{E}\left\{|x_\mu(n)|^2\right\} \sum_{i=N_\mu}^{\infty} |h_{\mu,i}(n)|^2 . \qquad (9.116)$$

The coefficients of the subband impulse responses $h_{\mu,i}(n)$ were computed in Appendix A. As in Section 3.2.2, it is assumed that the excitation signal is white noise, that no local signals are present $[n_\mu(n) = 0]$ and that the adaptive filters did converge $[\widehat{h}_\mu(n) = h_\mu(n)]$. Furthermore, all noncausal parts of the impulse responses are neglected.

Equation 9.116 shows that the average subband error power $\mathrm{E}\{|e_\mu(n)|^2\}$ is dependent on the excitation power $\mathrm{E}\{|x_\mu(n)|^2\}$ in subband μ. If speech is the excitation signal, the subband powers will differ largely among high- and low-frequency subbands. In the middle part of Fig. 3.1 the average power spectral density of a speech signal is depicted. It is clearly visible that low-frequency speech components have larger (average) power than do high-frequency components. In the speech sequence depicted in Fig. 3.1, the ratio between speech components around the pitch frequency (100–200 Hz) and half of the sampling rate (4000 Hz) is about 30–35 dB. Even if this is just an average ratio, it is the first hint for emphasizing the importance of low-frequency components in subband echo cancellation systems.

A more detailed analysis of speech signals provide short-term power spectral densities. In this case the input signal is split into short (e.g., 20-ms) frames and an estimate of the current power spectral density is computed via a DFT for each frame. Also with this more detailed analysis the dominance of the low frequencies shows up. Thus, the distribution of the subband powers in speech signals is a first motivation to spend more coefficients for the low-frequency subbands than for the high-frequency bands. Nevertheless, room acoustics should also be taken into account when choosing the subband filter orders.

Also the materials at the walls and the furniture of the room where the hands-free system is installed impair the optimal distribution of the subband filter orders. For the direct sound the locations of the loudspeaker and the microphone are most important. When placing both devices on the left and the right of a computer monitor, for instance, a better acoustic decoupling is achieved as if both were placed on the same side. Even if both electroacoustic transducers have to be installed in the same housing, one should try to separate them as much as possible. The diffuse sound can be affected by the materials of the enclosure. Porous materials like textile fibers (carpets, curtains, etc.) or plasterboards (ceiling suspensions) absorb high-frequency sound energy faster than low-frequency energy. In rooms equipped with these materials the same echo reductions can be achieved in high-frequency subbands with much less coefficients than in low-frequency bands.

When considering all arguments discussed so far, the main part of the available memory and computational power should be used to cancel low-frequency echoes. If the orders of the adaptive filters are chosen exclusively according to the minimization of the broadband error power, short high-frequency echoes, such as those caused by fricative excitation, lead to a degradation of the subjective speech quality of the outgoing signal. Broadband power minimization does not take the masking properties [250] of the human ear into account; when we listen to speech signals, strong low-frequency components are able to mask high-frequency signals. According to the echo cancellation in low subbands, the masking effect is reduced considerably. For this reason, one should keep a few coefficients also in the high-frequency subbands in order to cancel at least the direct sound.

If neither the room in which the hands-free system will be installed nor the placement of the loudspeaker and the microphone is known a priori, only guiding values can be given for the distribution of the subband filter orders. Table 9.3 shows such guiding values for a subband system with $M = 16$ channels (only 9 of the 16 subbands have to be computed due to spectral symmetries) and a subsampling rate $r = 12$ for four different memory sizes. The values have been found by various real-time tests with different loudspeakers and microphones. Variations up to 10–15% were not audible as long as the orders remain above a limit of about 20 coefficients (in order to cancel the direct sound). For better visualization the orders of the subband cancellers (memory size = 1000 words) are depicted as bars within the time–frequency analysis of a room impulse response in Fig. 9.30. The length of the bars represents the order of the subband filter and the y-axis locations of the bars represent the related frequency bins.

Table 9.3 Orders of the subband echo cancellation filters for different memory sizes

	Available memory (only for the filter coefficients of the subband echo cancellers)			
Subband number	700 words	800 words	900 words	1000 words
0	65	70	75	80
1	85	95	110	115
2	65	70	75	85
3	40	50	60	65
4	40	45	45	55
5	35	40	45	50
6	25	30	35	40
7	20	25	30	35
8	20	20	25	30

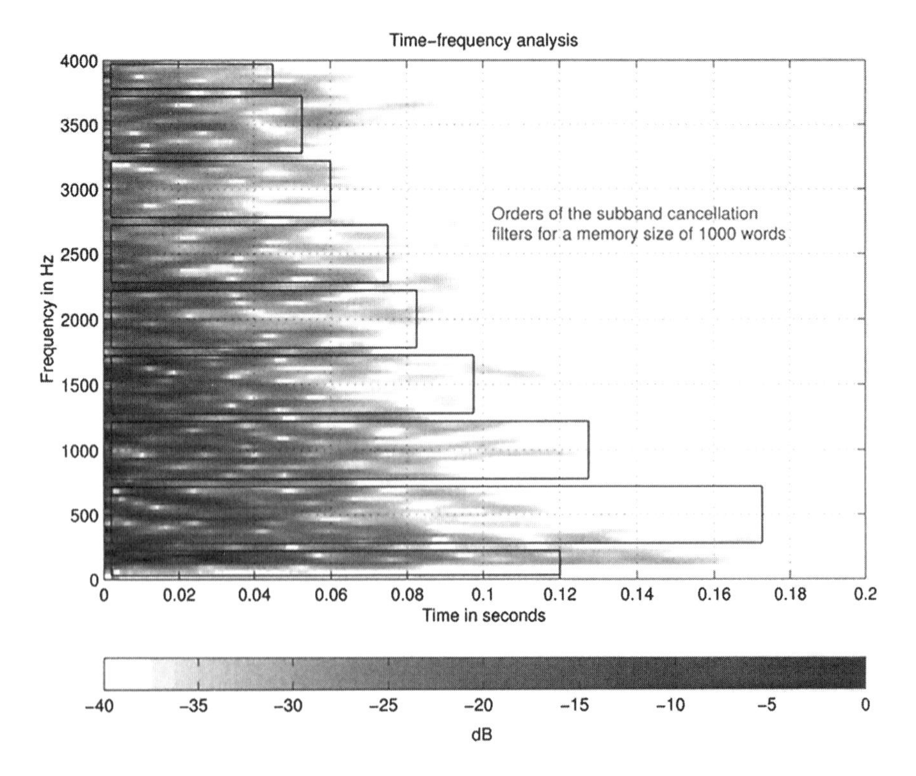

Fig. 9.30 Orders of the subband echo cancellation filters. Depicted is a time–frequency analysis of a room impulse response (300 ms reverberation time) and nine bars representing the orders of the subband echo cancellation filters. The y-axis locations of the bars represent the related frequency bins (with permission from [25]).

9.2 STEREOPHONIC AND MULTICHANNEL ECHO CANCELLATION

Monophonic hands-free telephone systems do not provide the local listener with spatial sound information. Sterophonic or even multichannel (with more than two channels) systems provide a more realistic acoustic presence of the far-end partner(s). Unfortunately, the system identification problem is no longer unique if the loudspeaker signals are pairwise correlated. We will explain this nonuniqueness problem in this section and present a short overview of solutions. We will start with stereophonic systems and give a brief outlook on system with more than two loudspeaker channels at the end of this section.

9.2.1 Stereophonic Acoustic Echo Cancellation

Stereophonic devices have two separate acoustic transmission channels that allow listeners to distinguish between different speakers. Furthermore, these systems are able to enhance the sound realism.

A stereophonic structure exhibits two loudspeakers as well as two microphones, in both the local and the remote rooms. As a direct consequence, four echo paths (i.e., impulse responses) have to be identified individually for each room. Thus, an increased amount of processing power has to be devoted to this task when compared to a monophonic hands-free system.

For the sake of clarity, only one half of the echo path system in the local room is shown in Fig. 9.31. It is assumed that neither local noise nor local speech signals are present. In the remote room, one speaker's voice in conjunction with ambient noise (which is not depicted in Fig. 9.31) is picked up by two microphones. The disturbed speech signals $x_1(n)$ and $x_2(n)$ are transmitted in two distinct paths to the local room and radiated by two loudspeakers. Two adaptive filters with impulse responses $\hat{\boldsymbol{h}}_1(n)$ and $\hat{\boldsymbol{h}}_2(n)$, respectively, try to imitate the echo signal $y(n)$ by appropriate outputs $\hat{d}_1(n)$ and $\hat{d}_2(n)$, in order to cancel the echo.

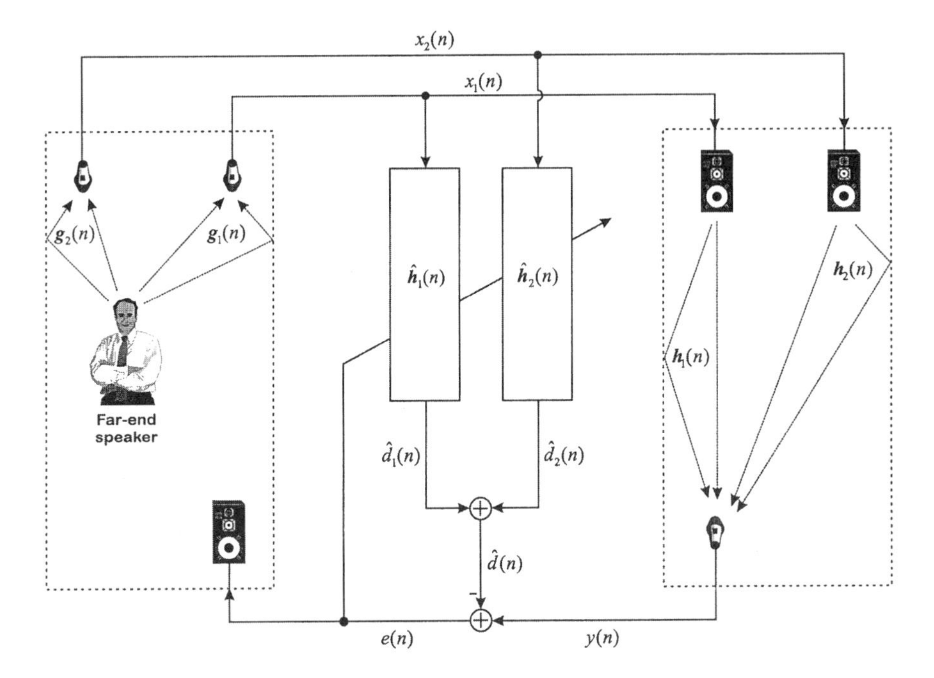

Fig. 9.31 Basic scheme for stereophonic acoustic echo cancellation. A speaker (source) in the far-end room is active. Due to acoustic coupling between loudspeakers and microphone, an echo signal $y(n)$ is coupled back from the local to the far-end room. Two filters with impulse responses $\hat{\boldsymbol{h}}_1(n)$ and $\hat{\boldsymbol{h}}_2(n)$, respectively, are adapted in order to cancel the echo.

9.2.1.1 Problems

In analogy to the situation for a single-channel device, the echo cancellation filters have finite-length impulse responses $\hat{\boldsymbol{h}}_1(n)$ and $\hat{\boldsymbol{h}}_2(n)$. With such confined sets of filter coefficients, it is impossible to perfectly model the infinite impulse responses $\boldsymbol{h}_1(n)$ and $\boldsymbol{h}_1(n)$ of the echo paths. Therefore, any solution of the identification problem is biased or "misaligned," due to undermodelization.

In addition, a dual-channel structure as depicted in Fig. 9.31 poses the following problems with regard to the adaptation of the echo cancellation filters [11]:

- The signals $x_1(n)$ and $x_2(n)$ are strongly cross-correlated since they are filtered versions of a common source signal. For this reason, the correlation matrix[10]

$$\begin{aligned} \boldsymbol{R}_{\boldsymbol{xx}}(n) &= \mathrm{E}\left\{ \begin{bmatrix} \boldsymbol{x}_1(n) \\ \boldsymbol{x}_2(n) \end{bmatrix} \left[\boldsymbol{x}_1^T(n)\; \boldsymbol{x}_2^T(n) \right] \right\} \\ &= \begin{bmatrix} \boldsymbol{R}_{\boldsymbol{x}_1\boldsymbol{x}_1}(n) & \boldsymbol{R}_{\boldsymbol{x}_1\boldsymbol{x}_2}(n) \\ \boldsymbol{R}_{\boldsymbol{x}_2\boldsymbol{x}_1}(n) & \boldsymbol{R}_{\boldsymbol{x}_2\boldsymbol{x}_2}(n) \end{bmatrix} \end{aligned} \tag{9.117}$$

  is ill-conditioned and consequently, the performance of the adaptation algorithm is degraded. However, noise components in the excitation signals $x_1(n)$ and $x_2(n)$ help reduce this effect to a certain but limited amount.

- With an ill-conditioned correlation matrix $\boldsymbol{R}_{\boldsymbol{xx}}(n)$, the misalignment of the echo cancellation filters is much worse for the dual-channel case than for the single-channel case.

- Since $\boldsymbol{x}_1(n)$ and $\boldsymbol{x}_2(n)$ are correlated, the optimal impulse responses $\hat{\boldsymbol{h}}_1(n)$ and $\hat{\boldsymbol{h}}_2(n)$ are not uniquely defined. Consequently the filter adaptations may have to follow dislocations of the remote signal source.

9.2.1.2 Proposed Methods

In general, any imaginable method should avoid affecting stereophonic perception. Several proposals for a solution of the problems listed immediately above have been made. According to Benesty et al. [11], any adaptive algorithm should be backed by a decorrelating component, in order to enhance the conditioning of the correlation matrix $\boldsymbol{R}_{\boldsymbol{xx}}(n)$ and to get a robust solution that is no longer dependent on remote room properties.

9.2.1.2.1 Decorrelation of the Excitation Signals by Inserting Noise

The addition of independent random noise to each channel may help reduce the correlation of the stereo signals and therefore enhance the performance of the echo cancellation filters (see Fig. 9.32).

[10] For simplicity, all signals are assumed to be real.

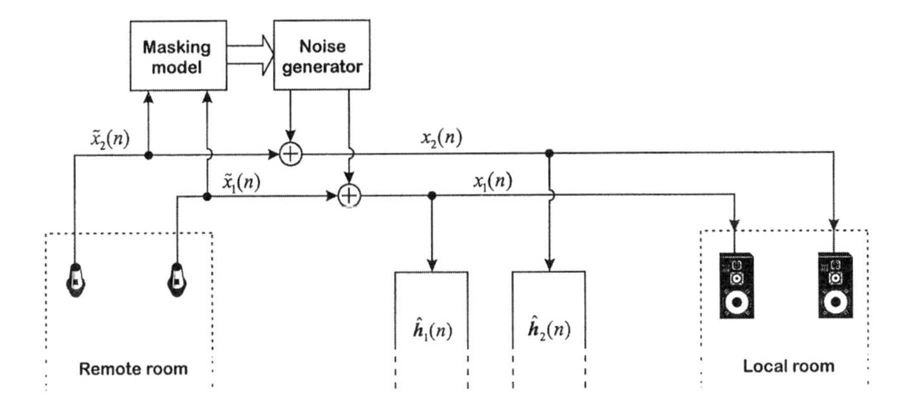

Fig. 9.32 Decorrelation of the excitation signals by inserting noise.

Unfortunately, an improvement of the conditioning of the correlation matrix requires rather high and thus disturbing noise levels [218]. A possible solution to this dilemma may be to use the auditory masking properties of the human ear. The idea consists of adding spectrally shaped random noise that is masked by the loudspeaker input signals to each channel [83]. The additional costs for this method can be kept relatively low when frequency-domain adaptive algorithms for the echo cancellation filters are applied.

9.2.1.2.2 Nonlinear Transformations

Further research has been directed toward the *application of nonlinear transformations* [11]. In principle, a small signal derived by a nonlinear function, such as half-wave rectification, of the signal itself is added to one of the excitation signals (see Fig. 9.33):

$$x_1(n) = \tilde{x}_1(n)\,, \tag{9.118}$$

$$x_2(n) = h_{\mathrm{HP},n} * \begin{cases} \tilde{x}_2(n), & \text{if } \tilde{x}_2(n) > 0\,, \\ (1+\alpha)\,\tilde{x}_2(n), & \text{else}\,. \end{cases} \tag{9.119}$$

For small amplifications ($0 \geq \alpha > 0.3$), the distortion introduced by the use of such nonlinearities is hardly perceptible. Half-wave rectification and weighted addition according to Eq. 9.119 introduc an offset. For compensation of this effect, a convolution with a highpass filter $h_{\mathrm{HP},i}$ is applied.

In combination with complementary combfiltering, the approach of nonlinearly transforming the excitation signals provides improved convergence of the echo cancellation filters without compromising spatial realism or increasing computational complexity [12]. This method is based on the observation that, in general, the signal energy below 1 kHz is decisive for auditory localization and that complementary combfiltering in frequency regions above 1 kHz does not distort stereophonic perception. In the frequency band below 1 kHz, a nonlinear transformation is carried

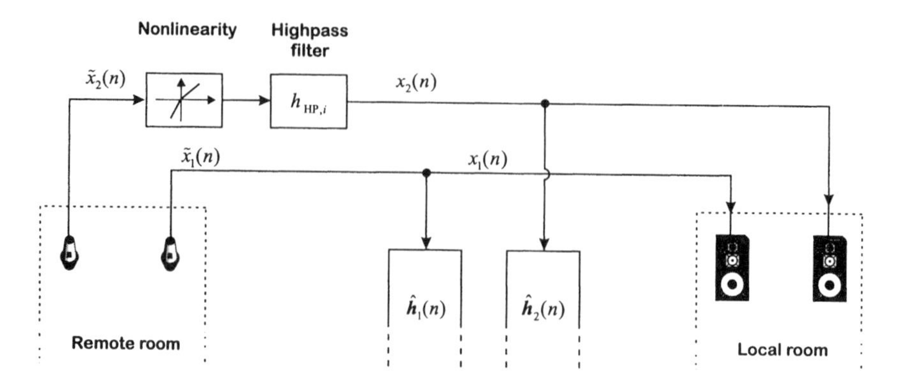

Fig. 9.33 Decorrelation of the excitation signals by nonlinear transformation.

out. Subsampling in this spectral range allows the rapidly converging frequency-domain RLS algorithm to be applied with a modest computational complexity. In the high-frequency range, two complementary combfilters perfectly cancel the correlation of the stereo signals. Therefore, the NLMS algorithm is appropriate in this frequency range.

9.2.1.2.3 Time-Variant Filtering

Another method for decorrelating stereo signals proposes the periodic delay of the excitation signal in one channel by a simple filter with time-variable coefficients [129]. The time-variant filter consists of a 2-tap FIR filter whose coefficients are controlled by a periodic function $c(n)$ with period Q. For the first $Q/2$ iterations $c(n) = 1$ and the filter output $x_2(n)$ is equal to the input signal $\tilde{x}_2(n)$. When $c(n) = 0$ for the following $Q/2$ iterations, the filter output is a one-sample delayed version $x_2(n-1)$ of the input signal. This alternating procedure is repeated every Q samples. A block diagram of the *sliding unit* is depicted in Fig. 9.34.

Unfortunately, the sudden change of the coefficient $c(n)$ every $Q/2$ iterations leads to audible distortions, clicks, in the processed signal $x_2(n)$. To avoid these clicks, $c(n)$ can be varied smoothly between zero and one over L samples. For a sampling rate of 8 kHz, the parameters Q and L were suggested to be $Q = 2000$ and $L = 200$ [221].

9.2.1.3 Dual-Channel Adaptive Algorithms

In stereo applications, the time-domain NLMS algorithm is sensitive to the conditioning of the correlation matrix $\boldsymbol{R}_{\boldsymbol{xx}}(n)$. In order to improve the speed of convergence the RLS algorithm in the frequency domain is a better candidate for a stereophonic system.

The RLS algorithm may be viewed as a performance reference [82]. With this algorithm, the mean square error converges independently of the conditioning of the

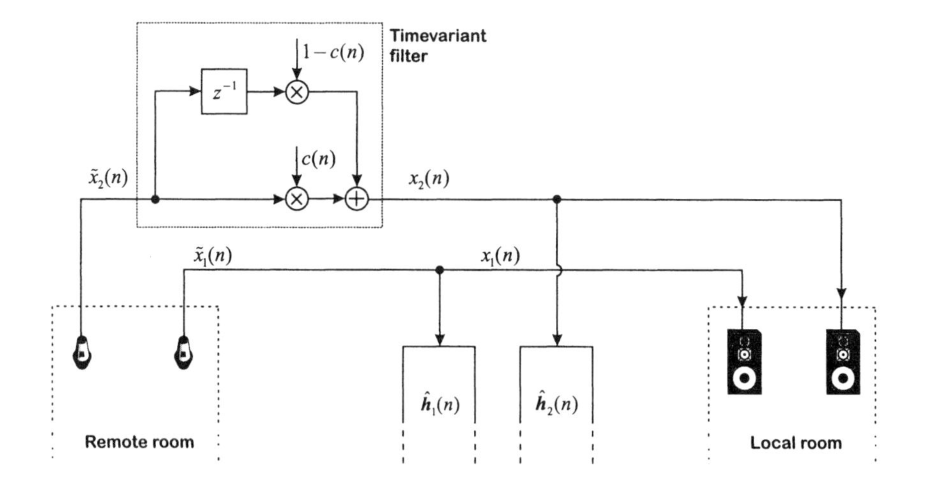

Fig. 9.34 Time-variant filter for identification of the true echo paths.

correlation matrix $\boldsymbol{R}_{\boldsymbol{xx}}(n)$. Its major drawback is high computational complexity even if it is applied in the frequency domain.

An LMS-like algorithm, the so-called extended LMS (XLMS) algorithm, can be derived from the two-channel RLS algorithm [10]. It inherently provides a partial decorrelation of the inputs $x_1(n)$ and $x_2(n)$, while roughly approximating the correlation matrix $\boldsymbol{R}_{\boldsymbol{xx}}(n)$. Therefore, this algorithm is less sensitive to correlated input signals than is the conventional LMS algorithm or the NLMS algorithm. By the introduction of leakage, which is proposed for stabilizing systems and alleviating "stalling" of adaptive coefficients, the filter update equation of the leaky XLMS algorithm may be denoted as [119]

$$\begin{bmatrix} \widehat{\boldsymbol{h}}_1(n+1) \\ \widehat{\boldsymbol{h}}_2(n+1) \end{bmatrix} = (1-\gamma) \begin{bmatrix} \widehat{\boldsymbol{h}}_1(n) \\ \widehat{\boldsymbol{h}}_2(n) \end{bmatrix} + \mu\, \boldsymbol{M}^{-1}(n) \begin{bmatrix} \boldsymbol{x}_1(n) \\ \boldsymbol{x}_2(n) \end{bmatrix} e(n), \quad (9.120)$$

with the leakage factor γ and the stepsize μ. The XLMS algorithm explicitly takes the cross-correlation between the two input signals $x_1(n)$ and $x_2(n)$ into account. This can be observed by the expansion of the matrix $\boldsymbol{M}(n)$, a (rather simplified) two-channel correlation matrix, defined by the following equation

$$\begin{aligned} \boldsymbol{M}(n) &= \begin{bmatrix} \boldsymbol{x}_1^{\mathrm{T}}(n)\,\boldsymbol{x}_1(n)\,\boldsymbol{I} & \rho\,\boldsymbol{x}_1^{\mathrm{T}}(n)\,\boldsymbol{x}_2(n)\,\boldsymbol{I} \\ \rho\,\boldsymbol{x}_1^{\mathrm{T}}(n)\,\boldsymbol{x}_2(n)\,\boldsymbol{I} & \boldsymbol{x}_2^{\mathrm{T}}(n)\,\boldsymbol{x}_2(n)\,\boldsymbol{I} \end{bmatrix} \\ &= \begin{bmatrix} p_{11}(n)\,\boldsymbol{I} & \rho\, r_{12}(n)\,\boldsymbol{I} \\ \rho\, r_{12}(n)\,\boldsymbol{I} & p_{22}(n)\,\boldsymbol{I} \end{bmatrix}, \end{aligned} \quad (9.121)$$

where ρ represents a fixed correlation coefficient that scales the sum squared cross-channel coefficient $r_{12}(n) = \boldsymbol{x}_1^{\mathrm{T}}(n)\,\boldsymbol{x}_2(n)$ and $p_{ii}(n) = \boldsymbol{x}_i^{\mathrm{T}}(n)\,\boldsymbol{x}_i(n)$, $i = 1, 2$, is the sum of squared input data in the two channels. **I** denotes the unit matrix. For

$$0 < \mu < 1, \quad (9.122)$$

$$0 \leq \rho < 1, \quad (9.123)$$

$$0 \leq \gamma \ll 1, \quad (9.124)$$

the leaky XLMS algorithm is stable.

The decoupled update term for the coefficients of the first filter in Eq. 9.120 considered in Eq. 9.121 shows the decorrelation property of this algorithm:

$$\begin{aligned}\widehat{\boldsymbol{h}}_1(n+1) &= (1-\gamma)\,\widehat{\boldsymbol{h}}_1(n) \\ &+ \frac{\mu}{p_{11}(n)p_{22}(n) - \rho^2 r_{12}^2(n)} \left[p_{22}(n)\boldsymbol{x}_1(n) - \rho\, p_{11}(n)\boldsymbol{x}_2(n)\right] e(n). \quad (9.125)\end{aligned}$$

In terms of decorrelation, leakage can be interpreted as the addition of white noise to the input signals. Compared to the addition of white noise, leakage is implemented directly in the algorithm and will therefore not affect the characteristics of the input signals.

The overall complexity is kept at a modest number of $6N$ additions and $8N$ multiplications. Thus, this approach is particularly effective for large filter lengths N, although it leads to a slight $2N$ increase of the computational cost in comparison with the conventional XLMS algorithm.

Several authors investigate the applicability of projection algorithms to stereophonic devices. By exploiting the time variance of the short-time cross-correlation functions, projection algorithms in fullband or subbands provide efficient solutions with respect to the speed of convergence and the residual misalignment [158, 210]. Fast versions of projection algorithms accomplish a minimum complexity of $6N$ + O(P), where P represents the projection order.

9.2.2 Multichannel Systems

For applications such as home entertainment, computer games, or advanced teleconferencing systems, multichannel surround sound is often employed. If the sound source is, for instance, a DVD player with five independent broadband channels and one subwoofer channel and the DVD player must be controlled by voice, multichannel echo cancellation is a key technology. Fig. 9.35 shows the basic setup of such a system. If the position of the television viewer is known a priori fixed beamforming (see Chapter 11) can be applied in order to enhance the signal-to-noise ratio.

Before emitting the DVD channels over the loudspeakers, one of the decorrelation methods discussed before needs to be applied. The interchannel correlation might be monitored and the "strength" of the decorrelation scheme can be adjusted according to the correlation. Afterward multichannel echo cancellation can be used to remove most of the residual echoes within the output signal of the beamformer. The resulting signal can be fed into a speech recognition system in order to allow voice control of the DVD player.

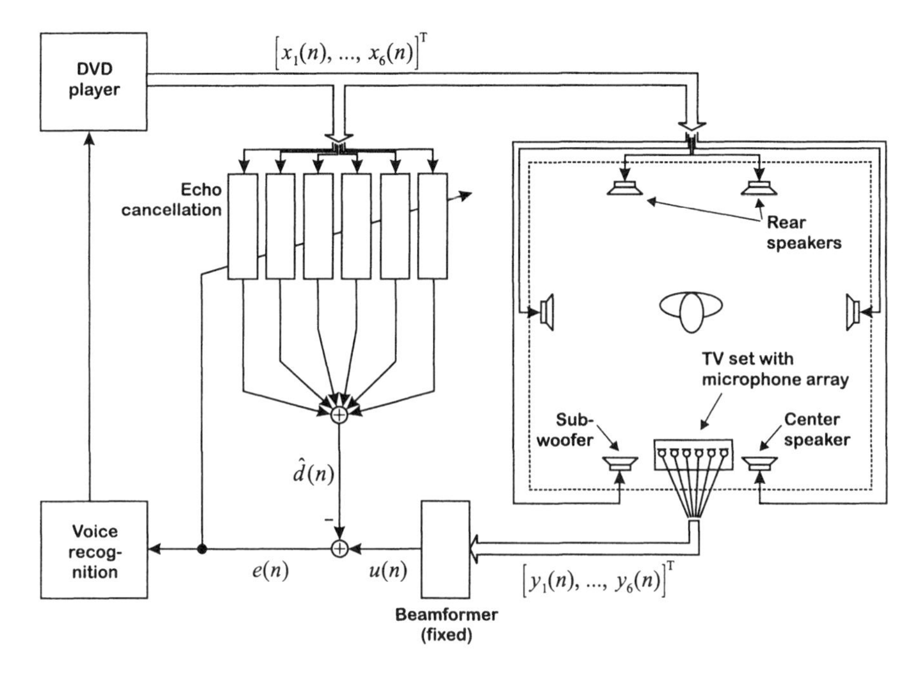

Fig. 9.35 General setup of a voice-controlled DVD player.

Applications like this might enter our lives in the near future. Even if most stereo and TV sets are controlled by a remote control panel, for some applications a voice-controlled communication between the user and the electronic device might be desirable. Whenever security aspects become important, voice control by means of speech recognition is the favorite choice for human–machine communication. Examples of such applications are dialing telephone numbers while driving a car or controlling the (audio) DVD player of a car. In these cases the visual attention of the driver can be kept at the traffic if voice control is applied. Multichannel echo cancellation allows one to "open" the speech recognizer even when music or other multichannel signals are played via the loudspeakers.

10
Residual Echo and Noise Suppression

In most applications the recording of a speech signal takes place in a noisy environment. In case of hands-free telephony the local speech signal is corrupted by background noise and echo components (see Fig. 10.1). The level of the background

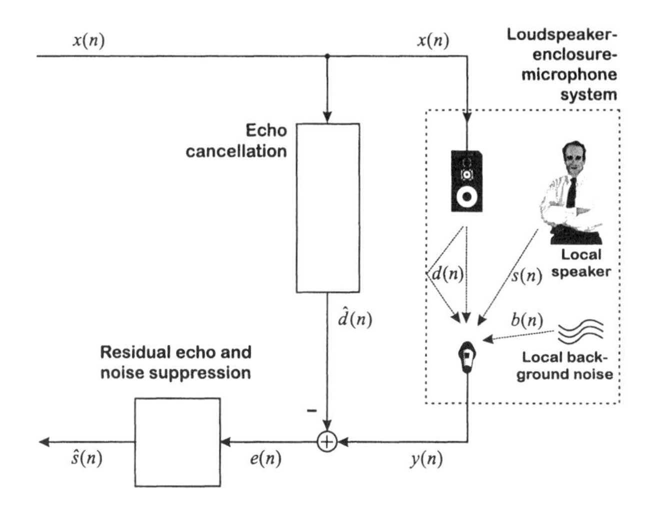

Fig. 10.1 Basic structure of a single-channel system for residual echo and noise suppression.

noise depends on the area of application. While only a moderate noise level can be assumed in quiet offices, a signal-to-noise ratio up to 0 dB might be expected if a phone call is made from a car, a train, or an airplane.

Even if echo components can be reduced by echo cancellation, as described in Chapter 9, the residual echo might still bother the remote conversation partner. In this chapter we will describe basic methods to reduce background noise as well as residual echoes. We will assume the single-channel case. Thus, we are addressing systems with only one microphone or systems that are connected to the output of a beamformer (see Chapter 11). Furthermore, only the basics of residual echo and noise suppression will be described in this chapter. Any type of control, including the estimation of the background noise power spectral density, will be described in Chapter 14.

10.1 BASICS

Before we start with the description of systems for residual echo and noise suppression, we will introduce some general aspects. The aim of these systems is to reduce the distorting components while keeping the local speech signal as natural as possible. Thus, the desired and the distorting components of the microphone $y(n)$ or the output signal of an echo cancellation stage $e(n)$, respectively, need to be separated (see Fig. 10.1). In order to achieve this, the input signal is split into overlapping blocks of appropriate size (e.g., 32 ms). Within such a block the local speech signal is assumed to be stationary. Each block is transformed via a filterbank or a DFT into the frequency domain. In order to remove the distorting echo and noise components, each subband or frequency bin is weighted with an attenuation factor, which depends on the current signal-to-noise ratio. Additionally, postprocessing such as pitch-adaptive filtering [224] or frequency shifting[1] [112] can be applied. The resulting representation of the enhanced signal spectrum is transformed back into the time domain (see Fig. 10.2).

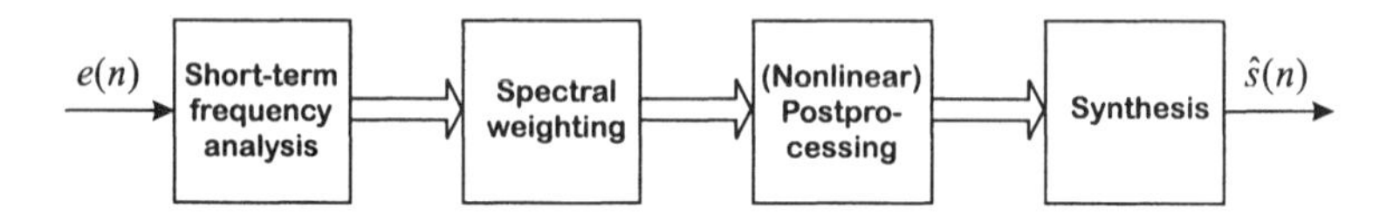

Fig. 10.2 Basic units of a noise suppression system.

This basic principle is common to most systems for residual echo and noise suppression. Thus, in Section 10.1.1 we will introduce appropriate transformations. For computing the spectral weighting coefficients, several characteristics have been proposed. Section 10.1.2 gives an overview about the most popular ones.

[1] In some hands-free telephone systems the bandwidth of the recorded signal is larger than the bandwidth of the transmission system. Instead of simply removing the surplus frequency components, a frequency shifting procedure can be applied. In case of fricative sounds like /s/ or /f/ the frequency range from, for example, 4 to 5 kHz is copied to the range between 3 and 4 kHz, resulting in a better speech intelligibility. For further details, see Section 10.3.1.3.

10.1.1 Echo and Noise Suppression in the Frequency Domain

For the analysis of the input signal $e(n)$ of a residual echo and noise suppression scheme two major types of short-term spectral decompositions have been proposed:

1. Schemes that are matched to the properties of the human auditory system, which might be regarded as the receiver (sink) of a noise reduced signal.

2. Schemes that are matched to the properties of speech signals. These analyses are optimized for the desired source of the noise reduction task.

For type 1, the frequency and time resolution is distributed nonuniformly. Often cascaded filterbanks as described in Section 9.1.3.2.2 or wavelet-based schemes are utilized. As in the human ear, high frequency and low time resolution are achieved at low frequencies and vice versa for high frequencies. Nonuniform frequency decompositions will be discussed in more detail in Section 10.1.1.2.

For analysis type 2, uniformly distributed time and frequency resolution are used. Typically, DFTs or polyphase filterbanks are utilized to achieve such resolutions. The application of uniform frequency decompositions can be motivated by the properties of speech signals.

During voiced sequences a speech spectrum shows strong peaks nearly equidistantly distributed over the frequency range. The distance between two peaks represents the pitch frequency. Fig. 10.3 depicts the short-term spectrum of a voiced speech sequence, spoken by a young woman, with a pitch frequency of about 280 Hz.

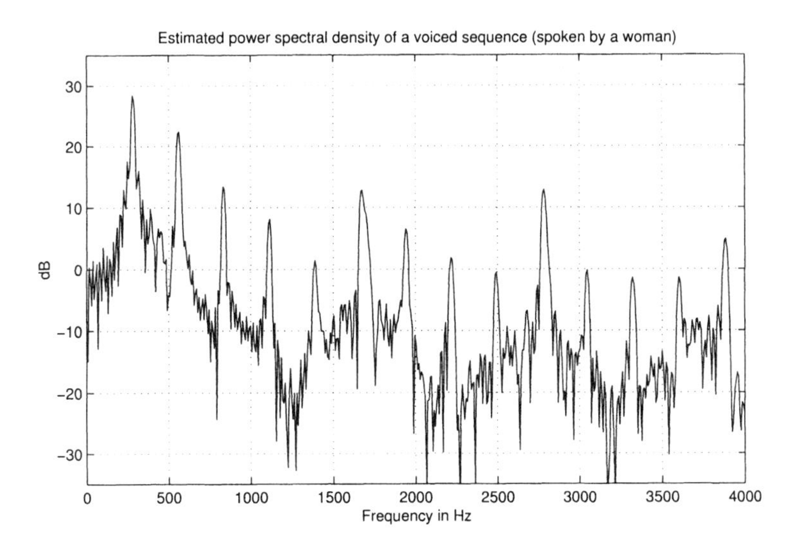

Fig. 10.3 Short-term spectrum of a voiced speech sequence.

In order to obtain a sufficient frequency resolution, the distance between two frequency supporting points should be at most half of the pitch frequency. Typical pitch frequencies are 80–350 Hz. Thus, often a frequency resolution of about 20–50 Hz is chosen. If the sample rate is $f_s = 8$ kHz a 256-point DFT results in a resolution of

$$\Delta f = \frac{f_s}{N_{\text{FFT}}} = \frac{8000\,\text{Hz}}{256} \approx 30\,\text{Hz}\,. \tag{10.1}$$

Note that Δf does not represent the "real" frequency resolution. Due to windowing, a lower resolution of about 2–3 times the DFT resolution can usually be achieved.

Uniform frequency decomposition schemes and their advantages and disadvantages with respect to residual echo and noise suppression are briefly described in the next section.

10.1.1.1 Analysis–Synthesis Schemes Matched to the Properties of the Source

The most popular uniform analysis–synthesis scheme performs periodically DFTs and inverse DFTs of overlapping signal segments. The basic structure of such a system is depicted in Fig. 10.4.

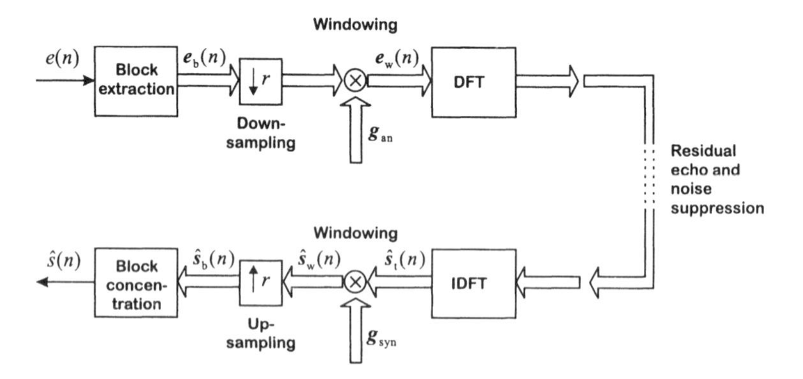

Fig. 10.4 Basic building blocks of a DFT-based analysis–synthesis system.

In the following we will describe briefly the building blocks of the system presented in Fig. 10.4 and the resulting restrictions for the choice of the subsampling rate r and the analysis and synthesis window functions $\mathbf{g}_{\text{An}}$ and $\mathbf{g}_{\text{Syn}}$, respectively. The segmentation can be described by extracting the last M samples of the input signal $e(n)$ and combining these samples to a vector

$$\mathbf{e}_{\text{b}}(n) = \big[e(n), e(n-1), \ldots, e(n-M+1)\big]^{\text{T}}. \tag{10.2}$$

Usually successive segments are overlapping by 50% or 75%. We will model this overlapping by subsampling the vectors with $r = M/2$ or $r = M/4$, respectively. Each vector is multiplied by a window function

$$\mathbf{g}_{\text{An}} = \big[g_{\text{An},0},\, g_{\text{An},1},\, \ldots,\, g_{\text{An},M-1}\big]^{\text{T}}. \tag{10.3}$$

For the resulting windowed and subsampled vectors $\boldsymbol{e}_\mathrm{w}(n)$, we obtain

$$\begin{aligned}\boldsymbol{e}_\mathrm{w}(n) &= \boldsymbol{e}_\mathrm{b}(nr) \otimes \boldsymbol{g}_\mathrm{An} \\ &= \begin{bmatrix} e(nr)\, g_{\mathrm{An},0} \\ e(nr-1)\, g_{\mathrm{An},1} \\ \vdots \\ e(n-M+1)\, g_{\mathrm{An},M-1} \end{bmatrix} .\end{aligned} \tag{10.4}$$

The symbol $\otimes$ denotes elementwise multiplication. Applying a DFT to the vector $\boldsymbol{e}_\mathrm{w}(n)$ results in M frequency supporting points of the short-term spectrum of the current frame:

$$E(\Omega_\mu, n) = \sum_{k=0}^{M-1} e(nr-k)\, g_{\mathrm{An},k}\, e^{-j\Omega_\mu k} . \tag{10.5}$$

The frequency supporting points Ω_μ are equidistantly distributed over the normalized frequency range:

$$\Omega_\mu = \frac{2\pi}{M}\mu \quad \text{with } \mu \in \{0\,\ldots,\, M-1\} . \tag{10.6}$$

After weighting the frequency bins $E(\Omega_\mu, n)$ with appropriate attenuation factors in order to perform a residual echo and noise reduction we obtain the enhanced spectral coefficients $\widehat{S}(\Omega_\mu, n)$. For synthesizing the output signal first an inverse DFT is performed:

$$\hat{s}_{\mathrm{t},i}(n) = \frac{1}{M} \sum_{\mu=0}^{M-1} \widehat{S}(\Omega_\mu, n)\, e^{j\Omega_\mu i} . \tag{10.7}$$

The resulting signals $\hat{s}_{\mathrm{t},i}(n)$ are combined to a vector

$$\hat{\boldsymbol{s}}_\mathrm{t}(n) = \left[\hat{s}_{\mathrm{t},0}(n), \hat{s}_{\mathrm{t},1}(n)), \ldots, \hat{s}_{\mathrm{t},M-1}(n)\right]^\mathrm{T} . \tag{10.8}$$

As in the analysis stage a window function $\boldsymbol{g}_\mathrm{Syn}$ can be applied:

$$\begin{aligned}\hat{\boldsymbol{s}}_\mathrm{w}(n) &= \hat{\boldsymbol{s}}_\mathrm{t}(nr) \otimes \boldsymbol{g}_\mathrm{Syn} \\ &= \begin{bmatrix} \hat{s}_{\mathrm{t},0}(n)\, g_{\mathrm{Syn},0} \\ \hat{s}_{\mathrm{t},1}(n)\, g_{\mathrm{Syn},1} \\ \vdots \\ \hat{s}_{\mathrm{t},M-1}(n)\, g_{\mathrm{Syn},M-1} \end{bmatrix} .\end{aligned} \tag{10.9}$$

In many applications synthesis windowing is omitted ($g_{\mathrm{Syn},i} = 1$). In case of time-invariant weighting coefficients, such as for equalization purposes in the DFT domain, the synthesis windowing does not have any positive effect; on the contrary, due to synthesis windowing, the spectral resolution of the analysis window is reduced. This effect will become clear when we formulate the conditions for perfect reconstruction of the through-connected analysis–synthesis system. Nevertheless, in case of time-varying spectral weighting coefficients synthesis windowing has the positive effect of reducing distortions at frame boundaries [222], especially when the average spectral attenuation has changed significantly from one frame to another.

For deriving the conditions for perfect reconstruction we assume all coefficients of the residual echo and noise suppression to be one. Thus, we have

$$\widehat{S}(\Omega_\mu, n) = E(\Omega_\mu, n)\,. \tag{10.10}$$

Computing a DFT and afterwards its inverse does not change the input vector. For this reason, we also have

$$\hat{\mathbf{s}}_{\mathrm{t}}(n) = \mathbf{e}_{\mathrm{w}}(n)\,. \tag{10.11}$$

In order to achieve perfect reconstruction the sum of shifted versions of the product of the analysis and synthesis window should always be one. If we define all coefficients $g_{\mathrm{An},i}$ and $g_{\mathrm{Syn},i}$ to be zero for $i < 0$ and $i \geq M$ we can denote the condition for perfect reconstruction as

$$\sum_{m=-\infty}^{\infty} g_{\mathrm{An},i\,+\,mr}\, g_{\mathrm{Syn},i\,+\,mr} = 1 \quad \text{for } i \in \{0,\, \ldots,\, M-1\}\,. \tag{10.12}$$

The parameter r is denoting the subsampling rate. If no synthesis windowing is applied and the subsampling rate is chosen as $r = M/4$ weighted Hanning windows according to

$$g_{\mathrm{An},i} = \begin{cases} \frac{1}{4}\left[1 - \cos\left(\frac{2\pi}{M} i\right)\right], & \text{for } i \in \{0,\, \ldots,\, M-1\}, \\ 0 & \text{else}, \end{cases} \tag{10.13}$$

$$g_{\mathrm{Syn},i} = \begin{cases} 1, & \text{for } i \in \{0,\, \ldots,\, M-1\}, \\ 0 & \text{else}, \end{cases} \tag{10.14}$$

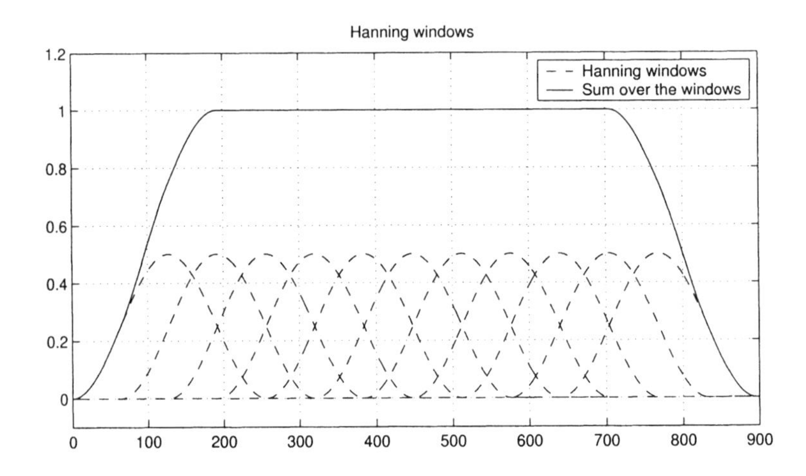

Fig. 10.5 Weighted Hanning windows and the corresponding sum.

fulfill the perfect reconstruction condition. The sum over a few weighted Hanning windows of length $M = 256$ is depicted in Fig. 10.5. At the beginning and at the end of the sequence the block processing results in a smoothed fade-in and a smoothed fade-out over two adjacent frames.

If synthesis windowing should be applied, the square root of a weighted Hanning window can be utilized for the analysis as well as for the synthesis stage:

$$g_{\mathrm{An},i} = g_{\mathrm{Syn},i} = \begin{cases} \frac{1}{2}\sqrt{1 - \cos\left(\frac{2\pi}{M} i\right)} & \text{for } i \in \{0, \ldots, M-1\}, \\ 0 & \text{else.} \end{cases} \tag{10.15}$$

In this case the subsampling rate has to be $r = M/4$. Further window functions that can be applied for this choice are weighted Hamming windows. If a larger subsampling rate is required, weighted Bartlett windows and $r = M/2$ can be used [109].

When applying synthesis windowing, one should be aware of the reduced frequency resolution within the analysis stage. Taking the square root of the analysis window results in the frequency domain in a narrower main lobe and also in smaller attenuation outside it. For visualizing this effect, a Hanning window as well as its square root and the corresponding frequency responses are depicted in Fig. 10.6.

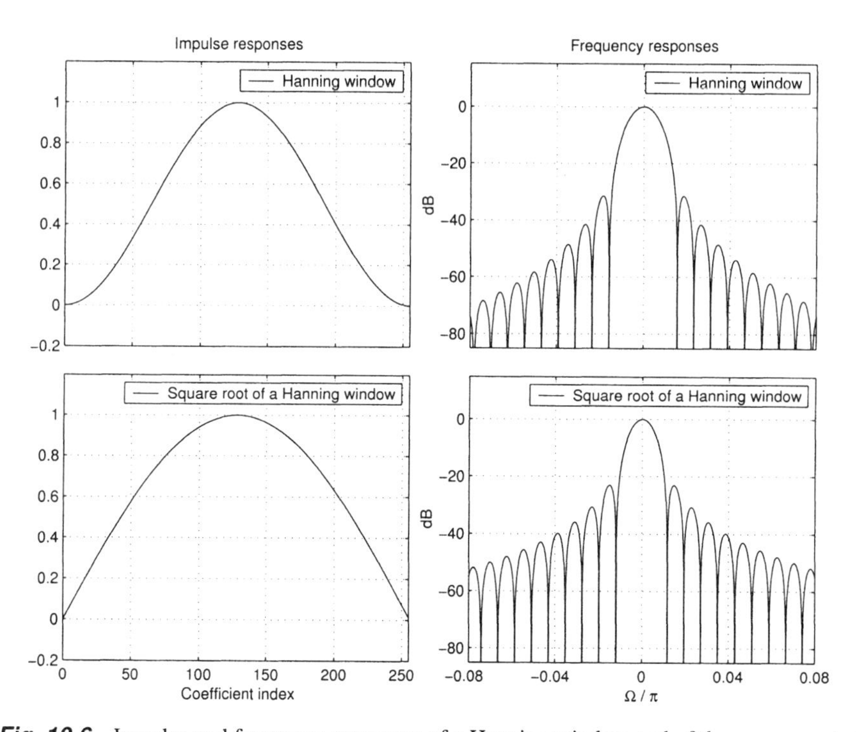

Fig. 10.6 Impulse and frequency responses of a Hanning window and of the square root of this window. Note that the impulse responses have been normalized to the range from 0 to 1 and the frequency responses were normalized to 0 dB at $\Omega = 0$.

When comparing the frequency responses (e.g., at $\Omega = 0.05\,\pi$), the attenuation decreases from 55 dB without synthesis windowing to 45 dB with synthesis windowing. The distance $\Omega = 0.05\,\pi$ corresponds approximately to 2 times the distance between adjacent frequency supporting points of a DFT with 256 input samples.

If synthesis windowing should be applied and the frequency selectivity of the analysis should be enhanced at the same time, the DFT and its inverse can be extended to polyphase filterbanks. To achieve this, only preceding frames need to be added before performing the DFT. Furthermore, the window function g_{An} has to be replaced by a so-called prototype lowpass filter. Fig. 10.7 gives an overview of both structures.

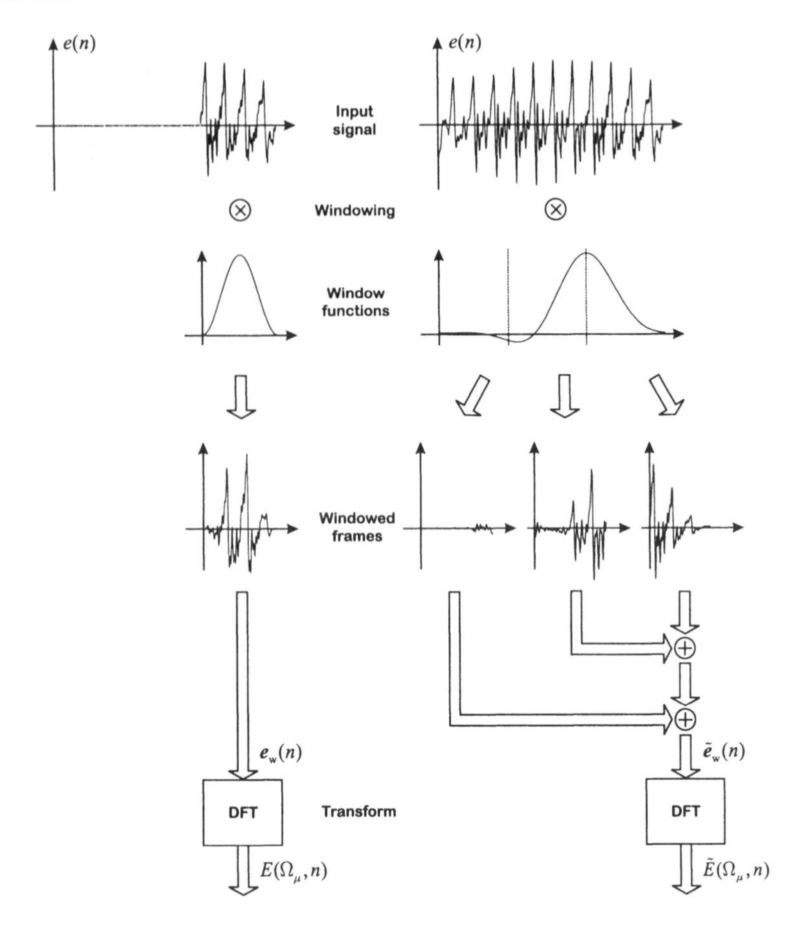

Fig. 10.7 Analysis performed by a DFT (left) and by a polyphase filterbank (right).

A detailed description of DFT-modulated polyphase filterbanks as well as the derivation of the structure presented in Fig. 10.7 and its inverse can be found in Section 9.1.3.2.3. The design of prototype lowpass filters is described in detail in Appendix B.

Using more than one frame for the current spectral analysis leads to a better frequency resolution. In the right part of Fig. 10.8 the frequency responses of a Hanning window with $M = 256$ coefficients and of a prototype lowpass filter are depicted. To be comparable to the simple DFT structure the lowpass filter was designed for a

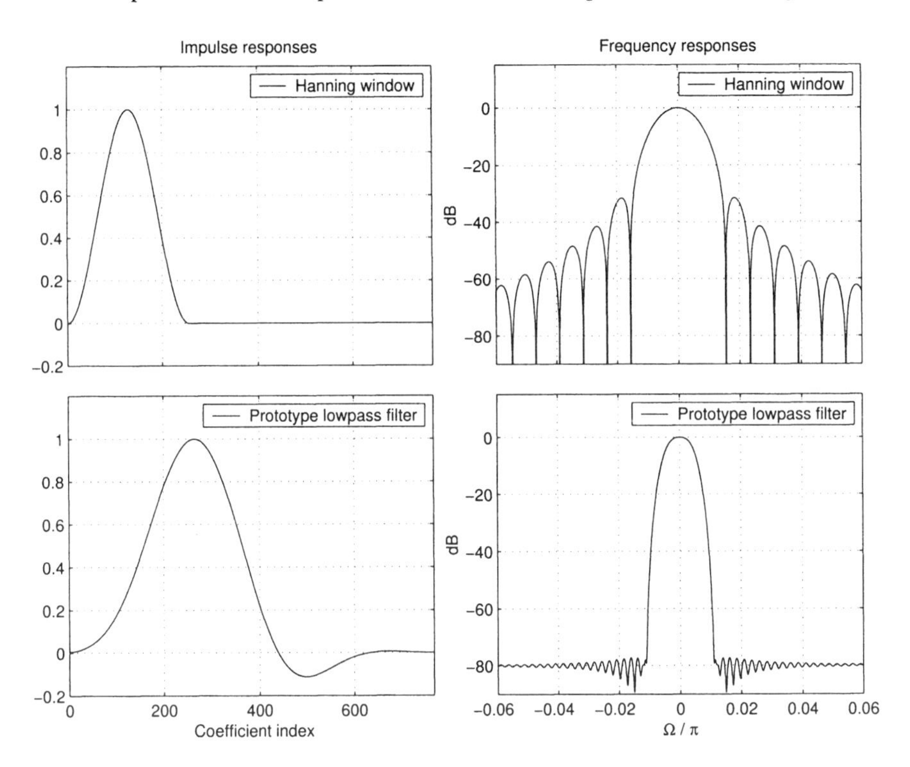

Fig. 10.8 Impulse and frequency responses of a Hanning window and of a prototype lowpass filter. The prototype filter was designed for a polyphase filterbank, having the same frequency resolution as the DFT preceding the Hanning window ($M = 256$). Note that the impulse responses have been normalized to 1 and the frequency responses were normalized to 0 dB at $\Omega = 0$.

filterbank with $M = 256$ channels. Its length was set to $N_{\text{LP}} = 768$, which means that three adjacent (nonoverlapping) frames are used for analyzing the signal. The impulse responses of the Hanning window and of the prototype filter are depicted in the left part of Fig. 10.8.

When comparing the frequency responses, the advantages of the filterbank become clearly visible:

- The spectrum of the prototype filter has a narrower mainlobe.
- Sidelobes are negligible.

Nevertheless, the filterbank structure has also disadvantages. As we see in the left part of Fig. 10.8, the impulse response of the prototype filter is much longer than the one of the Hanning window. This results in a reduced time resolution of the filterbank. If the input signal changes its spectral characteristic within the time corresponding to the memory size of the analysis stage of the filterbank, the short-time spectrum is smoothed. For very long prototype filters (e.g., $N_{\text{LP}} > 1000$ at 8 kHz sampling rate) the spectral smoothing results in *postechoes* [52].

Postechoes appear at sudden changes of the coefficients of the residual echo and noise suppression filter. Due to a long prototype filter, the synthesis also exhibits a long memory. Filling this memory with large amplitudes during a speech sequence and inserting only very small amplitudes afterward leads to artifacts during the output of the stored large samples.

Anyhow, if the length of the prototype lowpass filter is chosen not too large, filterbanks are a good candidate for analysis–synthesis systems with uniform frequency resolution.

10.1.1.2 Analysis–Synthesis Schemes Matched to the Properties of the Sink

The main motivation of applying analysis–synthesis schemes with nonuniform frequency resolution is the perception property of the human ear. The human auditory system is able to separate low frequencies more precisely than high frequencies. Applying short analysis windows for the high-frequency components and high-order windows for the lower frequencies leads to a high spectral resolution in the lower bands and to low resolutions at higher frequencies. Since the human perception of audio signals also works in a logarithmic scale, this way of signal analysis matches the properties of the sink (a human listener) of a noise reduction scheme.

The resolution of human auditory systems is described by the critical band tuning curves of the inner ear. Based on psychoacoustic experiments [250], the frequency range is divided into *critical bands*. Fig. 10.9 shows the width of the critical bands in dependence of the center frequency. Nonuniform frequency analysis schemes are tuned mostly to achieve the resolution depicted in Fig. 10.9.

The concept of critical bands leads to a nonlinearly warped frequency scale ξ called the *Bark scale.* [2] The unit of this frequency scale is Bark, where each critical band has a bandwidth of 1 Bark. The transformation of the frequency f into the Bark scale is approximately given by [250]

$$\frac{\xi}{\text{Bark}} = 13 \arctan\left(\frac{0.76\,f}{\text{kHz}}\right) + 3.5 \arctan\left[\left(\frac{f}{7.5\,\text{kHz}}\right)^2\right]. \qquad (10.16)$$

One possibility to obtain a nonuniform frequency resolution according to the Bark scale is to modify the polyphase filterbank presented in Section 9.1.3.2.3 by replacing the delay elements of the input and the output delay chain by allpass filters [90, 130]. The structure of the allpass-transformed filterbank is shown in Fig. 10.10 .

[2]The term *Bark* was chosen in memory of the scientist Heinrich Barkhausen (1881–1956), who introduced the *phon*: a value describing the loudness level.

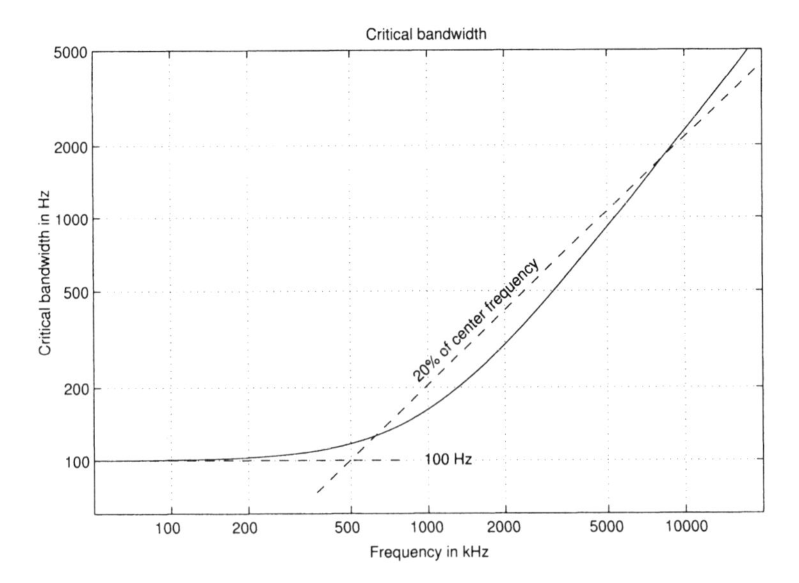

Fig. 10.9 Critical bandwidth as a function of frequency. At lower frequencies (below 500 Hz) the widths of the critical bands can be approximated by 100 Hz. At higher frequencies (above 500 Hz) critical bands have a width of about 20% of their center frequency.

The coefficients g_i (see Fig. 10.10) belong to the so-called *prototype lowpass filter*. This filter can be designed in the same manner as for the uniform frequency resolution. Details about the filter design can be found in Appendix B.

Exchanging the delay elements z^{-1} by allpass filters of first order

$$z^{-1} \Longrightarrow \frac{\alpha z + 1}{z + \alpha} = H_{\mathrm{AP}}(z) \tag{10.17}$$

results in a transformation of the abscissa Ω of the frequency response. However, the attenuation properties of the bandpass filters are not changed. In order to obtain a stable recursive filter $H_{\mathrm{AP}}(z)$ the constant α is bounded according to

$$-1 < \alpha < 1\,. \tag{10.18}$$

Substituting z by $e^{j\Omega}$ in Eq. 10.17, the transformation of the frequency axis can be expressed as

$$\Omega_{\mathrm{new}} = 2\,\arctan\left[\frac{1+\alpha}{1-\alpha}\,\tan\left(\frac{\Omega_{\mathrm{old}}}{2}\right)\right]. \tag{10.19}$$

For $\alpha = 0$ the allpass is only a delay element and the frequency axis is not changed. If α is chosen positive ($0 < \alpha < 1$), the bandwidths of bandpass filters with input $e(n)$ and output $e_\mu(n)$ decrease monotonously with increasing center frequencies. In the opposite case, $-1 < \alpha < 0$, the bandwidths increase with rising center frequencies.

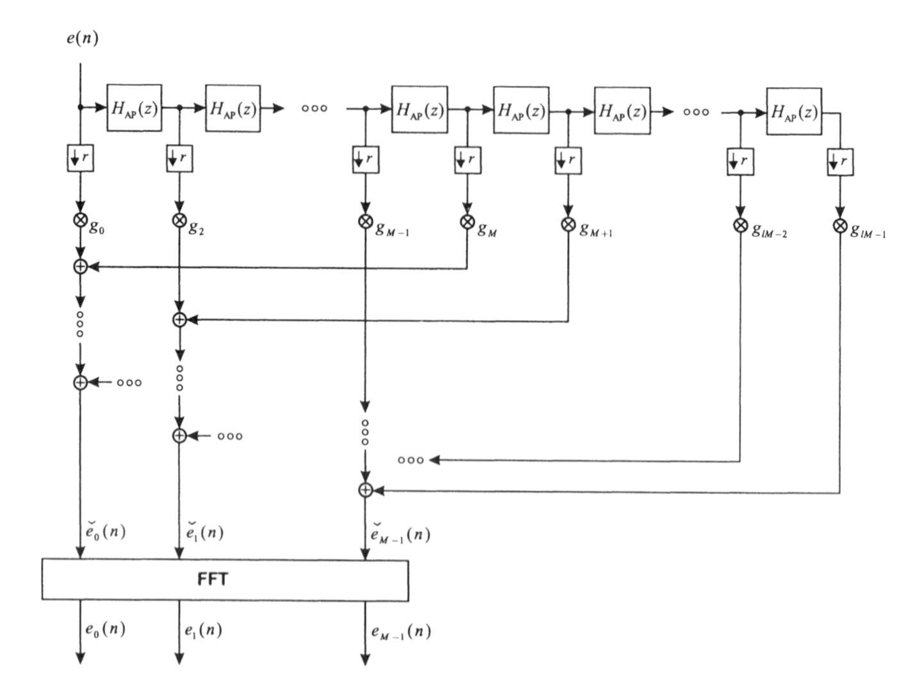

Fig. 10.10 Structure of an allpass-transformed polyphase filterbank.

In case of no subsampling $r = 1$ and using only delay elements (instead of allpass filters) the bandpass transfer functions can be expressed as

$$G_\mu\left(e^{j\Omega}\right)\Big|_{\alpha=0} = \sum_{k=0}^{N_{\mathrm{LP}}-1} g_k \left(e^{-j\Omega}\right)^k e^{-j\frac{2\pi}{M}\mu k}, \tag{10.20}$$

where N_{LP} denotes the length of the prototype lowpass filter (see Appendix B). When replacing the delay elements by allpass filters according to Eq. 10.17, the transfer functions are changed to

$$G_\mu\left(e^{j\Omega}\right) = \sum_{k=0}^{N_{\mathrm{LP}}-1} g_k \left(\frac{\alpha\, e^{j\Omega}+1}{e^{j\Omega}+\alpha}\right)^k e^{-j\frac{2\pi}{M}\mu k}. \tag{10.21}$$

In Fig. 10.11 frequency resolutions of three allpass transformed polyphase filterbanks with $M = 16$ channels are depicted. All filterbanks have been designed with the same prototype lowpass filter with $N_{\mathrm{LP}} = 128$ coefficients. In the upper diagram α was set to zero, leading to a uniform frequency resolution. In the lower two diagrams the choices $\alpha = 0.5$ and $\alpha = -0.5$, respectively, are depicted. For a sampling rate of 11.025 kHz (a quarter of the sampling rate of CD players), for instance, a transformation with $\alpha = -0.49$ results in a good approximation of the bandwidths corresponding to critical bands.

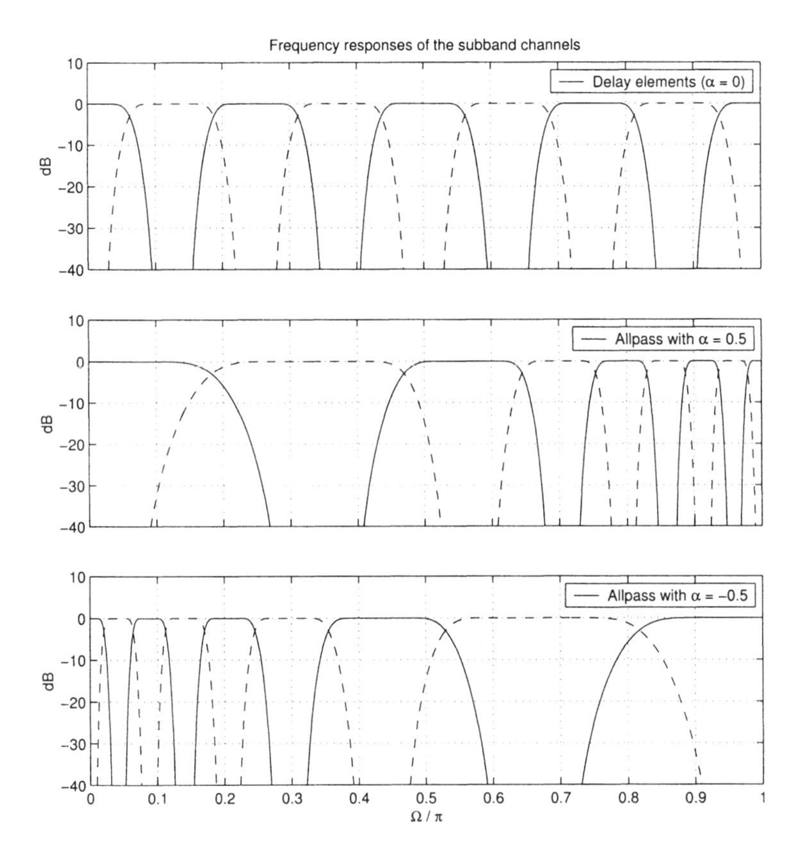

Fig. 10.11 Transfer functions of allpass-transformed polyphase filterbanks.

The synthesis stage of the filterbank is derived in the same manner than the analysis stage–all delay elements of the structure depicted in Fig. 9.25 are replaced by allpass filters.

Unfortunately, allpass-transformed filterbanks have two drawbacks:

- The cascade of allpass filters of order one (a through-connected analysis–synthesis system assumed) produces an output signal $\widehat{s}(n)$ whose phase is disturbed. The correction of the phase is possible only in an approximate manner. Furthermore, the phase correction introduces a large delay in the signal path.

- By replacing the delay elements with allpass filters, it is no longer possible to realize the delay line by a (computational effective) circular buffer. The outputs of the allpass filters have to be computed at the full sampling rate resulting in a large computational complexity.

Nevertheless, it has been reported [90] that the application of this type of filterbank (compared to DFT analysis) in combination with spectral subtraction noise reduction rules (see Section 10.1.2.2) increases the quality of background noise suppression schemes.

Another possibility to obtain different bandwidths of the subbands is to use a two-stage cascaded polyphase filterbank according to Fig. 10.12. First, the signal is split

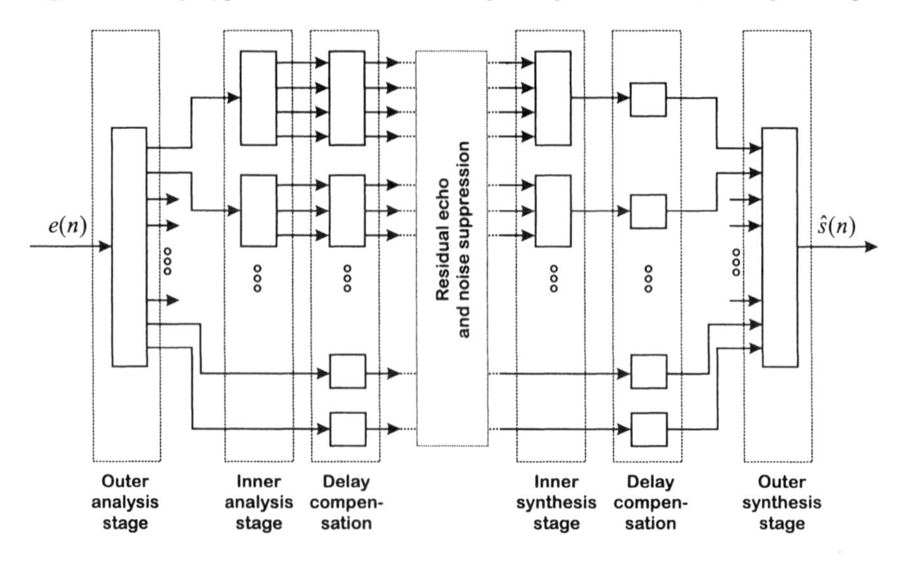

Fig. 10.12 Two-stage cascaded filterbank.

up into a small number of subbands. After downsampling the subbands are split up a second time using filterbanks with different frequency resolutions. The number of subbands of the second stage is chosen with respect to the Bark scale.

Because of the different frequency resolutions of the inner filterbank, the delay introduced in each subband is also different. A possible way to compensate for this would be to add an appropriate number of delay elements behind the inner synthesis filterbanks. Unfortunately, noise reduction schemes often take adjacent subbands into account, such as for applying psychoacoustic models. Thus, one has to ensure that each frequency component of the input signal $e(n)$ undergoes the same delay before reaching the residual echo and noise suppression stage. For this reason, we have introduced a two-stage delay compensation: the first one for compensating different delays within the inner analysis stage and the second one for equalizing the delay of the inner synthesis stage (see Fig. 10.12). Equal gain of each subband decomposition must also be considered.

The design of each filterbank can be performed according to Appendix B. Dreiseitel and Puder [55] suggested the following decomposition scheme for a sampling rate of 8 kHz:

- For the outer stage a polyphase filterbank as described in Section 9.1.3.2.3 with 16 channels was used. As the input signal $e(n)$ is real, the first stage delivers symmetric subbands. Thus, only the lower half of the channels has to be processed.

- The individual subbands are decomposed within the second stage into 32 channels at low frequencies. At high frequencies no further splitting is applied. The decomposition parameters are listed in Table 10.1 .

Table 10.1 Frequency resolutions of the inner filterbanks (sampling rate of the entire system is 8 kHz)

Frequency range in Hz	Bandwidth in Hz	Number of subbands
0– 250	16	32
250– 750	16	32
750–1250	31	16
1250–1750	31	16
1750–2250	62	8
2250–2750	125	4
2750–3250	125	4
3250–3750	500	1
3750–4000	250	1

The computational complexity of cascaded filterbanks is much lower than the complexity of allpass-transformed filterbanks, since very efficient ways exist to implement polyphase structures (see Section 9.1.3.2.3). However, the delay introduced by cascaded filterbanks is not negligible. It consists of the window length of the outer stage plus the largest window length of the inner stage multiplied by the outer subsampling rate. The resulting delay is comparable to a uniformly spaced frequency decomposition having the finest resolution of the cascaded filterbank.

Nevertheless, cascaded filterbanks show advantages if also echo cancellation should be performed. In this case the outer filterbank can be designed such that aliasing within each outer subband is kept very small. This leads to a good performance of the echo cancellation filters (see Section 9.1.3). Subtraction of the estimated echo components can be performed between the outer and the inner analysis stage (see Fig. 10.13). The aliasing requirements for the inner filterbanks are not critical any more. If the delay of the inner filterbank should be kept small, a DFT/IDFT structure with large overlapping can be applied. If no subsampling is applied, no further delay is introduced. In this case the DFT can be computed recursively by sliding window techniques [209].

10.1.2 Filter Characteristics

Before distinguishing between residual echo and background noise suppression (see Sections 10.2 and 10.3, respectively), we will introduce several filter characteristics

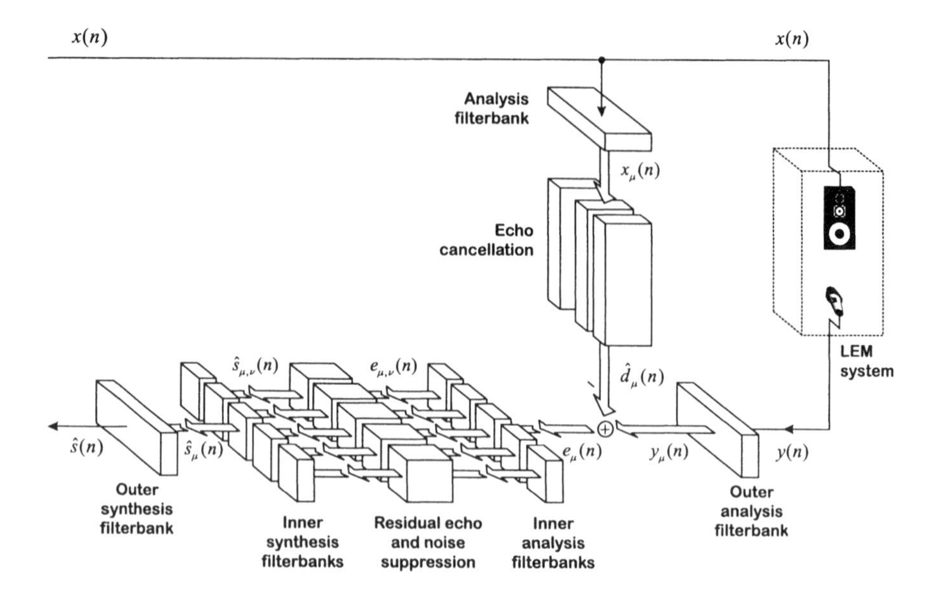

Fig. 10.13 Structure of a cascaded filterbank system utilized for echo cancellation and residual echo and noise suppression.

which can be applied for both types of suppression. Finally, the combination of residual echo and background noise suppression will be presented in Section 10.4.

The filter characteristics will differ in their "aggressiveness," their robustness against speech distortions and residual noise components, and their complexity. Before applying one of the characteristics, extensions such as overestimation of the noise spectrum and limitation of the attenuation should be applied. These extensions—which are especially suited for background noise suppression—are described in Chapter 14. Furthermore, several methods for estimating the power spectral density of background noise are presented in Chapter 14. The aim of this section is to describe how the filter characteristics can be applied, assuming that all necessary quantities were already estimated.

The input signal for a residual echo and noise reduction scheme is usually the error signal $e(n)$ of an echo cancellation unit (see Fig. 10.1). As described in the previous chapters, $e(n)$ is obtained by subtracting the estimated echo signal $\widehat{d}(n)$ from the microphone signal $y(n)$:

$$e(n) = y(n) - \widehat{d}(n)\,. \tag{10.22}$$

The microphone signal consists of the local speech signal $s(n)$, local background noise $b(n)$, as well as echo components $d(n)$:

$$y(n) = s(n) + b(n) + d(n)\,. \tag{10.23}$$

By subtracting the estimated echo signal $\widehat{d}(n)$ from the real echo signal $d(n)$, we obtain the residual echo signal $e_\mathrm{u}(n)$:

$$e_\mathrm{u}(n) = d(n) - \widehat{d}(n)\,. \tag{10.24}$$

By inserting Eqs. 10.23 and 10.24 into Eq.10.22 we can separate the input signal $e(n)$ of a suppression scheme into three independent components: speech, background noise, and residual echo:

$$e(n) = s(n) + b(n) + e_\mathrm{u}(n)\,. \tag{10.25}$$

For derivation of the filter characteristics within the next few sections, we will not distinguish between the two distortions $b(n)$ and $e_\mathrm{u}(n)$. Thus, both components are combined as

$$\tilde{n}(n) = b(n) + e_\mathrm{u}(n)\,. \tag{10.26}$$

10.1.2.1 Wiener Filter

One of the most basic filter characteristics has been derived by Wiener (see Chapter 5). Assuming all signals to be stationary and minimizing $\mathrm{E}\left\{\tilde{n}^2(n)\right\}$ leads to

$$\begin{aligned} H_\mathrm{W}\left(e^{j\Omega}\right) &= \frac{S_{ss}(\Omega)}{S_{ee}(\Omega)} \\ &= 1 - \frac{S_{bb}(\Omega) + S_{e_u e_u}(\Omega)}{S_{ee}(\Omega)} \\ &= 1 - \frac{S_{\tilde{n}\tilde{n}}(\Omega)}{S_{ee}(\Omega)}\,. \end{aligned} \tag{10.27}$$

For deriving the last line in Eq. 10.27 we assume the local background noise $b(n)$, the residual echo signal $e_\mathrm{u}(n)$, and the local speech signal $s(n)$ to be mutually orthogonal. By introducing the frequency-selective signal-to-noise ratio according to

$$SNR(\Omega) = \frac{S_{ss}(\Omega)}{S_{\tilde{n}\tilde{n}}(\Omega)} \tag{10.28}$$

we can rewrite Eq. 10.27 as

$$H_\mathrm{W}\left(e^{j\Omega}\right) = \frac{SNR(\Omega)}{SNR(\Omega)+1}\,. \tag{10.29}$$

The Wiener characteristic attenuates frequency components that have a low signal-to-noise ratio. Spectral components with large signal-to-noise ratios are barely attenuated.

Due to the nonstationary nature of speech signals, the signal-to-noise ratio according to Eq. 10.28 can only be estimated on a limited time basis. Thus, when computing the filter according to Eq. 10.27, the power spectral densities $S_{ee}(\Omega)$ and $S_{\tilde{n}\tilde{n}}(\Omega)$ are replaced by their short-term estimates $\widehat{S}_{ee}(\Omega,n)$ and $\widehat{S}_{\tilde{n}\tilde{n}}(\Omega,n)$, respectively. Furthermore, the power spectral densities are estimated only at discrete frequencies:

$$\Omega_\mu = \frac{2\pi}{M}\mu \quad \text{with } \mu \in \{0,\, \ldots,\, M-1\}\,. \tag{10.30}$$

Due to errors within the estimation process, $\widehat{S}_{\tilde{n}\tilde{n}}(\Omega_\mu,n)$ might be temporarily larger than $\widehat{S}_{ee}(\Omega_\mu,n)$. In this case the frequency characteristic $H_{\mathrm{W}}\left(e^{j\Omega_\mu},n\right)$ would become negative. To avoid this effect, it is bounded by zero:

$$\widehat{H}_{\mathrm{W}}\left(e^{j\Omega_\mu},n\right) = \max\left\{1 - \frac{\widehat{S}_{\tilde{n}\tilde{n}}(\Omega_\mu,n)}{\widehat{S}_{ee}(\Omega_\mu,n)},\, 0\right\}. \tag{10.31}$$

The attenuation of the Wiener filter in relation to the signal-to-noise ratio is depicted in Fig. 10.14. Among all filter characteristics presented within this chapter, the Wiener filter has the highest attenuation for a given signal-to-noise ratio.

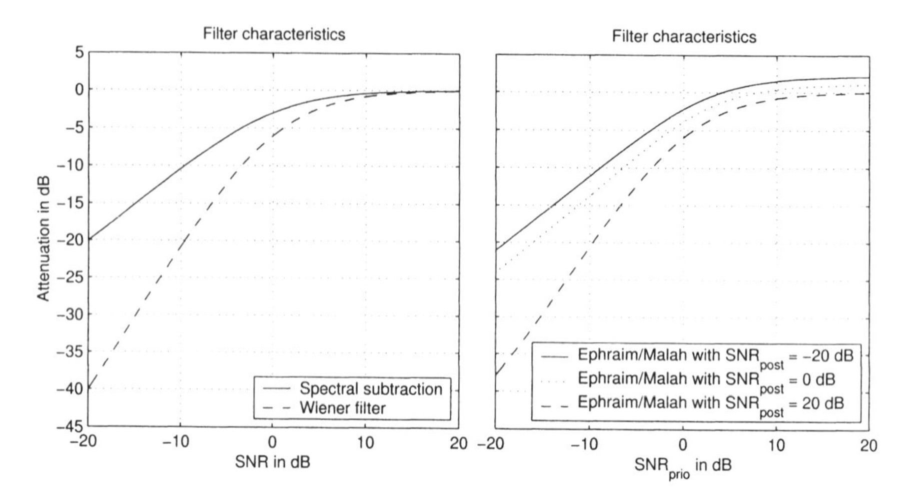

Fig. 10.14 Filter characteristics.

10.1.2.2 Spectral Subtraction

Another well-known noise reduction method is called *spectral subtraction* [18], an approach based on the subtraction of the distorting spectral components. As in the Wiener approach, we assume mutual orthogonality of all components of the error signal $e(n)$. Thus, we can write for the power spectral density of the error signal

$$\begin{aligned} S_{ee}(\Omega) &= S_{ss}(\Omega) + S_{bb}(\Omega) + S_{e_u e_u}(\Omega) \\ &= S_{ss}(\Omega) + S_{\tilde{n}\tilde{n}}(\Omega)\,. \end{aligned} \tag{10.32}$$

Rearranging the sum in the last line of Eq. 10.32 leads to

$$S_{ss}(\Omega) = \left(1 - \frac{S_{\tilde{n}\tilde{n}}(\Omega)}{S_{ee}(\Omega)}\right) S_{ee}(\Omega)\,. \tag{10.33}$$

The term in brackets can be interpreted as the squared magnitude of a frequency response $H_{\mathrm{SS},1}(e^{j\Omega})$:

$$S_{ss}(\Omega) = \left|H_{\mathrm{SS},1}\left(e^{j\Omega}\right)\right|^2 S_{ee}(\Omega)\,. \tag{10.34}$$

By comparing Eqs. 10.33 and 10.34 we obtain the frequency response of the spectral subtraction filter $H_{\mathrm{SS},1}(e^{j\Omega})$. Except for the square root, the filter is equal to the Wiener solution:

$$\begin{aligned} H_{\mathrm{SS},1}\left(e^{j\Omega}\right) &= \sqrt{1 - \frac{S_{\tilde{n}\tilde{n}}(\Omega)}{S_{ee}(\Omega)}} \\ &= \sqrt{H_{\mathrm{W}}\left(e^{j\Omega}\right)}\,. \end{aligned} \tag{10.35}$$

Rearranging the power spectral densities, we can express $H_{\mathrm{SS},1}(e^{j\Omega})$ in dependence of the signal-to-noise ratio $SNR(\Omega)$:

$$H_{\mathrm{SS},1}\left(e^{j\Omega}\right) = \sqrt{\frac{SNR(\Omega)}{SNR(\Omega)+1}}\,. \tag{10.36}$$

As in the derivation of the Wiener characteristic, the power spectral densities have to be replaced by their short-term estimates. To accommodate estimation errors, the estimated (short-term) power spectral density of the speech signal is bounded by zero. Thus, the initial assumption (Eq. 10.32) is modified to

$$\widehat{S}_{ss}(\Omega_\mu, n) = \max\left\{\widehat{S}_{ee}(\Omega_\mu, n) - \widehat{S}_{\tilde{n}\tilde{n}}(\Omega_\mu, n), 0\right\}\,. \tag{10.37}$$

By applying this limitation also within the derivation of the time-varying filter characteristic, we get

$$\widehat{H}_{\mathrm{SS},1}\left(e^{j\Omega_\mu}, n\right) = \max\left\{\sqrt{1 - \frac{\widehat{S}_{\tilde{n}\tilde{n}}(\Omega_\mu, n)}{\widehat{S}_{ee}(\Omega_\mu, n)}},\, 0\right\}\,. \tag{10.38}$$

In the left part of Fig. 10.14 the filter characteristics of the Wiener solution and spectral subtraction are presented as functions of the signal-to-noise ratio. By comparing both characteristics, we see that the Wiener characteristic attenuates twice as much as spectral subtraction (due to the square root within the characteristic). For comparison, the right part of the figure shows the characteristics of a method by Ephraim and Malah [61, 62], which will be explained in the next section.

Alternatively a subtraction rule can be formulated for the absolute values of a spectrum. By starting with

$$\left|\widehat{S}\left(e^{j\Omega_\mu}, n\right)\right| = \max\left\{\left|E\left(e^{j\Omega_\mu}, n\right)\right| - \left|\tilde{N}\left(e^{j\Omega_\mu}, n\right)\right|, 0\right\} \tag{10.39}$$

we obtain for the frequency response of the subtraction rule for absolute values [18]:

$$\widehat{H}_{\text{SS},2}\left(e^{j\Omega_\mu}, n\right) = \max\left\{1 - \sqrt{\frac{\widehat{S}_{\tilde{n}\tilde{n}}(\Omega_\mu, n)}{\widehat{S}_{ee}(\Omega_\mu, n)}},\, 0\right\}. \tag{10.40}$$

10.1.2.3 Filtering According to Ephraim and Malah

A second filter characteristic (besides the Wiener solution) that is optimal in the mean square error (MSE) sense is the solution proposed by Ephraim and Malah [61, 62]. In contrast to the Wiener solution, which minimizes the broadband mean square error $\mathrm{E}\left\{\,(\hat{s}(n) - s(n))^2\,\right\}$, here the mean square amplitude error

$$\mathrm{E}\left\{\left[f\left(\left|\widehat{S}\left(e^{j\Omega_\mu}, n\right)\right|\right) - f\left(\left|S\left(e^{j\Omega_\mu}, n\right)\right|\right)\right]^2\right\} \longrightarrow \min \tag{10.41}$$

is minimized in the frequency domain. The expression $f(x)$ represents an arbitrary (invertible) function. If a DFT/IDFT structure is applied and the speech and noise signal in all spectral bins can be modeled as Gaussian random processes and furthermore $f(n)$ is chosen as $f(x) = x$, the result of the minimization is

$$\begin{aligned}\widehat{H}_{\text{EM},1}\left(e^{j\Omega_\mu}, n\right) &= \frac{\sqrt{\pi}}{2}\left(\frac{1}{1 + \widehat{SNR}_{\text{post}}(\Omega_\mu, n)}\right)^{1/2}\left(\frac{\widehat{SNR}_{\text{prio}}(\Omega_\mu, n)}{1 + \widehat{SNR}_{\text{prio}}(\Omega_\mu, n)}\right)^{1/2} \\ &\quad F\left(\frac{\left(1 + \widehat{SNR}_{\text{post}}(\Omega_\mu, n)\right)\widehat{SNR}_{\text{prio}}(\Omega_\mu, n)}{1 + \widehat{SNR}_{\text{prio}}(\Omega_\mu, n)}\right).\end{aligned} \tag{10.42}$$

The function $F(x)$ abbreviates

$$F(x) = e^{-\frac{x}{2}}\left[(1 + x)\, I_0\left(\frac{x}{2}\right) + x\, I_1\left(\frac{x}{2}\right)\right], \tag{10.43}$$

where $I_0(x)$ and $I_1(x)$ are modified Bessel functions of orders zero and one, respectively [4]. The term $\widehat{SNR}_{\text{post}}(\Omega_\mu, n)$ describes the a posteriori signal-to-noise ratio of the current frame and is computed as

$$\widehat{SNR}_{\text{post}}(\Omega_\mu, n) = \frac{\widehat{S}_{ee}(\Omega_\mu, n) - \widehat{S}_{\tilde{n}\tilde{n}}(\Omega_\mu, n)}{\widehat{S}_{\tilde{n}\tilde{n}}(\Omega_\mu, n)}. \tag{10.44}$$

One of the most important features of the characteristic proposed by Ephraim and Malah is the introduction (and utilization) of a smoothed version of the a posteriori signal-to-noise ratio. The resulting quantity, called the a priori signal-to-noise ratio, is computed according to

$$\begin{aligned}\widehat{SNR}_{\text{prio}}(\Omega_\mu, n) &= \gamma\left|\widehat{H}_{\text{EM},1}\left(e^{j\Omega_\mu}, n-1\right)\right|^2\left(\widehat{SNR}_{\text{post}}(\Omega_\mu, n-1) + 1\right) \\ &\quad + (1 - \gamma)\max\left\{\widehat{SNR}_{\text{post}}(\Omega_\mu, n),\, 0\right\}\end{aligned} \tag{10.45}$$

where the subscripts "post" and "prio" denote "a posteriori" and "a priori," respectively. Smoothing constants γ chosen from the interval $\gamma \in [0.98,\ 0.998]$ have been proven to achieve subjectively good performances [33]. Furthermore, smoothing according to Eq. 10.45 reduces *musical noise* (see Section 14.2) significantly. In the right part of Fig. 10.14 the filter characteristic is depicted in terms of the a priori signal-to-noise ratio $\widehat{SNR}_{\text{prio}}(\Omega_\mu, n)$ for several values of the a posteriori ratio $\widehat{SNR}_{\text{post}}(\Omega_\mu, n)$.

If $f(x) = \log_{10}(x)$ instead of $f(x) = x$ is inserted in Eq. 10.41, logarithmic spectral differences are applied to obtain an optimal filter characteristic. For the resulting frequency response, one gets [62][3]

$$\widehat{H}_{\text{EM},2}\left(e^{j\Omega_\mu}, n\right) = \frac{\widehat{SNR}_{\text{prio}}(\Omega_\mu, n)}{1 + \widehat{SNR}_{\text{prio}}(\Omega_\mu, n)} \exp\left\{ \frac{1}{2} \int\limits_{v(\Omega_\mu, n)}^{\infty} \frac{e^{-t}}{t}\, dt \right\}. \tag{10.46}$$

The lower limit of the integral is defined as

$$v(\Omega_\mu, n) = \frac{\widehat{SNR}_{\text{prio}}(\Omega_\mu, n)\, \widehat{SNR}_{\text{post}}(\Omega_\mu, n)}{1 + \widehat{SNR}_{\text{prio}}(\Omega_\mu, n)}. \tag{10.47}$$

For practical realizations the modified Bessel function in Eq. 10.43 as well as the integral in Eq. 10.46 are calculated in advance for several supporting points within the interval [-20 dB, 20 dB] for $\widehat{SNR}_{\text{prio}}(\Omega_\mu, n)$ and $\widehat{SNR}_{\text{post}}(\Omega_\mu, n)$. The resulting values are stored in a lookup table. In this case, the computational complexity is not much higher than for the Wiener or for the spectral subtraction characteristic.

10.1.2.4 Further Filter Characteristics

Besides the characteristics presented until now, various other filter characteristics can be applied. Common to all approaches should be that the characteristic should have values close to one if the signal-to-noise ratio is rather high. In case of a small signal-to-noise ratio, the characteristic should introduce an attenuation.

A very simple characteristic that fulfills these restrictions is a binary attenuation strategy according to

$$H_{\text{BS}}\left(e^{j\Omega}\right) = \begin{cases} 1, & \text{if } S_{ss}(\Omega) > K_0\, S_{\tilde{n}\tilde{n}}(\Omega), \\ H_{\text{min}}, & \text{else}, \end{cases} \tag{10.48}$$

with $0 < H_{\text{min}} < 1$. Introducing short-term estimations for the power spectral densities and rewriting Eq. 10.48 using the definition of the signal-to-noise ratio (see Eq. 10.28) leads to

$$\widehat{H}_{\text{BS}}\left(e^{j\Omega}, n\right) = \begin{cases} 1, & \text{if } \widehat{SNR}(\Omega_\mu, n) > K_0, \\ H_{\text{min}}, & \text{else}. \end{cases} \tag{10.49}$$

[3]To improve readability, e^x has been written as $\exp\{x\}$ in Eq. 10.46.

The constant K_0 should be chosen according to the expected average signal-to-noise ratio during speech activity. Values of about $K_0 = 3...20$ have been proved to achieve good results in case of low or moderate noise levels. Even if the binary strategy is very simple, it should be applied only if the available processing power is extremely limited. Especially the abrupt transition from H_{min} to 1 produces audible artifacts. Other characteristics with a smoother transition are, for instance, fourth order power series

$$\widehat{H}_{\text{PC},1}\left(e^{j\Omega}, n\right) = \min\left\{0.1 + \widehat{SNR}^4(\Omega_\mu, n),\, 1\right\}, \tag{10.50}$$

or power series of even higher order

$$\widehat{H}_{\text{PC},2}\left(e^{j\Omega}, n\right) = \min\left\{0.2 + \widehat{SNR}^8(\Omega_\mu, n),\, 1\right\} \tag{10.51}$$

are able to reduce these artifacts. In the left part of Fig. 10.15 the binary characteristic as well as the attenuation according to Eq. 10.50 are depicted. Both approaches

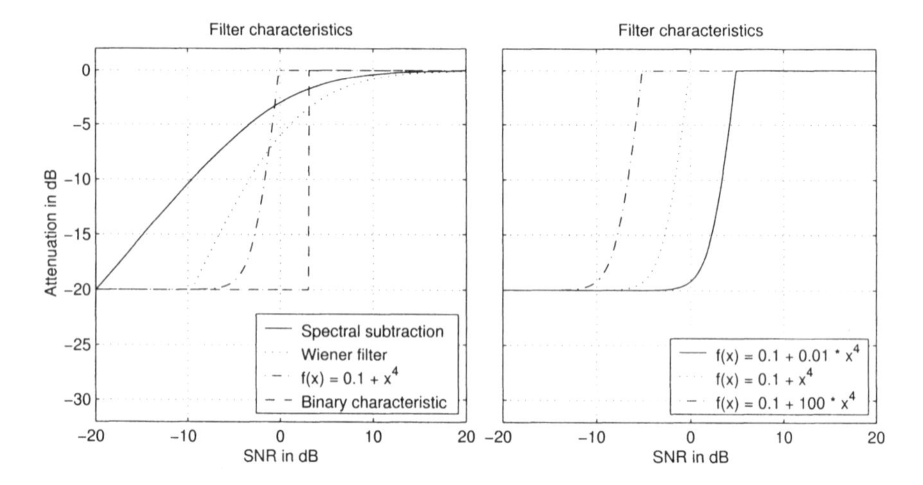

Fig. 10.15 Further filter characteristics (the Wiener filter was bounded by a maximum attenuation of -20 dB).

(Eqs. 10.49 and 10.50) share two common characteristics:

1. they have a smaller attenuation than does the Wiener or the spectral subtraction approach at large signal-to-noise ratios and
2. a larger attenuation is introduced for small signal-to-noise ratios.

By introducing an overestimation parameter $\beta(n)$ within the potential series according to

$$H_{\text{PC},3}\left(e^{j\Omega}, n\right) = \min\left\{0.1 + \beta(n)\, \widehat{SNR}^4(\Omega_\mu, n),\, 1\right\}, \tag{10.52}$$

the "aggressiveness" of the noise suppression scheme can be adjusted. A small value for $\beta(n)$ [e.g., $\beta(n) = 0.01$], leads to larger attenuation and a large value [e.g., $\beta(n) = 100$], reduces the attenuation. Both cases are depicted in the right part of Fig. 10.15. The parameter $\beta(n)$ can be controlled with respect to the detected speech activity. Details about control strategies will be presented in Chapter 14.

10.1.2.5 Concluding Remarks

Before we will differentiate between suppressing residual echoes and reducing local background noise a few final remarks on filter characteristics should be given. Usually each filter characteristic is modified in several ways:

- Most often the maximal attenuation is bounded by values larger than zero in order to reduce artifacts:

$$H\left(e^{j\Omega}, n\right) = \max\left\{..., H_{\text{min}}\right\}. \tag{10.53}$$

  The limiting value H_{min} is called the *spectral floor*. This parameter is usually chosen from the range -6 to -20 dB for background noise suppression and from -30 to -60 dB for residual echo suppression. In background noise suppression schemes the effect of this limitation is a residual noise floor. This noise floor is considerably smaller than in the non-processed signal, but its presence also gives information of the acoustical scenario to the far-end listener. The spectral floor parameter can also be controlled adaptively and can be chosen frequency selective [185]

$$H_{\text{min}} \longrightarrow H_{\text{min}}\left(e^{j\Omega}, n\right). \tag{10.54}$$

- Another modification which is often applied is overestimation of the power spectral density of the noise

$$\widehat{S}_{\tilde{n}\tilde{n}}(\Omega_\mu, n) \longrightarrow \beta(\Omega_\mu, n)\, \widehat{S}_{\tilde{n}\tilde{n}}(\Omega_\mu, n). \tag{10.55}$$

  Due to the overestimation, the musical noise phenomena (see Section 14.2) can be reduced. Like the filter limitation, the overestimation parameter $\beta(\Omega_\mu, n)$ can be chosen time and frequency selective. Details about this type of filter modification will be presented in Chapter 14.

Finally, it should be noted that different filter characteristics can be utilized within different subbands. Especially in the field of background noise suppression, characteristics with only a moderate "aggressiveness" are applied in the lower frequency range while more aggressive (larger overestimation, higher attenuation) characteristics are utilized at higher frequencies.

10.2 SUPPRESSION OF RESIDUAL ECHOES

In this section we will assume that the signal components that are to be removed consist only of the residual echo

$$\tilde{n}(n) = e_{\text{u}}(n)\,. \tag{10.56}$$

This would be the case in quiet offices or in cars that are not moving and when the engine is switched off. In these situations a small amount of local background noise is still present, but it is not necessary to remove this component. In the next section we will discuss the opposite case (only local background noise). The combination of residual echo and background noise suppression will be presented in Section 10.4. Furthermore, we will assume that echo cancellation according to Chapter 9 has been performed and only the remaining components of the echo have to be suppressed.

The impact of the echo cancellation filter on the acoustical echo is limited by at least two facts:

- Only echoes due to the linear part of the transfer function of the LEM system can be canceled.
- The order of the adaptive filter typically is much smaller than the order of the LEM system.

Therefore, a residual echo suppression filter [7, 20, 60, 228] is used to reduce the echo further. For the transfer function of this filter, a Wiener characteristic is often applied:

$$H_{\text{RE}}\left(e^{j\Omega_\mu}, n\right) = \frac{\widehat{S}_{es}(\Omega_\mu, n)}{\widehat{S}_{ee}(\Omega_\mu, n)}\,. \tag{10.57}$$

In good agreement with reality, one can assume that the undisturbed error $e_{\text{u}}(n)$ and the local speech signal $s(n)$ are orthogonal. Under this assumption the frequency response of the filter reduces to

$$H_{\text{RE}}\left(e^{j\Omega_\mu}, n\right) = 1 - \frac{\widehat{S}_{e_{\text{u}}e_{\text{u}}}(\Omega_\mu, n)}{\widehat{S}_{ee}(\Omega_\mu, n)}\,. \tag{10.58}$$

Since the signals involved are highly nonstationary, the short-term power spectral densities have to be estimated for time intervals not longer than 10–20 ms. The overwriting problem, however, is that the locally generated speech signal $s(n)$ is observable only during the absence of the remote excitation signal $x(n)$. It should be noted, however, that any impact of the residual echo suppression filter on residual echoes also impacts the local speech signal $s(n)$ and, thus, reduces the quality of the speech output of hands-free telephones or speech recognition systems.

The coupling between the loudspeaker and the microphone is reduced in proportion to the degree of the match of the echo cancellation filter $\widehat{\boldsymbol{h}}(n)$ to the original system $\boldsymbol{h}(n)$. In real applications, since a perfect match (over all times and all situations) cannot be achieved, the remaining signal $e(n)$ still contains echo components.

When applying Eq. 10.58, the estimated power spectral densities $\widehat{S}_{ee}(\Omega_\mu, n)$ and $\widehat{S}_{e_u e_u}(\Omega_\mu, n)$ contain estimation errors. Therefore, the quotient may become larger than one. Consequently, the filter exhibits a phase shift of π. To prevent that, the filter transfer function can be extended according to the modifications presented in Section 10.1.2.5

$$H_{\text{RE}}\left(e^{j\Omega_\mu}, n\right) = \max\left[1 - \beta\,\frac{\widehat{S}_{e_u e_u}(\Omega_\mu, n)}{\widehat{S}_{ee}(\Omega_\mu, n)},\, H_{\min}\right], \tag{10.59}$$

where $H_{\min}$ determines the maximal attenuation of the filter. The overestimation parameter β can be used to control the "aggressiveness" of the filter. Details can be found in, for example, the book by Quatieri [191].

In order to estimate the short-term power spectral density of the error signal $S_{ee}(\Omega_\mu, n)$ first-order IIR smoothing of the squared magnitudes of the subband error signals $e_\mu(n)$ is applied in order to estimate the short-term power spectral density:[4]

$$\widehat{S}_{ee}\left(\Omega_\mu, n\right) = (1-\gamma)\,\left|e_\mu(n)\right|^2 + \gamma\,\widehat{S}_{ee}\left(\Omega_\mu, n-1\right). \tag{10.60}$$

In contrast to the short-term power estimation presented in Chapter 8, only a fixed time constant γ is utilized here. Because the undistorted error signal $e_u(n)$ is not accessible, the estimation of the short-term power spectral density $\widehat{S}_{e_u e_u}(\Omega_\mu, n)$ cannot be approximated in the same manner as $\widehat{S}_{ee}(\Omega_\mu, n)$.

According to our model of the LEM system, the undisturbed error $e_u(n)$ can be expressed by a convolution of the excitation signal $x(n)$ and the system mismatch[5] $h_{\Delta,i}(n)$ (see Chapter 7):[6]

$$e_u(n) = \sum_{i=0}^{N-1} h_{\Delta,i}(n)\, x(n-i). \tag{10.61}$$

Hence, the power spectral density of the undisturbed error signal can be estimated by multiplying the short-term power spectral density of the excitation signal with the squared magnitude spectrum of the estimated system mismatch:

$$\widehat{S}_{e_u e_u}\left(\Omega_\mu, nr\right) = \widehat{S}_{xx}\left(\Omega_\mu, n\right)\left|\widehat{H}_\Delta\left(e^{j\Omega_\mu}, n\right)\right|^2. \tag{10.62}$$

The short-term power spectral density of the excitation signal is estimated in the same manner the one of the error signal (according to Eq. 10.60). For estimating the

[4]If a DFT/IDFT structure is applied for short-term spectral analysis and synthesis the subband signal $e_\mu(n)$ has also been denoted as $E(e^{j\Omega_\mu}, n)$.

[5]Note also that the system mismatch is an unknown quantity. Thus, the estimation problem is not solved yet.

[6]It is assumed that the excitation signal and all other involved signals are real. Hence, the system mismatch vector is real, too.

system mismatch, a *double-talk detector* and a *detector for enclosure dislocations* are required. A detailed description of such detectors would go beyond the scope of this chapter—the interested reader is referred to Chapter 13, where several methods for the solution of such detection problems are described. If the local background noise $b(n)$ has only moderate power, the spectrum of the system mismatch can be estimated by smoothing the ratio of the excitation and the error spectrum:

$$\left|\widehat{H}_\Delta\left(e^{j\Omega_\mu},n\right)\right|^2 = \begin{cases} (1-\gamma_\Delta)\,\dfrac{\widehat{S}_{ee}\left(\Omega_\mu,n\right)}{\widehat{S}_{xx}\left(\Omega_\mu,n\right)} + \gamma_\Delta\left|\widehat{H}_\Delta\left(e^{j\Omega_\mu},n-1\right)\right|^2, \\ \qquad\qquad \text{if condition 10.64 is true,} \\ \left|\widehat{H}_\Delta\left(e^{j\Omega_\mu},n-1\right)\right|^2, \qquad \text{else}. \end{cases} \tag{10.63}$$

The recursive smoothing is updated only if one of the following conditions are true:

$$\begin{aligned} &\{\text{Remote single-talk is detected.}\} \vee \\ &\qquad\qquad \{\text{Enclosure dislocations are detected.}\}\,. \end{aligned} \tag{10.64}$$

The suppression of residual echoes can be performed either in the time or in the frequency domain. In the first case the frequency response of the suppression filter is transformed periodically via inverse DFT into the time domain

$$h_{\mathrm{RE}}(n) = \begin{cases} \mathrm{IDFT}\left\{H_{\mathrm{RE}}\left(e^{j\Omega_\mu},\frac{n}{r}\right)\right\}, & \text{if } n \bmod r \equiv 0\,, \\ h_{\mathrm{RE}}(n-1), & \text{else}\,. \end{cases} \tag{10.65}$$

An example of residual echo suppression is presented in Fig. 10.16 . In the two upper diagrams the excitation signal $x(n)$ as well as the sum of all local signals $n(n)$ are depicted. The bottom diagram shows the short-term power of the microphone signal $y(n)$, of the error signal (after echo cancellation) $e(n)$, and of the signal $\hat{s}(n)$ after echo cancellation and suppression. The echo cancellation was performed using the NLMS algorithm. In remote single-talk situations the residual echo signal is attenuated by 12 dB, which was the maximum attenuation according to the parameter H_{min}. During the double-talk situation (3.5–6.2 seconds) only a small attenuation of about 3 dB was introduced. The overestimation factor was chosen to be $\beta = 3$.

Beside the advantages of residual echo suppression also the disadvantages should be mentioned here. In contrast to echo cancellation, echo suppression schemes are introducing attenuation within the sending path of the local signal. Therefore a compromise between attenuation of residual echoes and reduction of the local speech quality always has to be made. In case of severe background noise, a *modulation effect* due to the attenuation of the noise in remote single-talk situations appears. In such scenarios the echo suppression should always be combined with a comfort noise generator, which will be presented in the next section.

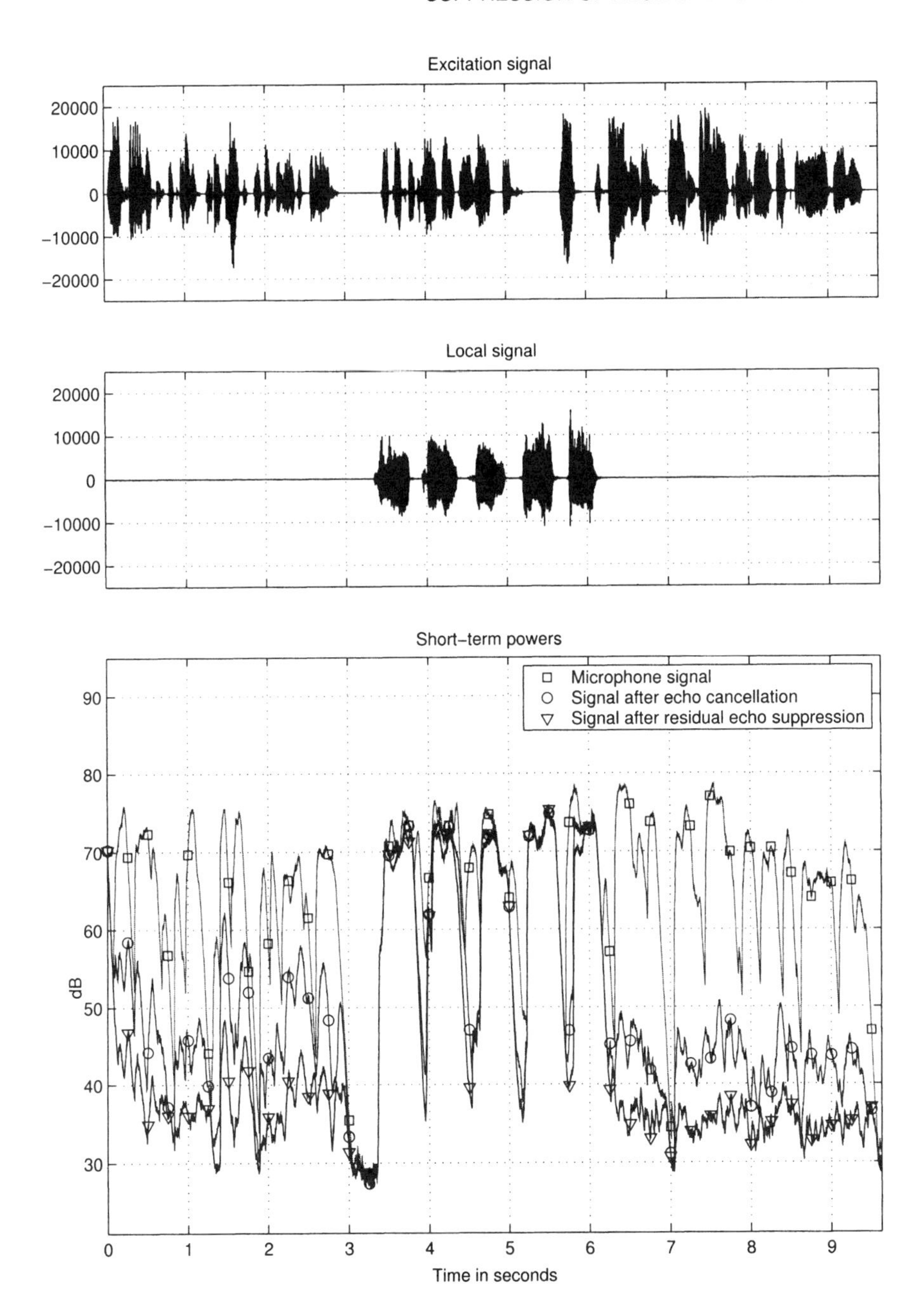

Fig. 10.16 Suppression of residual echoes. In the upper two diagrams the excitation signal $x(n)$ as well as the sum of all local signals $n(n)$ are depicted. The bottom diagram shows the short-term power of the microphone signal $y(n)$, of the error signal (after echo cancellation) $e(n)$, and of the signal after echo suppression $\hat{s}(n)$. In remote single-talk situations the residual echo signal is attenuated by 12 dB, which was the maximum attenuation according to the parameter $H_{\min}$ (with permission from [100]).

10.2.1 Comfort Noise

An additional attenuation of only 12 dB—as applied in the last section—is usually not sufficient to suppress residual echoes completely. If a smaller limit, for example, 40 or 50 dB, is utilized, the local background noise is erased, too. For the remote conversation partner the line seems disconnected whenever this person speaks. To overcome this problem *comfort noise injection* can be applied [106, 140]. Whenever the short-term output power of the residual echo suppression is smaller than the background noise level, the output signal is substituted by artificially generated noise. This noise has the same power and spectral characteristics as the local background noise. The structure of a residual echo suppression system with comfort noise injection is depicted in Fig. 10.17.

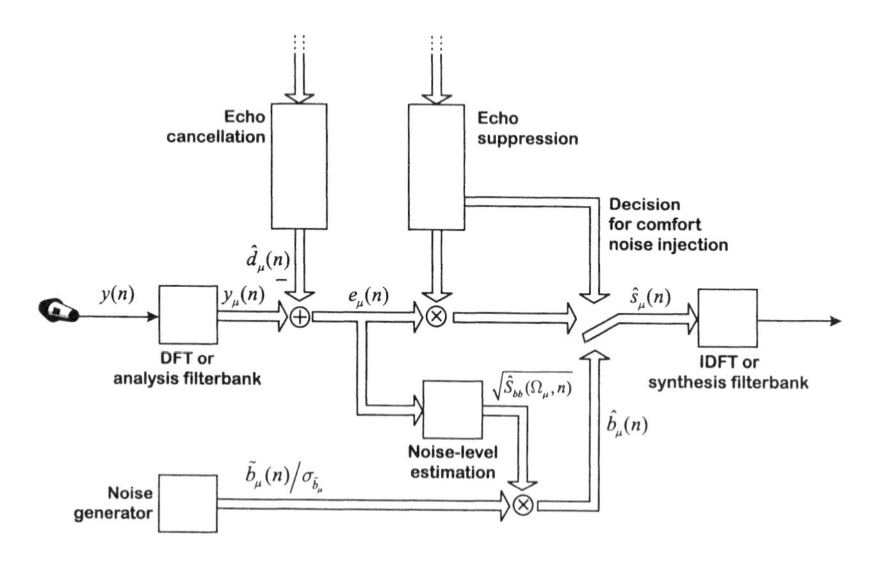

Fig. 10.17 Structure of a residual echo suppression system with comfort noise injection.

The rule for replacing the attenuated error signal by artificial noise $\hat{b}_\mu(n)$ within subband μ can be denoted as

$$\hat{s}_\mu(n) = \begin{cases} e_\mu(n)\, H_{\mathrm{RE}}\left(e^{j\Omega_\mu}, n\right), & \text{if} \left|e_\mu(n)\, H_{\mathrm{RE}}\left(e^{j\Omega_\mu}, n\right)\right|^2 > \widehat{S}_{bb}\left(\Omega_\mu, n\right), \\ \hat{b}_\mu(n), & \text{else}. \end{cases} \tag{10.66}$$

For the estimation of the power spectral density of the background noise, one of the methods presented in Chapter 14 can be applied. Usually, the noise generator produces zero-mean white noise $\tilde{b}_\mu(n)$ with variance $\sigma^2_{\tilde{b}_\mu}$. The artificial noise signal is generated by normalizing $\tilde{b}_\mu(n)$ by its standard derivation and multiplying it with the square root of the estimated power spectral density of the background noise:

$$\hat{b}_\mu(n) = \frac{\tilde{b}_\mu(n)}{\sigma_{\tilde{b}_\mu}} \sqrt{\widehat{S}_{bb}(\Omega_\mu, n)}. \tag{10.67}$$

A simple method for generating the noise $\tilde{b}_\mu(n)$ will be presented in the next section.

In order to show the advantages of comfort noise injection, the outputs of two noise suppression systems are depicted in Fig. 10.19. The difference between the two systems was that in one system the comfort noise injection was activated. All other parameters of both systems were equal.

In the upper two diagrams of Fig. 10.19 the excitation signal as well as the local signal are presented. The lower two diagrams show time–frequency analyses of the output signals. During the single-talk periods (first to third second and sixth second to end) large attenuation at the pitch harmonics at low frequencies and large (overall) attenuation at higher frequencies (white areas within the third diagram) can be observed. By applying comfort noise injection, these areas are filled with artificial noise that cannot be distinguished from the real background noise (see start to first second). The insertion of comfort noise improves the overall signal quality considerably.

10.2.1.1 Noise Generation

A very simple and computationally effective procedure of noise generation consists of binary shift registers with feedback paths. The basic structure of such a system is depicted in Fig. 10.18. The output of this type of noise generator is binary

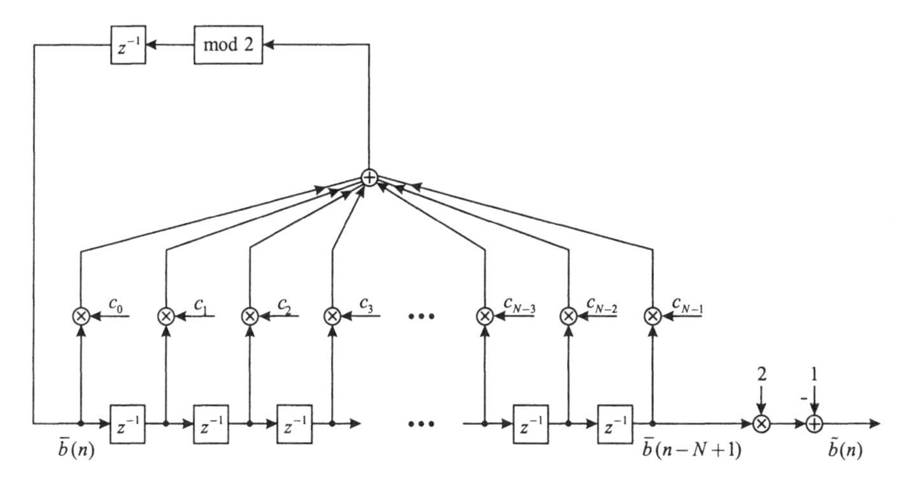

Fig. 10.18 Structure of a binary pseudo–noise generator.

and (nearly) white. The term "nearly" is used because the output signal is periodic. However, it exhibits a very long period even for medium memory sizes, for example, around 6 days for a register length of 32 and a sampling frequency of 8 kHz.

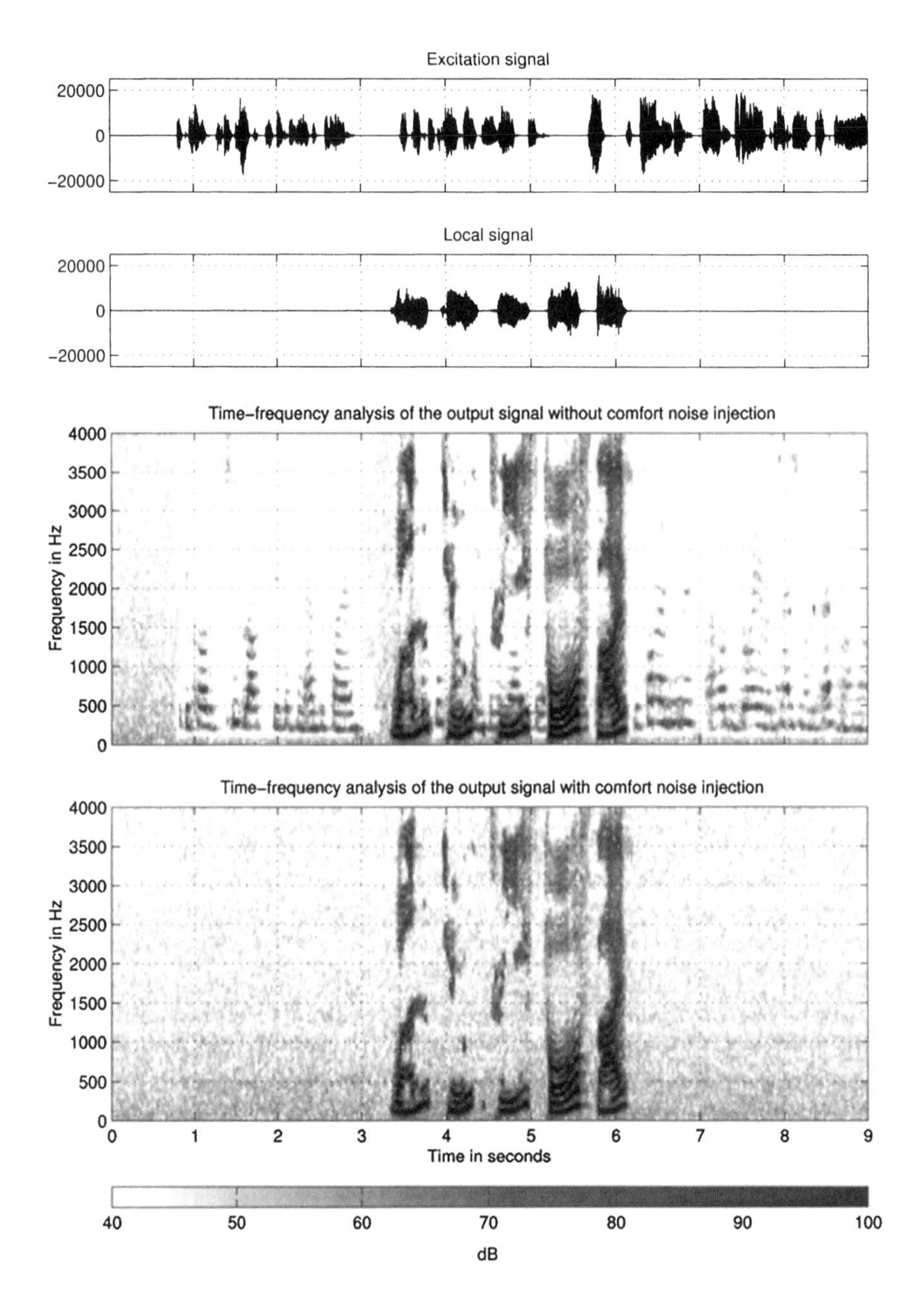

Fig. 10.19 Simulation examples for comfort noise injection. In the upper two diagrams the excitation signal $x(n)$ as well as the local signal $n(n)$ are presented. The lower two diagrams show time–frequency analyses of the output signals (third diagram without comfort noise injection, fourth diagram with comfort noise injection).

Each delay element in Fig. 10.18 stores one bit of the noise generator. A few of these bits are coupled back in order to excite the generator. The feedback coefficients as well as the internal signals are binary values:

$$\bar{b}(n),\ c_i \in \{0,\ 1\}\ . \tag{10.68}$$

Before feeding the result of the summation back to the input of the register, a modulo-2 operation is performed:

$$\bar{b}(n+1) = \left(\sum_{i=0}^{N-1} \bar{b}(n-i)\, c_i \right) \bmod 2\,. \tag{10.69}$$

The period length of the output $\bar{b}(n-N+1)$ of the register depends on the choice of the feedback coefficients c_i. Only for a few choices of coefficient combinations the maximum period length is achieved. In Table 10.2 the indices of the nonzero feedback coefficients as well as the periods of the resulting maximum-length pseudonoise sequence are listed for several register lengths.

In order to have a zero-mean noise sequence the output of the shift register is modified according to

$$\tilde{b}(n) = 2\,\bar{b}(n-N+1) - 1\,. \tag{10.70}$$

The resulting signal $\tilde{b}(n)$ consists only of the values +1 and -1. Thus, its variance is

$$\sigma_{\tilde{b}}^2 = 1\,. \tag{10.71}$$

10.3 SUPPRESSION OF BACKGROUND NOISE

In analogy to Section 10.2, we will assume now that the signal component to be suppressed is only local background noise. Thus, we have

$$\tilde{n}(n) = b(n) \tag{10.72}$$

and

$$\widehat{S}_{\tilde{n}\tilde{n}}\left(\Omega_\mu, n\right) = \widehat{S}_{bb}\left(\Omega_\mu, n\right)\,, \tag{10.73}$$

respectively. Estimation of the power spectral density $\widehat{S}_{bb}(\Omega_\mu, n)$ is one of the major problems of noise suppression schemes. Usually this quantity is estimated during speech pauses. In order to reduce the variance of the estimation process, temporal smoothing is applied. During periods of speech activity the estimation procedure is stopped and the values computed at the end of the last speech pause are utilized for computing the coefficients of the noise suppression filter. A more detailed description of schemes for estimating the power spectral density of the background noise will be presented in Chapter 14.

Table 10.2 Indices of the nonzero feedback coefficients for maximum-length pseudonoise sequences [244]

Register width	Period length at 8 kHz sampling rate	Feedback paths
2	0.4 ms	{1, 0}
3	0.9 ms	{2, 0}
4	1.9 ms	{3, 0}
5	3.9 ms	{4, 1}
6	7.9 ms	{5, 0}
7	15.9 ms	{6, 0}
8	31.9 ms	{7, 5, 4, 0}
9	63.9 ms	{8, 3}
10	127.9 ms	{9, 2}
11	0.3 s	{10, 1}
12	0.5 s	{11, 6, 3, 2}
13	1 s	{12, 3, 2, 0}
14	2 s	{13, 11, 10, 0}
15	4.1 s	{14, 0}
16	8.2 s	{15, 4, 2, 1}
17	16.4 s	{16, 2}
18	32.8 s	{17, 6}
19	1.1 min	{18, 5, 4, 0}
20	2.2 min	{19, 2}
21	4.4 min	{20, 1}
22	8.7 min	{21, 0}
23	17.5 min	{22, 4}
24	35 min	{23, 3, 2, 0}
25	1.2 h	{24, 2}
26	2.3 h	{25, 7, 6, 0}
27	4.7 h	{26, 7, 6, 0}
28	9.3 h	{27, 2}
29	18.6 h	{28, 1}
30	1.6 d	{29, 15, 14, 0}
31	3.1 d	{30, 2}
32	6.2 d	{31, 27, 26, 0}
33	12.4 d	{32, 12}
34	24.9 d	{33, 14, 13, 0}

An example of noise suppression is depicted in Fig. 10.20. The power spectral density of the noise is computed during the first second (a speech pause) of the sequence and is kept until the end of the speech interval. The Wiener characteristic according to Eq. 10.31 has been utilized to adjust the filter coefficients. In order to reduce artifacts, the attenuation factors $\widehat{H}_{\mathrm{W}}(e^{j\Omega_\mu}, n)$ have been limited to a maximum attenuation of 12 dB and an overestimation factor $\beta = 2$ has been utilized.

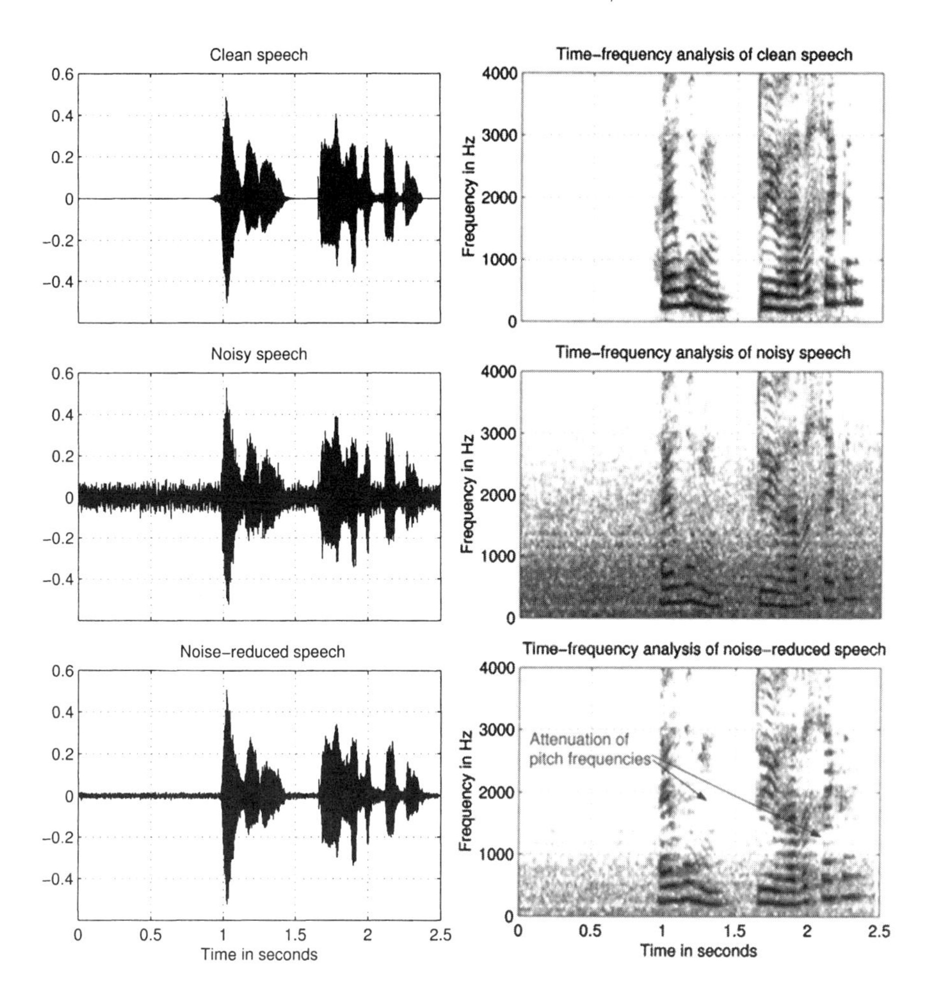

Fig. 10.20 Example of a background noise suppression. The Wiener characteristic according to Eq. 10.31 has been applied and the attenuation factors were limited to a maximum of 12 dB.

The noisy signal was generated artificially by adding car noise recorded at a speed of about 100 km/h and a clean speech signal. The signal-to-noise ratio of the noisy speech signal (depicted in the center part of Fig. 10.20), which is computed simply by dividing the energies of the two signals, is about 14 dB. By applying the filter characteristic the background noise can be reduced considerably as shown in the lowest two diagrams of Fig. 10.20.

Nevertheless, in periods of only low signal-to-noise ratio, as in the example after 1.3 seconds around 2000 Hz and after 2.2 seconds around 1200 Hz, standard noise suppression characteristics introduce audible distortions. In the example presented in Fig. 10.20, several pitch harmonics are attenuated. In order to reduce these artifacts, extensions have been proposed. Before we describe such extensions in Section 10.3.1, a few comments on the choice of the filter characteristic are given.

First, we have to say that it is not possible to name an optimal characteristic in general. The best choice depends on the involved signals, especially on the type of background noise, on the evaluation setup, and on the individual preferences of the listener. The filter characteristic according to Ephraim and Malah (see Section 10.1.2.3), for example, shows a good performance concerning artifacts during noise-only periods but sounds slightly reverberant during speech periods. If one listens to the output signal of this characteristic with headphones, the reverberant impression is much more audible than listening to the output using loudspeakers.

When evaluating the quality of the characteristic, several criteria have to be mentioned. Examples are

- Attenuation of the background noise
- Speech quality
- Naturalness of the residual noise

Dreiseitel [56] interviewed approximately 30 people regarding the importance of these criteria. Fig. 10.21 presents the results of the survey.

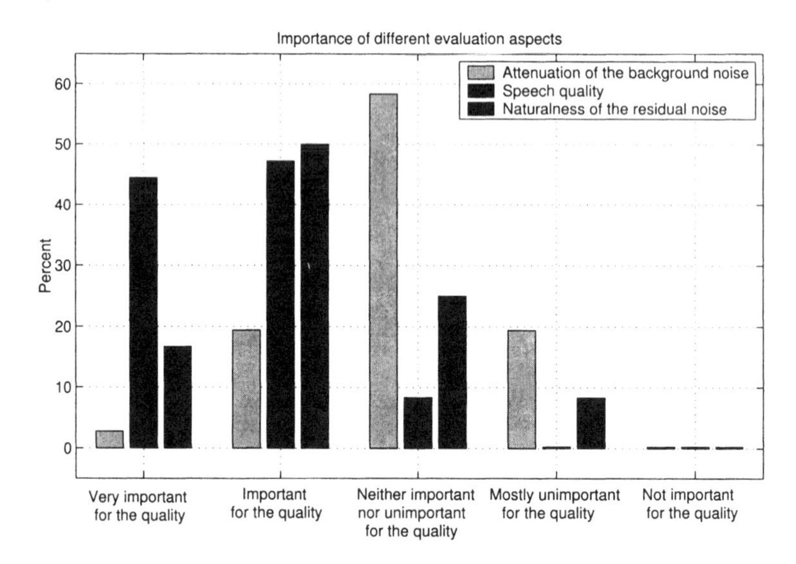

Fig. 10.21 Results of a survey on the importance of several aspects for evaluation of a noise suppression scheme [56].

To evaluate the quality of a noise suppression scheme a variety of processed speech examples should be presented to a small group (10–30) of persons. The playback setup should be as close as possible to the final application. By varying several parameters and algorithmic parts (e.g., the filter characteristics[7]), the average quality rating of the test subjects should be utilized to optimize the noise suppression scheme.

10.3.1 Postprocessing

After applying spectral attenuation factors in the frequency or subband domain, several approaches for further enhancement of the denoised speech signal have been proposed. We will mention three of these methods:

- Automatic gain control in order to achieve the same average loudness in noisy and in quiet environments
- Pitch-adaptive postfiltering for attenuating the signal below the pitch frequency and in between its higher-order harmonics
- Fricative spreading, in order to copy high-frequency parts of the spectra of fricatives into regions below the cutoff frequency of the transmission channel

Each approach is described in more detail in Sections 10.3.1.1–10.3.1.3.

10.3.1.1 Automatic Gain Control

Any person who speaks in a noisy environment will automatically alter the speech characteristics in order to increase the efficiency of communication over the noisy channel. This is known as the *Lombard effect* [149]. The effect can be described in more detail by the following statements [105]:

- The average bandwidth of most phonemes decreases.
- The formant frequencies of vowels are increased.
- The first formant frequency of most phonemes increases.
- The formant amplitudes increase, leading to an increased spectral tilt.

As a result of the last point, the overall speech level rises with increasing background noise power. Rates of about 0.3–1 dB of speech power increment per decibel of background noise increment have been reported [105]. Thus, the power of the same speech sequence recorded in a car can vary by about 10–30 dB, depending on whether the car is parked or moving with high speed.

In order to reduce the level variations of the output of a noise suppression system, an automatic gain control should be applied. The basic parts of such an algorithm have already been described in Section 8.3.1. In combination with a noise suppression scheme, a frame-based gain control architecture is often preferred. Instead of

[7]Note that it is not necessary to use the same filter characteristic in each subband.

smoothing the absolute value of the output signal $\hat{s}(n)$[8] (see Eqs. 8.31 and 8.34), Parseval's relation [49] can be applied and the sum of the squared magnitudes of the current frame can be smoothed in a fast and a slow version:

$$\overline{|x_{\mathrm{S}}(n)|} = (1 - \gamma_{\mathrm{S}}(n))\, P_{\hat{s}}(n) + \gamma_{\mathrm{S}}(n)\, \overline{|x_{\mathrm{S}}(n-1)|}\,, \tag{10.74}$$

$$\overline{|x_{\mathrm{F}}(n)|} = (1 - \gamma_{\mathrm{F}}(n))\, P_{\hat{s}}(n) + \gamma_{\mathrm{F}}(n)\, \overline{|x_{\mathrm{F}}(n-1)|}\,, \tag{10.75}$$

with

$$P_{\hat{s}}(n) = \sqrt{\sum_{\mu=0}^{M-1} |\hat{s}_\mu(n)|^2}\,. \tag{10.76}$$

The square root was taken in order to reduce the dynamics of the estimation. With these modified (compared to Section 8.3.1) slow and fast short-term power estimations, an average peak power can be estimated. Comparing this peak power with a desired value results in a slowly varying gain factor. The computation of this gain factor is described in detail in Section 8.3.1. The only difference is that all time constants have to be exchanged according to subsampling rate r of the frame-based processing. In general this can be achieved by setting

$$\gamma_{\mathrm{sb}} = \sqrt[r]{\gamma_{\mathrm{fb}}}\,. \tag{10.77}$$

The subscripts "sb" and "fb" denote subband and fullband quantities, respectively.

10.3.1.2 Pitch-Adaptive Postfiltering

Single-channel background noise suppression improves the signal-to-noise ratio of noisy speech signals at the expense of speech distortions or residual noise. Especially nonstationary noise components cause problems because their power spectral density cannot be estimated properly during speech activity. Musical noise can be reduced by limiting the attenuation to a maximum, called the spectral floor. However, the more the attenuation is limited, the more noise remains within the cleaned speech signal.

In a car environment the background noise usually has a lowpass characteristic and often the enhanced signal still contains noise components below the pitch frequency and in between low-order harmonics of the pitch. For further enhancement of the speech quality, *pitch-adaptive postfiltering* can be applied. This can be realized in either the time or frequency domain.

10.3.1.2.1 Time-Domain Filtering

Time-domain postfiltering is applied in several speech coding standards, such as in the enhanced full-rate speech codec for GSM systems [125]. The

[8] In Section 8.3.1 this signal was denoted by $x(n)$.

postfilter can be designed as the long-term part of an adaptive predictor error filter [37]. The frequency response of such a filter is given by

$$H_{\mathrm{t}}\left(e^{j\Omega},n\right) = \frac{1-\alpha}{1+\beta}\,\frac{1+\beta\,e^{-jp_{0}(n)\Omega}}{1-\alpha\,e^{-jp_{0}(n)\Omega}}\,. \tag{10.78}$$

The filter is applied to the output signal of a conventional noise suppression scheme. The maximum filter attenuation and the steepness of the filter slopes can be controlled using the coefficients α and β. The quantity $p_0(n)$ is an estimate of the current pitch period. The filter $H_{\mathrm{t}}(e^{j\Omega},n)$ introduces minima at approximately odd-integer multiples of half of the pitch frequency $\Omega_{\mathrm{p}}(n)$:

$$\Omega_{\mathrm{min},i}(n) \approx (2i+1)\,\frac{\Omega_{\mathrm{p}}(n)}{2}\,. \tag{10.79}$$

No attenuation at all ($H_{\mathrm{t}}(e^{j\Omega},n)=1$) is achieved at frequencies

$$\Omega_{\mathrm{max},i}(n) \approx i\,\Omega_{\mathrm{p}}(n)\,. \tag{10.80}$$

The drawback of this approach is that the minima and the maxima of the filter occur only approximately at the desired frequencies. There are mainly two reasons for it:

- The filter characteristic according to Eq. 10.78 is realized with delay elements in the time domain. Thus, only integer values for $p_0(n)$ are allowed. For this reason, $p_0(n)$ is obtained by rounding the real pitch period $p_{\mathrm{real}}(n)$ to the nearest integer of a sample period.

- Furthermore, the pitch frequency can only be estimated. Due to estimation errors, the minima and maxima deviate from the calculated ones. The deviation becomes larger with increasing frequency.

The power spectral density of a voiced speech sequence and the frequency response of the corresponding pitch-adaptive postfilter are depicted in Fig. 10.22. Because of a small estimation error, the maxima of the filter do not fit the true higher-order harmonics of the pitch frequency.

Another drawback of postfiltering according to Eq. 10.78 is that the attenuation below half of the pitch frequency decreases to 0 dB. A large attenuation would be desirable in order to suppress low-frequency background noise, especially in automotive applications.

10.3.1.2.2 Frequency-Domain Filtering

To overcome the drawbacks of time-domain postfiltering at only moderate computational complexity, frequency-domain approaches were investigated [224]. These approaches directly modify the spectrum $\hat{S}(e^{j\Omega_\mu},n)$ in the DFT domain or the subband signal $\hat{s}_\mu(n)$ by use of a pitch-adaptive weighting function $H_{\mathrm{f}}(e^{j\Omega_\mu},n)$.

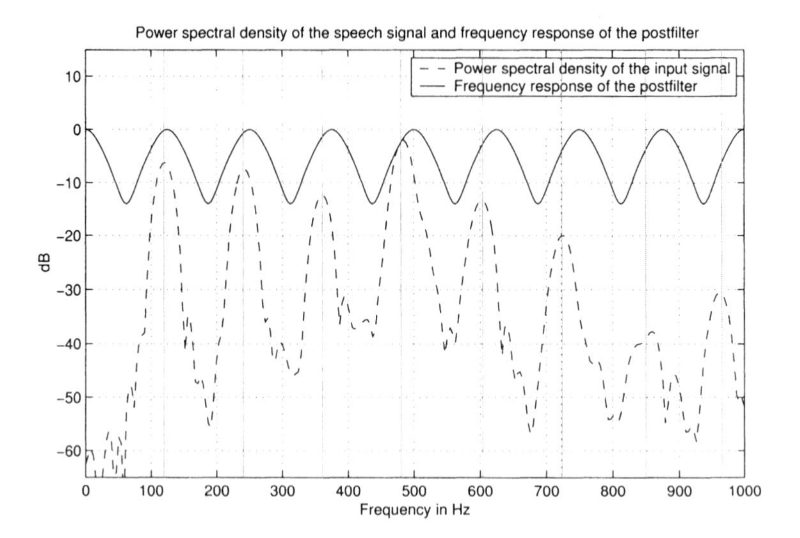

Fig. 10.22 Mismatch between maxima of the postfilter and harmonics of clean, voiced speech. The postfilter was designed with $\alpha = 0.5$ and $\beta = 0.25$.

In addition, below half of the estimated pitch frequency $\Omega_{\text{p}}(n)$ the filter has constant magnitude

$$H_{\text{f}}\left(e^{j\Omega_\mu}, n\right) = H_{\text{min}}, \quad \text{for } |\Omega_\mu| < \frac{\Omega_{\text{p}}(n)}{2}. \tag{10.81}$$

Above half of the estimated pitch frequency a modified version of the time-domain filter $H_{\text{t}}(e^{j\Omega}, n)$ is applied. The modification consists of a multiplication with a logarithmic ramp function starting at $\Omega = 0$ with the value one and ending with a value larger than one at $\Omega = \pi$:

$$H_{\text{r}}\left(e^{j\Omega}, n\right) = \Omega^{K_0}. \tag{10.82}$$

To avoid spectral amplification, the resulting filter is bounded by one:

$$H_{\text{f}}\left(e^{j\Omega_\mu}, n\right) = \begin{cases} H_{\text{min}}, & \text{for } |\Omega_\mu| < \frac{\Omega_{\text{p}}(n)}{2}, \\ \max\left\{\left|H_{\text{t}}\left(e^{j\Omega_\mu}, n\right)\right| H_{\text{r}}\left(e^{j\Omega_\mu}, n\right), 1\right\} & \text{else}. \end{cases} \tag{10.83}$$

Fig. 10.23 shows the frequency response of the modified postfilter as well as the power spectral density of the same voiced speech sequence as in Fig. 10.22. The filter according to Eq. 10.83 is more robust against pitch estimation errors, and also the attenuation below half of the pitch frequency is visible.

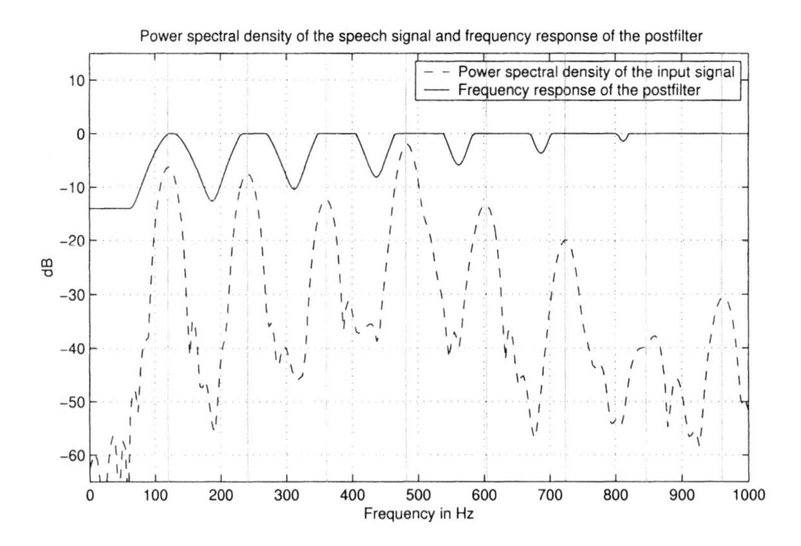

Fig. 10.23 Frequency response of the postfilter designed in the frequency domain.

10.3.1.2.3 Application of the Postfilter

Applying the postfilter requires two further detection mechanisms. First, the postfilter will be introduced only during voiced sequences. Thus, a scheme for detecting those periods is needed. A variety of methods have been established [5, 32, 224]; three of them are listed below:

- *Broadband energy*: Voiced sequences usually have larger energy than unvoiced sequences. By comparing the energy of the current frame with a threshold, voiced sequences can be detected.

- *Highpass / lowpass power ratio*: Unvoiced sounds have more energy at high frequencies than at low frequencies. The opposite is true for voiced sounds. A comparison can be utilized for a voiced/unvoiced classification.

- *Zero-crossing rate*: Unvoiced sounds are changing the sign of succeeding samples more often than voiced sequences. Thus, a large zero-crossing rate is an indicator for unvoiced sounds.

Several methods are known for determining the pitch frequency. We will list only a few of them and refer the interested reader to the literature [115, 190, 207]:

- *Autocorrelation function*: During voiced sequences the first maximum of the autocorrelation function (at nonzero delay) occurs at a lag corresponding to the pitch period. Monitoring this lag can be used to estimate the fundamental frequency of a speech signal.

- *Cepstral analysis*: If a source-filter model is assumed, voiced speech can be described as a convolution of a periodic excitation with the impulse response of the vocal tract. In the spectral domain, this relation becomes a product of the respective transforms. Applying the logarithm turns this product into a simple addition. After another transformation back into the time (cepstral) domain, it is possible to separate the spectrally slow changing vocal tract information from the fundamental frequency of the excitation signal.

- *Harmonic product spectrum*: Another method that sometimes outperforms the autocorrelation and the cepstral analysis in case of (telephone) bandlimited speech (300–3400 Hz) is the harmonic product spectrum [192, 206]. By computing the product over several equidistant frequency supporting points of the spectrum, the pitch frequency can be estimated. Only when the frequency increment corresponds to the pitch frequency (or a multiple of it) the resulting product becomes large.

If a frame is classified as voiced, the frequency response of the entire noise suppression filter is computed by multiplying one of the filter characteristics described in Section 10.1.2 with the frequency response of the postfilter[9] $H_{\mathrm{f}}(e^{j\Omega_\mu}, n)$. In case of unvoiced frames or silence only the standard characteristic is applied:

$$H_{\mathrm{BN}}\left(e^{j\Omega_\mu}, n\right) = \begin{cases} \max\left\{H_{\mathrm{FC}}\left(e^{j\Omega_\mu}, n\right),\, H_{\mathrm{min,vo}}\right\} H_{\mathrm{f}}\left(e^{j\Omega_\mu}, n\right), & \text{in case of voiced frames,} \\ \max\left\{H_{\mathrm{FC}}\left(e^{j\Omega_\mu}, n\right),\, H_{\mathrm{min,uv}}\right\}, & \text{else.} \end{cases} \tag{10.84}$$

$H_{\mathrm{FC}}(e^{j\Omega_\mu}, n)$ is denoting any of the filter characteristics presented in Section 10.1.2. During unvoiced sequences the filter is limited to a value $H_{\mathrm{min,uv}}$ (e.g., -12 dB). This parameter is smaller than the limiting value $H_{\mathrm{min,vo}}$ for voiced frames. A typical choice for $H_{\mathrm{min,vo}}$ might be -6 dB. This means that the amount of noise suppression introduced by the standard characteristic is decreased during voiced periods. Thus, artifacts are reduced but also the background noise level is higher compared to unvoiced frames. Due to the postfilter, an additional amount of background noise attenuation is introduced. The ratio of the two limiting variables $H_{\mathrm{min,vo}}$ and $H_{\mathrm{min,uv}}$ should be in the same range as the ratio of the peaks and valleys of the pitch-adaptive postfilter. In this case it is possible to avoid loudness variations of the residual background noise level within the processed signal. Finally, it should be noted that the quality of postfilter approaches depends crucially on the reliability of the pitch estimation and the voiced/unvoiced classification.

10.3.1.3 Fricative Spreading

In most current voice transmission systems, the bandwidth is still limited to 3.4 or 4 kHz (depending on the transmission system). This bandwidth is sufficiently large for vowels as spoken by a majority of speakers. However, for consonants,

[9] It is assumed that the frequency-domain approach according to Section 10.3.1.2.2 has been utilized.

especially for fricatives like /s/ or /f/, this is no longer true because their spectral energy is often located above 4 kHz. In Fig. 10.24 a time–frequency analysis of the consecutive numbers "one, two, ..., ten" spoken by a male person is depicted. The spectral energy of fricatives (e.g., /s/ within "six" or "seven") above 4 kHz is clearly visible.

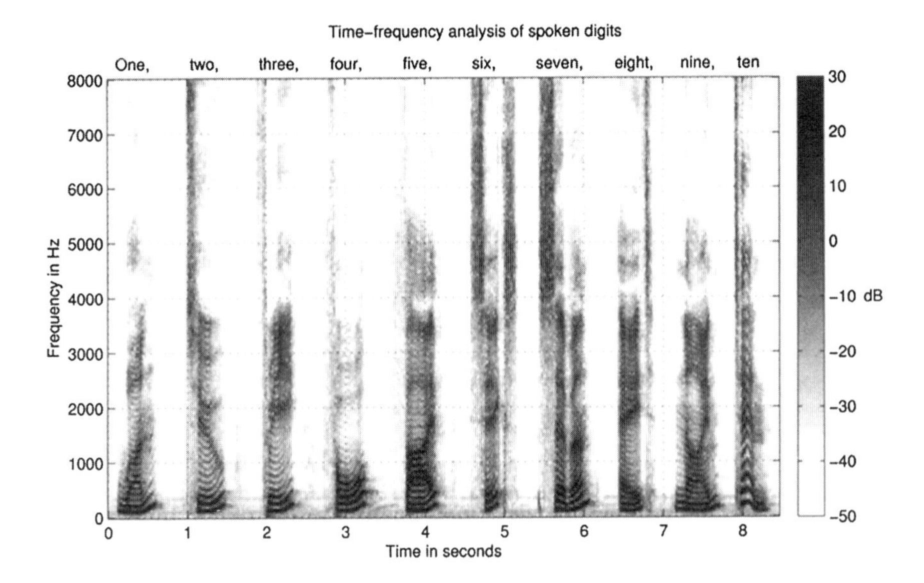

Fig. 10.24 Time–frequency analysis of the spoken digits "one" to "ten." The spectral energy of fricatives, for example, /s/ within "six" or "seven," above 4 kHz is clearly visible.

To improve the quality of the narrowband signal, techniques to shift the spectrum of fricatives below the cutoff frequency of the transmission system can be applied [112]. To achieve this, the signal is recorded first at a high sample rate (e.g., 11.025 or 16 kHz) and noise suppression is applied to the wideband signal. The resulting signal is decomposed into subbands of approximately 1 kHz bandwidth and a short-term energy $P_\nu(n)$ is computed for each subband. If the noise suppression scheme is performed within the DFT or subband domain with, for example, $M = 256$ channels at a sampling rate of $f_s = 16$ kHz, the short-term energy computation can be performed by adding the squared amplitudes of groups of subbands or DFT bins:

$$P_\nu(n) = \sum_{\mu=\nu\frac{M}{16}}^{(\nu+1)\frac{M}{16}-1} |\hat{s}_\mu(n)|^2 \,. \tag{10.85}$$

After computing the short-term energies, the subband with the maximum energy is searched:

$$\nu_{\text{cen}}(n) = \text{argmax}\,\{P_\nu(n)\} \,. \tag{10.86}$$

Denoting the spectral index of this subband by $\nu_{\mathrm{cen}}(n)$, the following spectral translation rules can be applied:

- If the centroid $\nu_{\mathrm{cen}}(n)$ is less than or equal to 2, no spectral translation is applied.
- If $\nu_{\mathrm{cen}}(n) = 3$, add all components of the frequency range from 4 to 5 kHz to the spectrum between 3 and 4 kHz.
- If $\nu_{\mathrm{cen}}(n) = 4$, add all components of the frequency range from 4 to 5 kHz as well as the of the range from 5 to 6 kHz to the spectrum between 3 and 4 kHz.
- If $\nu_{\mathrm{cen}}(n) = 5$, add all components of the frequency range from 4 to 5 kHz, from 5 to 6 kHz, and from 6 to 7 kHz to the spectrum between 3 and 4 kHz.

The rules given above were originally presented by Heide and Kang [112]. A sampling frequency of $f_s = 16$ kHz was assumed and the (single-sided) bandwidth of the transmission channel was assumed to be 4 kHz. An increase of the speech intelligibility from 90.2% (without fricative spreading) to 94.3% (with fricative spreading) was measured [112]. Furthermore, a MELP (mixed excitation linear predictor) coding scheme has been utilized before evaluating the speech intelligibility. In Fig. 10.25 time–frequency analyses of a bandlimited signal without (top) and with (bottom) applying fricative spreading are depicted. The numbers "six" and "seven" have been spoken. The increase of high-frequency energy during fricatives, such as /s/ in "six" and /s/ in "seven," is clearly visible.

The success of this method depends on the channel bandwidth. If frequencies up to 4 kHz can be transmitted, fricative spreading is able to enhance the speech intelligibility of most speakers. Nevertheless, if the transmission system is able to support frequencies up to only 3.4 kHz (analog telephone lines), enhancement of the speech quality is rather small or even counterproductive.

10.4 COMBINING BACKGROUND NOISE AND RESIDUAL ECHO SUPPRESSION

In Sections 10.2 and 10.3 residual echo and background noise suppression have been described separately. If both should be combined, some minor modifications are necessary. In a first stage for both types of noise components, separate filter frequency responses should be computed according to one of the methods presented before. Each filter characteristic should be limited by its maximum attenuation:

$$H_{\mathrm{BN}}\left(e^{j\Omega_\mu}, n\right) = \max\left\{..., H_{\mathrm{min,B}}\right\}, \tag{10.87}$$

$$H_{\mathrm{RE}}\left(e^{j\Omega_\mu}, n\right) = \max\left\{..., H_{\mathrm{min,E}}\right\}, \tag{10.88}$$

where the subscripts"BN" and "RE" denote background noise suppression and residual echo suppression, respectively. For the exact computation (dots within Eqs. 10.87

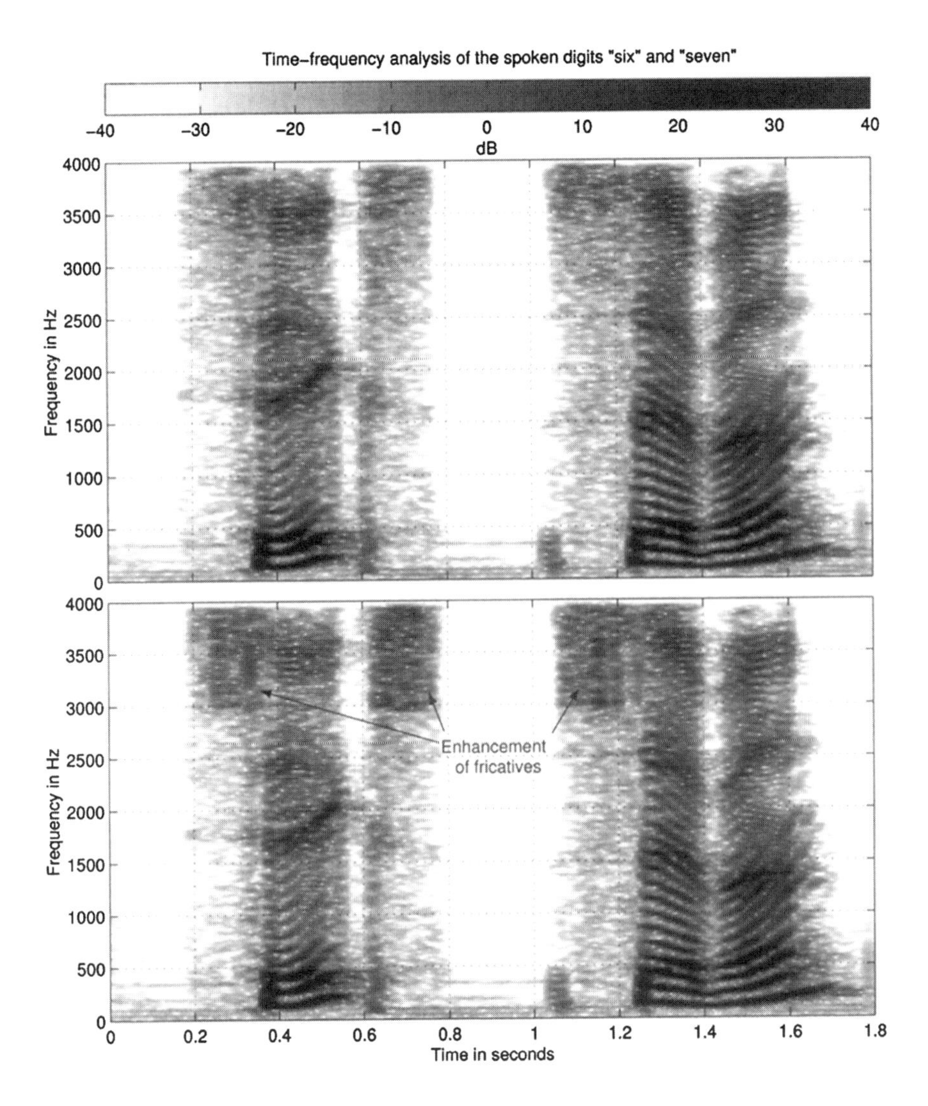

Fig. 10.25 Time-frequency analysis of the spoken digits "six" and "seven" without (top) and with (bottom) the application of fricative spreading.

and 10.88) of the filter characteristics the reader is referred to the beginning of this chapter.

As already described, the attenuation limit of the residual echo suppression can be chosen much smaller (from a linear perspective) than the maximum attenuation of the background noise suppression:

$$H_{\text{min,B}} \gg H_{\text{min,E}} \,. \tag{10.89}$$

Next a preliminary common filter characteristic $H(e^{j\Omega_\mu}, n)$ is computed by multiplication of both subfilters

$$H\left(e^{j\Omega_\mu}, n\right) = H_{\text{BN}}\left(e^{j\Omega_\mu}, n\right) H_{\text{RE}}\left(e^{j\Omega_\mu}, n\right) . \tag{10.90}$$

Before the filters are applied to the subband signals $e_\mu(n)$, it is determined whether the power of the resulting signal would be smaller than the residual background noise level:

$$\left|e_\mu(n)\right|^2 \left|H\left(e^{j\Omega_\mu}, n\right)\right|^2 > \widehat{S}_{bb}\left(\Omega_\mu, n\right) \left|H_{\text{min,B}}\right|^2 . \tag{10.91}$$

Whenever this is the case, the output signal is not computed by multiplication of the subband input signals with the frequency response $H(e^{j\Omega_\mu}, n)$. This would result in a modulation of the residual background noise level, which would be very annoying to the remote conversation partner. Instead, the resulting signal is generated by insertion of artificial noise:

$$\hat{s}_\mu(n) = \begin{cases} e_\mu(n)\, H\left(e^{j\Omega_\mu}, n\right), & \text{if condition 10.91 is true}, \\ \hat{b}_\mu(n)\, H_{\text{min,B}}, & \text{else}. \end{cases} \tag{10.92}$$

To show the cooperation of noise reduction and residual echo suppression, a simulation example is presented in Fig. 10.26. The microphone signal (see second and fourth diagrams of Fig. 10.26) consists of stationary background noise, echo, and of local speech. A large part of the echo signal is removed by an echo cancellation unit. Nevertheless, residual echoes would still be audible without further processing, especially at the beginning of the remote speech period (see upper part of Fig. 10.26). The residual echo and background noise suppression unit has removed nearly all echo components, and the background noise level is reduced by a large amount (here 12 dB). When examining the time–frequency analysis of the output signal (lowest diagram in Fig. 10.26), it is not possible to distinguish between real residual background noise (first five seconds) and artificial noise (e.g. from sixth to eighth second).

Nevertheless, during double-talk situations the postfilter introduces small artifacts. If the residual echo suppression part of the postfilter is adjusted too aggressive the local speech is attenuated. In the other case soft echoes might be audible for the remote conversation partner. These artifacts can be kept at a very low level by exploiting the masking properties of the human ear [91, 92, 93].

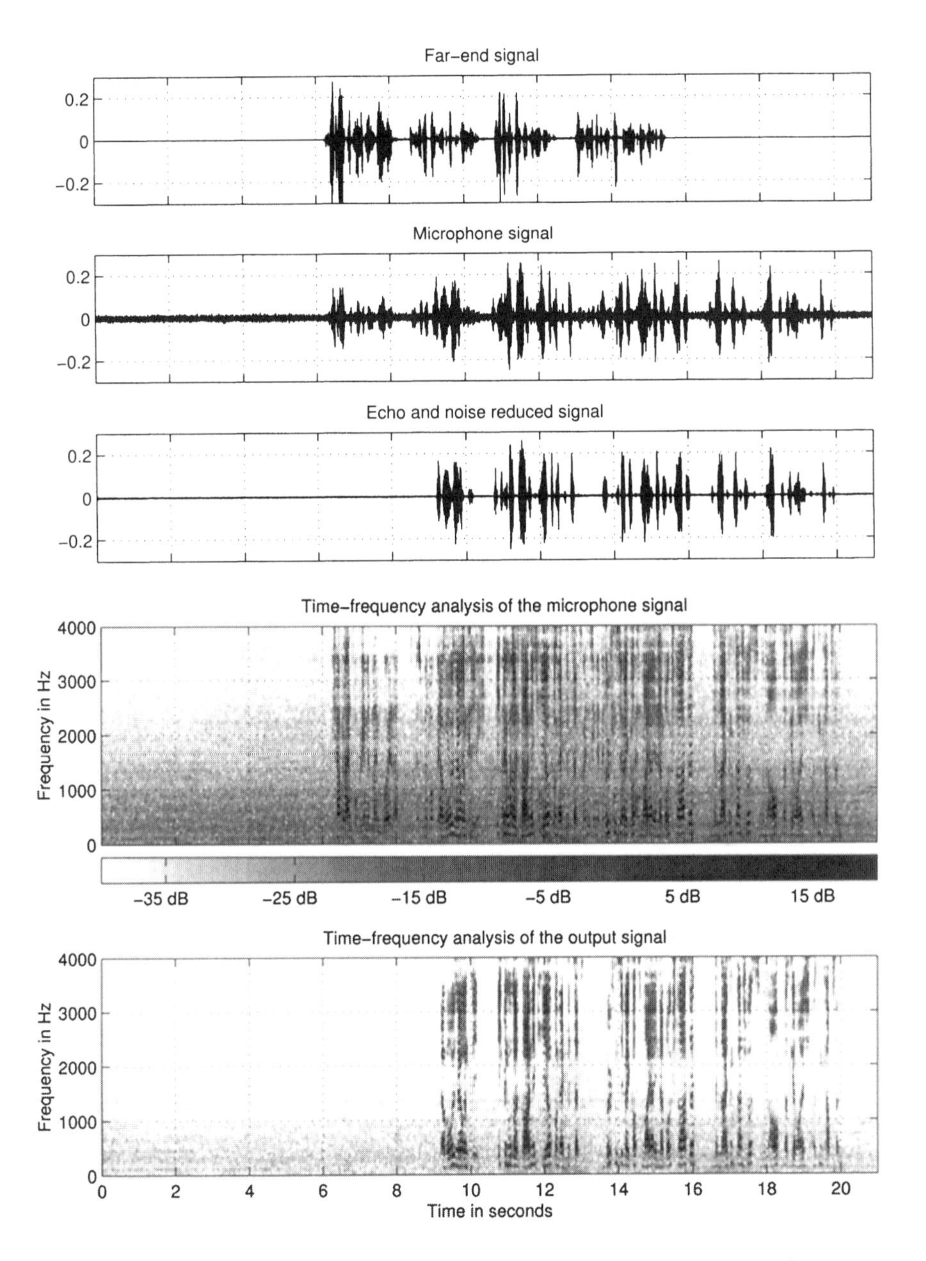

Fig. 10.26 Simulation example for a combined residual echo and noise suppression system.

11

Beamforming

An alternative to the algorithms based on spectral attenuation in order to suppress background noise or residual echoes is the usage of a microphone array. Here, the output signals of several microphones are combined in such a way that the signals coming from a predefined source direction (e.g., the direction pointing to the local speaker) are added in phase. Signals from other directions are combined only diffusely or are even destructively, which leads to an attenuation of signals from these directions. In case of nonadaptive combination of the microphone signals, no time-variant distortions are introduced in the signal path. This is one of the most important advantages of beamformers over conventional single-channel noise suppression systems.

Beamforming was first developed in the field of radar, sonar, and electronic warfare [234]. In these applications the ratio of bandwidth and center frequency of the band of interest is very small. In audio application this ratio is rather large,[1] leading to the problem of *grating lobes* (see Section 11.1). In contrast to earlier publications, we will use the terms *beamformer* and *microphone array* in an equivalent manner.

Before we describe several aspects of beamforming, we will start with an example that motivates the application. The command correctness of a speech recognition system [107] is depicted in Fig. 11.1. Several single- or double-word commands, like "radio on" or "radio off," were spoken by different persons. The total number of commands was $N_{\mathrm{Com}} = 30$. Several recordings for evaluating the speech recognition system have been performed within a car. The rear mirror of this car was equipped

[1] In audio applications the bandwidth of interest is located symmetrically around 0 Hz. Thus, the center frequency is 0 Hz, which leads to an infinitely large ratio between bandwidth and center frequency.

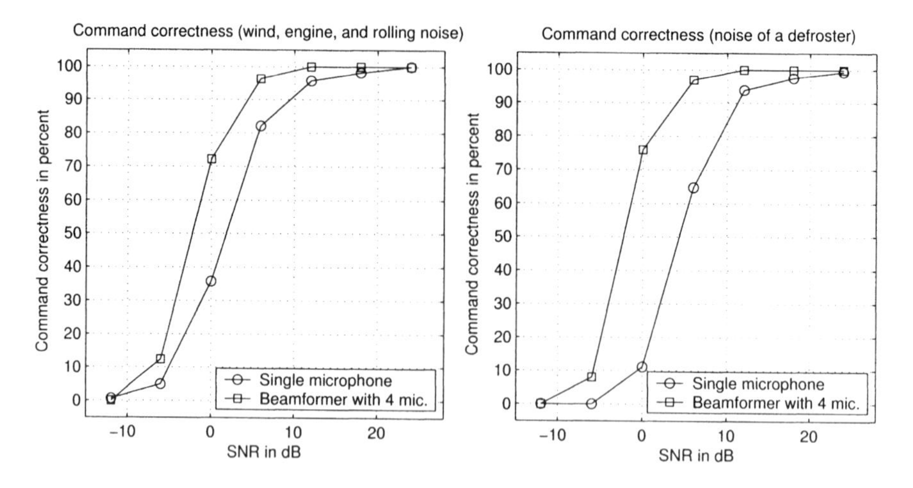

Fig. 11.1 Influence of beamforming on the recognition rate of speech recognition systems (mic. = microphones).

with a microphone array consisting of four sensors. Furthermore, several recordings of driving noise (wind, engine, and rolling noise) and of noise produced by a defroster have been taken with the same microphone array. During the evaluation process the speech and the noise components were added with different gains in order to achieve several signal-to-noise ratios.[2] The output of a single microphone and the output of the beamformer (design according to Section 11.4) were connected to a speech recognition system (for details, see Ref. [202]). The achieved command correctnesses are depicted for both types of noise and for several signal-to-noise ratios in Fig. 11.1.

Even if the beamformer improves–in the case of wind, motor, and rolling noise–the average signal-to-noise ratio only about 6 dB (compared to the single microphone), the recognition rates can be increased considerably. Due to the beamforming, the reverberant character the recorded speech signals is reduced, which is also a reason for enhanced recognition performance. In case of a directional distortion, like the defroster, the beamformer is even more effective. Especially for low signal-to-noise ratios (0–10 dB) the command correctness can be improved considerably.

A second feature of multimicrophone processing should also be considered even at this early point in the chapter. By using more than one microphone, it becomes possible to estimate the direction of a sound source. With this information it is possible to distinguish between different sound sources (e.g., the local speaker and the loudspeaker that emits the signal from the remote speaker) that are very similar

[2] With this setup, the Lombard effect (see Section 10.3.1.1) was not modeled. Nevertheless, the experiment gives a first hint of the importance of beamforming if a speech recognition system should perform well in noisy environments.

in their statistical properties. This spatial information can be exploited for enhanced system control (see Part IV of this book).

The chapter is organized as follows. First the fundamentals of sound propagation are briefly described. In order to evaluate the performance of beamformers, several analyses and performance measurements are introduced in Section 11.2. Afterward, delay-and-sum as well as filter-and-sum structures are described. The chapter closes with the introduction of adaptive beamformers.

Due to multipath propagation of sound in small enclosures or sensor imperfections, it is necessary to control the adaptation of beamformers. As in the previous chapters, this will be treated separately—the interested reader is referred to Chapter 15.

11.1 BASICS

In order to understand and to develop algorithms for array processing, it is necessary to understand the basic temporal and spatial properties of sound propagation. Sound propagation of moderately loud sound sources can be described with sufficient accuracy by the linear wave equation [127] (written in cartesian notation)

$$\frac{\partial^2 p(t,\boldsymbol{r})}{\partial x^2} + \frac{\partial^2 p(t,\boldsymbol{r})}{\partial y^2} + \frac{\partial^2 p(t,\boldsymbol{r})}{\partial z^2} = \frac{1}{c^2}\frac{\partial^2 p(t,\boldsymbol{r})}{\partial t^2}, \tag{11.1}$$

where $p(t,\boldsymbol{r})$ denotes the sound pressure at position $\boldsymbol{r} = [x,\, y,\, z]^{\mathrm{T}}$ and at time t. The constant c represents the sound velocity. Furthermore, it is assumed that the medium in which the sound propagates is homogeneous and nondispersible. If the wave equation should be solved for a point source located at the origin and that excites the sound field according to

$$p(t,\boldsymbol{0}) = A\, e^{j\omega t}, \tag{11.2}$$

it is favorable to transform Eq. 11.1 into spherical coordinates by using the following transformations:

$$\begin{aligned} r &= \sqrt{(x^2+y^2+z^2)}, & (11.3)\\ \theta &= \arccos\left(\frac{x}{\sqrt{x^2+y^2}}\right), & (11.4)\\ \phi &= \arccos\left(\frac{z}{\sqrt{x^2+y^2+z^2}}\right). & (11.5) \end{aligned}$$

If we assume that the resulting sound field exhibits spherical symmetry

$$\frac{\partial p(t,r,\theta,\phi)}{\partial \theta} = \frac{\partial p(t,r,\theta,\phi)}{\partial \phi} = 0, \tag{11.6}$$

we obtain for the sound equation in spherical coordinates:

$$\frac{\partial^2 \big(r\, p(t,r)\big)}{\partial r^2} = \frac{1}{c^2} \frac{\partial^2 \big(r\, p(t,r)\big)}{\partial t^2} . \tag{11.7}$$

By solving this equation for the excitation according to Eq. 11.2, we obtain the following for the sound pressure outside the origin

$$p(t,r) = \frac{A}{r}\, e^{j\omega\left(t-\frac{r}{c}\right)} . \tag{11.8}$$

This equation means that with increasing distance, the sound pressure decreases. Doubling the distance between a source and a sensor results in only half of the amplitude. Thus, the recorded power is reduced by 6 dB if the source–sensor–distance is doubled. On the other hand, if the distance between two sensors is rather small compared to the distance to the source

$$r_1,\, r_2 \gg |r_1 - r_2| , \tag{11.9}$$

the signal of the second sensor can be approximated from the signal of the first sensor by simple multiplication with an appropriate exponential term:

$$p(t,r_1) = \frac{A}{r_1}\, e^{j\omega(t-r_1/c)} , \tag{11.10}$$

$$p(t,r_2) = \frac{A}{r_2}\, e^{j\omega(t-r_2/c)} , \tag{11.11}$$

$$\approx p(t,r_1)\, e^{j\omega(r_1-r_2)/c} . \tag{11.12}$$

This approximation can be interpreted as a plane wave arriving at the microphones. The amplitude difference at the microphone positions is dependent only on the distance between the microphones (and not on the distance to the sound source). We will make use of this approximation whenever we analyze the spatial properties of microphone arrays.

11.1.1 Spatial Sampling of a Sound Field

For temporal sampling of a signal the Nyquist criterion

$$f_s \geq 2\, f_{\max} \tag{11.13}$$

has to be fulfilled. This means that the sampling rate f_s should be chosen at least twice as high as the maximum frequency $f_{\max}$ of the desired signal. For spatial sampling of a sound field an appropriate criterion can be derived from the Nyquist rule and the dispersion equation

$$c = \lambda f , \tag{11.14}$$

where λ denotes the wavelength and f the frequency of the wave. Inserting Eq. 11.14 into the Nyquist criterion leads to the spatial sampling criterion

$$\lambda_s \leq \frac{\lambda_{\min}}{2} = \frac{c}{2\, f_{\max}}. \tag{11.15}$$

This means that the sampling wavelength, which is the distance between adjacent microphones, should be at most half of the minimal wavelength of the signal. If criterion 11.15 is violated, spatial aliasing appears. In this case spatial ambiguities result in unwanted maxima within the receiving characteristic of beamformers; these are called *grating lobes*.

For visualization of such grating lobes the beampatterns[3] of two beamformers are depicted in Fig. 11.2. Both beamformers consist of two omnidirectional micro-

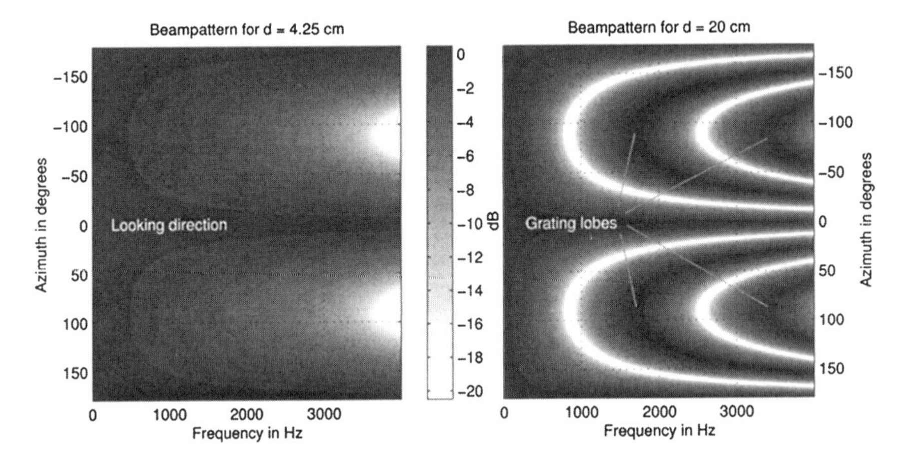

Fig. 11.2 Beampatterns of two beamformers. Each beamformer consists of two sensors. For the left example the distance between the microphones is 4.25 cm; for the right example both sensors were placed at a distance of 20 cm.

phones. For the first beamformer (left part of Fig. 11.2) the distance between the microphones is 4.25 cm; for the second (right part of Fig. 11.2), the microphones are placed 20 cm away from each other. The beamformer was designed to block all signals except those arriving from an azimuth angle of 0°. The maximum signal frequency is 4 kHz—thus the distance between the microphones should not be larger than 4.25 cm (the speed of sound is assumed to be 340 m/s) if spatial aliasing should be avoided.

When comparing the receiving characteristics of both beamformers, the grating lobes in the right beampattern are clearly visible. On the other hand, if the spatial sampling criterion is fulfilled, only a poor spatial selectivity at low frequencies can be

[3]The computation of beampatterns will be explained in detail in Section 11.2.1.

achieved. For this reason a compromise between spatial aliasing and low-frequency selectivity is often preferred.

11.2 CHARACTERISTICS OF MICROPHONE ARRAYS

In this section we will describe the basic aspects of microphone arrays. The sensors of an array can be placed in a linear (one-dimensional) manner, within a plane (planar or two-dimensionally), or even three-dimensionally. For reasons of simplicity we will focus only on linear arrays with uniformly spaced microphones. Figure 11.3 shows such an array with $L = 4$ sensors. Each microphone has an omnidirectional characteristic, and the distance between two adjacent sensors is $d_{\mathrm{mic}} = 4$ cm. Furthermore, it is assumed that a speaker is located at an angle $\phi_0 = 35°$ in front of the array.

If a wavefront coming from the source or looking direction reaches all sensors at the same time ($\phi_0 = 0°$ according to the coordinate system of Fig. 11.3), the array is called a *broadside* array. On the other hand, if a wavefront coming from the source direction reaches the sensors of a linear array with maximum delay ($\phi_0 = 90°$ or $\phi_0 = 270°$), the array is called an *endfire* array.

The performance of beamformers can be analyzed by a variety of methods. Two of these analyses (a polar plot and a beampattern) are depicted in the right part of Fig. 11.3. For both methods a *spatial frequency response* (derived in the following paragraphs) is required.

The output signal $u(n)$ of the beamformer is given by the addition of the (FIR) filtered microphone signals:[4]

$$\begin{aligned} u(n) &= \sum_{l=0}^{L-1} \sum_{i=0}^{N-1} y_l(n-i)\, g_{l,i}(n) \\ &= \sum_{l=0}^{L-1} \boldsymbol{y}_l^T(n)\, \boldsymbol{g}_l(n). \end{aligned} \tag{11.16}$$

The quantities L and N in Eq. 11.16 are denoting the number of microphones and the length of the FIR filters, respectively. We will assume for the rest of this chapter that the electroacoustic transducers produce an output signal proportional to the sound pressure (minus an offset according to the nominal sound pressure p_0):

$$y_l(n) = K\big(p(nT_0, \boldsymbol{r}_l) - p_0\big). \tag{11.17}$$

T_0 represents the sampling time of the transducer, which is equal to the inverse of the sampling frequency:

$$T_0 = \frac{1}{f_s}. \tag{11.18}$$

[4]For the reason of simplicity real input signals $y_l(n)$ and real filter impulse responses $g_{l,i}(n)$ are assumed.

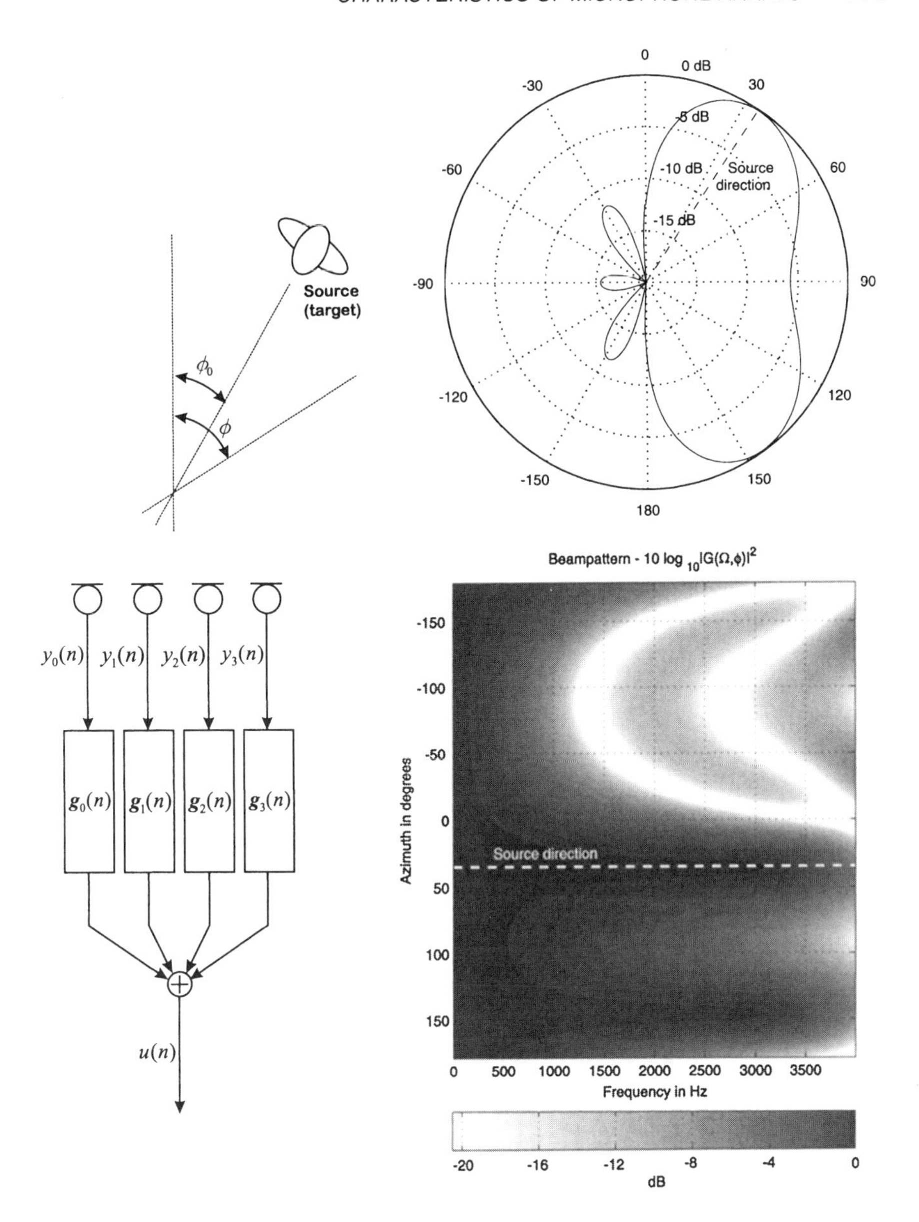

Fig. 11.3 Characteristics of a microphone array. In the right part of the figure a polar plot for the frequency 1500 Hz (top) and a beampattern (bottom) are depicted.

The vector $\boldsymbol{y}_l(n)$ contains the last N samples of the lth microphone signal

$$\boldsymbol{y}_l(n) = \left[y_l(n),\, y_l(n-1),\, \ldots,\, y_l(n-N+1)\right]^{\mathrm{T}} \tag{11.19}$$

and the coefficients $g_{l,\,i}(n)$ (with $i \in \{0,\,\ldots,\,N-1\}$) of the FIR filter make up the vector

$$\boldsymbol{g}_l(n) = \left[g_{l,\,0}(n),\, g_{l,\,1}(n),\, \ldots,\, g_{l,\,N-1}(n)\right]^{\mathrm{T}}. \tag{11.20}$$

In case of fixed beamformers, where the filter coefficients will not change over time

$$g_{l,\,i}(n) = g_{l,\,i}(n-1) = g_{l,\,i}, \tag{11.21}$$

the spectrum of the beamformer output is given by

$$U\left(e^{j\Omega}\right) = \sum_{l=0}^{L-1} Y_l\left(e^{j\Omega}\right) G_l\left(e^{j\Omega}\right). \tag{11.22}$$

For the investigation of the directivity of a microphone array, we will assume that a plane wave comes into the array (see Section 11.1). The direction of arrival $\boldsymbol{r}$ is given according to the coordinate system depicted in Fig.11.4.

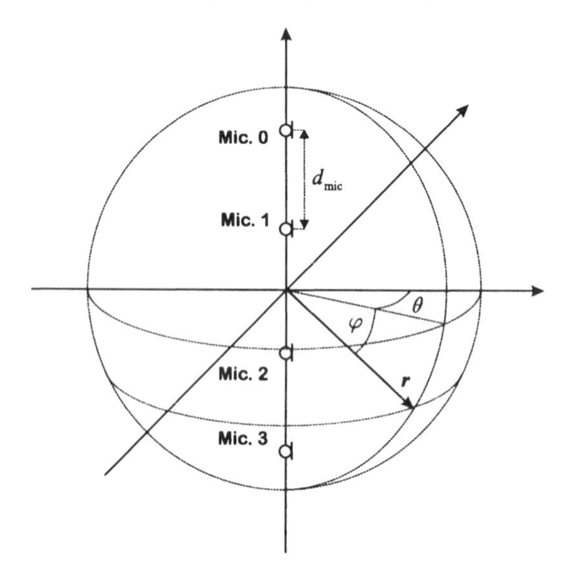

Fig. 11.4 Coordinate system utilized for deriving the spatial transfer function of linear beamformers.

The origin of the coordinate system is chosen such that all position vectors of the microphones sum up to zero:

$$\sum_{l=0}^{L-1} \boldsymbol{r}_l = 0. \tag{11.23}$$

This point (the origin) is also called the *phase center* of an array. The length of the vector pointing into the direction of arrival $\boldsymbol{r}$ is defined to be

$$\|\boldsymbol{r}\| = 1. \tag{11.24}$$

If we assume a plane wave $S(e^{j\Omega})$ arriving from direction $\boldsymbol{r}$ at the array each microphone signal is a delayed (positively or negatively) version of the wavefront arriving at the phase center of the array. According to the coordinate system of Fig. 11.4, the delay τ_l[5] of the lth microphone can be denoted as

$$\tau_l = \frac{f_s}{c}\,\boldsymbol{r}_l^T \boldsymbol{r}\,. \tag{11.25}$$

Furthermore, we assume that each microphone has an individual receiving characteristic. This can be modeled also by spatial frequency responses $H_l(e^{j\Omega}, \boldsymbol{r})$. For the spectra of the microphone signals,[6] we obtain

$$Y_l\big(e^{j\Omega}, \boldsymbol{r}\big) = S\big(e^{j\Omega}\big)\ H_l\big(e^{j\Omega}, \boldsymbol{r}\big)\ e^{j\Omega\tau_l}\,. \tag{11.26}$$

Inserting this result into Eq. 11.22, we get for the spectrum of the beamformer output

$$U\big(e^{j\Omega}, \boldsymbol{r}\big) = S\big(e^{j\Omega}\big) \sum_{l=0}^{L-1} H_l\big(e^{j\Omega}, \boldsymbol{r}\big)\ G_l\big(e^{j\Omega}\big)\ e^{j\Omega\tau_l}\,. \tag{11.27}$$

The spatial frequency response of the entire beamformer (including all microphone tolerances) is given by

$$G_{\mathrm{BF}}\big(e^{j\Omega}, \boldsymbol{r}\big) = \frac{U\big(e^{j\Omega}, \boldsymbol{r}\big)}{S\big(e^{j\Omega}\big)} = \sum_{l=0}^{L-1} H_l\big(e^{j\Omega}, \boldsymbol{r}\big)\ G_l\big(e^{j\Omega}\big)\ e^{j\Omega\tau_l}\,. \tag{11.28}$$

In case of equidistantly spaced microphones with a distance d_{mic} between each neighboring pair of sensors, the delay τ_l of the lth microphone is given by

$$\tau_l = \left(\frac{L-1}{2} - l\right) \frac{d_{\mathrm{mic}}\, f_s}{c}\, \sin\varphi\,. \tag{11.29}$$

The independence of the angle θ is a result of the spatial symmetry (see Fig. 11.4). Inserting Eq. 11.29 into Eq. 11.28 leads to the spatial frequency response of linear arrays with equidistantly spaced microphones:[7]

$$G_{\mathrm{BF}}\big(e^{j\Omega}, \varphi\big) = \sum_{l=0}^{L-1} H_l\big(e^{j\Omega}, \boldsymbol{r}\big)\ G_l\big(e^{j\Omega}\big)\ \exp\left[j\Omega\left(\frac{L-1}{2} - l\right)\frac{d_{\mathrm{mic}}\, f_s}{c}\, \sin\varphi\right]. \tag{11.30}$$

In most textbooks on beamforming and microphone arrays the influence of microphone characteristics and of microphone tolerances is not included [22]. Thus,

[5]Note that τ_l denotes the (noninteger) number of samples and not the continuous delay time.

[6]According to the spatial sound model introduced in this section, the vector $\boldsymbol{r}$ has also been added as a variable.

[7]For better readability, e^x is written as $\exp(x)$.

our definition of the spatial transfer function differs from other texts. However, in most consumer audio applications cardioid or hypercardioid microphones are used. The inclusion of $H_l(e^{j\Omega}, \boldsymbol{r})$ influences the result of several optimization procedures (e.g., when optimizing the gain of an array; see Section 11.2.2).

If the reader prefers the conventional definition, optimal omnidirectional microphones can be assumed. Their receiving characteristics can be described by

$$H_l\left(e^{j\Omega}, \boldsymbol{r}\right)\Big|_{\text{ideal omnidir. mic.}} = 1\,. \tag{11.31}$$

Under this assumption the spatial frequency response of an arbitrary array simplifies to

$$G_{\text{BF}}\left(e^{j\Omega}, \boldsymbol{r}\right)\Big|_{\text{ideal omnidir. mic.}} = \sum_{l=0}^{L-1} G_l\left(e^{j\Omega}\right)\, e^{j\Omega\tau_l}\,. \tag{11.32}$$

11.2.1 Directional Pattern

An illustrative quantity used to visualize the receiving characteristics of an array is its *directional pattern*, which is defined as the squared magnitude of the spatial frequency response:

$$\Phi\left(\Omega, \boldsymbol{r}\right) = \sum_{l=0}^{L-1} \left|H_l\left(e^{j\Omega}, \boldsymbol{r}\right)\, G_l\left(e^{j\Omega}\right)\, e^{j\Omega\tau_l}\right|^2\,. \tag{11.33}$$

We have already utilized this quantity for visualizing the characteristics of a basic delay-and-sum array (see Fig. 11.3) and for visualizing grating lobes (see Fig. 11.2). If Eq. 11.33 is examined for a single frequency Ω_0, so-called polar plots are utilized to visualize the spatial properties of an array. In case of linear arrays the spatial symmetry concerning the array axis can be exploited and two-dimensional diagrams as depicted in Fig. 11.3 can be utilized. For arbitrary array geometries, pseudo-three-dimensional plots have to be used.

If all sensors have an identical receiving characteristic

$$H_l\left(e^{j\Omega}, \boldsymbol{r}\right) = H\left(e^{j\Omega}, \boldsymbol{r}\right)\,, \tag{11.34}$$

it is possible to separate the beampattern of the microphone part and the beampattern of the filtering part:

$$\begin{aligned}
\Phi\left(\Omega, \boldsymbol{r}\right) &= \left|H\left(e^{j\Omega}, \boldsymbol{r}\right) \sum_{l=0}^{L-1} G_l\left(e^{j\Omega}\right)\, e^{j\Omega\tau_l}\right|^2 \\
&= \left|H\left(e^{j\Omega}, \boldsymbol{r}\right)\right|^2 \left|\sum_{l=0}^{L-1} G_l\left(e^{j\Omega}\right)\, e^{j\Omega\tau_l}\right|^2 \\
&= \Phi_{\text{mic}}\left(\Omega, \boldsymbol{r}\right)\, \Phi_{\text{fil}}\left(\Omega, \boldsymbol{r}\right)\,.
\end{aligned} \tag{11.35}$$

The first factor in Eq. 11.35, namely

$$\Phi_{\text{mic}}(\Omega, \boldsymbol{r}) = \left|H\left(e^{j\Omega}, \boldsymbol{r}\right)\right|^2, \tag{11.36}$$

represents the beampattern of the microphones. The second factor, namely

$$\Phi_{\text{fil}}(\Omega, \boldsymbol{r}) = \left|\sum_{l=0}^{L-1} G_l\left(e^{j\Omega}\right) e^{j\Omega\tau_l}\right|^2, \tag{11.37}$$

denotes the spatial properties that result from the filtering part of the beamformer. The choice of the primary sensors has large influence on the entire beampattern. To demonstrate this two beampatterns are depicted in Fig. 11.5. The filtering part

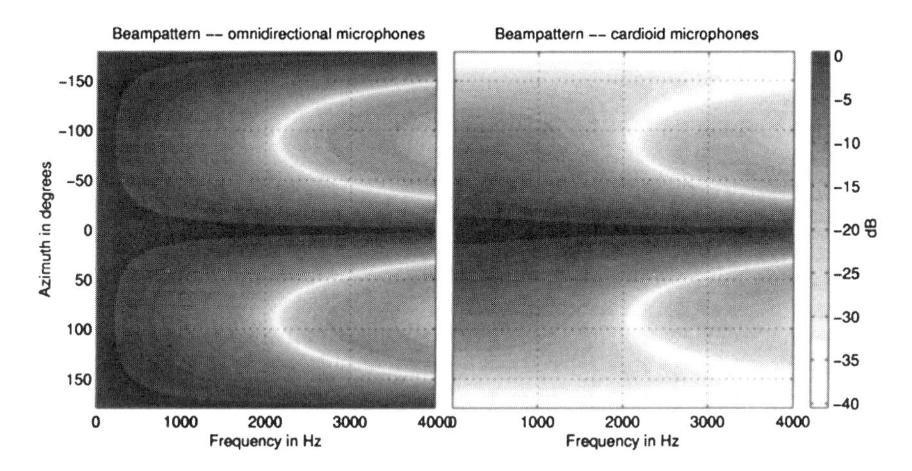

Fig. 11.5 Beampatterns of arrays with different sensor types. For the beamformer corresponding to the left beampattern, omnidirectional microphones were utilized; for the right one, sensors with a cardioid beampattern were chosen. The filtering parts of both beamformers were equal.

is equal in both cases; each filter $g_{l,i}$ consists of only a single coefficient which is $g_{l,0} = 0.25$. The steering direction of the array is $\varphi = 0°$. Four microphones are used in both cases. All microphones are located on a line with $d_{\text{mic}} = 4$ cm neighboring sensor distance. For the left beampattern omnidirectional microphones have been used—for the right analysis ideal cardioid sensors [58] with a spatial frequency response

$$H_{\text{card., id.}}\left(e^{j\Omega}, \varphi\right) = \frac{1}{2}\left(1 + \cos\varphi\right) \tag{11.38}$$

were applied. The differences (especially the backside rejection properties) in both beampatterns are clearly visible.

11.2.2 Array Gain

Another criterion besides the beampattern for evaluating beamformers is the directional gain of an array. This is the gain of a microphone array over a single omnidirectional sensor in a noise field. A common quantity used to measure this gain is the *directivity factor*

$$Q(\Omega, \boldsymbol{r}_s) = \frac{\Phi(\Omega, \boldsymbol{r}_s)}{\frac{1}{4\pi} \int\limits_{A_s} \Phi(\Omega, \boldsymbol{r})\, dA} \tag{11.39}$$

or its logarithmic counterpart, the *directivity index* [16, 58]:

$$Q_{\log}(\Omega, \boldsymbol{r}_s) = 10 \log_{10}\left[Q(\Omega, \boldsymbol{r}_s)\right] . \tag{11.40}$$

Within the denominator of Eq. 11.39 the integral is taken over a spherical surface with unit radius. The vector $\boldsymbol{r}_s$, $\|\boldsymbol{r}_s\| = 1$, is denoting the steering direction of the array.

The directivity factor is also used for the definition of so-called superdirective microphone arrays. An array is called *superdirective* whenever its directivity factor is larger than that of a delay-and-sum beamformer (see Section 11.3.1) with the same array geometry [16].

11.2.3 Further Performance Criteria

Besides the beampattern and the directivity factor (or index), a variety of other performance criteria have been established. The drawback of the directivity measures is that a predefined steering direction is required. In some applications, such as audio and videoconferences or music recordings, such a direction cannot be explicitly given. Here the *front-to-back ratio* according to

$$R_{\mathrm{FB}}(\Omega) = \frac{\int\limits_{A_f} \Phi(\Omega, \boldsymbol{r})\, dA}{\int\limits_{A_b} \Phi(\Omega, \boldsymbol{r})\, dA} \tag{11.41}$$

might be the better choice for evaluating the performance of an array, particularly for optimizing the filter weights. In Eq. 11.41 A_f denotes that half of a spherical surface with unit radius located in front of the array and A_b denotes the other half located behind the beamformer.

Another criterion that is especially suitable for enhancing the robustness according to sensor imperfections and mounting inaccuracies is the *white noise gain* [46]

$$R_{\mathrm{WN}}(\Omega) = \frac{\left| \sum\limits_{l=0}^{L-1} G_l(e^{j\Omega})\, e^{-j\Omega\tau_l} \right|^2}{\sum\limits_{l=0}^{L-1} \left| G_l(e^{j\Omega}) \right|^2} . \tag{11.42}$$

When applying the white noise gain as a side condition during the design process of a beamformer, such as

$$Q(\Omega, \boldsymbol{r}_\mathrm{s}) \longrightarrow \max \quad \text{with respect to} \quad R_\mathrm{WN}(\Omega) = \delta^2, \tag{11.43}$$

one has to take the upper limit of white noise gain

$$R_\mathrm{WN}(\Omega) \leq L \tag{11.44}$$

into consideration in order to keep the optimization self-consistent.

11.3 FIXED BEAMFORMING

In this section basics of the design of nonadaptive beamformers will be described. We will start with simple delay-and-sum beamformers, where all microphone signals are time-aligned for a predefined direction. This type of beamformer will serve as a reference for enhanced designs that will be described toward the end of this section.

In most design procedures, computation of the filter weights $g_{l,i}$ is split into two steps:

1. Delay compensation filters $g_{\mathrm{D},l,i}$ are designed.

2. The delay-compensated microphone signals are fed into a second (independent) stage of the beamformer. For this stage broadside optimization procedures can be applied in order to find the coefficients of the filters $g_{\mathrm{F},l,i}$ and to achieve superdirectivity.

The frequency response of the entire filter $G_l(e^{j\Omega})$ within the lth microphone channel makes up to

$$G_l\left(e^{j\Omega}\right) = G_{\mathrm{D},l}\left(e^{j\Omega}\right) G_{\mathrm{F},l}\left(e^{j\Omega}\right). \tag{11.45}$$

11.3.1 Delay-and-Sum Beamforming

The basic structure of a delay-and-sum beamformer is depicted in Fig. 11.6. The filters $g_{\mathrm{D},l}$ (here in FIR structure) have allpass character. The filters of the second stage $G_{\mathrm{F},l}(e^{j\Omega})$ can be implemented as simple multiplications

$$G_{\mathrm{F},l}\left(e^{j\Omega}\right) = \frac{1}{L}. \tag{11.46}$$

The filters of the first stage $G_{\mathrm{D},l}(e^{j\Omega})$ should delay the incoming signals $y_l(n)$ by fixed (noninteger) numbers of samples τ_l. These delays are not necessarily a multiple of a sampling interval. Thus, it is not possible to write the delaying operation as a simple difference equation with only one input and one output signal in the discrete time domain. The problem can be solved by regarding a noninteger delay as a resampling process. The desired solution can be obtained in the frequency domain by

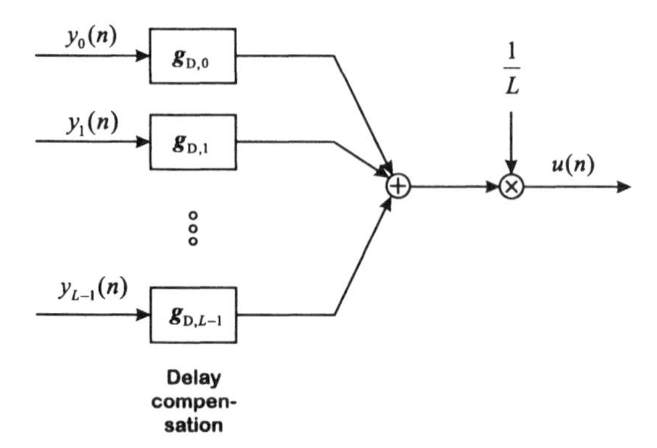

Fig. 11.6 Structure of the delay-and-sum beamformer.

reconstructing the bandlimited signal in the continuous time domain and resampling it after applying the delay [138]. In this case we obtain the following relation for the frequency response of ideal delay compensation filters:

$$G_{D,l}\left(e^{j\Omega}\right) = e^{-j\Omega\tau_l}\,. \tag{11.47}$$

The delays τ_l can be obtained by analyzing the array geometry. For linear arrays with equidistant microphone positioning, it is rather simple to compute the required delays in advance (see Fig. 11.7). If the phase center of the array is located according

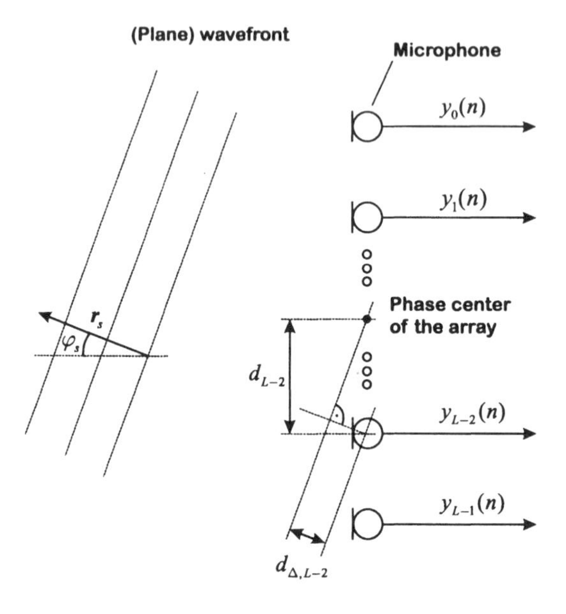

Fig. 11.7 Geometry of a linear beamformer with equidistant microphone positions.

to Eq. 11.23, the distance from the lth microphone to the phase center is

$$d_l = \|\boldsymbol{r}_l\| = \left(l - \frac{L-1}{2}\right) d_{\text{mic}} \,. \tag{11.48}$$

As already mentioned, d_{mic} denotes the (constant) distance between adjacent microphones. If a plane wave arrives from the steering direction $\boldsymbol{r}_{\text{s}}$ of the array, described by the angle φ_{s}, the plane wave reaches the lth sensor with a delay of

$$t_l = \frac{d_l \, \sin\varphi_{\text{s}}}{c} = \left(l - \frac{L-1}{2}\right) \frac{d_{\text{mic}} \, \sin\varphi_{\text{s}}}{c} \tag{11.49}$$

compared to the phase center of the array. Multiplying Eq. 11.49 with the sampling frequency f_{s} leads to the delay in terms of samples for each microphone:

$$\tau_l = t_l \, f_{\text{s}} = \left(l - \frac{L-1}{2}\right) \frac{d_{\text{mic}} \, \sin\varphi_{\text{s}}}{c} \, f_{\text{s}} \,. \tag{11.50}$$

Note that Eqs. 11.49 and 11.50 also denote negative (noncausal) delays. For the realizations presented in the next two sections, this is unproblematic. It will be more important that the maximum and the minimum required delays do not differ considerably. If this is the case, first the "traditional" (integer) delays realized by ring buffers should be implemented and only the residual delays should be realized by one of the methods presented below.

11.3.1.1 Frequency–Domain Implementation

A simple and computational effective way of implementing delay-and-sum beamformers are DFT or filterbank structures. The general setup of such structures is depicted in Fig. 11.8. Each microphone signal is split into overlapping blocks of appropriate size. The resulting signal blocks are transformed via a DFT or a filterbank into the frequency or subband domain. Delaying the microphone signals corresponds in the frequency domain to the multiplication with fixed coefficient vectors:

$$\boldsymbol{d}_l = \left[1, \, e^{-j\frac{2\pi}{M}\tau_l}, \, e^{-j\frac{2\pi}{M}2\tau_l}, \, \ldots, \, e^{-j\frac{2\pi}{M}(M-1)\tau_l}\right]^{\text{T}} . \tag{11.51}$$

The vectors $\boldsymbol{d}_l$ (with $l \in \{0, \ldots, L-1\}$) result from sampling of the ideal frequency response (see Eq. 11.47) at M equidistant frequency supporting points. If uniform polyphase filterbanks according to Section 9.1.3.2.3 with M channels are utilized, the vectors according to Eq. 11.51 can also be used. After multiplication with the delay compensation vectors $\boldsymbol{d}_l$, all microphone channels are added and an inverse transformation (IDFT or synthesis filterbank) is applied. In order to obtain a unit spatial frequency response in the steering direction of the beamformer, the output signal of the inverse transform is multiplied by the factor $1/L$.

In case of large block sizes, such as $M = 256$ samples, and no excessive delays, for instance, $|\tau_l| < 3$ samples, this processing structure has sufficient accuracy. On

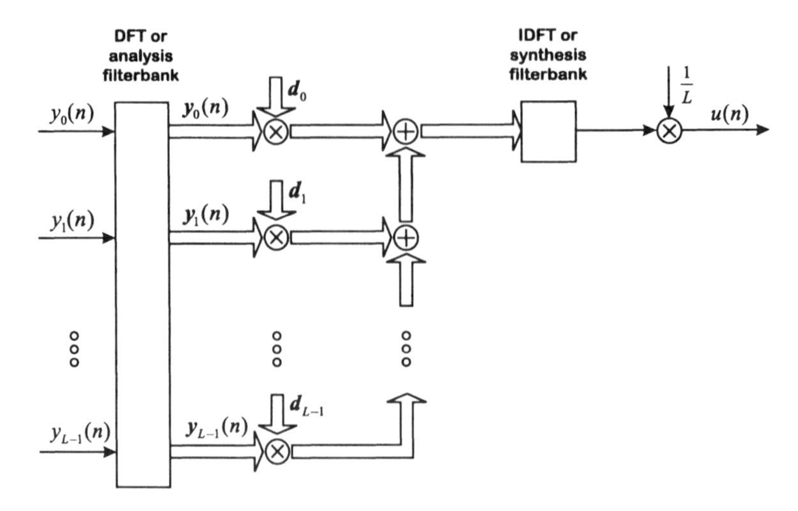

Fig. 11.8 Structure of a delay-and-sum beamformer implemented in the DFT or subband domain.

the other hand, the drawback of block processing structures should also be considered. Due to the necessary input and output buffers, a delay is always introduced in the signal path.

11.3.1.2 Design of Time-Domain Fractional Delay Filters

When realizing the delay compensation stage in the time domain FIR structures are usually applied. By transforming the frequency response of the ideal filters (see Eq. 11.47) into the time domain [138], we obtain

$$g_{\mathrm{D},l,i} = \begin{cases} \dfrac{\sin\Big(\pi\left(i-\tau_l\right)\Big)}{\pi\left(i-\tau_l\right)}, & \text{for } \tau_l - i \neq 0\,, \\ 1, & \text{else}\,. \end{cases} \tag{11.52}$$

Unfortunately, the resulting impulse response is both infinite and noncausal for noninteger delays τ_l. In order to obtain solutions that can be realized by FIR filters with N_{D} coefficients, a variety of methods have been proposed. We will describe only three basic design schemes (for further details and design procedures, the interested reader is referred to the article by Laakso et al. [138]):

- A straightforward solution is to minimize the average squared spectral difference between the frequency response of an optimal delay compensation filter according to Eq. 11.47 and the FIR filter $\bar{G}_{\mathrm{D},l}(e^{j\Omega})$ that should be designed:

$$E_{\mathrm{LS}} = \frac{1}{\pi} \int\limits_{\Omega=0}^{\pi} \left| G_{\mathrm{D},l}\left(e^{j\Omega}\right) - \bar{G}_{\mathrm{D},l}\left(e^{j\Omega}\right)\right|^2 d\Omega \longrightarrow \min\,. \tag{11.53}$$

The solution to this minimization criterion is—for linear phase FIR filters—simply a truncation of the ideal impulse response to N_D coefficients [179]. For small delays $|\tau_l| < 0.5$, the impulse response is shifted by $N_\mathrm{D}/2$ coefficients in order to obtain a causal solution:

$$\bar{g}_{\mathrm{D},l,i} = \begin{cases} g_{\mathrm{D},l,i-\frac{N_\mathrm{D}}{2}}, & \text{for } 0 \le i \le N_\mathrm{D}-1, \\ 0, & \text{else}. \end{cases} \tag{11.54}$$

- Although the design according to Eq. 11.54 is optimal in the least square sense, the well-known Gibbs phenomenon appears and causes ripple in the magnitude response (see lower part of Fig. 11.9). For reducing these ripples, window functions such as Hamming or Hanning windows [109] can be applied:

$$\tilde{g}_{\mathrm{D},l,i} = \begin{cases} g_{\mathrm{D},l,i-\frac{N_\mathrm{D}}{2}}\, g_{\mathrm{han},i-\frac{N_\mathrm{D}}{2}}, & \text{for } 0 \le i \le N_\mathrm{D}-1, \\ 0, & \text{else}. \end{cases} \tag{11.55}$$

The term $g_{\mathrm{han},i}$ in Eq. 11.55 denotes a Hanning window, which is defined as

$$g_{\mathrm{han},i} = \begin{cases} K\left[1+\sin\left(\frac{2\pi}{N_\mathrm{D}}\left(i-\frac{N_\mathrm{D}}{2}\right)\right)\right], & \text{for } i \in \{0,\ldots,N_\mathrm{D}-1\}, \\ 0 & \text{else}. \end{cases} \tag{11.56}$$

The constant K is chosen such that the average low-frequency magnitude of the windowed filter satisfies the condition[8]

$$\frac{1}{\Omega_1}\int_{\Omega=0}^{\Omega_1} \tilde{G}_{\mathrm{D},l}\left(e^{j\Omega}\right) d\Omega = 1, \quad \text{with} \quad \Omega_1 = \frac{\pi}{2}. \tag{11.57}$$

Due to the multiplication with the impulse response of a lowpass filter, the oscillations within the transfer function can be reduced considerably (see lower part of Fig. 11.9).

- As a last design method, *Lagrange interpolation* is described. Applying this procedure to design the delay compensation filters leads to a maximally flat frequency response at a certain frequency (here $\Omega_0 = 0$). This means that the first $N_\mathrm{D}-1$ derivatives of the frequency domain error

$$E\left(e^{j\Omega}\right) = \check{G}_{\mathrm{D},l}\left(e^{j\Omega}\right) - G_{\mathrm{D},l}\left(e^{j\Omega}\right) \tag{11.58}$$

are zero at $\Omega = 0$:

$$\left.\frac{d^n\, E\left(e^{j\Omega}\right)}{(d\,\Omega)^n}\right|_{\Omega=0} = 0, \quad \text{for } n \in \{0, 1, \ldots, N_\mathrm{D}-1\}. \tag{11.59}$$

[8] The upper integration limit Ω_1 should be chosen in dependence of the desired approximation quality. The better the approximation is, the higher should be the choice of Ω_1.

After differentiation and inserting the results at $\Omega = 0$, a system of equations of order N_D is obtained. Transforming these equations into the time domain leads to

$$\sum_{i=0}^{N_\mathrm{D}-1} i^n \, \breve{g}_{\mathrm{D},l,i} = \tau_l^n, \quad \text{for } n \in \{0, 1, \ldots, N_\mathrm{D} - 1\}. \tag{11.60}$$

This time-domain system of equations can be solved explicitly [138], and we obtain for the filter coefficients

$$\breve{g}_{\mathrm{D},l,i} = \prod_{n=0,\, n\neq i}^{N_\mathrm{D}-1} \frac{\tau_l - n}{i - n}, \quad \text{for } i \in \{0, 1, \ldots, N_\mathrm{D} - 1\}. \tag{11.61}$$

In Fig. 11.9 group delays (top) and frequency responses (bottom) of three delay compensation filters—one for each of the design criteria described above—are depicted. All filters have $N_\mathrm{D} = 21$ coefficients and were designed to delay the incoming signal by 0.4 sample. In order to avoid noncausality, the filters were shifted such that the "center of gravity" is around coefficient index $i = 10$. Thus, the desired group delay is 10.4 samples. The Lagrange filter shows the smallest ripple at low frequencies. However, the performance in terms of group delay and frequency response deviation at high frequencies is rather poor. The large ripple of the direct approach by simply truncating the impulse response of the ideal filter is also clearly visible. The windowing approach seems to be a good compromise between small ripple and good performance even at higher frequencies.

11.3.1.3 Results

In order to provide an impression of the performance of delay-and-sum beamforming, two measured signals are presented in Fig. 11.10. The recordings were made in a car driving with a speed of about 90 km/h. The rear mirror was equipped with $L = 4$ microphones (distance between adjacent sensors: 5 cm). The recorded signals consist of typical car background noise (wind, rolling, and engine noise) as well as speech from the driver. In the upper diagram the time-domain signal of the second microphone is depicted.

The second diagram shows the output of a time-domain delay-and-sum beamformer. The angle between the driver and the array axis was about 65°. FIR filters with $N_\mathrm{D} = 21$ coefficients were used for implementing the required delays. Hanning windowed $\sin(x)/x$ functions (according to the design procedure of the previous section) have been utilized to compute the filter coefficients. A visual comparison of both signals does not really show much difference between the signals. However, when listening to both signals, the beamformer output consists of less noise components at medium and high frequencies. Furthermore, the beamformer output signal sounds "clearer" (less reverberant) compared to a single-microphone signal. The enhancement concerning signal-to-noise ratio can be visualized by comparing the power spectral densities that have been estimated during the first 6 seconds of

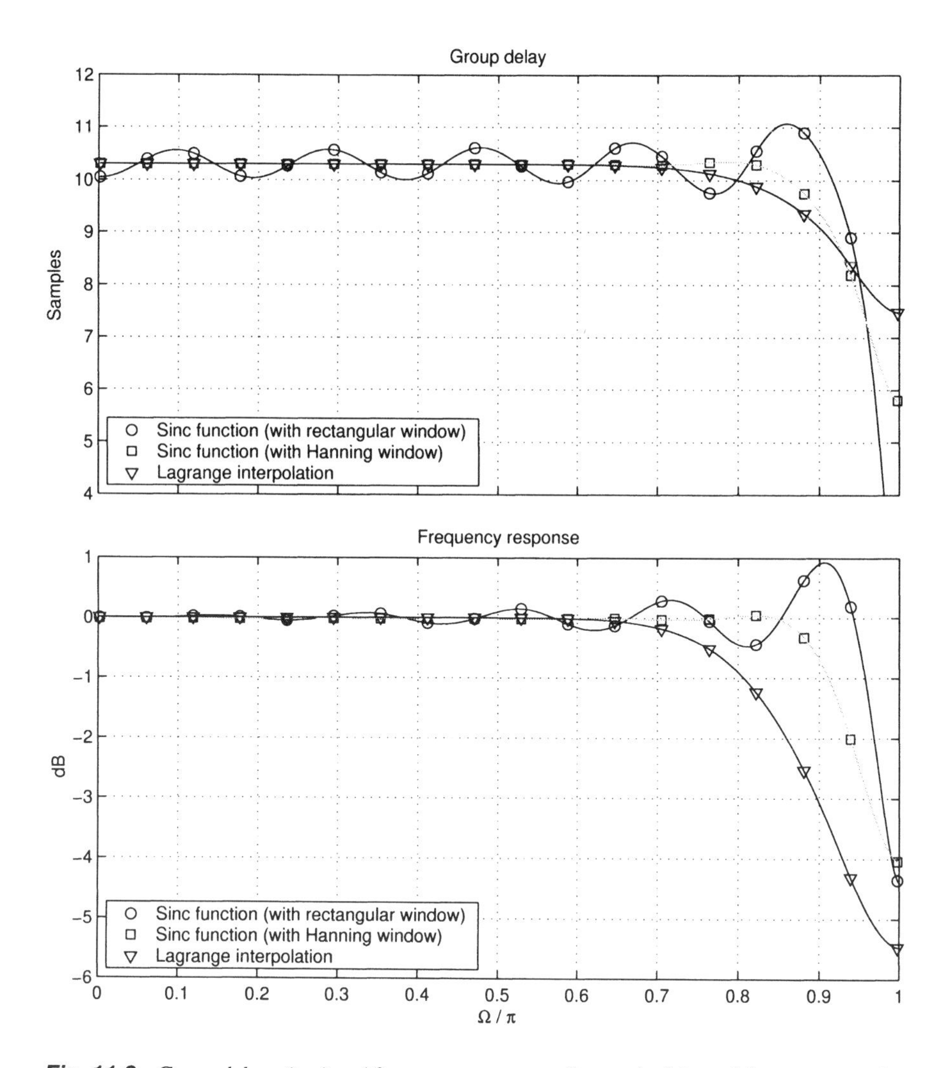

Fig. 11.9 Group delays (top) and frequency responses (bottom) of three delay compensation filters.

the recording. Both power spectral densities are depicted in the lowest diagram of Fig. 11.10.

Car noise can be modeled with sufficient accuracy as a diffuse noise field [59]. In such noise fields the improvement of the signal-to-noise ratio can be approximated at high frequencies by 3 dB per doubling of the number of microphones. The measurement presented in Fig. 11.10 verifies this approximation quite well. The enhancement up to 6 dB at higher frequencies is clearly visible.

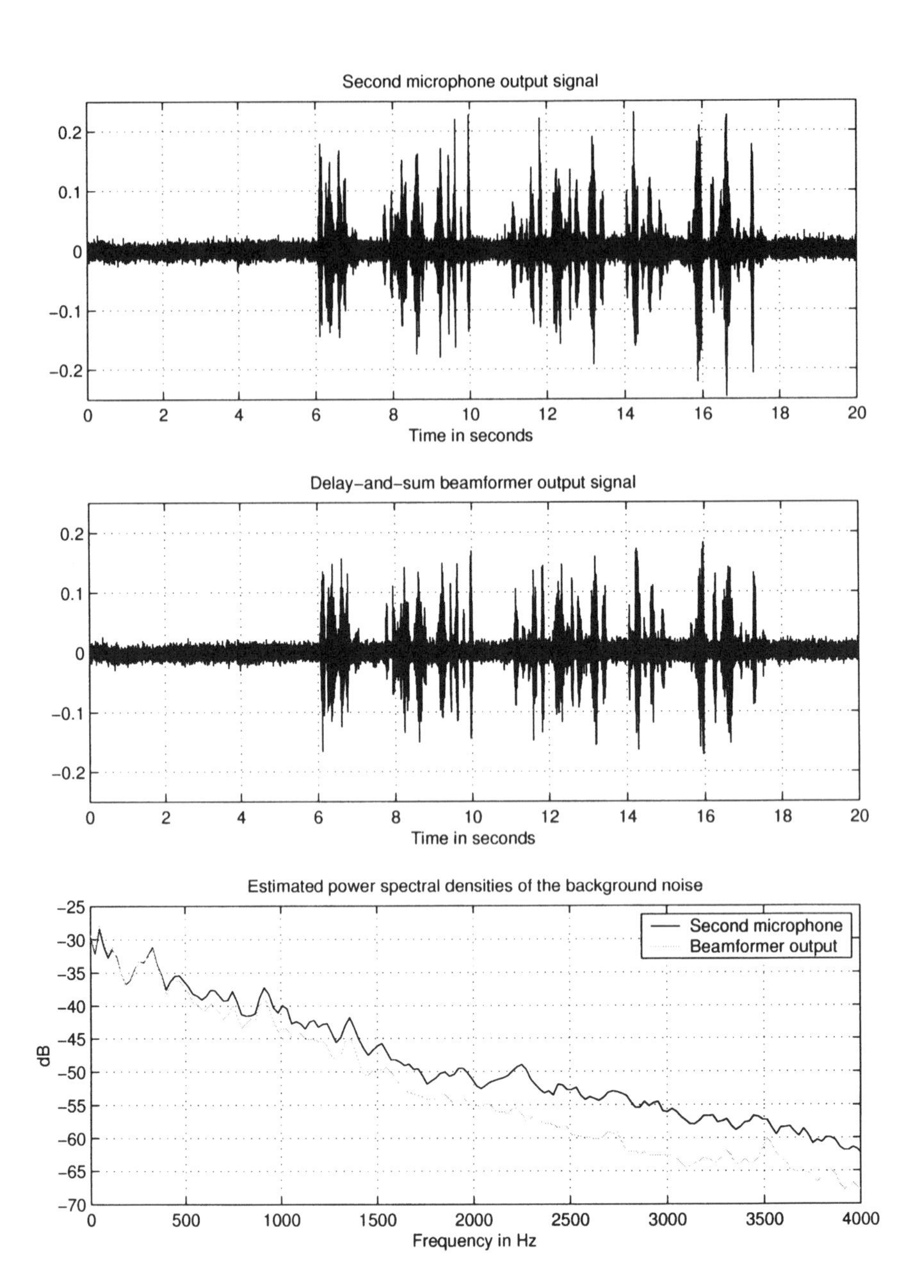

Fig. 11.10 Comparison of a single-microphone signal and an output signal of a delay-and-sum beamformer. The signals were measured in a car driving at a speed of about 90 km/h.

11.3.2 Filter-and-Sum Beamforming

The design of a filter-and-sum beamformer can be formulated as a constrained minimization [71]. In order to obtain a simple formulation of the optimization problem, the filters g_l driven by the lth microphone signal are split into delay compensating filters $g_{D,l}$ with unit gain and *free* filters $g_{F,l}$

$$g_{l,n} = \sum_{m=-\infty}^{\infty} g_{D,l,\,n\,-\,m}\, g_{F,l,\,m}\,. \tag{11.62}$$

The structure of a filter-and-sum beamformer with separate delay compensation and filtering part is depicted in Fig. 11.11. If we assume that the filters $g_{D,l}$ perform a perfect delay compensation, the optimization problem can be stated as

$$\mathrm{E}\left\{\, u^2(n) \,\right\} \longrightarrow \min \tag{11.63}$$

with respect to

$$\sum_{l=0}^{L-1} g_{F,l} = c\,. \tag{11.64}$$

The solution to this optimization problem is called the linearly constrained minimum variance (LCMV) beamformer. The constraint vector c represents a desired impulse response in the looking direction of the beamformer. The elements of c are usually zeros, except for a single weight, which is chosen to be one. This nonzero weight is located in the center of the vector:

$$c = \left[0,\, \ldots,\, 0,\, 1,\, 0,\, \ldots,\, 0\right]^{\mathrm{T}}\,. \tag{11.65}$$

Without the constraint, the trivial solution $g_{F,l} = \mathbf{0}$ would also be allowed. If the filters $g_{F,l,\,n}$ are assumed to be FIR filters of length N_F, the constraint according to

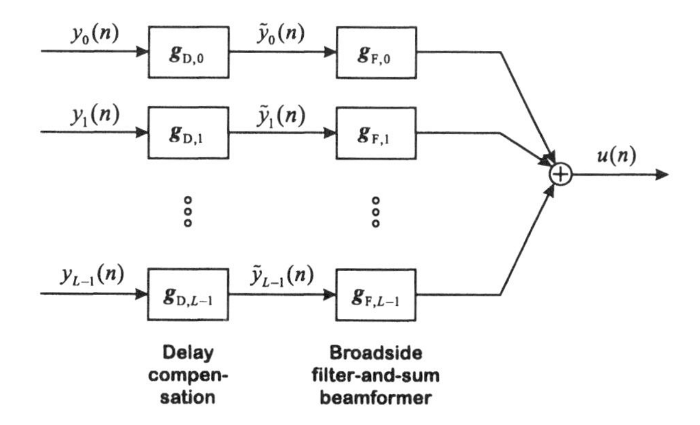

Fig. 11.11 Structure of the filter-and-sum beamformer.

Eq. 11.64 represents N_F linear equations for the coefficients $g_{\text{F},l,n}$. Thus, $(L-1)\,N_\text{F}$ degrees of freedom are left for minimizing the power of the beamformer output. In order to obtain a compact solution, some definitions are introduced. First a vector containing the coefficients of all filters is introduced as

$$\boldsymbol{g}_\text{F} = \left[\boldsymbol{g}_{\text{F},0}^\text{T},\, \boldsymbol{g}_{\text{F},1}^\text{T},\, \ldots,\, \boldsymbol{g}_{\text{F},L-1}^\text{T}\right]^\text{T}. \tag{11.66}$$

Furthermore a signal vector $\tilde{\boldsymbol{y}}(n)$ is defined[9] such that

$$u(n) = \boldsymbol{g}_\text{F}^\text{T}\, \tilde{\boldsymbol{y}}(n)\,. \tag{11.67}$$

In this case the signal vector is defined as

$$\tilde{\boldsymbol{y}}(n) \;=\; \left[\tilde{\boldsymbol{y}}_0^\text{T}(n),\, \tilde{\boldsymbol{y}}_1^\text{T}(n),\, \ldots,\, \tilde{\boldsymbol{y}}_{L-1}^\text{T}(n)\right]^\text{T}, \tag{11.68}$$

with

$$\tilde{\boldsymbol{y}}_l(n) \;=\; \left[\tilde{y}_l(n),\, \tilde{y}_l(n-1),\, \ldots,\, \tilde{y}_l(n-N_\text{F}+1)\right]^\text{T}. \tag{11.69}$$

Using these definitions, we obtain for the output power of the beamformer

$$\begin{aligned} \text{E}\left\{\, u^2(n) \,\right\} &= \text{E}\left\{\, \boldsymbol{g}_\text{F}^\text{T}\, \tilde{\boldsymbol{y}}(n)\, \tilde{\boldsymbol{y}}^\text{T}(n)\, \boldsymbol{g}_\text{F} \,\right\} \\ &= \boldsymbol{g}_\text{F}^\text{T}\, \text{E}\left\{\, \tilde{\boldsymbol{y}}(n)\, \tilde{\boldsymbol{y}}^\text{T}(n) \,\right\}\, \boldsymbol{g}_\text{F}\,. \end{aligned} \tag{11.70}$$

In the following we will abbreviate the correlation matrix of the delay compensated microphone signals $\tilde{y}_l(n)$ by

$$\text{E}\left\{\, \tilde{\boldsymbol{y}}(n)\, \tilde{\boldsymbol{y}}^\text{T}(n) \,\right\} = \boldsymbol{R}_{\tilde{\boldsymbol{y}}\tilde{\boldsymbol{y}}}\,. \tag{11.71}$$

Inserting this abbreviation into Eq. 11.70, we get

$$\text{E}\left\{\, u^2(n) \,\right\} = \boldsymbol{g}_\text{F}^\text{T}\, \boldsymbol{R}_{\tilde{\boldsymbol{y}}\tilde{\boldsymbol{y}}}\, \boldsymbol{g}_\text{F}\,. \tag{11.72}$$

Finally, the constraint of the minimization criterion (Eq. 11.64) can also be written in matrix–vector notation. For this, the summation process on the left side of Eq. 11.64 can be transformed into a matrix vector multiplication:

$$\boldsymbol{C}\, \boldsymbol{g}_\text{F} = \boldsymbol{c}\,. \tag{11.73}$$

The $L\, N_\text{F} \times N_\text{F}$ matrix $\boldsymbol{C}$ has the form

$$\boldsymbol{C} = \begin{bmatrix} 1 & 0 & 0 & \ldots & 0 & \ldots & 1 & 0 & 0 & \ldots & 0 \\ 0 & 1 & 0 & \ldots & 0 & \ldots & 0 & 1 & 0 & \ldots & 0 \\ 0 & 0 & 1 & \ldots & 0 & \ldots & 0 & 0 & 1 & \ldots & 0 \\ & & & & & & & & & & \\ & & \vdots & & & \ddots & & & \vdots & & \\ & & & & & & & & & & \\ 0 & 0 & 0 & \ldots & 1 & \ldots & 0 & 0 & 0 & \ldots & 1 \end{bmatrix}. \tag{11.74}$$

[9] Real input data are assumed here.

This constraint minimization problem—described by Eqs. 11.72 and 11.73—can be solved by the method of Lagrange multipliers [151]. We choose

$$F(\boldsymbol{g}_{\text{F}}) = \frac{1}{2}\, \boldsymbol{g}_{\text{F}}^{\text{T}}\, \boldsymbol{R}_{\tilde{\boldsymbol{y}}\tilde{\boldsymbol{y}}}\, \boldsymbol{g}_{\text{F}} + \boldsymbol{\lambda}^{\text{T}} \left(\boldsymbol{C}\, \boldsymbol{g}_{\text{F}} - \boldsymbol{c}\right) \tag{11.75}$$

as a cost function, where $\boldsymbol{\lambda}$ denotes an N_{F}-dimensional vector containing the Lagrange multipliers. For the gradient of the cost function with respect to $\boldsymbol{g}_{\text{F}}$, we obtain

$$\nabla_{\boldsymbol{g}_{\text{F}}} F(\boldsymbol{g}_{\text{F}}) = \boldsymbol{R}_{\tilde{\boldsymbol{y}}\tilde{\boldsymbol{y}}}\, \boldsymbol{g}_{\text{F}} + \boldsymbol{C}^{\text{T}}\, \boldsymbol{\lambda}\,. \tag{11.76}$$

For optimality, the gradient has to be zero, which leads to the wanted filter coefficient vector

$$\boldsymbol{g}_{\text{F, opt}} = -\boldsymbol{R}_{\tilde{\boldsymbol{y}}\tilde{\boldsymbol{y}}}^{-1}\, \boldsymbol{C}^{\text{T}}\, \boldsymbol{\lambda}\,. \tag{11.77}$$

The correlation matrix $\boldsymbol{R}_{\tilde{\boldsymbol{y}}\tilde{\boldsymbol{y}}}$ is assumed to be positive definite, which guarantees that its inverse exists. The optimal coefficient vector has to fulfill the constraint described by Eq. 11.73

$$\boldsymbol{C}\left(-\boldsymbol{R}_{\tilde{\boldsymbol{y}}\tilde{\boldsymbol{y}}}^{-1}\, \boldsymbol{C}^{\text{T}}\, \boldsymbol{\lambda}\right) = \boldsymbol{c}\,, \tag{11.78}$$

which can be utilized to compute the Lagrange multipliers:

$$\boldsymbol{\lambda} = -\left[\boldsymbol{C}\, \boldsymbol{R}_{\tilde{\boldsymbol{y}}\tilde{\boldsymbol{y}}}^{-1}\, \boldsymbol{C}^{\text{T}}\right]^{-1} \boldsymbol{c}\,. \tag{11.79}$$

The existence of $[\boldsymbol{C}\, \boldsymbol{R}_{\tilde{\boldsymbol{y}}\tilde{\boldsymbol{y}}}^{-1}\, \boldsymbol{C}^{\text{T}}]^{-1}$ is ensured since $\boldsymbol{R}_{\tilde{\boldsymbol{y}}\tilde{\boldsymbol{y}}}$ is assumed to be positive definite and $\boldsymbol{C}$ has full rank. Inserting this result into Eq. 11.77 leads to the optimum constraint filter vector in the least mean square sense:

$$\boldsymbol{g}_{\text{F, opt}} = \boldsymbol{R}_{\tilde{\boldsymbol{y}}\tilde{\boldsymbol{y}}}^{-1}\, \boldsymbol{C}^{\text{T}} \left[\boldsymbol{C}\, \boldsymbol{R}_{\tilde{\boldsymbol{y}}\tilde{\boldsymbol{y}}}^{-1}\, \boldsymbol{C}^{\text{T}}\right]^{-1} \boldsymbol{c}\,. \tag{11.80}$$

To show the advantages of filter-and-sum beamforming over delay-and-sum beamforming, two beampatterns are depicted in Fig. 11.12. In both cases a linear array with $L = 4$ cardioid microphones has been utilized. The upper diagram shows the beampattern of a simple delay-and-sum beamformer. The delay compensation filters were designed using the windowing approach described in Section 11.3.1.2. FIR filters of length $N_{\text{D}} = 21$ and a Hamming window [203] have been utilized. In the lower diagram the beampattern of a filter-and-sum beamformer is depicted. Besides the same delay compensation filters that were applied in the delay-and-sum approach, additional FIR filters $\boldsymbol{g}_{\text{F},l}$ with $N_{\text{F}} = 32$ coefficients were utilized. For the computation of the filter coefficients the correlation matrix of a diffuse noise field [59] has been applied.

The advantages of the additional filtering—in terms of better directivity—appear mostly at low frequencies (see Fig. 11.12). Because speech has most of its energy at low frequencies, this advantage can improve the quality of speech enhancement and speech recognition systems considerably.

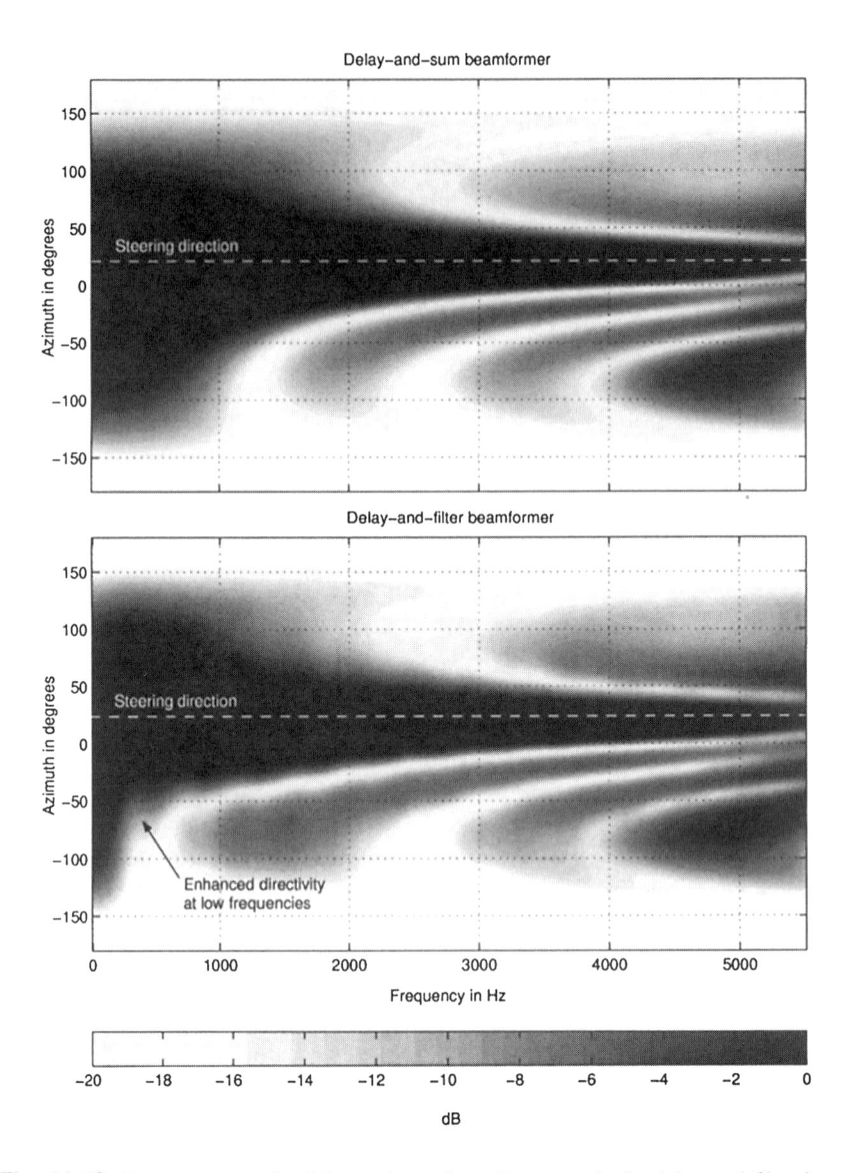

Fig. 11.12 Beampatterns of a delay-and-sum beamformer and of a delay-and-filter beamformer.

The direct constraint according to Eq. 11.64 is not the only possible side condition. Also additional a priori knowledge can be introduced via constraints. Furthermore, several constraints can be introduced in order to numerically stabilize the optimization process [29, 46, 235].

11.4 ADAPTIVE BEAMFORMING

The major drawback of fixed filter-and-sum beamforming—as described in Section 11.3.2—is that the correlation matrix $\boldsymbol{R}_{\tilde{\boldsymbol{y}}\tilde{\boldsymbol{y}}}$ of the input signals has to be known a priori. In many applications speech of a desired speaker is corrupted by a second speaker (as well as by background noise). Often the position of the second (unwanted) source is not known in advance. In this case adaptive beamforming algorithms can be applied. Such an algorithm, called the *constrained least mean square* algorithm [71], will be described in the following paragraphs.

The algorithm uses the same structure as a filter-and-sum beamformer (see Fig. 11.11). The vector containing the filter weights $\boldsymbol{g}_{\mathrm{F}}(n)$[10] is corrected at each iteration in the negative direction of the constrained gradient (see Eq. 11.76):

$$\begin{aligned} \boldsymbol{g}_{\mathrm{F}}(n+1) &= \boldsymbol{g}_{\mathrm{F}}(n) - \alpha\, \boldsymbol{\nabla}_{\boldsymbol{g}_{\mathrm{F}}} F\left(\boldsymbol{g}_{\mathrm{F}}(n)\right) \\ &= \boldsymbol{g}_{\mathrm{F}}(n) - \alpha \left[\boldsymbol{R}_{\tilde{\boldsymbol{y}}\tilde{\boldsymbol{y}}}(n)\, \boldsymbol{g}_{\mathrm{F}}(n) + \boldsymbol{C}^{\mathrm{T}}\, \boldsymbol{\lambda}(n)\right]. \end{aligned} \tag{11.81}$$

The cost function $F(\boldsymbol{g}_{\mathrm{F}}(n))$ is chosen as stated in Eq. 11.75. The parameter α is denoting the stepsize, which guarantees—if chosen sufficiently small—the convergence of the algorithm. At this early point of the derivation, it is assumed that the correlation matrix $\boldsymbol{R}_{\tilde{\boldsymbol{y}}\tilde{\boldsymbol{y}}}(n)$ at time index n is known. We will return to this point later on. The Lagrange multipliers $\boldsymbol{\lambda}(n)$ are chosen such that the filter vector $\boldsymbol{g}_{\mathrm{F}}(n+1)$ satisfies constraint 11.73:

$$\boldsymbol{c} = \boldsymbol{C}\, \boldsymbol{g}_{\mathrm{F}}(n+1)\,. \tag{11.82}$$

Inserting the filter recursion (see Eq. 11.81) leads to

$$\boldsymbol{c} = \boldsymbol{C}\, \boldsymbol{g}_{\mathrm{F}}(n) - \alpha\, \boldsymbol{C}\, \left[\boldsymbol{R}_{\tilde{\boldsymbol{y}}\tilde{\boldsymbol{y}}}(n)\, \boldsymbol{g}_{\mathrm{F}}(n) + \boldsymbol{C}^{\mathrm{T}}\, \boldsymbol{\lambda}(n)\right]. \tag{11.83}$$

By separation of $\boldsymbol{\lambda}(n)$ in Eq. 11.83 and inserting the result into Eq. 11.81, we obtain

$$\begin{aligned} \boldsymbol{g}_{\mathrm{F}}(n+1) &= \boldsymbol{g}_{\mathrm{F}}(n) - \alpha \left(\boldsymbol{I} - \boldsymbol{C}^{\mathrm{T}} \left[\boldsymbol{C}\,\boldsymbol{C}^{\mathrm{T}}\right]^{-1} \boldsymbol{C}\right) \boldsymbol{R}_{\tilde{\boldsymbol{y}}\tilde{\boldsymbol{y}}}(n)\, \boldsymbol{g}_{\mathrm{F}}(n) \\ &\quad + \boldsymbol{C}^{\mathrm{T}} \left[\boldsymbol{C}\,\boldsymbol{C}^{\mathrm{T}}\right]^{-1} \left[\boldsymbol{c} - \boldsymbol{C}\, \boldsymbol{g}_{\mathrm{F}}(n)\right]. \end{aligned} \tag{11.84}$$

In numerically robust versions of the algorithm the last factor in Eq. 11.84, $\boldsymbol{c} - \boldsymbol{C}\,\boldsymbol{g}_{\mathrm{F}}(n)$, is not assumed to be zero. This would be the case only if the coefficient vector precisely satisfies the constraint according to Eq. 11.82 during each iteration. However, due to numerical inaccuracies, the constraint is seldom fulfilled in real systems. Not neglecting the last term in Eq. 11.84 prevents error accumulation and error growth and makes the algorithm suitable for implementations on processors with fixed-point arithmetic. For a more compact notation, the vector

$$\tilde{\boldsymbol{c}} = \boldsymbol{C}^{\mathrm{T}} \left[\boldsymbol{C}\,\boldsymbol{C}^{\mathrm{T}}\right]^{-1} \boldsymbol{c} \tag{11.85}$$

[10] The filter vector $\boldsymbol{g}_{\mathrm{F}}(n)$ is defined as in Eq. 11.66. A time index n has been added here.

and the matrix

$$P = I - C^{\mathrm{T}} \left[C\, C^{\mathrm{T}}\right]^{-1} C \tag{11.86}$$

are introduced. Now, the algorithm can be written as

$$g_{\mathrm{F}}(n+1) = P \left[\, g_{\mathrm{F}}(n) - \alpha\, R_{\tilde{y}\tilde{y}}(n)\, g_{\mathrm{F}}(n) \,\right] + \tilde{c}\,. \tag{11.87}$$

For computing the negative gradient direction, knowledge of the input correlation matrix $R_{\tilde{y}\tilde{y}}(n)$ is required. A simple approximation for $R_{\tilde{y}\tilde{y}}(n)$ at the nth iteration is the outer product of the signal vector $\tilde{y}(n)$ with itself:

$$R_{\tilde{y}\tilde{y}}(n) \longrightarrow \tilde{y}(n)\, \tilde{y}^{\mathrm{T}}(n)\,. \tag{11.88}$$

In this case the matrix vector product $R_{\tilde{y}\tilde{y}}(n)\, g_{\mathrm{F}}(n)$ can be simplified as

$$R_{\tilde{y}\tilde{y}}(n)\, g_{\mathrm{F}}(n) \longrightarrow \tilde{y}(n)\, \tilde{y}^{\mathrm{T}}(n)\, g_{\mathrm{F}}(n) = \tilde{y}(n)\, u(n)\,. \tag{11.89}$$

Furthermore, the stepsize factor α is usually normalized to the squared norm of the input vector:

$$\alpha = \frac{\mu}{\|\tilde{y}(n)\|^2}\,. \tag{11.90}$$

The norm of the input vector can be computed efficiently by using the recursion

$$\|\tilde{y}(n)\|^2 = \|\tilde{y}(n-1)\|^2 + \sum_{l=0}^{L-1} \tilde{y}_l^2(n) - \sum_{l=0}^{L-1} \tilde{y}_l^2(n - N_{\mathrm{F}})\,. \tag{11.91}$$

Substituting the estimation of the correlation matrix and inserting the stepsize normalization leads to the stochastic constrained NLMS algorithm:

$$u(n) = \tilde{y}^{\mathrm{T}}(n)\; g_{\mathrm{F}}(n) \tag{11.92}$$

$$g_{\mathrm{F}}(n+1) = P \left[\, g_{\mathrm{F}}(n) - \mu\, \frac{\tilde{y}(n)}{\|\tilde{y}(n)\|^2}\, u(n) \,\right] + \tilde{c}\,. \tag{11.93}$$

In order to fulfill the constraint already at the first iteration an initialization of the filter vector according to

$$g_{\mathrm{F}}(0) = \tilde{c} \tag{11.94}$$

should be applied. Finally, a remark on the complexity of the algorithm is given. Even if Eq. 11.93 looks computationally expensive (due to the multiplication with P) the entire algorithm is rather effective and requires only twice as much operations as does the standard NLMS algorithm. To be precise, for updating the $L\, N_{\mathrm{F}}$ filter coefficients only $2\, L\, N_{\mathrm{F}}$ additions, $2\, L\, N_{\mathrm{F}}$ multiplications, and one division are required. The low complexity results mainly from the structure of the matrix P. Details about efficient time-domain realizations can be found in Frost's article [71].

11.4.1 Generalized Sidelobe Canceler

An alternative implementation of the linearly constrained adaptive beamforming algorithm—as presented in the last section—that achieves the same performance but in a simpler manner is the *generalized sidelobe canceler* [89], whose structure is depicted in Fig. 11.13. Essentially, the generalized sidelobe canceler is a structure that transforms the constraint minimization problem into an unconstrained form.

The algorithm consists of two substructures, which are depicted in the upper and lower part, respectively, of Fig. 11.13.

- The upper part is a conventional delay-and-sum beamformer with fixed delay compensation filters $\boldsymbol{g}_{\mathrm{D},l}$.
- The lower part is called the *sidelobe cancellation path*. It consists of a *blocking matrix* $\boldsymbol{B}$ followed by a set of adaptive filters $\boldsymbol{g}_{\mathrm{B},l}(n)$.

The output signals of the adaptive filters are subtracted from a delayed version of the output of the fixed beamformer. Thus, the output signal $u(n)$ of the entire structure can be written as

$$u(n) = \tilde{u}(n) - \hat{n}(n)\,, \tag{11.95}$$

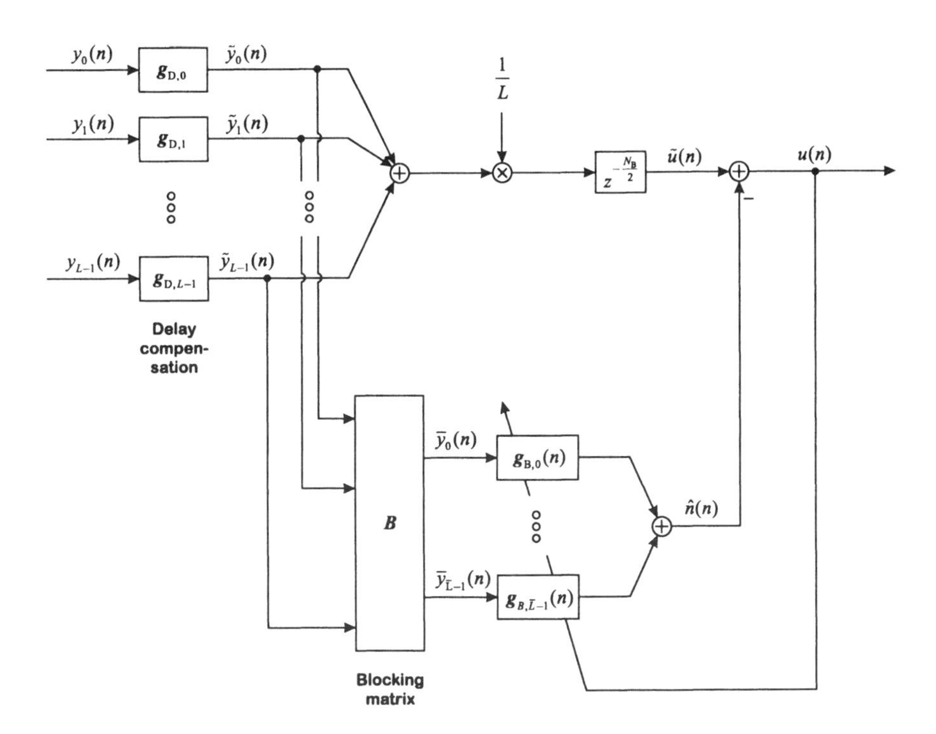

Fig. 11.13 Structure of the generalized sidelobe canceler.

where $\hat{n}(n)$ denotes the sum of the output signals of the adaptive filters

$$\hat{n}(n) = \sum_{l=0}^{\bar{L}-1} \bar{\boldsymbol{y}}_l^{\mathrm{T}}(n)\, \boldsymbol{g}_{\mathrm{B},l}(n)\,. \tag{11.96}$$

The aim of the matrix $\boldsymbol{B}$ is to block the desired signal from the lower path of the structure depicted in Fig. 11.13. The input signals of the adaptive filter $\bar{y}_l(n)$ are computed out of the delay compensated microphone signals as

$$\begin{bmatrix} \bar{y}_0(n) \\ \bar{y}_1(n) \\ \vdots \\ \bar{y}_{\bar{L}-1}(n) \end{bmatrix} = \boldsymbol{B} \begin{bmatrix} \tilde{y}_0(n) \\ \tilde{y}_1(n) \\ \vdots \\ \tilde{y}_{L-1}(n) \end{bmatrix}. \tag{11.97}$$

To ensure that no components of the desired signal enter the adaptive filtering part, all elements within a row of the matrix $\boldsymbol{B}$ should sum up to zero

$$\sum_{i=0}^{L-1} b_{l,i} = 0\,, \quad \text{for} \quad l \in \{0,\, \bar{L}-1\}\,. \tag{11.98}$$

Furthermore, all rows should be linearly independent. As a result, the row dimension of $\boldsymbol{B}$, which is introduced as $\bar{L}$, has to be $L-1$ or less. In many applications a very simple blocking matrix that has only two entries, 1 and -1, in each row is utilized. For the case of $L = 4$ microphones and $\bar{L} = 3$ adaptive filters the matrix is defined as

$$\boldsymbol{B} = \begin{bmatrix} 1 & -1 & 0 & 0 \\ 0 & 1 & -1 & 0 \\ 0 & 0 & 1 & -1 \end{bmatrix}. \tag{11.99}$$

Applying such a blocking matrix means that the differences between adjacent (and delay-aligned) sensor signals enter the adaptive filter stage. Unconstrained algorithms can be applied within the adaptive stage because the blocking matrix prevents the desired signal from entering this path. In case of using the NLMS algorithm, for instance, the update of the filter coefficients can be described by

$$\boldsymbol{g}_{\mathrm{B},l}(n+1) = \boldsymbol{g}_{\mathrm{B},l}(n) + \mu \frac{\bar{\boldsymbol{y}}_l(n)}{\|\bar{\boldsymbol{y}}(n)\|^2}\, u(n)\,. \tag{11.100}$$

Note that the normalization has to be performed to the squared norm of the entire filter vector

$$\bar{\boldsymbol{y}}(n) = \left[\bar{\boldsymbol{y}}_0^{\mathrm{T}}(n),\, \bar{\boldsymbol{y}}_1^{\mathrm{T}}(n),\, \ldots,\, \bar{\boldsymbol{y}}_{\bar{L}-1}^{\mathrm{T}}(n),\, \right]^{\mathrm{T}} \tag{11.101}$$

and not to the squared norm of each individual signal vector $\bar{\boldsymbol{y}}_l(n)$ [89]. In order to also model noncausal impulse responses with the filters $\boldsymbol{g}_{\mathrm{B},l}(n)$, the desired signal, which is the output of the delay-and-sum beamformer, is delayed by half of the adaptive filter length $N_{\mathrm{B}}/2$ (see Fig. 11.13). Due to its simplicity, the generalized sidelobe structure is the most frequently applied structure for adaptive beamforming. Further improvement and enhanced robustness can be achieved if the blocking matrix is designed adaptively [118].

Nevertheless, it should be noted that the performance of an adaptive beamformer relies crucially on the control of the adaptive filters. Therefore, control issues will be investigated in Chapter 15.

Part IV

Control and Implementation Issues

12

System Control—Basic Aspects

The basic algorithms and processing structures for acoustic echo and noise suppression systems were described in Chapters 4–11. Until now control aspects have not been treated. This will be the aim in this part. Chapters 13–15 describe various estimation and detection schemes as well as possibilities of how to combine these schemes to entire control units. Before we start with details about control structures, some general aspects will be considered.

12.1 CONVERGENCE VERSUS DIVERGENCE SPEED

In order to highlight the importance of control in acoustic echo and noise control systems, we will start with a simple example. We assume to have an adaptive filter for echo cancellation purposes with $N = 1000$ coefficients. If we further assume to have white noise as remote excitation as well as no local speech and no background noise

$$s(n) = 0, \; b(n) = 0, \tag{12.1}$$

the convergence properties of the NLMS algorithm are easily predictable (see Section 7.1.2). Furthermore, the NLMS algorithm shows (under these idealized circumstances) nearly the same convergence behavior as for all other algorithms presented for adaptive filters.

If we adapt the filter with a fixed stepsize $\mu = 0.1$ and without any regularization $\Delta = 0$, the expected system distance $\mathrm{E}\{\|\boldsymbol{h}_\Delta(n)\|^2\}$ can be predicted using the theory presented in Chapter 7:

$$\mathrm{E}\left\{ \|\boldsymbol{h}_\Delta(n+1)\|^2 \right\} \approx \left(1 - \frac{\mu\,(2-\mu)}{N}\right) \mathrm{E}\left\{ \|\boldsymbol{h}_\Delta(n)\|^2 \right\} + \frac{\mu^2}{N}\,\frac{\sigma_n^2}{\sigma_x^2}\,. \tag{12.2}$$

Using Eq. 12.2, it can be computed straightforward that 12,000 iterations are necessary in order to decrease the system distance from -20 to -30 dB. If—after the convergence process—a double-talk situation appears with $\sigma_n^2 = \sigma_x^2$ and the filter is still adapted with stepsize $\mu = 0.1$, it takes only 1000 iterations until the filter diverges from -30 to -20 dB (see Fig. 12.1). This simple example shows that the divergence rate is much faster (in our example more than 10 times) than convergence rate.

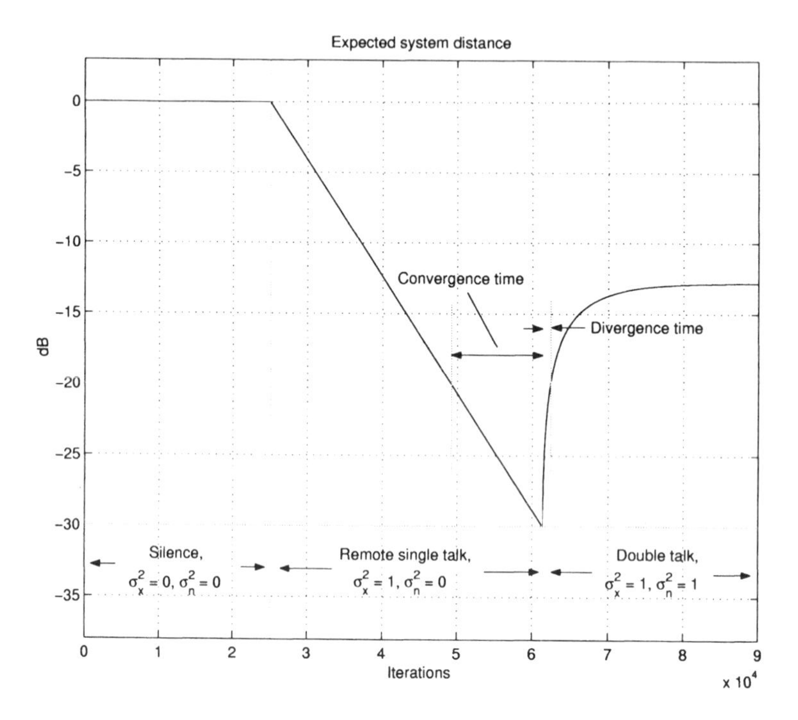

Fig. 12.1 Convergence and divergence of an adaptive filter. The filter with $N = 1000$ coefficients was adapted using the NLMS algorithm with a fixed stepsize $\mu = 0.1$ and without regularization $\Delta = 0$. Mutually independent white noise was used for remote excitation as well as local distortion. The figure shows the expected system distance during a silence period (the adaptation was stopped), a remote single-talk situation, and a double-talk interval.

For this reason, situations with small signal-to-noise ratio have to be detected quickly and reliably and the control parameters have to be adjusted according to the state of the system. The quality of acoustic echo and noise control systems relies crucially on the performance of their control mechanisms. When using fast algorithms (e.g., RLS or affine projection), the need for reliable control is even increased.

For acoustic echo cancellation and system identification problems in general, it is possible to derive pseudooptimal expressions for the control parameters. The term "pseudo" is used because a variety of approximations and assumptions will be made during the derivation of the parameters (see Chapter 13). For residual echo and noise suppression it is not as easy to derive optimal control parameters like overestimation factors or spectral floors. In these algorithmic parts more complicated cost functions than "simple" mean square error minimization[1] are usually applied. In most cases psychoacoustic criteria or even listening tests only are used in order to find parameter sets. A variety of proposals have been published in recent years, and we will try to mention the basic ideas of a few of them in Chapter 14.

12.2 SYSTEM LEVELS FOR CONTROL DESIGN

Before we start describing the details of control structures for echo and noise control systems, some basic aspects of control schemes should be mentioned. It is possible to design control structures on several system levels:

- The simplest method is to design units that control only specific parts of the system. This means that the echo cancellation, for instance, has its own control unit and this unit does not *communicate* with control units of other algorithmic parts. The advantages of this design method are simple verification and validation possibilities. These benefits have to be paid by larger computational load (some detectors, e.g., for double-talk detection, have to be computed more than once) as well as smaller reliability compared to the methods presented next.

- If a limited amount of communication or even interaction between the subunits is allowed, the performance (in terms of false alert rates, etc.) can be increased and the computational complexity can be decreased. An example for the latter is the estimation of the local background noise level. If the short-term power spectral density $\widehat{S}_{bb}(\Omega, n)$ of the background noise is already estimated for noise reduction, it can be reused within the echo cancellation, the loss control, the residual echo suppressing, and/or the generation of comfort noise. An example of increased reliability is the interaction of beamforming and echo cancellation. The beamformer is able to estimate the (spatial) direction of sound sources. This means that it can be distinguished between signals coming from the direction of a local speaker and signals coming from the direction of a loudspeaker (assuming that the locations of both are known a priori). When controlling the echo cancellation, this spatial information can be used for enhanced distinguishing between increased error power caused by double talk or changes within the local room.

[1]This would lead to the Wiener solution. Due to errors within the power spectral density estimation of the background noise, the Wiener solution is not applicable without modification in practice.

- Another possibility for control design is to have only one global control for the entire system. In this case all detectors and estimators are linked mutually and also thresholds, and so on should be allowed to be changed by higher control levels of the system. If, for example, a question is asked by a speech dialog system the probability of double talk rises at the end of the question. For this reason, the parameters of a double-talk detector should be modified in such a manner that the detector is more "watchful" at the end of the question . The combination of the outputs of all detection and estimation schemes in order to compute the required control parameters can be done either heuristically [154] or by means of neural networks [24, 153, 155] or fuzzy systems [26, 27]. Even if this control approach offers the highest potential for both reliability and computational complexity,[2] testing and verification of the system as well as predicting the system behavior under changed boundary conditions is much more complicated as in the two previous schemes.

In general, it is not possible to decide which of the presented control strategies would be the best choice. The decision depends crucially on the boundary conditions of each specific application as well as on the available processing power. Also, hardware aspects should be mentioned when designing the control part of acoustic echo and noise control systems. If, for example, the input level is not sufficiently high, roundoff errors caused by fixed-point architecture start becoming important and the adaptation process should be stopped. Also overloading of the AD converters, the microphones, the loudspeakers, and the amplifiers should be detected. In these cases the assumption that the loudspeaker–enclosure–microphone system can be modeled by a linear system is not true any more, and the system control should freeze all adaptive parts of the system.

[2]The computational effort required for training of the neural networks is not considered here.

13

Control of Echo Cancellation Systems

In Chapter 7 adaptive algorithms for system identification purposes have been introduced. In Chapter 9 these algorithms have been applied to acoustic echo cancellation. In both chapters two competitive requirements were made on the stepsize and the regularization parameter:

- In order to achieve a fast initial convergence or a fast readaptation after system changes, a large stepsize ($\mu = 1$) and a small regularization parameter ($\Delta = 0$) should be used.
- For achieving a small final misadjustment ($\|\boldsymbol{h}_\Delta(n)\|^2 \to 0$), a small stepsize $\mu \to 0$ and/or a large regularization parameter $\Delta \to \infty$ is necessary.

Both requirements cannot be fulfilled with fixed (time-invariant) control parameters. Therefore, a time-variant stepsize and a time-variant regularization parameter

$$\begin{aligned} \mu \quad &\to \quad \mu(n) \\ \Delta \quad &\to \quad \Delta(n) \end{aligned}$$

are introduced in this section. In the previous chapters the normalized version of the LMS algorithm has been utilized as a basis for all other algorithms. We will keep this line of proceeding and derive in the next section pseudooptimal control parameters for the NLMS algorithm. The term *pseudo* is utilized because of several approximations as well as simplifications during the derivation. Afterward the derivation is extended for the affine projection algorithm.

Unfortunately, the computation of the optimal control parameters requires knowledge of system and signal parameters that cannot be measured directly. To overcome this problem, several estimation and detection schemes will be presented in Section 13.4. These schemes differ in their computational complexity and their reliability. At the end of this chapter possible combinations and examples for combined control methods are given.

13.1 PSEUDOOPTIMAL CONTROL PARAMETERS FOR THE NLMS ALGORITHM

In most system designs that utilize the NLMS algorithm only stepsize or only regularization control is applied. For this reason, in the first two subsections pseudooptimal choices for both control parameters are derived. Because of the scalar regularization, both control strategies can easily be exchanged, which is shown in Section 13.1.3. Nevertheless, their practical implementations often differ. Even if in most cases only stepsize control is implemented, in some situations control by regularization or a mixture of both may be a good choice as well. Therefore, in the last part of this chapter some hints concerning the choice of the control structure are given.

13.1.1 Pseudooptimal Stepsize

In the following, a pseudooptimal stepsize for the nonregularized NLMS algorithm is derived. As suggested in Section 7.1, the optimization criterion will be the minimization of the expected system distance. The adaptation rule of this form of the NLMS algorithm can be expressed as follows:

$$\hat{\boldsymbol{h}}(n+1) \;=\; \hat{\boldsymbol{h}}(n) + \mu(n)\,\frac{e(n)\,\boldsymbol{x}(n)}{\|\boldsymbol{x}(n)\|^2}\,. \tag{13.1}$$

For reasons of simplicity in Eq. 13.1 as well as in the following paragraphs, real input signals and real systems are assumed. Supposing that the system is time–invariant ($\boldsymbol{h}(n+1) = \boldsymbol{h}(n)$), the recursion of the expected system distance can be denoted as follows (see Eq. 7.24, for $\Delta = 0$):

$$\begin{aligned}\mathrm{E}\left\{\|\boldsymbol{h}_\Delta(n+1)\|^2\right\} \;=\;& \mathrm{E}\left\{\|\boldsymbol{h}_\Delta(n)\|^2\right\} - 2\,\mu(n)\,\mathrm{E}\left\{\frac{e(n)\,e_\mathrm{u}(n)}{\|\boldsymbol{x}(n)\|^2}\right\} + \\ &+ \mu^2(n)\,\mathrm{E}\left\{\frac{e^2(n)}{\|\boldsymbol{x}(n)\|^2}\right\}.\end{aligned} \tag{13.2}$$

In Eq. 13.2 the undistorted error $e_\mathrm{u}(n)$ has been utilized, which is defined as

$$e_\mathrm{u}(n) = \boldsymbol{x}^\mathrm{T}(n)\,\boldsymbol{h}_\Delta(n)\,. \tag{13.3}$$

For further explanations of this signal, see Section 7.1.2.1. For determining an optimal stepsize [246], the cost function, which is the squared norm of the system mismatch vector, should decrease (to be precise: should not increase) on average for every iteration step:

$$\mathrm{E}\left\{\|\boldsymbol{h}_\Delta(n+1)\|^2\right\} - \mathrm{E}\left\{\|\boldsymbol{h}_\Delta(n)\|^2\right\} \leq 0\,. \tag{13.4}$$

Inserting the recursive expression for the expected system distance (Eq. 13.2) in relation 13.4 leads to

$$\mu^2(n)\,\mathrm{E}\left\{\frac{e^2(n)}{\|\boldsymbol{x}(n)\|^2}\right\} - 2\,\mu(n)\,\mathrm{E}\left\{\frac{e(n)\,e_\mathrm{u}(n)}{\|\boldsymbol{x}(n)\|^2}\right\} \leq 0\,. \tag{13.5}$$

Thus, the stepsize $\mu(n)$ has to fulfill the condition

$$0 \leq \mu(n) \leq 2\,\frac{\mathrm{E}\left\{\dfrac{e(n)\,e_\mathrm{u}(n)}{\|\boldsymbol{x}(n)\|^2}\right\}}{\mathrm{E}\left\{\dfrac{e^2(n)}{\|\boldsymbol{x}(n)\|^2}\right\}}\,. \tag{13.6}$$

The largest decrease of the system distance is achieved in the middle of the defined interval. To prove this statement, we differentiate the expected system distance at time index $n+1$. Setting this derivation to zero leads—due to the quadratic surface of the cost function—to the optimal stepsize:

$$\left.\frac{\partial\,\mathrm{E}\left\{\|\boldsymbol{h}_\Delta(n+1)\|^2\right\}}{\partial\,\mu(n)}\right|_{\mu(n)=\mu_\mathrm{opt}(n)} = 0\,. \tag{13.7}$$

We assume that the stepsizes at different time instants are uncorrelated. Therefore, the expected system distance at time index n is not dependent on the stepsize $\mu(n)$ (only on $\mu(n-1)$, $\mu(n-2)$, ...). Using this assumption and the recursive computation of Eq. 13.2, we can solve Eq. 13.7:

$$\begin{aligned}
-2\,\mathrm{E}\left\{\frac{e(n)\,e_\mathrm{u}(n)}{\|\boldsymbol{x}(n)\|^2}\right\} + 2\,\mu_\mathrm{opt}(n)\,\mathrm{E}\left\{\frac{e^2(n)}{\|\boldsymbol{x}(n)\|^2}\right\} &= 0 \\
\mu_\mathrm{opt}(n)\,\mathrm{E}\left\{\frac{e^2(n)}{\|\boldsymbol{x}(n)\|^2}\right\} &= \mathrm{E}\left\{\frac{e(n)\,e_\mathrm{u}(n)}{\|\boldsymbol{x}(n)\|^2}\right\} \\
\mu_\mathrm{opt}(n) &= \frac{\mathrm{E}\left\{\dfrac{e(n)\,e_\mathrm{u}(n)}{\|\boldsymbol{x}(n)\|^2}\right\}}{\mathrm{E}\left\{\dfrac{e^2(n)}{\|\boldsymbol{x}(n)\|^2}\right\}}\,.
\end{aligned} \tag{13.8}$$

If we assume that the squared norm of the excitation vector $\|\boldsymbol{x}(n)\|^2$ can be approximated by a constant, as well as that the local signals $n(n) = b(n) + s(n)$[1] and the far-end signal $x(n)$ (and therefore also $n(n)$ and $e_\mathrm{u}(n)$) are uncorrelated, the optimal stepsize can be simplified as follows:

$$\mu_\mathrm{opt}(n) \approx \frac{\mathrm{E}\left\{e_\mathrm{u}^2(n)\right\}}{\mathrm{E}\left\{e^2(n)\right\}}. \tag{13.9}$$

In case of complex signals and systems (e.g., for control within subband structures), an equivalent derivation leads to

$$\mu_\mathrm{opt}(n) \approx \frac{\mathrm{E}\left\{|e_\mathrm{u}(n)|^2\right\}}{\mathrm{E}\left\{|e(n)|^2\right\}}. \tag{13.10}$$

In the absence of local speech and noise ($n(n) = 0$), the distorted error signal $e(n)$ equals the undistorted error signal $e_\mathrm{u}(n)$ and the optimal stepsize is one. If the adaptive filter is accurately adjusted to the impulse response of the unknown system, the power of the residual error signal $e_\mathrm{u}(n)$ is very small. In the presence of local background noise or local speech, the power of the distortion $n(n)$ and therefore also the power of the distorted error signal $e(n)$ increases. In this case, the numerator of Eq. 13.8 is much smaller than the denominator, resulting in a stepsize close to zero—so the filter is not or is only marginally changed in such situations. Both examples show that Eq. 13.9 is (at least for these two boundary cases) a useful approximation.

To show the advantages of a time-variant stepsize control, the simulation example of Section 7.1.2.3 (see Fig. 7.11) is repeated. This time a fourth convergence curve, where the stepsize is estimated by

$$\hat{\mu}_\mathrm{opt}(n) = \frac{\overline{e_\mathrm{u}^2(n)}}{\overline{e^2(n)}}, \tag{13.11}$$

is added. The terms $\overline{e_\mathrm{u}^2(n)}$ and $\overline{e^2(n)}$ denote short-term smoothing (first-order IIR filters) of the squared input signals:

$$\overline{e^2(n)} = \gamma\, \overline{e^2(n-1)} + (1-\gamma)\, e^2(n), \tag{13.12}$$

$$\overline{e_\mathrm{u}^2(n)} = \gamma\, \overline{e_\mathrm{u}^2(n-1)} + (1-\gamma)\, \big(e(n) - n(n)\big)^2. \tag{13.13}$$

The time constant was set to $\gamma = 0.995$. The resulting stepsize $\hat{\mu}_\mathrm{opt}(n)$ is depicted in the lower diagram of Fig. 13.1. At the beginning of the simulation the short-term power of the undistorted error is much larger than that of the local noise. Therefore, the stepsize $\hat{\mu}_\mathrm{opt}(n)$ is close to one and a very fast initial convergence (comparable to the case $\Delta = 0$ and $\mu = 1$) can be achieved (see upper diagram of Fig. 13.1). The better the filter converges, the smaller is the undistorted error. With decreasing

[1] For details on the definition of the signals $n(n)$, $b(n)$, and similar terms, see Fig. 9.1.

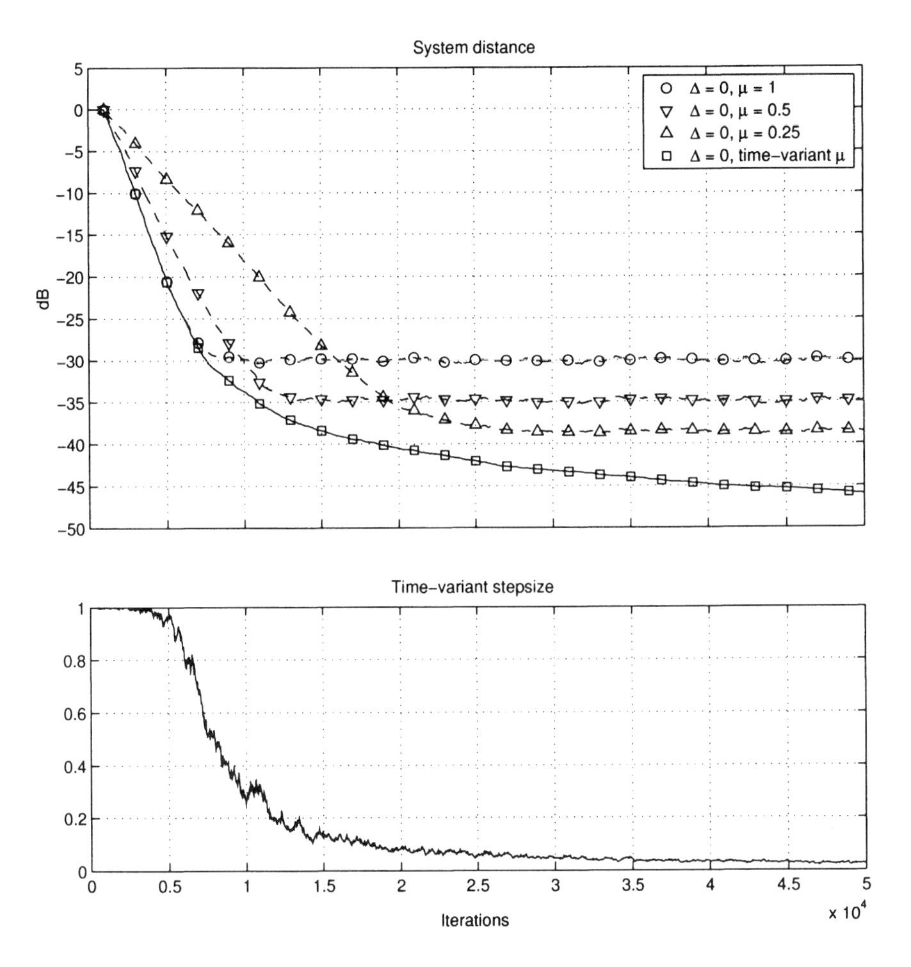

Fig. 13.1 Convergence using a time-variant stepsize. To show the advantages of a time-variant stepsize control, the simulation example of Section 7.1.2.3 is repeated. This time a fourth convergence curve, where the stepsize is estimated as proposed in Eq. 13.11, is added. The resulting stepsize is depicted in the lower diagram. If we compare the three curves with fixed stepsizes (dotted lines) and the convergence curve with the time-variant stepsize (solid line), the advantage of a time-variant stepsize control is clearly visible (with permission from [102]).

error power, the influence of the local noise in Eq. 13.13 increases. This leads to a decrease of the stepsize parameter $\mu(n)$.

Due to the reduction of the stepsize, the system distance can be further reduced. If we compare the convergence curves of fixed stepsizes and the curve with a time-variant stepsize (see upper diagram of Fig. 13.1), the advantage of a time-variant control is clearly visible. A stepsize control based on Eq. 13.9 is able to achieve a fast initial convergence (and also a fast readaptation after system changes) as well as a good steady-state performance.

Unfortunately, neither the undistorted error signal $e_{\mathrm{u}}(n)$ nor its power is accessible in real system identification problems. In Section 13.4, methods for estimating the power of this signal and therefore also the optimal stepsize are presented for the application of acoustic echo cancellation.

13.1.2 Pseudooptimal Regularization

After deriving an approximation of the optimal stepsize in the last section, we will do the same in this section for the only regularized adaptation process. The only regularized NLMS algorithm has the following update equation:

$$\hat{\boldsymbol{h}}(n+1) = \hat{\boldsymbol{h}}(n) + \frac{e(n)\,\boldsymbol{x}(n)}{\|\boldsymbol{x}(n)\|^2 + \Delta(n)}. \tag{13.14}$$

The regularization parameter $\Delta(n) \geq 0$ increases the denominator of the update term and can therefore also be used for control purposes. With this update equation, the system mismatch vector can be computed recursively:

$$\boldsymbol{h}_\Delta(n+1) = \boldsymbol{h}_\Delta(n) - \frac{e(n)\,\boldsymbol{x}(n)}{\|\boldsymbol{x}(n)\|^2 + \Delta(n)}. \tag{13.15}$$

It was assumed that the system is time-invariant ($\boldsymbol{h}(n+1) = \boldsymbol{h}(n)$). In order to determine an optimal regularization, the same cost function as in the optimal stepsize derivation—the average of the squared norm of the system mismatch vector (expected system distance)—can be used. The expected system distance at time index $n+1$ can be recursively expressed as follows:

$$\begin{aligned}\mathrm{E}\left\{\|\boldsymbol{h}_\Delta(n+1)\|^2\right\} &= \mathrm{E}\left\{\|\boldsymbol{h}_\Delta(n)\|^2\right\} - 2\,\mathrm{E}\left\{\frac{e(n)\,e_u(n)}{\|\boldsymbol{x}(n)\|^2 + \Delta(n)}\right\} \\ &\quad + \mathrm{E}\left\{\frac{e^2(n)\,\|\boldsymbol{x}(n)\|^2}{\left(\|\boldsymbol{x}(n)\|^2 + \Delta(n)\right)^2}\right\}.\end{aligned} \tag{13.16}$$

As in the derivation in Section 13.1.1, the definition of the undistorted error signal $e_{\mathrm{u}}(n) = \boldsymbol{h}_\Delta^T(n)\,\boldsymbol{x}(n)$ was inserted. Differentiation of the cost function leads to

$$\begin{aligned}\frac{\partial\,\mathrm{E}\left\{\|\boldsymbol{h}_\Delta(n+1)\|^2\right\}}{\partial\,\Delta(n)} &= 2\,\mathrm{E}\left\{\frac{e_u(n)\,e(n)}{\left(\|\boldsymbol{x}(n)\|^2 + \Delta(n)\right)^2}\right\} \\ &\quad - 2\,\mathrm{E}\left\{\frac{\|\boldsymbol{x}(n)\|^2\,e^2(n)}{\left(\|\boldsymbol{x}(n)\|^2 + \Delta(n)\right)^3}\right\}.\end{aligned} \tag{13.17}$$

As done in the last section, we approximate the squared norm of the excitation vector $\|\boldsymbol{x}(n)\|^2$ by a constant, and we assume the signals $n(n)$ and $x(n)$—and therefore also $n(n)$ and $e_u(n)$)—to be orthogonal. Therefore, the derivative simplifies as follows:

$$\frac{\partial \mathrm{E}\left\{\|\boldsymbol{h}_\Delta(n+1)\|^2\right\}}{\partial \Delta(n)} \approx 2 \frac{\mathrm{E}\left\{e_u^2(n)\right\}}{\left(\|\boldsymbol{x}(n)\|^2 + \Delta(n)\right)^2} - 2 \frac{\|\boldsymbol{x}(n)\|^2}{\left(\|\boldsymbol{x}(n)\|^2 + \Delta(n)\right)^3} \mathrm{E}\left\{e^2(n)\right\}. \quad (13.18)$$

By setting the derivative to zero

$$\left.\frac{\partial \mathrm{E}\left\{\|\boldsymbol{h}_\Delta(n+1)\|^2\right\}}{\partial \Delta(n)}\right|_{\Delta(n)=\Delta_{\text{opt}}(n)} = 0 \quad (13.19)$$

and using approximation 13.18, we can find an approximation for the optimal regularization parameter $\Delta_{\text{opt}}(n)$:

$$\Delta_{\text{opt}}(n) \approx \frac{\left(\mathrm{E}\left\{e^2(n)\right\} - \mathrm{E}\left\{e_u^2(n)\right\}\right) \|\boldsymbol{x}(n)\|^2}{\mathrm{E}\left\{e_u^2(n)\right\}}. \quad (13.20)$$

Because of the orthogonality of the distortion $n(n)$ and the excitation signal $x(n)$, the difference between the two expectation operators can be simplified:

$$\Delta_{\text{opt}}(n) \approx \frac{\mathrm{E}\left\{n^2(n)\right\} \|\boldsymbol{x}(n)\|^2}{\mathrm{E}\left\{e_u^2(n)\right\}}. \quad (13.21)$$

If the excitation signal is white noise and the excitation vector and the system mismatch vector are assumed to be uncorrelated, the power of the undistorted error signal can be simplified, too:

$$\begin{aligned}
\mathrm{E}\left\{e_u^2(n)\right\} &= \mathrm{E}\left\{\boldsymbol{h}_\Delta(n)^T \boldsymbol{x}(n)\, \boldsymbol{x}^T(n)\, \boldsymbol{h}_\Delta(n)\right\} \\
&= \mathrm{E}\left\{\boldsymbol{h}_\Delta(n)^T \mathrm{E}\left\{x^2(n)\right\} \boldsymbol{h}_\Delta(n)\right\} \\
&= \mathrm{E}\left\{x^2(n)\right\} \mathrm{E}\left\{\|\boldsymbol{h}_\Delta(n)\|^2\right\}. \quad (13.22)
\end{aligned}$$

Inserting this result in approximation 13.21 as well as assuming that (a large filter length $N \gg 1$ is taken for granted)

$$N\,\mathrm{E}\left\{x^2(n)\right\} \approx \|\boldsymbol{x}(n)\|^2 \quad (13.23)$$

leads to a simplified approximation of the optimal regularization parameter

$$\Delta_{\text{opt}}(n) \approx N \frac{\mathrm{E}\left\{n^2(n)\right\}}{\mathrm{E}\left\{\|\boldsymbol{h}_\Delta(n)\|^2\right\}}. \quad (13.24)$$

As during the derivation of the pseudooptimal stepsize, real signals and systems were assumed. Dropping this assumption leads to

$$\Delta_{\text{opt}}(n) \approx N \frac{\text{E}\left\{|n(n)|^2\right\}}{\text{E}\left\{\|\boldsymbol{h}_\Delta(n)\|^2\right\}}. \tag{13.25}$$

As in the last section, the control based on approximation 13.24 will be compared with fixed regularization control approaches via a simulation. The power of the local signals was estimated using a first-order IIR smoothing filter:

$$\overline{n^2(n)} = \gamma \, \overline{n^2(n-1)} + (1-\gamma)\, n^2(n). \tag{13.26}$$

The time constant γ was set to $\gamma = 0.995$. Using this power estimation, the optimal regularization parameter was computed as follows:

$$\hat{\Delta}_{\text{opt}}(n) = N \frac{\overline{n^2(n)}}{\|\boldsymbol{h}_\Delta(n)\|^2}. \tag{13.27}$$

Furthermore, the expected system distance was approximated by its instantaneous value. In Fig. 13.2 the three examples of Fig. 7.12 with fixed regularization parameters as well as a convergence curve using Eq. 13.27 (time-variant regularization) are presented. All boundary conditions such as filter order, input signals, and initialization values were maintained.

In the lower part of Fig. 13.2 the regularization parameter resulting from Eq. 13.27 is depicted. With decreasing system distance, the regularization increases in order to improve the convergence state even if the power of the undistorted error signal is smaller than the one of the measurement noise.

13.1.3 "Relationship" between Both Control Methods

If only stepsize control or only regularization is applied (see Eqs. 13.1 and 13.14), the control parameters can easily be exchanged. Comparison of both update terms

$$\frac{\mu(n)\, e(n)}{\|\boldsymbol{x}(n)\|^2} = \frac{e(n)}{\|\boldsymbol{x}(n)\|^2 + \Delta(n)} \tag{13.28}$$

can be used to find the following converting rules:

$$\mu(n) = \frac{\|\boldsymbol{x}(n)\|^2}{\|\boldsymbol{x}(n)\|^2 + \Delta(n)}, \tag{13.29}$$

$$\Delta(n) = \|\boldsymbol{x}(n)\|^2 \, \frac{1-\mu(n)}{\mu(n)}. \tag{13.30}$$

Even if both control approaches are theoretically equivalent, their practical implementations often differ. In most real systems only stepsize control is applied. The

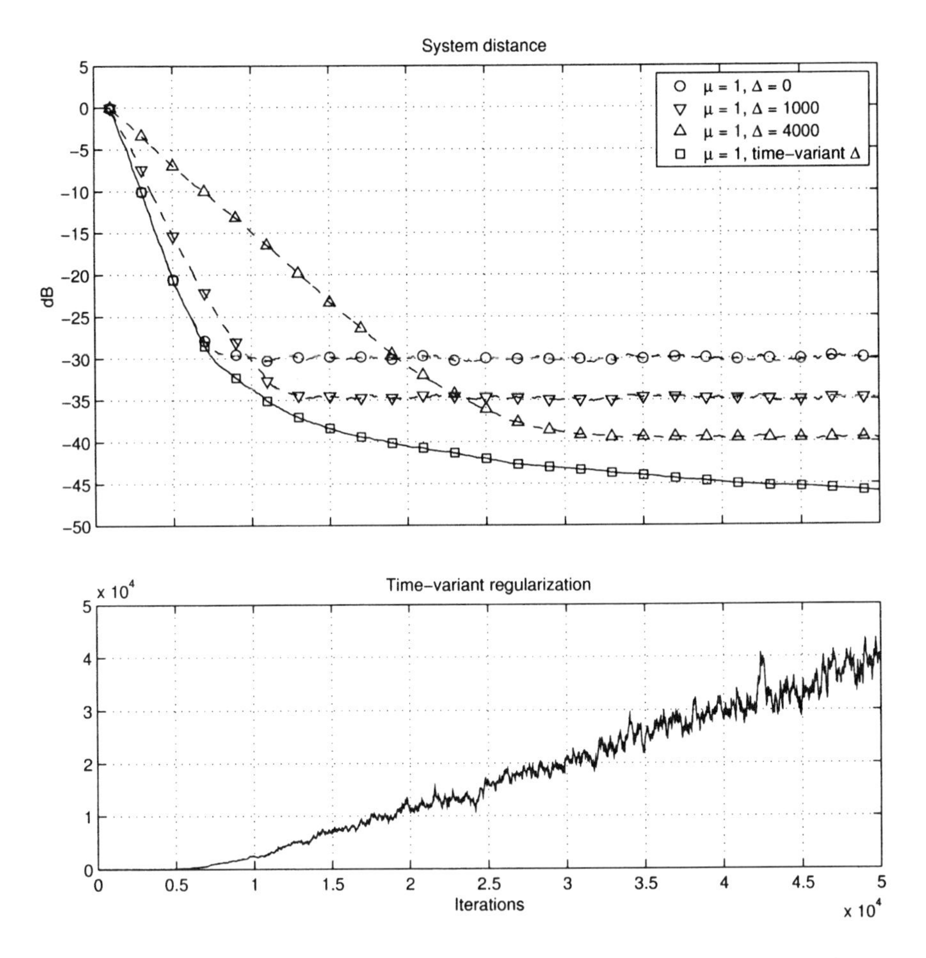

Fig. 13.2 Convergence using a time-variant regularization parameter. The simulation example of Section 7.1.2.4 is repeated. The interrupted lines show the progression of the system distance for different fixed regularization parameters. Additionally, a fourth convergence curve where the regularization parameter is adjusted according to Eq. 13.27 is presented (solid line). The progression of the control value $\hat{\Delta}_{\text{opt}}(n)$ is depicted in the lower diagram. The advantages of time-variant adaptation control are clearly visible (with permission from [102]).

most important reason for this is the limited range of values that is required for the stepsize ($\mu \in [0, 1]$). In contrast to this, the normalized range of the regularization parameter has no upper bound ($\Delta \in [0, \infty)$). Especially for implementations on signal processors with fixed-point arithmetic, this is a major obstacle.

Regularization control is often used for algorithms with matrix inversions like recursive least squares or affine projections. Originally regularization was applied in these algorithms for numerically stabilizing the solution. As a second step, this stabilization can be utilized for control purposes. But also for the scalar inversion case as in NLMS control, regularization might be superior to stepsize control in some situations.

For the pseudooptimal stepsize, short-term powers of two signals have to be estimated: $\sigma_e^2(n)$ and $\sigma_{e_u}^2(n)$. If first-order IIR smoothing of the squared values as presented in Section 13.1.1 or a rectangular window for the squared values is utilized for estimating the short-term power, the estimation process has its inherent "inertia." After a sudden increase or decrease of the signal power, the estimation methods follow only with a certain delay.

On the other hand, control by regularization as proposed in Eq. 13.27 requires only the power of the local signals. In applications with only time-invariant background noise but with time-variant excitation power, control by regularization should be preferred. In Fig. 13.3 a simulation with stationary measurement noise is presented. The excitation signal changes its power every 1000 iterations in order to obtain signal-to-noise ratios of 20 and -20 dB, respectively. The excitation signal and the local noise are depicted in the upper two diagrams of Fig. 13.3.

The impulse response of the adaptive filter as well as the one of the unknown system are causal and finite. The length of both was $N = 1000$. After 1000 iterations, the power of the excitation signal is enlarged by 40 dB. Both power estimations of the stepsize control need to follow this increase. The power of the distorted error starts from a larger initial value than the power estimation of the undistorted error. At the first few iterations, the resulting stepsize does not reach values close to one (optimal value). As a consequence, the convergence speed is also reduced. Because of the stationary behavior of the measurement noise, the regularization control—especially the power estimation for the measurement noise—does not have "inertia" problems.

After 2000 iterations, the power of the excitation signal is decreased by 40 dB. Both power estimations utilized in the stepsize computation decrease their value at the following first few iterations by nearly the same amount. This leads to a stepsize that is a little bit too large—resulting again in a reduced convergence speed.

In the lowest diagram of Fig. 13.3 the system distances resulting from stepsize as well as regularization control are depicted. The superior behavior of regularization control in this example is clearly visible. The advantages will disappear if the local noise also shows a nonstationary behavior.

13.2 PSEUDOOPTIMAL CONTROL PARAMETERS FOR THE AFFINE PROJECTION ALGORITHM

If fast algorithms such as affine projection (AP) or recursive least squares (RLS) are realized on hardware with limited accuracy (processors with fixed-point arithmetic), numerical stability problems arise. These problems are caused mainly by the inversion of ill-conditioned (autocorrelation) matrices.

Stabilizing the inversion process in terms of adding positive values to the main diagonal and controlling the adaptation process can be combined [169]. To avoid infinitely large regularization parameters a hybrid control strategy is often applied.

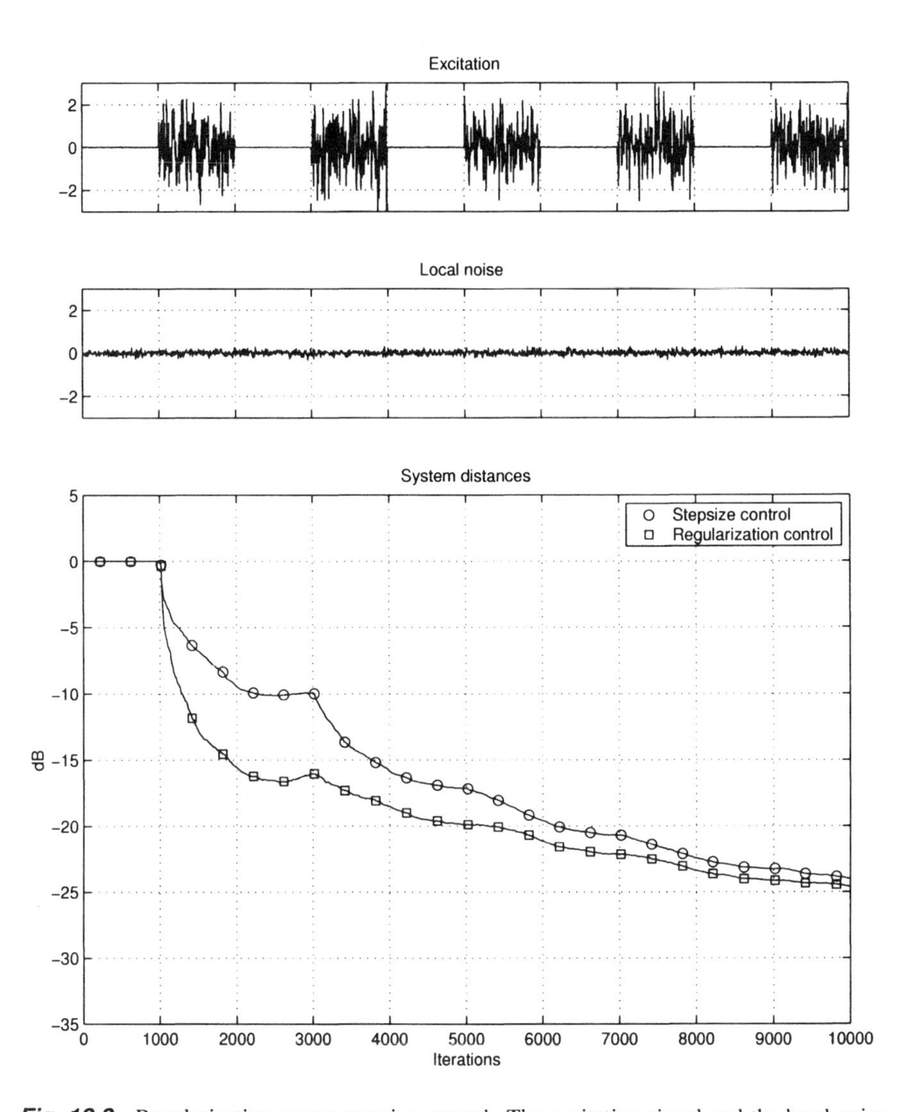

Fig. 13.3 Regularization versus stepsize control. The excitation signal and the local noise used in this simulations are depicted in the upper two diagrams. The excitation signal varies its power every 1000 iterations in order to have signal-to-noise ratios of 20 and -20 dB, respectively. In the lowest diagram the system distances resulting from stepsize as well as regularization control according to Eqs. 13.11 and 13.27 are depicted. The superior behavior of regularization control in this example is clearly visible. The advantages will disappear if the local signals also show a nonstationary behavior (with permission from [102]).

In a first stage, remote speech activity is detected. Only in periods with remote activity the adaptation process is "opened"—in all other periods the adaptation is stopped ($\mu(n) = 0$). During activity periods a stepsize $\mu(n) = 1$ is chosen and the actual control is performed by regularization. In the following we will derive a pseudooptimal regularization parameter for the affine projection algorithm and give an outlook (without derivation) on pseudooptimal stepsizes.

13.2.1 Pseudooptimal Regularization

For deriving a pseudooptimal regularization parameter, the expected norm of the system mismatch vector—as done in the previous sections—will be minimized:

$$\mathrm{E}\left\{ \|\boldsymbol{h}_\Delta(n)\|^2 \right\} \rightarrow \min . \tag{13.31}$$

Again the term "pseudo" is utilized because of several approximations that will be employed during the derivation. If the impulse response of the LEM system does not change, the system mismatch vector can be computed recursively [see Eq. 7.67 for real input data and for a stepsize $\mu(n) = 1$]:

$$\boldsymbol{h}_\Delta(n+1) = \boldsymbol{h}_\Delta(n) - \boldsymbol{X}(n)\big(\boldsymbol{X}^{\mathrm{T}}(n)\,\boldsymbol{X}(n) + \Delta(n)\boldsymbol{I}\big)^{-1}\boldsymbol{e}(n) . \tag{13.32}$$

The matrix $\boldsymbol{X}(n)$, containing the excitation signal, and the error vector[2] $\boldsymbol{e}(n)$ are defined as stated in Section 7.2. For white excitation and large filter length $N \gg 1$, the inverse matrix in Eq. 13.32 can be approximated by:

$$\big(\boldsymbol{X}^{\mathrm{T}}(n)\,\boldsymbol{X}(n) + \Delta(n)\boldsymbol{I}\big)^{-1} \approx \frac{1}{N\,\sigma_x^2(n) + \Delta(n)}\,\boldsymbol{I}. \tag{13.33}$$

Inserting this result into Eq. 13.32 leads to

$$\boldsymbol{h}_\Delta(n+1) \approx \boldsymbol{h}_\Delta(n) - \frac{1}{N\,\sigma_x^2(n) + \Delta(n)}\,\boldsymbol{X}(n)\boldsymbol{e}(n) . \tag{13.34}$$

The assumption of having white excitation is rarely fulfilled in real environments. Nevertheless, if fixed or adaptive prewhitening of the input signals according to Section 9.1.1.1 is employed, Approximation 13.33 is applicable. For the norm of the system mismatch vector, we find the following:

$$\begin{aligned} \|\boldsymbol{h}_\Delta(n+1)\|^2 \approx\; & \|\boldsymbol{h}_\Delta(n)\|^2 \\ & - \frac{2\,\boldsymbol{h}_\Delta^{\mathrm{T}}(n)\,\boldsymbol{X}(n)\,\boldsymbol{e}(n)}{N\,\sigma_x^2(n) + \Delta(n)} \\ & + \frac{\boldsymbol{e}^{\mathrm{T}}(n)\boldsymbol{X}^{\mathrm{T}}(n)\,\boldsymbol{X}(n)\,\boldsymbol{e}(n)}{\big(N\,\sigma_x^2(n) + \Delta(n)\big)^2} . \end{aligned} \tag{13.35}$$

[2]During the derivation of the affine projection algorithm it was distinguished between the a posteriori error vector $\boldsymbol{e}(n|n+1)$ and the a priori error vector $\boldsymbol{e}(n|n)$. For brevity, we will shorten $\boldsymbol{e}(n|n)$ to $\boldsymbol{e}(n)$ in this chapter.

Inserting the definition of the undistorted error vector

$$\boldsymbol{e}_{\mathrm{u}}(n) = \boldsymbol{X}^{\mathrm{T}}(n)\, \boldsymbol{h}_{\Delta}(n) \tag{13.36}$$

and using (as in the previous sections) the approximation

$$\boldsymbol{X}^{\mathrm{T}}(n)\, \boldsymbol{X}(n) \approx N\, \sigma_x^2(n)\, \boldsymbol{I} \tag{13.37}$$

leads to a simplified recursion of the system distance:

$$\begin{aligned} \|\boldsymbol{h}_{\Delta}(n+1)\|^2 \approx\ & \|\boldsymbol{h}_{\Delta}(n)\|^2 \\ & - \frac{2\, \boldsymbol{e}_{\mathrm{u}}^{\mathrm{T}}(n)\, \boldsymbol{e}(n)}{N\, \sigma_x^2(n) + \Delta(n)} \\ & + \frac{N\, \sigma_x^2(n)}{\left(N\, \sigma_x^2(n) + \Delta(n)\right)^2}\, \|\boldsymbol{e}(n)\|^2 . \end{aligned} \tag{13.38}$$

According to the minimization criterion, we will finally apply expectation operators to compute the system distance. Hereby, orthogonality of the remote and the local signals is assumed. Furthermore, the expected norm of the distorted error vector as well as the expected norm of the undistorted error vector will be approximated by

$$\mathrm{E}\left\{\|\boldsymbol{e}(n)\|^2\right\} \approx \sigma_{e_u}^2(n) + p\, \sigma_n^2(n), \tag{13.39}$$

$$\mathrm{E}\left\{\|\boldsymbol{e}_{\mathrm{u}}(n)\|^2\right\} \approx \sigma_{e_u}^2(n), \tag{13.40}$$

where p denotes the order of the affine projection algorithm. If the filter $\hat{\boldsymbol{h}}(n)$ is updated during each adaptation with the optimal control parameters, only the first element of the undistorted error vector $\boldsymbol{e}_{\mathrm{u}}(n)$ contains significant energy. Thus, the expected squared norm of the undistorted error vector is approximated only by $\sigma_{e_u}^2(n)$ and not by $p\, \sigma_{e_u}^2(n)$. The distorted error vector $\boldsymbol{e}(n)$ consists of undistorted error components and of local noise components $n(n) = b(n) + s(n)$.[3] The local signal is present in each of the p vector elements. For this reason, its expected squared norm is approximated by $\sigma_{e_u}^2(n) + p\, \sigma_n^2(n)$. After inserting approximations 13.39 and 13.40, we obtain the following equation for the system distance:

$$\begin{aligned} \mathrm{E}\left\{\|\boldsymbol{h}_{\Delta}(n+1)\|^2\right\} \approx\ & \mathrm{E}\left\{\|\boldsymbol{h}_{\Delta}(n)\|^2\right\} \\ & - \frac{2\, \sigma_{e_u}^2(n)}{N\, \sigma_x^2(n) + \Delta(n)} \\ & + \frac{N\, \sigma_x^2(n)\left(\sigma_{e_u}^2(n) + p\, \sigma_n^2(n)\right)}{\left(N\, \sigma_x^2(n) + \Delta(n)\right)^2} . \end{aligned} \tag{13.41}$$

In order to get an optimal regularization parameter, the expected norm of the system mismatch vector will be differentiated partially toward $\Delta(n)$. We assume that

[3] Here, $s(n)$ denotes local speech and $b(n)$ describes local background noise.

the regularization parameters at different time instants are uncorrelated. Therefore, the system distance at time index n is not dependent on $\Delta(n)$ [only on $\Delta(n-1)$, $\Delta(n-2)$, ...]. Using this assumption, we get

$$\frac{\partial \mathrm{E}\left\{\|\boldsymbol{h}_\Delta(n+1)\|^2\right\}}{\partial \Delta(n)} \approx \frac{2\,\sigma_{e_u}^2(n)}{\left(N\,\sigma_x^2(n)+\Delta(n)\right)^2} - \frac{2\,N\,\sigma_x^2(n)\left(\sigma_{e_u}^2(n)+p\,\sigma_n^2(n)\right)}{\left(N\,\sigma_x^2(n)+\Delta(n)\right)^3}. \tag{13.42}$$

By setting the derivation to zero

$$\left.\frac{\partial \mathrm{E}\left\{\|\boldsymbol{h}_\Delta(n+1)\|^2\right\}}{\partial \Delta(n)}\right|_{\Delta(n)=\Delta_{\text{opt}}(n)} = 0 \tag{13.43}$$

and assuming the denominator to be larger than zero, the optimal regularization parameter $\Delta_{\text{opt}}(n)$ can be separated:

$$\Delta_{\text{opt}}(n) \approx p\,\frac{N\,\sigma_x^2(n)\,\sigma_n^2(n)}{\sigma_{e_u}^2(n)}. \tag{13.44}$$

To show the advantages of time-variant regularization over fixed regularization, four convergences are depicted in Fig. 13.4. During the first three simulations, fixed regularization parameters $\Delta_1 = 0$, $\Delta_2 = 4000$, and $\Delta_3 = 16000$ have been utilized. For the fourth convergence a time-variant regularization according to Eq. 13.44 was applied. As in the previous sections the short-term powers $\sigma_x^2(n)$, $\sigma_n^2(n)$, and $\sigma_{e_u}^2(n)$ were estimated using first-order IIR smoothing filters. The advantages of time-variant control are clearly visible when analyzing the progression of the system distances (depicted in the upper diagram of Fig. 13.4).

13.2.2 Pseudooptimal Stepsize

When applying an equivalent derivation we obtain for the pseudooptimal stepsize of nonregularized affine projection algorithms

$$\mu_{\text{opt}}(n) = \frac{\mathrm{E}\left\{e_u^2(n)\right\}}{\mathrm{E}\left\{e_u^2(n)\right\} + p\;\mathrm{E}\left\{n^2(n)\right\}}. \tag{13.45}$$

Utilizing this stepsize (all expectations are replaced by first-order short-term smoothing filters) and performing simulation runs result in exactly the same performance as in the "only regularized" case (as depicted in Fig. 13.4). However, on hardware with limited numerical accuracy, control by regularization or a mixture of stepsize and regularization control shows a much more robust behavior than do only stepsize controlled affine projection adaptation processes. This behavior can be observed in all adaptive algorithms where (autocorrelation) matrix inversions are involved.

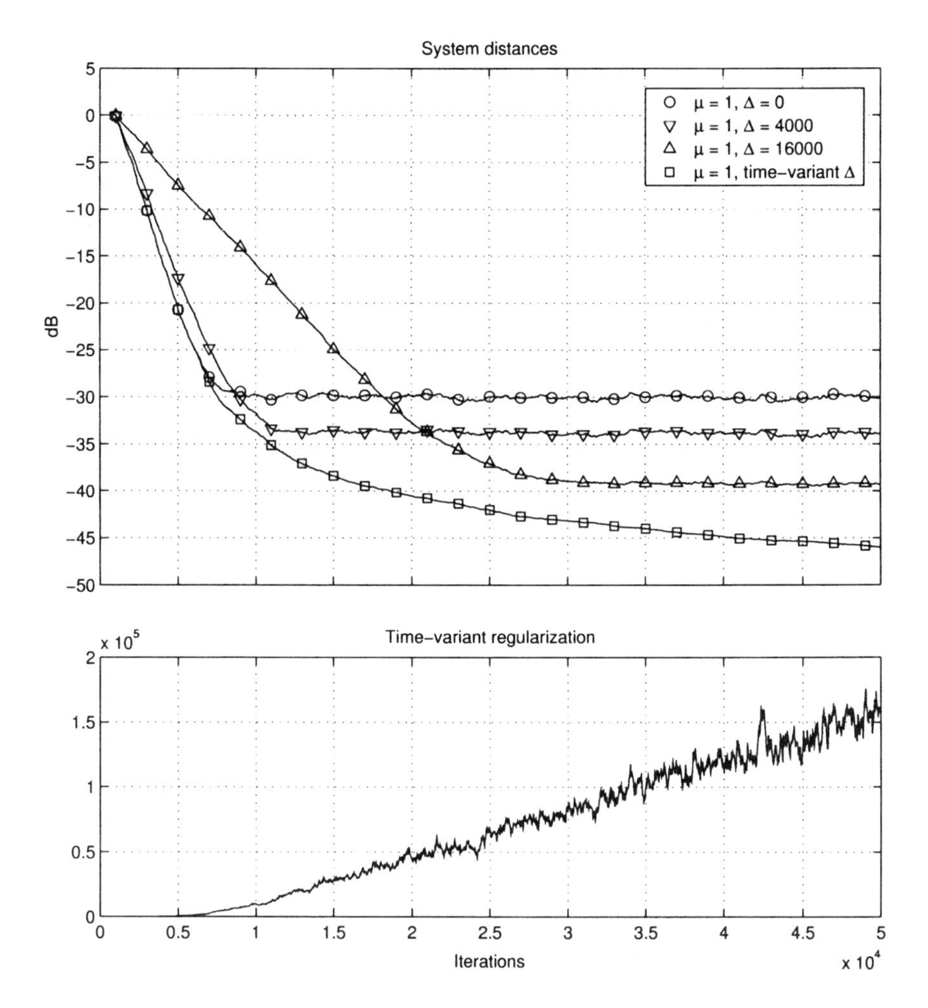

Fig. 13.4 Convergence using several fixed regularization parameters and one time-variant parameter. The time-variant regularization was adjusted according to Eq. 13.44 (with permission from [102]).

13.3 SUMMARY OF PSEUDOOPTIMAL CONTROL PARAMETERS

The pseudooptimal control parameters of the NLMS and of the affine projection algorithm are summarized in Table 13.1. As stated in Chapter 7, the NLMS algorithm can be interpreted as an affine projection of order $p = 1$. This relationship can also be observed when comparing the optimal control parameters.

The pseudooptimal stepsize for affine projection algorithms is p times smaller than the one for NLMS algorithms. If we consider that the system distance contracts on average only when the stepsize is chosen from an interval limited by zero on one side and by $2\,\mu_{\text{opt}}(n)$ on the other side, it turns out that controlling affine projection

Table 13.1 Pseudooptimal control parameters

Algorithm	Pseudooptimal stepsize	Pseudooptimal regularization
NLMS	$\dfrac{\mathrm{E}\{e_u^2(n)\}}{\mathrm{E}\{e_u^2(n)\} + \mathrm{E}\{n^2(n)\}}$	$\dfrac{\mathrm{E}\{n^2(n)\}\ \|\boldsymbol{x}(n)\|^2}{\mathrm{E}\{e_u^2(n)\}}$
AP	$\dfrac{\mathrm{E}\{e_u^2(n)\}}{\mathrm{E}\{e_u^2(n)\} + p\ \mathrm{E}\{n^2(n)\}}$	$p\ \dfrac{\mathrm{E}\{n^2(n)\}\ \|\boldsymbol{x}(n)\|^2}{\mathrm{E}\{e_u^2(n)\}}$

algorithms is more difficult than achieving a stable and fast convergence with the NLMS algorithm.

A similar relationship is observed when comparing the regularization parameters. Controlling affine projection algorithms turns out to be p times more sensitive to local signals than the NLMS algorithm. However, the affine projection and also other fast algorithms like RLS show faster convergence in case of nonwhite excitation signals like speech. This advantage has to be paid with a much larger sensitivity toward distorting components like local speech (double talk) and local background noise. In conclusion:

- One should spend more computational power on reliable double-talk detectors and on other enhanced detection and estimation schemes than would be necessary for standard NLMS algorithms.
- Additionally, one can apply more sophisticated inner cost functions that account for large error amplitudes as a possible result of double talk [73, 74].

13.4 DETECTION AND ESTIMATION METHODS

In the previous sections of this chapter, pseudooptimal control parameters have been derived. The results of the derivations contain second-order moments of signals that are not directly accessible. In this section, several estimation schemes as well as detection methods, which are required within the schemes, are introduced. Because of space limitations, only the basic principles or simulation examples, respectively, with input signals chosen to demonstrate the specific detection principles are presented.

Details about real-time implementation, computational complexity, and reliability can be found in the corresponding references.

All detectors and estimators can be grouped into four classes:

- As mentioned before, short-term power estimations (of accessible signals) are required. Even if there are several possibilities for estimating these terms, only one method is presented within thc first class.

- In Section 13.4.2, the basic principle for estimating the power of the undisturbed error signal $e_u(n)$ will be introduced and the necessity for estimating the system distance $\|\boldsymbol{h}_\Delta(n)\|^2$ will be explained. Methods for estimation of this parameter are grouped in the second class.

- The adaptation of several detection parameters (e.g., the expected system distance) should cease during activity of the local speaker. Principles for the detection of local speech activity are summarized in the third class.

- Finally, the principles for detecting enclosure dislocations, called "rescue detectors," are grouped in the fourth class.

At the end of this chapter several possibilities for combining the detectors and estimators to an entire control unit are presented and simulation results for some of the combinations are shown.

13.4.1 Short-Term Power Estimation

For all control parameters the short-term power of one or more signals is required. In this part we will address power estimations where the signal is either directly accessible all the time or is stationary and measurable within short periods.

13.4.1.1 First-Order IIR Smoothing

For estimation of the power of a signal, first-order IIR filters can be utilized to smooth the instantaneous power of the signal, as indicated in Fig. 13.5. To

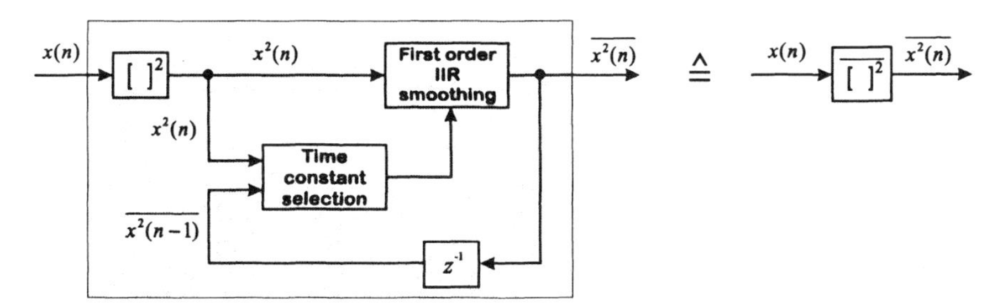

Fig. 13.5 Structure of the short-term power estimation applying a nonlinear first-order IIR filter.

detect rising signal powers (especially of the error signal) very rapidly, different smoothing constants for rising and falling signal edges are used ($\gamma_r < \gamma_f$):

$$\overline{x^2(n)} = (1 - \gamma(n))\, x^2(n) + \gamma(n)\, \overline{x^2(n-1)} \tag{13.46}$$

with

$$\gamma(n) = \begin{cases} \gamma_\mathrm{r}\,, & \text{if } x^2(n) > \overline{x^2(n-1)}, \\ \gamma_\mathrm{f}\,, & \text{otherwise.} \end{cases} \tag{13.47}$$

Besides taking the squared input signal [31], it is also possible to take the absolute value [114, 198]. The smoothed magnitudes and the smoothed squared values are related approximately via a form factor. When comparing the powers of different signals or computing their ratio, the knowledge of this form factor is not necessary if the same method was used for both power estimations.

It should be mentioned that the preceding estimation contains a small (multiplicative) bias due to the different smoothing constants. As with the form factor, this bias may be neglected if all power values to be compared are calculated with the same pair of smoothing constants.

An example for a short-term power estimation according to Eq. 13.46 (together with a background noise-level estimation, which will be introduced in the next section) is presented in Fig. 13.7. The smoothing constants were chosen as $\gamma_\mathrm{r} = 0.993$, $\gamma_\mathrm{f} = 0.997$, and the sampling rate was $f_\mathrm{s} = 8$ kHz.

13.4.1.2 Estimation of the Local Background Noise Power

For estimation of the background noise level a multitude of methods, such as minimum statistics [159] or IIR smoothing only during speech pauses, have been presented. For details, the interested reader is referred to Chapter 14. In this section we will describe only a very simple but robust scheme.

The input to this background noise level estimator is the short-term power of the microphone signal $\overline{y^2(n)}$ (computed according to Eq. 13.46). For estimating the background noise level, the minimum of the current short-term power and the previous output of the estimator is taken. To avoid freezing at a global minimum, the result of the minimum operator is multiplied by a constant slightly larger than one:

$$\overline{\hat{b}^2(n)} = \min\left\{\overline{y^2(n)},\, \overline{\hat{b}^2(n-1)}\right\}\, (1+\epsilon)\,. \tag{13.48}$$

The basic structure of the estimation scheme is depicted in Fig. 13.6. The constant ϵ is a small positive value that controls the maximum speed for increasing the estimated noise level. The quantity

$$B_\Delta(\epsilon) = f_\mathrm{s}\, 20\, \log_{10}(1+\epsilon) \tag{13.49}$$

describes the maximum power increase in decibels per second. For $\epsilon = 0.0002$ and a sampling rate $f_\mathrm{s} = 8000$ Hz, we obtain an upper limit for the incremental

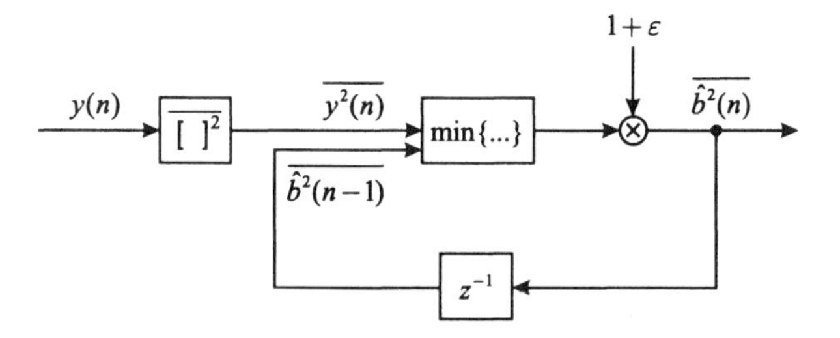

Fig. 13.6 Simple scheme for estimation of the local noise power.

speed of $B_\Delta(0.0002) = 13.9$ dB/second. Depending on the type of background noise, ϵ should be chosen such that $B_\Delta(\epsilon)$ is within the interval (0.5 dB/second, 3 dB/second).

Figure 13.7 depicts an example for short-term power estimation as well as background noise level estimation. The input signal was sampled at a rate of 8 kHz and the smoothing constants were chosen as $\gamma_r = 0.993$ and $\gamma_f = 0.997$. The noise increment constant ϵ was set to $\epsilon = 0.00002$.

13.4.2 Estimating the System Distance

For white remote excitation $x(n)$ and statistical independence of the excitation vector $\boldsymbol{x}(n)$ and the system mismatch vector $\boldsymbol{h}_\Delta(n)$, the estimated power of the undisturbed error signal can be noted as (see Eq. 13.3)

$$\begin{aligned} \mathrm{E}\left\{e_{\mathrm{u}}^2(n)\right\} &= \mathrm{E}\left\{\boldsymbol{h}_\Delta^{\mathrm{T}}(n)\,\boldsymbol{x}(n)\,\boldsymbol{x}^{\mathrm{T}}(n)\,\boldsymbol{h}_\Delta(n)\right\} \\ &= \mathrm{E}\left\{\boldsymbol{h}_\Delta^{\mathrm{T}}(n)\,\mathrm{E}\left\{\boldsymbol{x}(n)\,\boldsymbol{x}^{\mathrm{T}}(n)\right\}\,\boldsymbol{h}_\Delta(n)\right\} \\ &= \mathrm{E}\left\{\boldsymbol{h}_\Delta^{\mathrm{T}}(n)\,\mathrm{E}\left\{x^2(n)\right\}\,\boldsymbol{I}\,\boldsymbol{h}_\Delta(n)\right\} \\ &= \mathrm{E}\left\{x^2(n)\right\}\,\mathrm{E}\left\{\|\boldsymbol{h}_\Delta(n)\|^2\right\}. \end{aligned} \tag{13.50}$$

The second factor, the expected system distance, indicates the echo coupling

$$\beta(n) = \mathrm{E}\left\{\|\boldsymbol{h}_\Delta(n)\|^2\right\} = \mathrm{E}\left\{\left\|\boldsymbol{h}(n) - \hat{\boldsymbol{h}}(n)\right\|^2\right\} \tag{13.51}$$

after the echo cancellation. Fig. 13.8 illustrates the idea of replacing the parallel structure of the adaptive filter and the LEM system by a frequency-independent *coupling factor*. With this notation, the pseudooptimal stepsize for the NLMS algorithm, for instance, can be written as follows:

$$\mu(n) = \frac{\mathrm{E}\left\{e_{\mathrm{u}}^2(n)\right\}}{\mathrm{E}\left\{e^2(n)\right\}} = \frac{\mathrm{E}\left\{x^2(n)\right\}\,\beta(n)}{\mathrm{E}\left\{e^2(n)\right\}}. \tag{13.52}$$

Simulations, as well as real-time implementations, have shown that Eq. 13.52 also provides a good approximation in case of speech excitation. The problem of estimating the stepsize has thus been reduced to estimation of the power of the remote

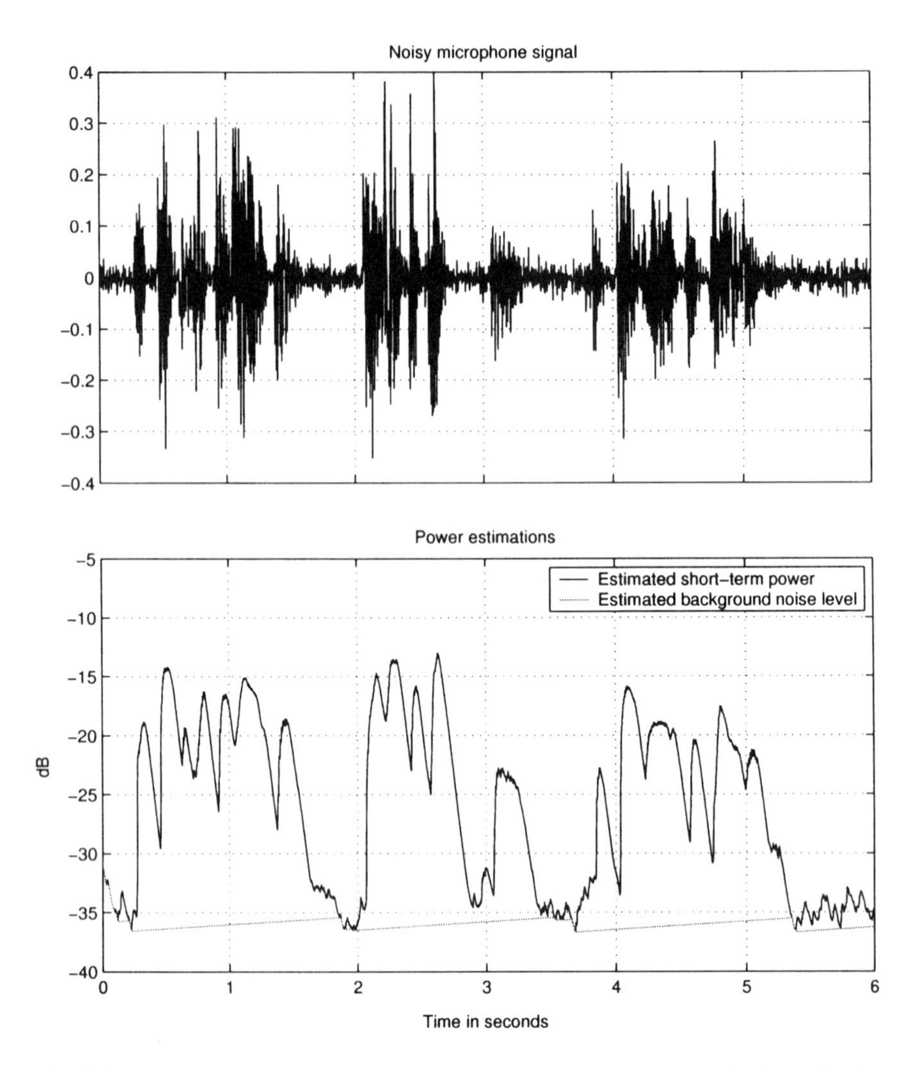

Fig. 13.7 Simulation example of the short-term power and the background noise estimation. The input signal was sampled at a rate of 8 kHz, and the smoothing constants were chosen as $\gamma_r = 0.993$ and $\gamma_f = 0.997$. The noise increment constant was set to $\epsilon = 0.00002$.

excitation signal and the power of the error signal (which can be computed according to Section 13.4.1.1), as well as estimation of the coupling factor. Since the excitation and the error signals are accessible, the major problem is to estimate the coupling factor.

In the following, we explain two different methods for estimating the coupling factor and illustrate their strong and weak properties with the help of simulations. The results depicted in Fig. 13.10 are compared with the true system distance available for simulations when the true LEM system impulse response $\boldsymbol{h}(n)$ is known.

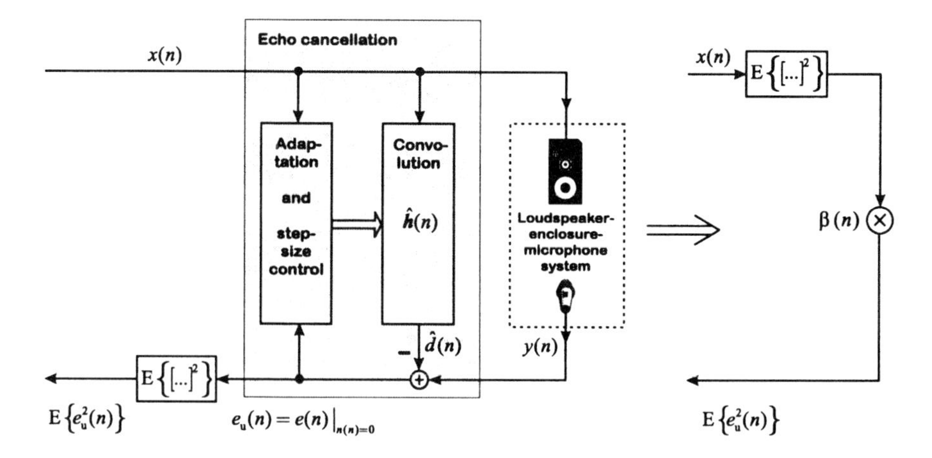

Fig. 13.8 Model of the parallel arrangement of the LEM system and the echo cancellation filter. To estimate the power of the undisturbed error signal, the parallel structure of the adaptive filter plus the LEM system is modeled by a coupling factor.

13.4.2.1 Estimation via "Delay" Coefficients

In general, the system distance according to

$$\left\|\boldsymbol{h}_\Delta(n)\right\|^2 = \left\|\boldsymbol{h}(n) - \hat{\boldsymbol{h}}(n)\right\|^2 \tag{13.53}$$

cannot be determined since the real LEM impulse response $\boldsymbol{h}(n)$ is unknown. If an additional artificial delay is introduced into the LEM system, this delay is also modeled by the adaptive filter. Thus, the section of the impulse response $\boldsymbol{h}(n)$ corresponding to the artificial delay is known to be zero and the system distance vector equals the negative filter vector for this section [246]

$$h_{\Delta,i}(n) = -\hat{h}_i(n) \qquad i \in [0, \ldots, N_\mathrm{D} - 1], \tag{13.54}$$

where N_D denotes the number of filter coefficients corresponding to the artificial delay. Utilizing the known property of adaptive algorithms to spread the filter mismatch evenly over all filter coefficients, the known part of the (real) system mismatch vector can be extrapolated to the system distance

$$\left\|\boldsymbol{h}_\Delta(n)\right\|^2 = \frac{N}{N_\mathrm{D}} \sum_{i=0}^{N_\mathrm{D}-1} \hat{h}_i^2(n), \tag{13.55}$$

where N denotes the length of the adaptive filter. The method is depicted in the upper diagram of Fig. 13.9. A stepsize control based on the estimation of the system distance with the "delay" coefficients generally shows good performance; when the power of the error signal increases as a result of a local excitation, the stepsize is reduced as the denominator in Eq. 13.52 increases. The reduced stepsize avoids a divergence of the filter.

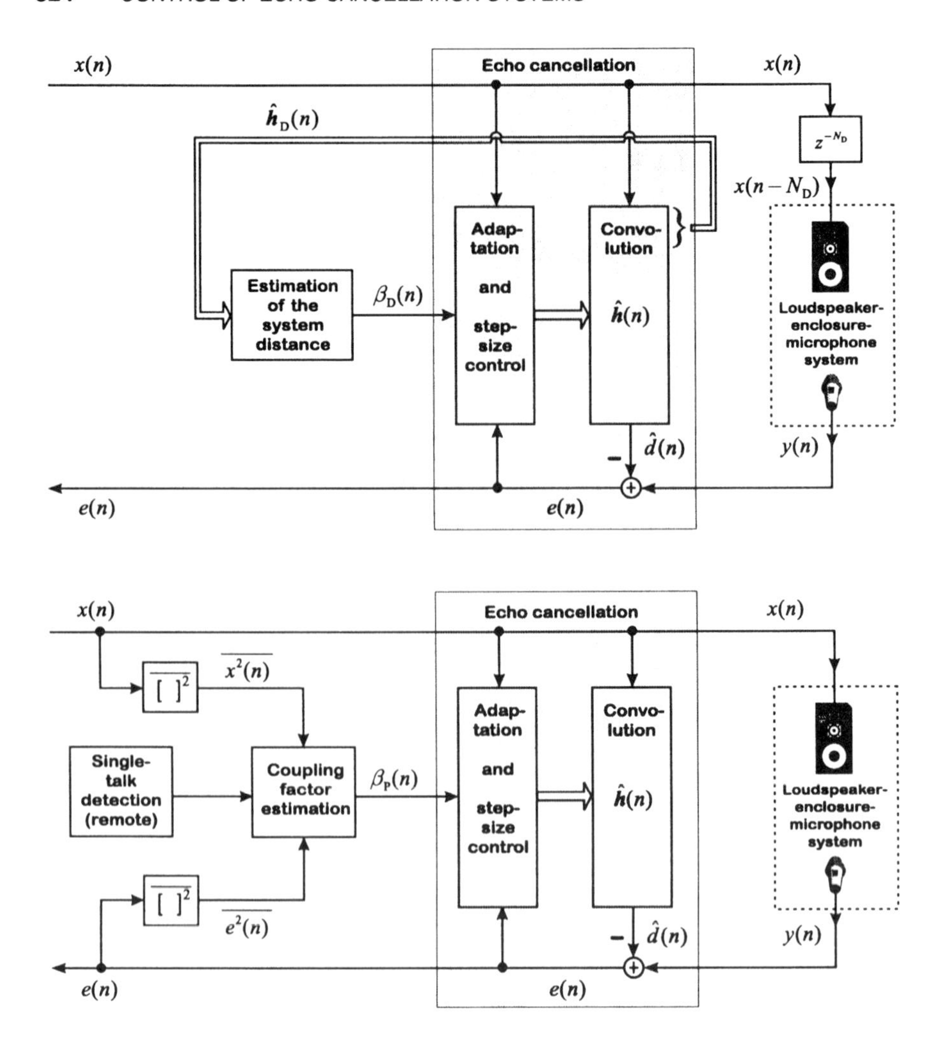

Fig. 13.9 Structures of the two principles for estimating the norm of the system mismatch vector. In the upper part, the method based on delay coefficients is depicted. In the lower diagram, the principle based on short-term power estimations of the excitation and the error signal, as well as on a remote single-talk detection is presented.

However, determination of the stepsize according to this method may lead to a freezing of the adaptation when the LEM system changes. This phenomenon can be observed, for example, in the lowest diagram of Fig. 13.10, after 60,000 iterations. Freezing occurs because a change in the LEM system also leads to an increase of the power of the error signal and consequently to a reduced stepsize. Thus, a new adaptation of the filter and the "delay" coefficients is prevented and the system freezes. To avoid this, we require an additional detector for LEM system changes that either sets the "delay" coefficients or the stepsize to larger values so that the filter can readapt.

13.4.2.2 Estimation via Power Estimation and Single-Talk Detection of the Remote Speaker

In this section a method to estimate the coupling factor based on power estimation is presented. The coupling factor (see Section 13.4.2) is determined by the ratio of the powers of the undisturbed error signal and the excitation signal. The undisturbed error signal $e_{\mathrm{u}}(n)$ can be approximated by the error signal if one ensures that sufficient excitation power is present and the local speaker is not active:

$$\beta_{\mathrm{P}}(n) = \frac{\overline{e_{\mathrm{u}}^2(n)}}{\overline{x^2(n)}} \approx \begin{cases} \gamma\beta_{\mathrm{P}}(n-1) + (1-\gamma)\dfrac{\overline{e^2(n)}}{\overline{x^2(n)}} & : \text{ if remote single talk is detected,} \\ \beta_{\mathrm{P}}(n-1) & : \text{ otherwise.} \end{cases} \tag{13.56}$$

Obviously, reliable single-talk detection (see lower section of Fig. 13.9) is necessary to ensure a good estimation with this method. However, even during remote single talk, the error signal still contains a disturbing component $t(n)$, due to the nonmodeled part of the impulse response, thus limiting the estimated coupling factor. Because of this limitation, the choice of stepsize is too large and it is not possible to obtain a convergence as good as that obtained with a stepsize based on the estimation with delay coefficients. The advantage of the stepsize control based on the coupling factor is that it does not lead to freezing of the adaptation when the LEM system changes. The dependency of the adaptation on reliable single-talk detection is explained explicitly with two wrong estimation results artificially introduced into the enabling signal in Fig. 13.10:

- Enabling the estimation during double talk (33,000–34,000 iterations) leads to an increase of the coupling factor and the stepsize and thus to a (slowly) diverging filter.
- Interpreting the LEM change as double talk (60,000–63,000 iterations) delays the tracking of the filter.

We conclude that both methods of estimating the system distance have strengths and weaknesses. Both methods need an additional detector. Estimation of the system distance with the delay coefficients requires a detector that can determine LEM changes, to avoid freezing of the filter. The estimation of the coupling factor requires a reliable remote single-talk detector.

13.4.3 Detection of Remote Single Talk

Detection of remote single talk is necessary for estimation of the coupling factor $\beta(n)$, as described in Section 13.4.2. It is obvious that detection of remote speech

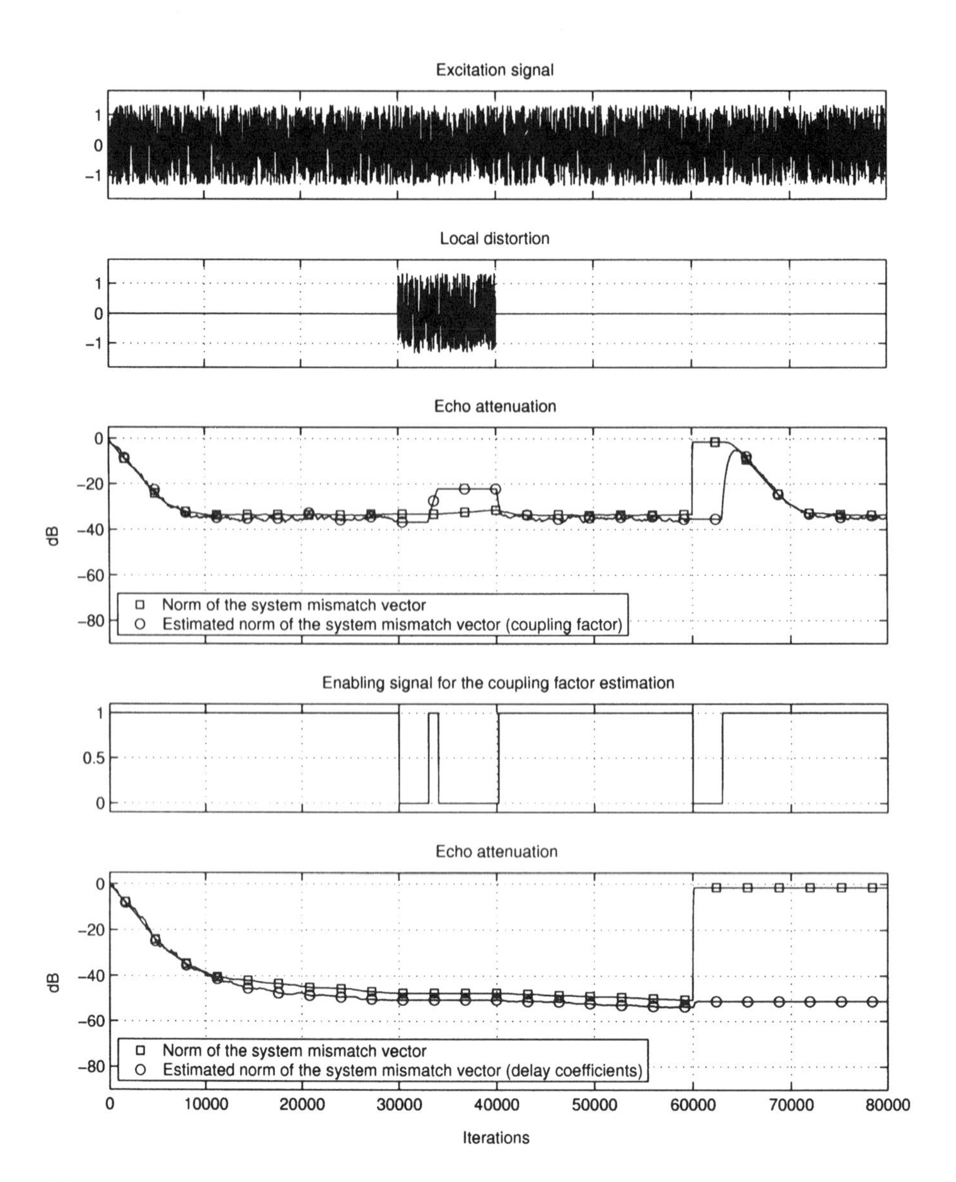

Fig. 13.10 Simulation examples for estimation of the system distance, or of the coupling factor, respectively. White noise was used for the excitation as well as for the local signal (see the two upper diagrams). Double talk takes place between iterations 30,000 and 40,000. After 60,000 iterations, an enclosure dislocation was simulated (a book was placed midway between the loudspeaker and the microphone). Details about the lower three diagrams are given in the text (with permission from [154]).

activity is required, since the coupling factor can be estimated only in periods of remote excitation. Additionally, the factor should be estimated only in periods when no local speech is present, because local signals disturb the determination of the coupling factor.

Combining these two conditions, remote single-talk detection is required for a reliable estimation of the coupling factor. As the remote activity can be easily detected by comparing the input signal power with a threshold, the major problem we will focus on in this section is the detection of local speech activity.

In the following, we will discuss two different detection methods. The first is based on power comparisons, whereas the second utilizes distance measures to determine the similarity of two signals.

13.4.3.1 Detection Principles Based on Power Comparisons

In general, voice activity of the local speaker can be detected from the error signal $e(n)$, since this signal is already freed of most parts of the incoming echo. Local activity is hereby detected if the short-term power of the error signal exceeds that of the estimated residual echo signal $\beta(n)\,\overline{x^2(n)}$. For a reliable estimation of $\beta(n)$, however, speech activity detection based on this value cannot be applied.[4] Instead, the power of the microphone signal and the power of its echo component estimated by the loudspeaker excitation, multiplied by the acoustic coupling $\beta_{\mathrm{L}}(n)$, are compared. The latter is dependent only on the loudspeaker, the enclosure, and the microphone. As it is independent of the convergence of the echo cancellation filter, movements of the local speaker and other changes of the echo path in the enclosure influence this coupling only marginally.

Thus, $\beta_{\mathrm{L}}(n)$ can be assumed to be approximately constant and is updated only during periods of large remote excitation:

$$\beta_{\mathrm{L}}(n) = \begin{cases} \gamma\beta_{\mathrm{L}}(n-1) + (1-\gamma)\dfrac{\overline{y^2(n)}}{\overline{x^2(n)}}\,, & \text{if large remote excitation is detected,} \\ \beta_{\mathrm{L}}(n-1)\,, & \text{otherwise.} \end{cases} \tag{13.57}$$

Due to the large smoothing, a distortion of the estimation during double talk may be neglected. In cases of a volume change of the loudspeaker, this factor can be updated directly, as volume information is available to most systems.

Local speech activity is detected if the power of the microphone signal exceeds that of the estimated echo signal by a threshold

$$\overline{y^2(n)} > K_{\mathrm{L}}\,\overline{x^2(n)}\,\beta_{\mathrm{L}}(n) \tag{13.58}$$

as indicated in Fig. 13.11. The detector was tested with speech signals depicted in the two upper diagrams of Fig. 13.12. During remote single talk, the powers of the

[4] This would cause a closed control loop with freezing and instability problems.

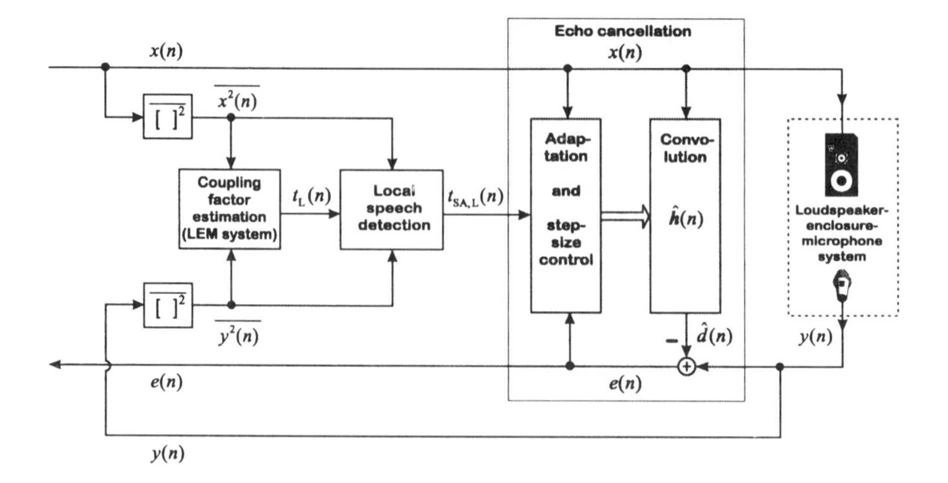

Fig. 13.11 Structure of local speech activity detection using the estimated coupling of the loudspeaker–enclosure–microphone (LEM) system as well as short-term power estimation of the excitation and the microphone signal.

echo signal and microphone signal are nearly identical. However, during double talk, the power of the microphone signal often exceeds the power of the echo signal and local activity can thus be detected. It is advisable to freeze the detection, for instance, for the duration of an average word, by decrementing a counter in order to detect the entire local excitation period. The local speech activity is detected $[t_{\mathrm{SA,L}}(n) = 1]$ if the counter exceeds a threshold. However, the detected period is then a little longer than the actual period of local activity.

The reliability of the detector decreases with increasing coupling between the loudspeaker and the microphone. But even for a large coupling, detection is still possible in periods of low remote excitation. However, if the beginning of double talk coincides with large remote activity, double talk may be detected only after a delay. To overcome this drawback modified cost functions, as presented by Gänsler and colleagues [73, 74], can be applied.

13.4.3.2 Detection Principles Based on Distance Measures

For speech coding and speech recognition purposes, several distance measures have been developed in order to determine the similarity of two signals [87]. In connection with controlling acoustic echo cancellation filters, distance measures are used to detect local speech activity. The degrees of similarity between the microphone signal $y(n)$ and the adaptive filter output signal $\hat{d}(n)$, or the loudspeaker signal $x(n)$, respectively, are determined.

All methods are based on the extraction and comparison of one feature of the signals. Besides these two distance measures, several other features of a speech

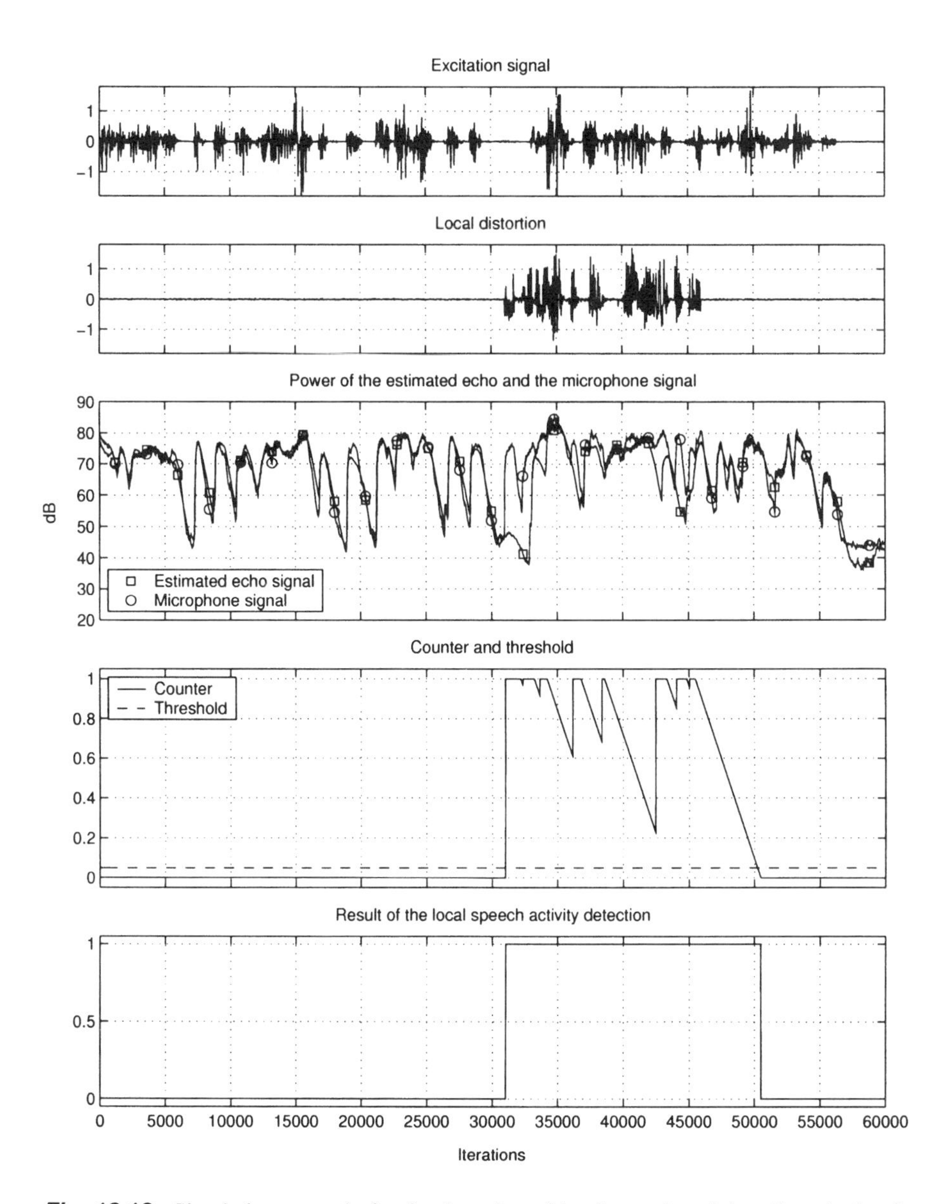

Fig. 13.12 Simulation example for the detection of local speech activity. Speech signals were used for both the excitation and the local signal. Double talk occurs during iterations 31,000 and 46,000. In the third diagram, the smoothed power of the excitation and of the microphone signal are depicted. The final detection of local speech activity and the counter on which this binary decision is based on are presented in the lower two diagrams (with permission from [154]).

signal, such as the pitch frequency [67], can also be utilized. However, we will describe only two examples in order to stay within the scope of this chapter.

13.4.3.2.1 Correlation Analysis

The aim of this detector is to measure the similarity between two signals by means of a correlation value [72, 114, 248]. Two different structures are practicable, as indicated in Fig. 13.13. The first approach—which we call *open-loop* structure—calculates a normalized correlation between the excitation signal $x(n)$ and the microphone signal $y(n)$. The second approach, denoted as a closed-loop structure, is based on a correlation measure between the excitation signal $x(n)$ and the output of the adaptive filter $\hat{d}(n)$. If the adaptive filter is adjusted sufficiently

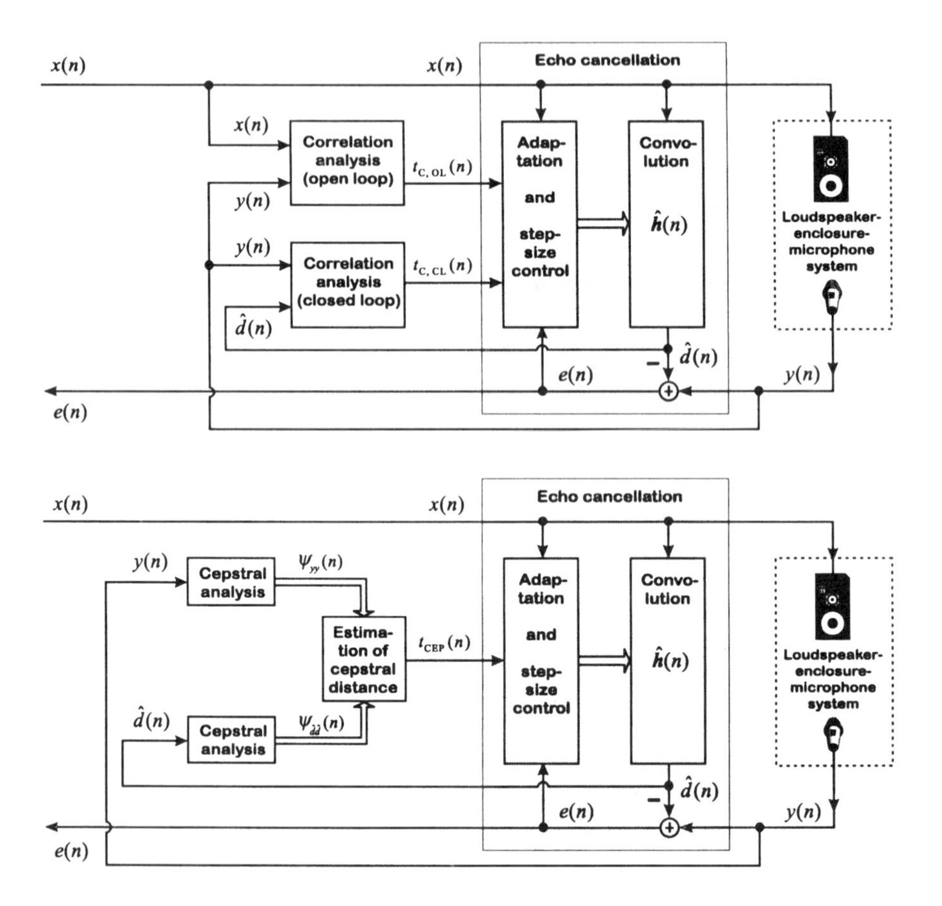

Fig. 13.13 Structures of remote single-talk detection methods based on distance measures. In the upper part an open and a closed-loop correlation analysis are depicted. A cepstral analysis (also a closed-loop structure) is presented in the lower diagram.

well, this structure yields better estimations because of the close similarity between the real echo $d(n)$ and the estimated echo $\hat{d}(n)$. However, the estimation quality depends on the convergence state of the adaptive filter, which is indicated by the term *closed-loop* structure. The other approach is independent of the adaptive filter and is therefore called open-loop structure.

The open-loop correlation factor

$$\rho_{\mathrm{OL}}(n) = \max_{l\in[0,L_{\mathrm{C}}]} \frac{\left|\sum\limits_{k=0}^{N_{\mathrm{C}}-1} x(n-k-l)\,y(n-k)\right|}{\sum\limits_{k=0}^{N_{\mathrm{C}}-1} \left|x(n-k-l)\,y(n-k)\right|} \tag{13.59}$$

has to be calculated for different delays l due to the time delay of the loudspeaker–microphone path. This is based on the assumption that the direct echo signal is maximally correlated with the excitation signal. In contrast, no delay has to be considered for the closed-loop correlation factor:

$$\rho_{\mathrm{CL}}(n) = \frac{\left|\sum\limits_{k=0}^{N_{\mathrm{C}}-1} \hat{d}(n-k)\,y(n-k)\right|}{\sum\limits_{k=0}^{N_{\mathrm{C}}-1} \left|\hat{d}(n-k)\,y(n-k)\right|}. \tag{13.60}$$

This is due to the fact that both signals are synchronous, if a sufficiently adjusted echo canceling filter is present. Both correlation values have to be calculated for a limited number of samples N_{C}, whereas a larger number ensures better estimation quality. However, there is a tradeoff between the estimation quality and the detection delay.

A decision for remote single talk $t_{\mathrm{C,OL}}(n)$, specifically $t_{\mathrm{C,CL}}(n)$ (see Fig. 13.13), can be easily generated by comparing the correlation value with a predetermined threshold (remote single talk is assumed if the correlation value is larger than the threshold). In Fig. 13.14, simulation results for the correlation values are shown. It is clear that the closed-loop structure ensures more reliable detection. However, in cases of misadjusted adaptive filters, this detector provides false estimations (e.g., at the beginning or after the enclosure dislocation at sample index $n = 60,000$).

Besides the time-domain-based approaches, frequency-domain structures for double-talk detection also exist. The interested reader is referred to the article by Benesty et al. [13].

13.4.3.2.2 Cepstral Analysis

The complex cepstrum of a signal $y(n)$ is defined as the inverse z transform of the logarithmic, normalized z transform of that signal [174]:

$$\log\left[\frac{Y(z)}{Y_0}\right] = \sum_{i=-\infty}^{\infty} \breve{y}(i)\,z^{-i}, \tag{13.61}$$

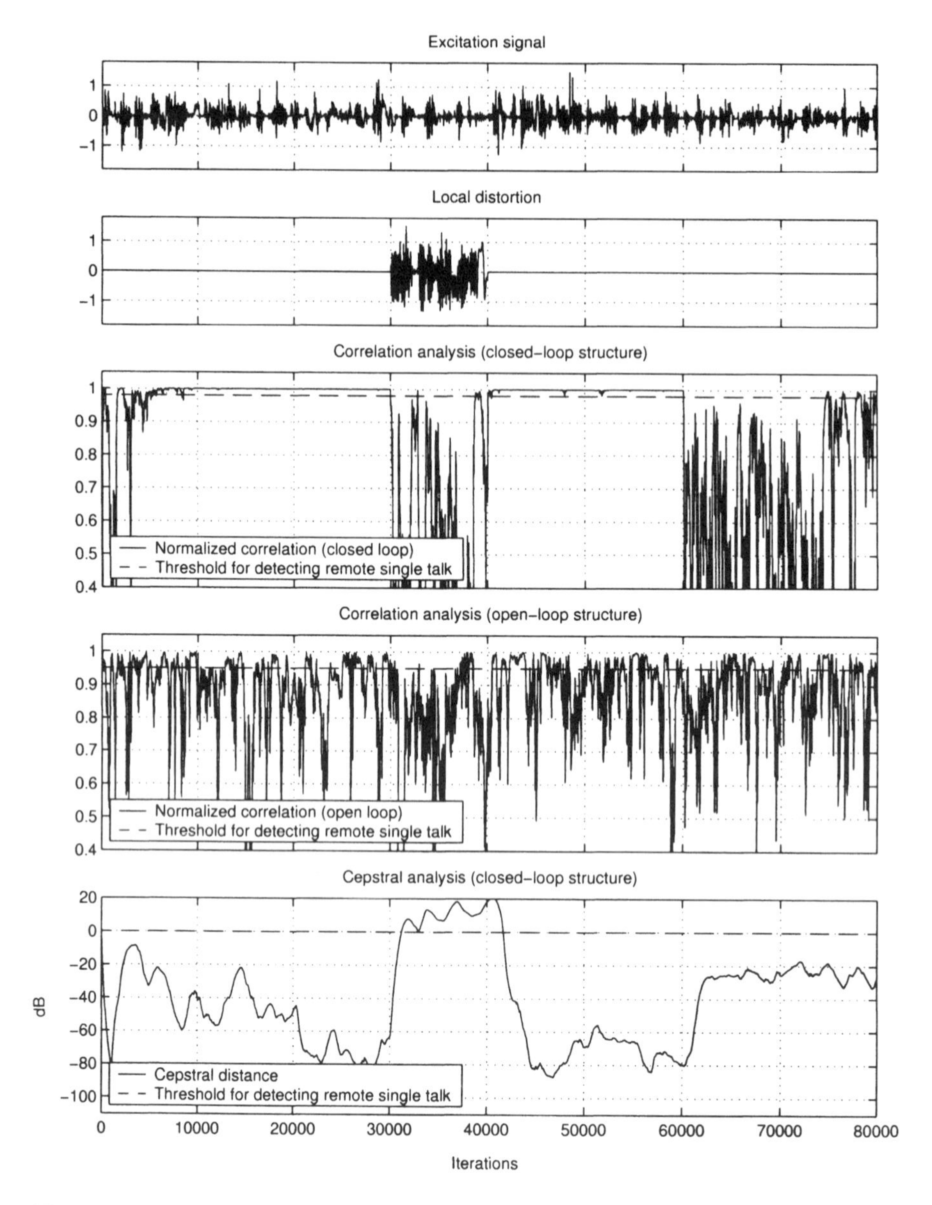

Fig. 13.14 Simulation example for detecting remote single talk with distance measure principles. Three methods—a closed- and an open-loop correlation analysis as well as a cepstral analysis—are depicted in the lower three diagrams. Speech signals were used for the excitation as well as for the local signal. Double talk occurs during iterations 30,000 and 40,000. At iteration 60,000, the impulse response of the LEM system was changed, leading to detection problems in the closed-loop correlation analysis (with permission from [154]).

where Y_0 is a suitable normalization and

$$Y(z) = \sum_{i=-\infty}^{\infty} y(i)\, z^{-i}. \tag{13.62}$$

The cepstrum exists if the quantity log $[Y(z)/Y_0]$ fulfills all conditions of the z transformation of a stable series. The cepstral distance measure, as defined by Gray and Markel [87], focuses on the problem of determining the similarity between two signals. A modified, truncated version adapted to acoustic echo control problems was proposed [69]

$$d_c(n) = \sum_{i=0}^{N_{\text{cep}}-1} \left(\psi_{yy,i}(n) - \psi_{\hat{d}\hat{d},i}(n)\right), \tag{13.63}$$

where $\psi_{yy,i}(n)$ and $\psi_{\hat{d}\hat{d},i}(n)$ denote the cepstra of the estimated (short-term) autocorrelation functions $\hat{s}_{yy,i}(n)$, respectively $\hat{s}_{\hat{d}\hat{d},i}(n)$, of the signals to be compared at time index n. The principle of the cepstral distance measure is to determine the spectral differences of two signals by calculating the difference of the logarithmic spectral densities. The truncation of the sum (Eq. 13.63) from infinity to $N_{\text{cep}}-1$ can be interpreted as smoothing the logarithmic spectral density functions. A variation of the LEM system does not affect the measurement since, typically, local dislocations vary only in the fine structure of the room frequency response. To avoid signal transformations when calculating the quantities, the signals can be modeled as AR processes of low order, and hence, the cepstral distance measure can be determined by a prediction analysis of the process parameters [193].

A structure for a detector incorporating the cepstral distance measure is depicted in Fig. 13.13. The cepstral distance is calculated between the microphone $y(n)$ and the estimated echo signal $\hat{d}(n)$, as proposed by Frenzel [69]. However, a distance measure between the microphone $y(n)$ and the excitation signal $x(n)$ is also possible, comparable to the open-loop correlation analysis. Remote single-talk detection can be performed similarly to the correlation analysis by implementing a predetermined threshold; thus, remote single talk is detected if the cepstral distance remains below the threshold.

Results of the simulation are depicted in Fig. 13.14. It is obvious that, for a good choice of the threshold, reliable detection of remote single talk is possible. The cepstral distance also rises when local dislocations are present, but not as much as in case of double talk.

13.4.4 Rescue Detectors

As mentioned before, methods estimating the system distance have to be combined with "rescue detectors," preserving the adaptive filter from longlasting misadjustment. Problems mainly arise due to the critical distinction between situations of double talk and enclosure dislocations.

In the following sections two examples for rescue detection methods are presented. Both methods ensure a reliable estimation of the system distance (see also Section 13.4.2).

13.4.4.1 Shadow Filter

The main idea of this detection principle is to implement a second filter (*shadow filter*) in parallel to the existing echo cancelling filter (*reference filter*), as depicted in Fig. 13.15.

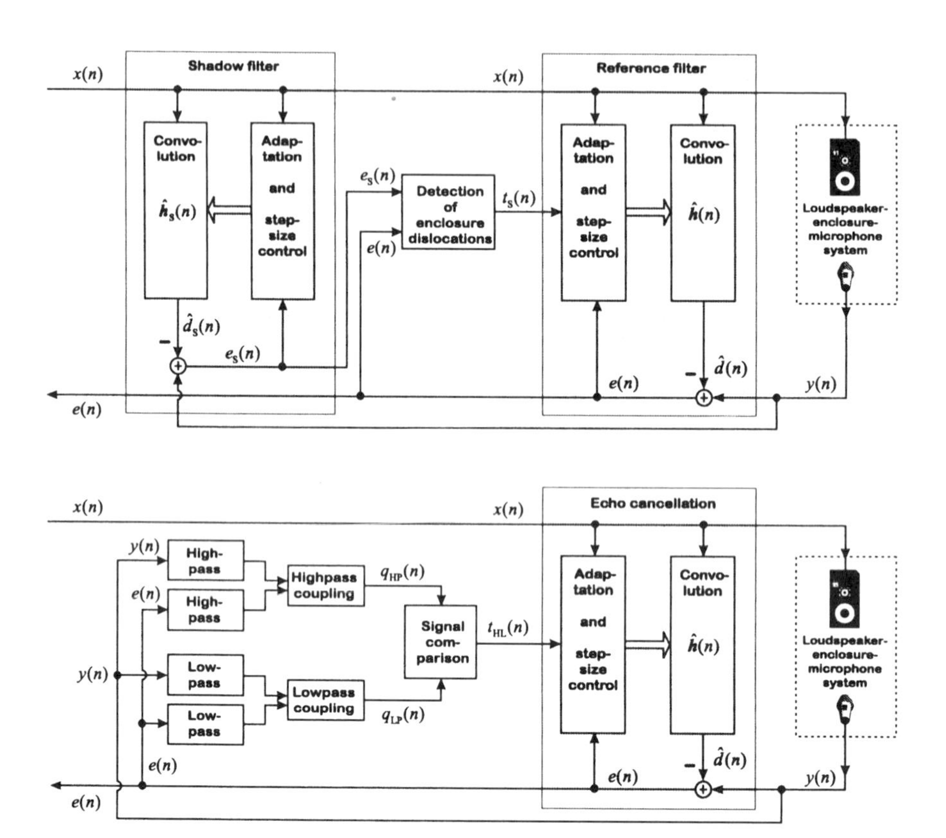

Fig. 13.15 Structures of two detection principles for enclosure dislocations. The upper part depicts a two-filter scheme (reference and shadow). Both filters are controlled independently. If one filter produces an error power much smaller than that of the other, either the filter coefficients can be exchanged or the parameters of the control mechanism can be reinitialized to make a new convergence possible. In the lower part, a detection scheme based on separate highpass and lowpass coupling analyses is depicted. Movements of persons only change the high-frequency characteristics of the LEM system, whereas activity of the local speaker also affects the low-frequency range. This relationship can be used to differentiate between increasing error powers due to double talk or to enclosure dislocations.

For this two-filter structure, different applications are possible; for instance, Haneda et al. [104] and Ochiai et al. [172] adapted only the shadow filter to the LEM impulse response. The reference filter used for echo cancellation has fixed coefficients. Coefficients are transferred from the shadow filter to the reference filter according to a transfer control logic, such as when the shadow filter gives a better approximation of the echo path impulse response than does the reference filter. Another approach is to adapt both the reference and the shadow filter, but with different stepsizes [216].

Here, the shadow filter is used for detecting enclosure dislocations [204]. The reference filter is adapted and controlled as in the single-filter case. The shadow filter is adapted similarly to the reference filter; however, its stepsize control is only excitation-based, that is, adaptation is stopped if the remote excitation falls below a predetermined threshold.

Furthermore, only half or less of the number of coefficients is used for the shadow filter, in comparison to the reference filter. These features ensure a high convergence speed of the shadow filter in the case of remote single talk. Of course, the filter diverges in case of local distortions. However, fast convergence after enclosure dislocations is ensured because the stepsize control is independent of the methods that can freeze the adaptive filter in these situations. Hence, the only situations in which the shadow filter is better adjusted to the LEM echo path than the reference filter are at enclosure dislocations. This is exploited to develop a detection mechanism; if the short-term error power of the shadow filter falls below the error power of the reference filter for a period of several iterations, enclosure dislocations are detected [in Fig. 13.15, $t_s(n)$ describes this detection result]. Consequently, the stepsize is enlarged to enable the adaptation of the reference filter toward the new LEM impulse response.

In Fig. 13.16, simulation results for this detector are shown. In the first graph, the powers of the error signal of both the reference and the shadow filter are pictured. Because the number of coefficients for the shadow filter is smaller than that for the reference filter, a faster convergence of the shadow filter is evident. However, a drawback of the decreased number of coefficients is the lower level of echo attenuation. After 62,000 iterations, when an enclosure dislocation takes place, fast convergence of the shadow filter can be observed, whereas the reference filter converges only slowly. Therefore an enclosure dislocation is detected (second graph in Fig. 13.16), which leads to a readjustment of the reference filter. At the beginning of the simulation, enclosure dislocations are also detected. However, this conforms with the requirements of the detector, because the beginning of the adaptation can also be interpreted as an enclosure dislocation due to misadjustment of the filter.

13.4.4.2 Low- and High-Frequency Power

The aim of this detector is to distinguish between two reasons for increasing power of the error signal $e(n)$: changes of the LEM impulse response or local speech activity. It was shown [163] that a typical change of the room impulse response, such as that caused by movements of the local speaker, affects mainly

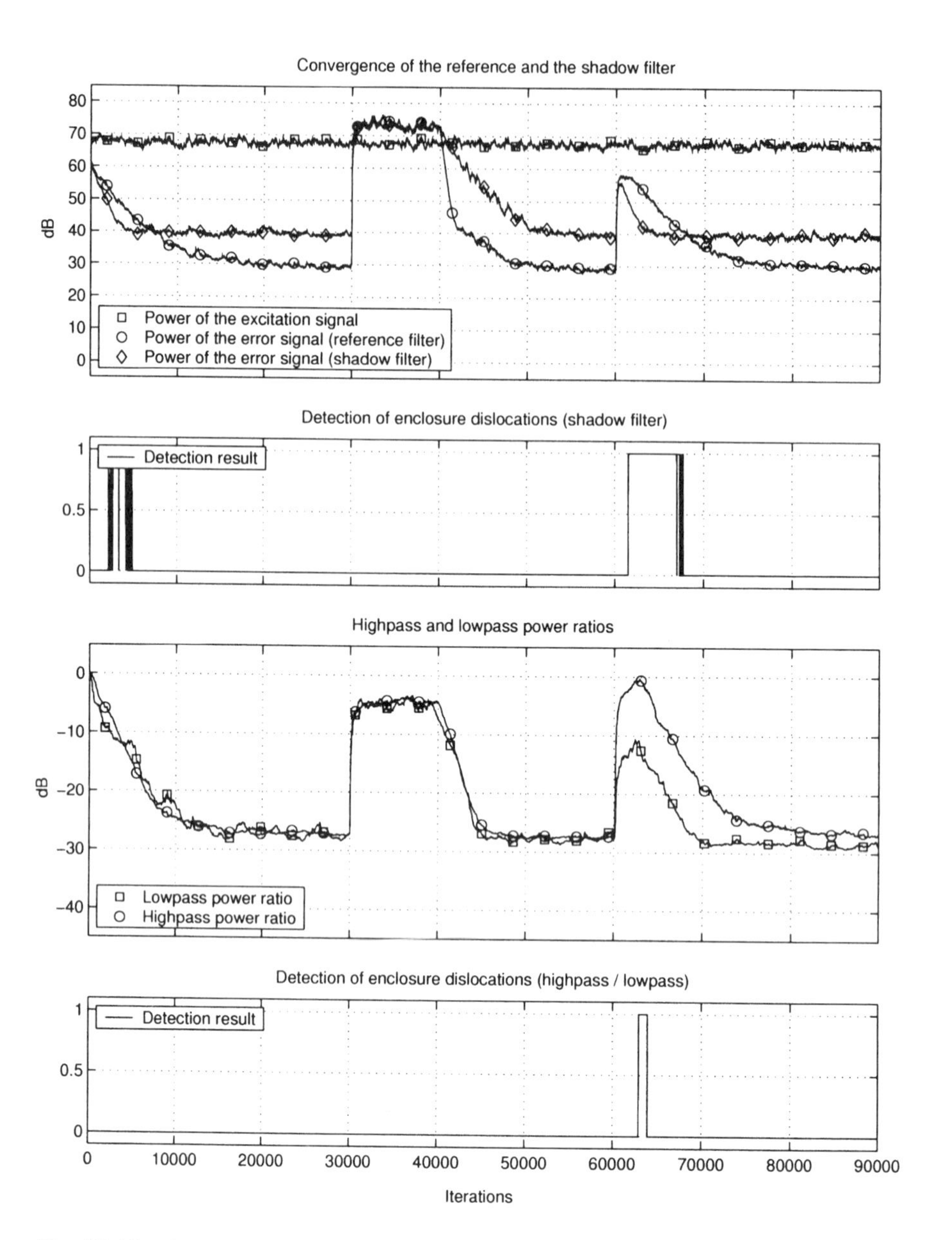

Fig. 13.16 Simulation examples for the detection of enclosure dislocations. Stationary noise with the same spectral characteristics as speech (linear predictive analysis of order 40) was used for the excitation signal as well as for the local signal. Double talk takes place during iterations 30,000 and 40,000. At iteration 62,000 the impulse response of the LEM system was changed. For both methods (shadow filter and separate highpass and lowpass coupling analyses), the detection result as well as the main analysis signal are depicted. Further details about the simulation are presented in the text (with permission from [154]).

the higher frequencies of the transfer function $H_\Delta(e^{j\Omega})$ corresponding to the system mismatch $\boldsymbol{h}_\Delta(n) = \boldsymbol{h}(n) - \hat{\boldsymbol{h}}(n)$. The reason for this characteristic is that movements of the local speaker may cause phase shifts up to 180° for high frequencies of the transfer function $H(e^{j\Omega})$ corresponding to the LEM system. In contrast, only slight phase shifts occur for low frequencies. The physical explanation is that the wavelengths of lower frequencies are large compared to typical enclosure dimensions, as the propagation of low-frequency sound waves is only marginally disturbed by local dislocations. Thus, the error signal generated by a system change exhibits smaller excitation for the low frequencies than for the high frequencies. Although these statements are valid mainly for white excitation and broadband LEM transfer functions, they can also be applied to speech excitation and real LEM systems.

In contrast to the error signal caused by the system mismatch considered above, the error signal generated by the local speaker's signal influences both the lower and the higher frequencies. This difference can be used to detect local dislocations as a larger increase of the power spectral density of the error signal for the high frequencies than for the low frequencies. In order to be independent of the shape of the LEM transfer function, the power spectral density of the error signal is normalized by the power spectral density of the microphone signal. (Note: the envelope of the transfer function is not influenced by local dislocations [163].) Therefore, the two ratios

$$q_{\mathrm{LP}}(n) = \frac{\int_0^{\Omega_g} S_{ee}(\Omega, n)\, d\Omega}{\int_0^{\Omega_g} S_{yy}(\Omega, n)\, d\Omega} \quad \text{and} \quad q_{\mathrm{HP}}(n) = \frac{\int_{\Omega_g}^{\pi} S_{ee}(\Omega, n)\, d\Omega}{\int_{\Omega_g}^{\pi} S_{yy}(\Omega, n)\, d\Omega} \tag{13.64}$$

are analyzed, where the short-term power spectral density is calculated by a recursively squared averaging. The cutoff frequency Ω_g should be chosen close to 700 Hz. A structure for the detector is proposed in Fig. 13.15.

There are different ways to finally generate the information about local dislocations. In [163], local dislocations are detected by processing differential values of $q_{\mathrm{LP}}(n)$ and $q_{\mathrm{HP}}(n)$ to detect a *change* of the LEM transfer function. However, if the peak indicating this change is not detected clearly, detection of the LEM change is totally missed. Another approach is based only on the current value of a slightly smoothed quotient $q_{\mathrm{LP}}(n)$ [26]. Our approach is to average the quotient $q_{\mathrm{HP}}(n)/q_{\mathrm{LP}}(n)$ by summing over the last 8000 samples. This procedure considerably increases the reliability of the detector but introduces a delay in the detection of enclosure dislocations.

Simulation results are depicted in Fig. 13.16. In the third graph, the lowpass and highpass power ratios, $q_{\mathrm{LP}}(n)$ and $q_{\mathrm{HP}}(n)$ respectively, are shown. It can be observed that both ratios rise close to -5 dB at $30,000$ iterations, during double-talk periods. In contrast, when a local dislocation occurs after $60,000$ samples, there is a clear increase in the highpass power ratio, whereas the lowpass power ratio is subject to only a small increase. The fourth graph shows the detection result of the sliding window for the quotient of the two power ratios. The enclosure dislocation is detected reliably, but with a small delay.

13.4.5 Concluding Remarks

In the previous four sections several estimation and detection schemes have been presented. Even if all procedures look very reliable within the presented simulation examples, their direct implementation in real scenarios shows several drawbacks. In general it is not possible to decide which of the methods of one class is superior over the others—this depends crucially on the specific type of application such as echo cancellation within cars (short impulse responses, high level of background noise) or hands-free systems in quiet offices (long impulse responses, low level of background noise).

13.5 DETECTOR OVERVIEW AND COMBINED CONTROL METHODS

After having described some of the most important detection principles in the previous section, we will now present an overview of the possibilities for combining these detectors to an entire stepsize or regularization control. As during the derivation of pseudooptimal control parameters, the NLMS algorithm will serve as a guiding example. An outlook is given on control for affine projection (AP) algorithms.

13.5.1 Stepsize Control for the NLMS Algorithm

In Fig. 13.17, the combination possibilities for building an entire stepsize control unit (without regularization) for the NLMS algorithm are depicted. The system designer has several choices, which differ in computational complexity, memory requirements, reliability, dependency of some types of input signals, and robustness in the face of finite word length effects.

Most of the proposed stepsize control methods are based on estimations of the short-term power of the excitation and the error signal. Estimating these quantities is, compared to the others remaining, relatively simple. As described in Section 13.4.1, in some applications only the absolute value of these signals is smoothed [113, 198], to reduce the required word length in fixed-point implementations. Also, smoothing filters other than (nonlinear) first-order IIR filters can be used.

Estimation of the current amount of echo attenuation is much more complicated. This quantity was introduced in Section 13.4.2 as echo coupling $\beta(n)$, which is an estimation of the norm of the system mismatch vector $\|\boldsymbol{h}_\Delta(n)\|^2$. A reliable estimation of this quantity is required not only for estimating the stepsize according to

$$\mu(n) = \frac{\overline{e_{\mathrm{u}}^2(n)}}{\overline{e^2(n)}} = \frac{\overline{x^2(n)}\ \overline{\beta(n)}}{\overline{e^2(n)}} \tag{13.65}$$

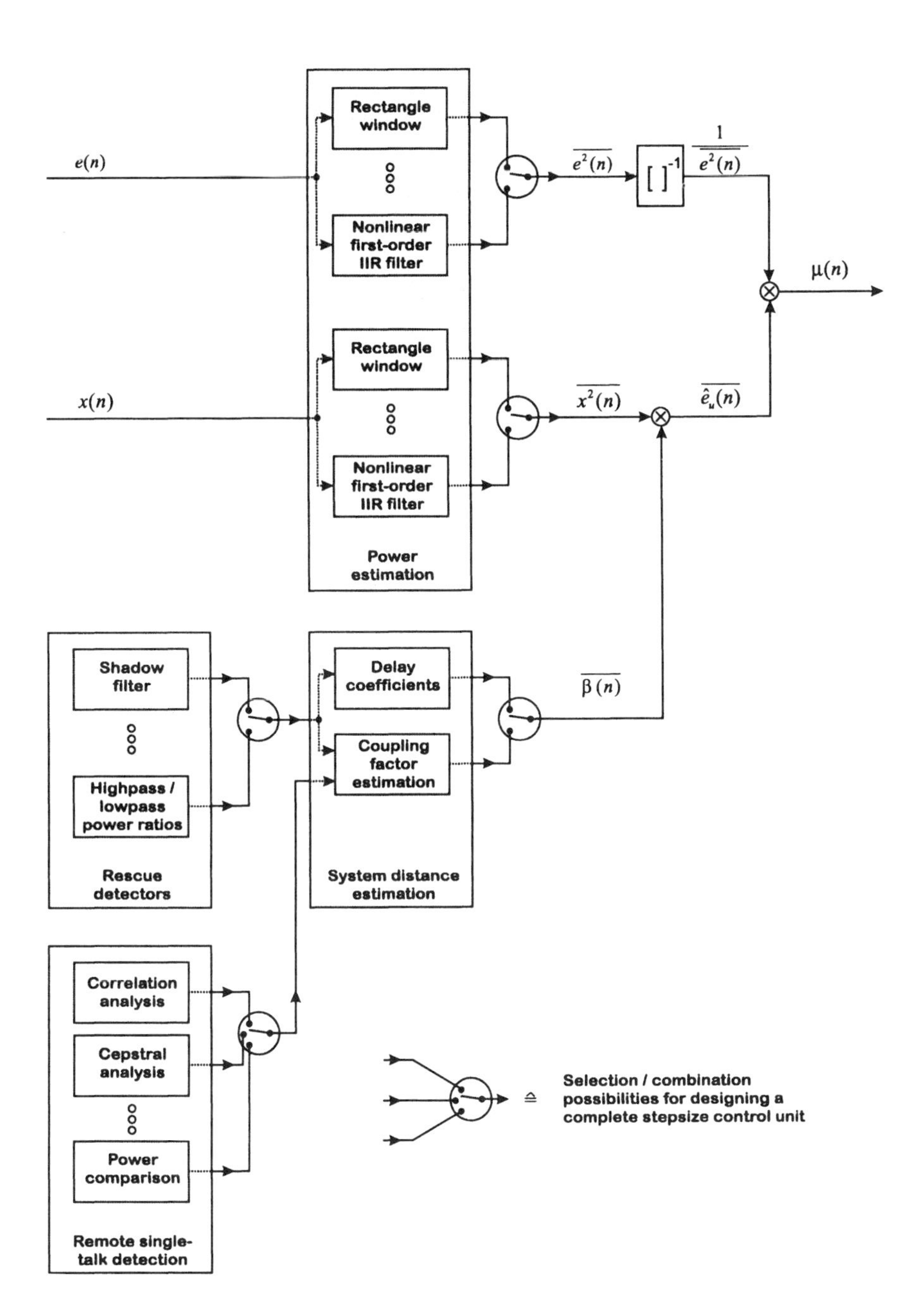

Fig. 13.17 Overview of possibilities for combining several detectors to a stepsize control unit.

(see also Eq. 13.52) but also for the interaction of the echo cancellation with other echo suppressing parts of an hands-free telephone, such as loss control [95, 96] and postfilter [98].

In Section 13.4.2, two methods for estimating the norm of the system mismatch vector were presented. Using the *delay coefficients* method has the advantage that no remote single-talk detection is required. Furthermore, the tail of the LEM impulse response, which is not canceled because of the limited order of the adaptive filter, does not affect this method. The disadvantage of this method is the extra delay that is necessary to generate the zero-valued coefficients of the LEM impulse response. If ITU-T or ETSI recommendations [64, 120] concerning the delay have to be fulfilled, the coupling factor estimation should be preferred, or a two-filter scheme [30] has to be implemented. A second drawback of the delay coefficients method is its insensitivity to enclosure dislocations. Without any rescuing mechanism, the estimation of the system distance would freeze and the filter would no longer converge. Finally, it should be noted that the computational load of the coupling factor method based on power estimations is only a fraction of that necessary for the delay coefficients method.

Two different methods for detecting enclosure dislocations were presented in Section 13.4.4. These detection principles also differ in their reliability, their computational complexity, and their memory requirements. Even though the shadow filter can operate in a reduced-frequency range, this detection method causes much higher computational load and requires more memory than does the detection principle based on the comparison of highpass and lowpass power ratios. Nevertheless, the shadow filter can detect a larger variety of enclosure dislocations. While movements of persons in the local room can be detected with both methods, an increase or decrease in the analog gain of the loudspeaker or the microphone, which is a frequency-independent gain modification of the LEM impulse response, can be detected only with the shadow filter principle.

If the coupling factor estimation is chosen, remote single-talk detection is required. Here, the system designer has also several combination alternatives. Methods based on feature extraction (correlation or cepstral analysis; see Section 13.4.3) and power-based detectors are possible candidates for this detection. Remote single-talk detection is often performed in a two-step procedure. The power of the excitation signal is first compared with a threshold. If remote speech activity is detected, one or more of the detectors mentioned above are used to correct the first decision in double-talk situations and to disable the adaptation of the coupling factor.

It should be mentioned that in real-time implementations, additional detectors are required that monitor the effects of limited arithmetical processor precision. For example, in the case of low remote excitation power, the stepsize $\mu(n)$ should also be reduced or even set to zero.

To conclude this section, we present one possibility for combining some of the detection principles mentioned above. For estimation of the system distance, the delay coefficients method was implemented with $N_{\mathrm{D}} = 40$ delay coefficients. Since this method was chosen, no remote single-talk detection was needed. The length of the adaptive filter was set to $N = 1024$, and speech signals were used for the remote excitation as well as for the local distortion. Both signals are depicted in the two upper sections of Fig. 13.18—a double-talk situation appears during iteration steps 30,000 and 40,000. After 62,000 iterations, an enclosure dislocation takes place (a book was placed between the loudspeaker and the microphone). To avoid freezing of the adaptive filter coefficients, a shadow filter of length $N_S = 256$ was implemented. If the power of the error signal of the shadow filter falls 12 dB or more below the error power of the reference filter, an enclosure dislocation is detected and the first N_D filter coefficients are reinitialized. In the center part of Fig. 13.18, the estimated and the true system distance are depicted. The rescue mechanism needs about 3000 iterations to detect the enclosure dislocation. During this time the stepsize $\mu(n)$ was set to very small values (see lower part of Fig. 13.18). After 65,000 iterations, the filter converges again. Even without detection of local speech activity, the stepsize was reduced during the double-talk situation.

13.5.2 Combined Stepsize and Regularization Control for the NLMS Algorithm

If echo cancellation is performed in quiet offices, it is sufficient to control the adaptation process only by utilizing a stepsize factor. However, in case of high background noise levels as they appear in automotive applications, it is advantageous to also apply control by regularization.

In this case the stepsize parameter is adjusted according to the local speech signal (double talk) and the regularization is modified according to the local background noise level. This is possible as long as all components of the error signal

$$e(n) = e_u(n) + s(n) + b(n) \tag{13.66}$$

are mutually orthogonal. The stepsize is adjusted slightly larger than in the control method presented in the last section. Compared to Eq. 13.65, the denominator is replaced by the sum of an estimate of the short-term power of the undistorted error $\overline{e_u^2(n)}$ and of the short-term power of the local speech signal $\overline{s^2(n)}$. This sum can be estimated in a simple manner by subtracting the power of the background noise $\overline{b^2(n)}$ from the power of the distorted error signal $\overline{e^2(n)}$. The resulting stepsize parameter can be computed as

$$\mu(n) = \frac{\overline{x^2(n)}\ \beta(n)}{\max\left\{\left(\overline{e^2(n)} - \overline{b^2(n)}\right), 0\right\}}. \tag{13.67}$$

The maximum operator within the denominator of Eq. 13.67 has been applied to make sure that the estimation

$$\overline{e_u^2(n)} + \overline{s^2(n)} \approx \overline{e^2(n)} - \overline{b^2(n)} \tag{13.68}$$

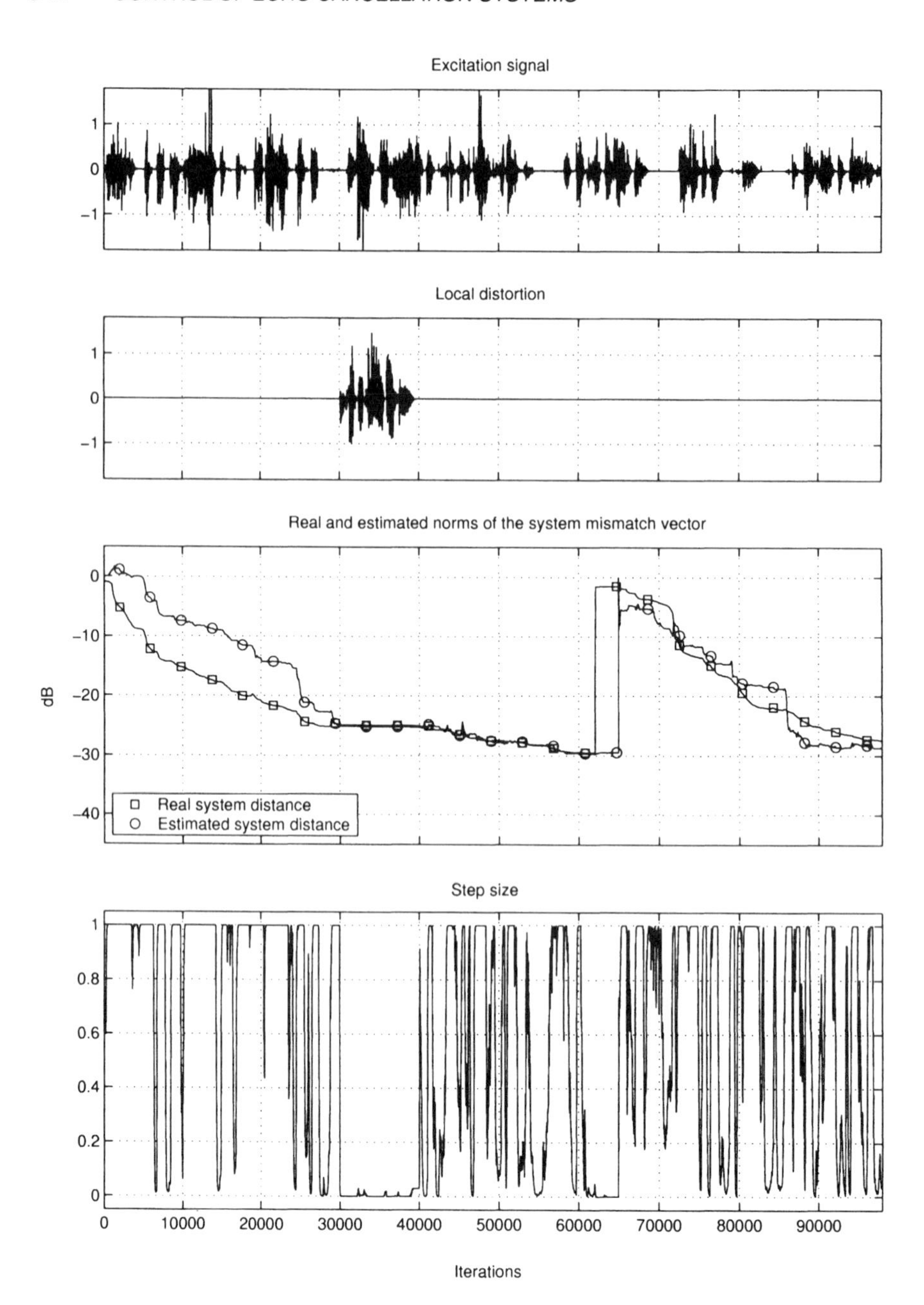

Fig. 13.18 Simulation example of a complete stepsize control unit. Speech signals were used for excitation as well as for local distortion (see first two diagrams). After 62,000 iterations, a book was placed between the loudspeaker and the microphone. In the third diagram, the measured and the estimated system distance are depicted. The lowest diagram shows the stepsize, based on the estimated norm of the system mismatch vector (with permission from [154]).

always has nonnegative results.[5] The regularization parameter is utilized to control the influence of the background noise. For this reason $\Delta(n)$ is chosen according to

$$\Delta(n) = N \frac{\overline{b^2(n)}}{\overline{\beta(n)}}. \tag{13.69}$$

As within the estimation of stepsize, the system designer has several choices for building the regularization control unit. Fig. 13.19 shows possible detector–estimator combinations.

As in the last section, a simulation example shows the performance of one of the possible detector combinations. All short-term powers of accessible signals were estimated using first-order IIR smoothing of the squared amplitudes. For estimating

[5] A better way of making the stepsize computation more robust against estimation errors would be to transform the regularization, as described in Eq. 13.69, into an equivalent stepsize (see Section 13.1.3). Afterward the stepsize according to Eq. 13.67 should be limited to ensure that the product of both stepsizes does not exceed the value 1.

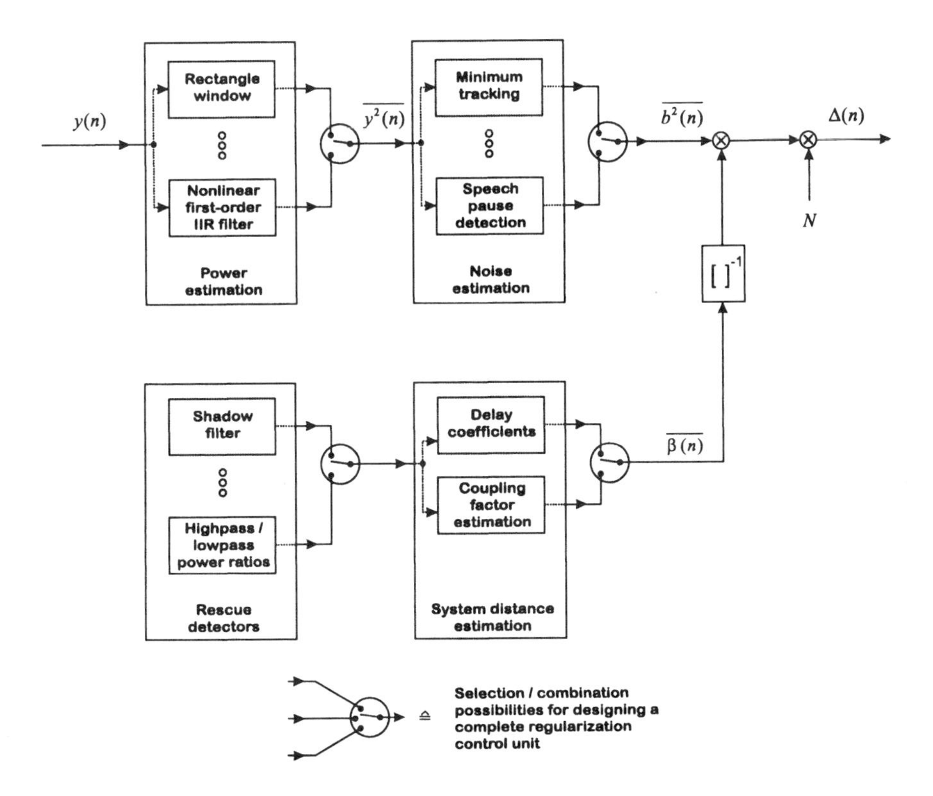

Fig. 13.19 Overview of possibilities for combining several detectors to a complete regularization control unit.

the background noise level, the method presented in Section 13.4.1.2 has been applied. All other quantities and states (system distance, enclosure dislocations, etc.) were estimated and detected, respectively, using the same methods as in the previous simulation example (see Section 13.5.1). Also the remote excitation as well as the local speech signal were the same as in the previous simulation. To simulate a car environment, local background noise—measured at a constant speed of 80 km/h—was added to the microphone signal. The upper two diagrams of Fig. 13.20 show the excitation and the microphone signal. The lower two diagrams show the stepsize (computed according to Eq. 13.67) and the regularization parameter (computed according to Eq. 13.69). Even if the resulting system distance (depicted in the center diagram of Fig. 13.20) does not converge as fast and as deep as in the simulation without background noise, the control strategy results in a robust double-talk behavior (see iterations 30,000–40,000) and in fast tracking of enclosure dislocations (which takes place after iteration 63,000).

13.5.3 Regularization Control for AP Algorithms

In this last section we will give an outlook on how the detection and estimation units presented in Sections 13.4.1–13.4.4 can be utilized to control algorithms faster than NLMS. The standard[6] affine projection (AP) algorithm is utilized as a guiding example.

As already explained earlier in this chapter, a combination of stepsize control and regularization should be applied in the same way for fast algorithms, such as AP or RLS. The latter type of control is rather important because it also numerically stabilizes the convergence process. In Section 13.2.1 a pseudooptimal regularization for the AP algorithm of order p was derived:

$$\Delta_{\mathrm{opt}}(n) \approx p \, \frac{N \, \sigma_x^2(n) \, \sigma_n^2(n)}{\sigma_{e_u}^2(n)} \,, \tag{13.70}$$

(see also Eq. 13.44). The quantity N denotes the length of the adaptive filter. Applying Eq. 13.70 is not possible yet, because neither the short-term power of the undistorted error signal $\sigma_{e_u}^2(n)$ nor the short-term power of the local signal $\sigma_n^2(n) = \sigma_s^2(n) + \sigma_b^2(n)$ is directly measurable. For a practical implementation, these quantities have to be estimated.

For white excitation and statistical independence of the excitation signal vector and the adaptive filter vector, the denominator of Eq. 13.70 can be estimated by

$$\sigma_{e_u}^2(n) = \sigma_x^2(n) \, \mathrm{E}\left\{ \|\boldsymbol{h}_\Delta(n)\|^2 \right\}. \tag{13.71}$$

If pre-whitening according to Section 9.1.1.1 is applied Eq. 13.71 can be used as an approximation for the power of the undistorted error $\sigma_{e_u}^2(n)$. The system distance

[6]The term *standard* should indicate that no fast version (see Section 7.2.2) of the algorithm is applied.

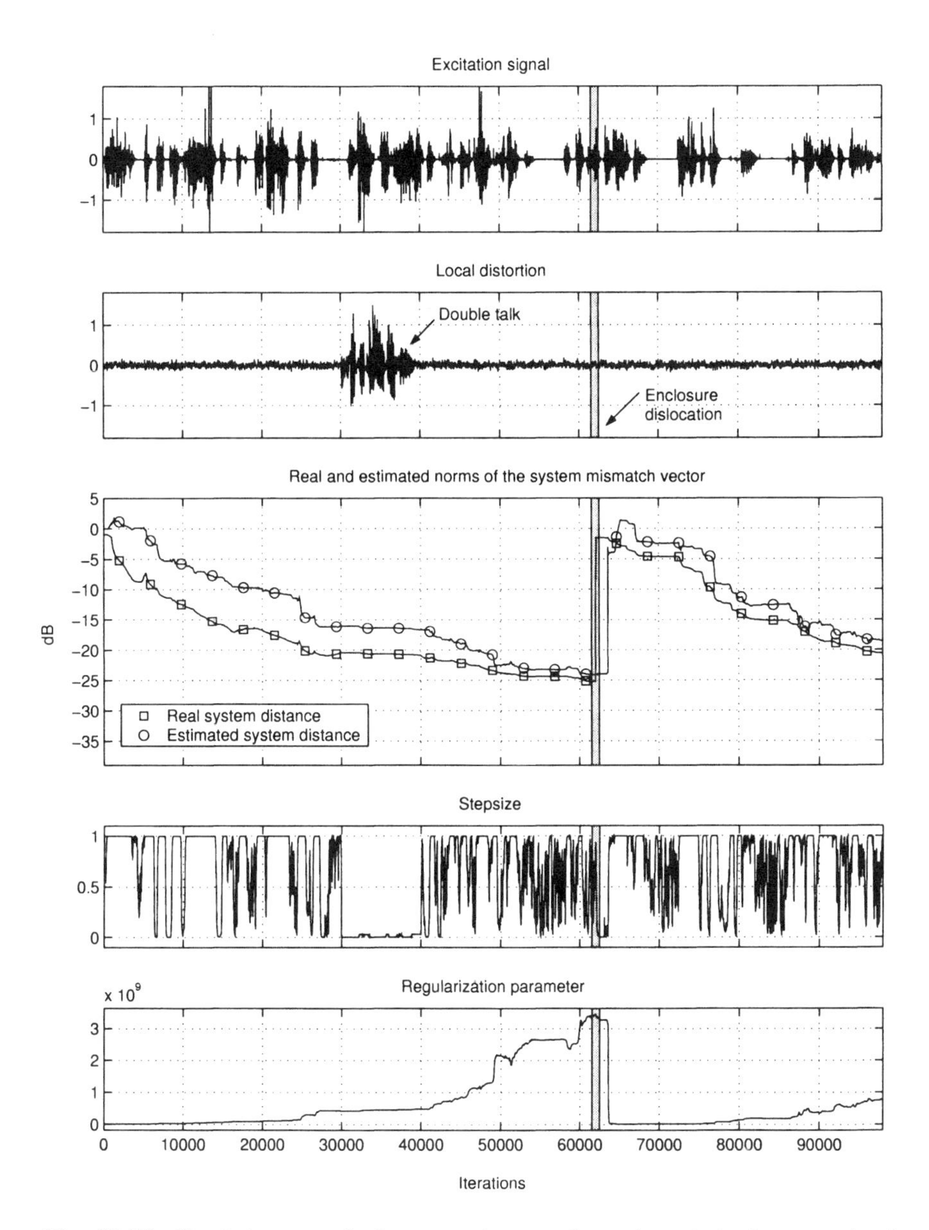

Fig. 13.20 Simulation example for a complete stepsize and regularization control unit. Speech signals were used for excitation as well as for local distortion (see first two diagrams). Additionally, the microphone signal consists of background noise (car noise, 80 km/h). After 62,000 iterations a book was placed between the loudspeaker and the microphone. In the third diagram, the measured and the estimated system distances are depicted. The lowest two diagrams show the stepsize and the regularization parameter; computed according to Eqs. 13.67 and 13.69, respectively (with permission from [102]).

can be estimated using, for instance, the delay coefficients methods or a coupling analysis:

$$\mathrm{E}\left\{\|\boldsymbol{h}_\Delta(n)\|^2\right\} \approx \overline{\beta(n)}\,. \tag{13.72}$$

If orthogonality of all signal sources (far-end and local speaker, as well as the background noise) is assumed, the short-term power of the local signal $\sigma_n^2(n)$ in Eq. 13.70 can be changed to

$$\sigma_n^2(n) = \sigma_s^2(n) + \sigma_b^2(n) = \sigma_e^2(n) - \sigma_{e_u}^2(n)\,. \tag{13.73}$$

Thus, the difference is at least as large as the power of the background noise. This power can be estimated using techniques known from noise reduction (see Section 13.4.1.2). For a practical implementation, we have limited the estimation of the sum of the local signals to the estimated background noise power $\overline{b^2(n)}$:

$$\Delta(n) = p\,\frac{N\,\max\left\{\overline{e^2(n)} - \overline{x^2(n)}\,\overline{\beta(n)},\,\overline{b^2(n)}\right\}}{\overline{\beta(n)}}\,. \tag{13.74}$$

The overbars here denote, as in the previous sections, first-order IIR smoothing. To show the performance of the method, a simulation in a realistic environment is presented. Speech signals were utilized, and for the echo path simulation, an impulse response measured in a car was used. In addition, background noise recorded in a car driven at a constant speed of about 130 km/h was added. Furthermore, fixed first-order prewhitening was applied to make the white noise assumptions more valid.

As depicted in the upper two diagrams of Fig. 13.21 longlasting double-talk situations and a high background noise level were simulated. The resulting regularization parameter $\Delta(n)$ is presented in the third diagram of Fig. 13.21. At the beginning of the simulation, $\Delta(n)$ increases because of the decreasing system distance (estimated using delay coefficients). Meanwhile, the noise estimate in the numerator remains stable. Due to local speech activity, the power of the error signal $\overline{e^2(n)}$ is enlarged, which in turn increases the resulting parameter $\Delta(n)$. An enclosure dislocation was simulated after 50,000 iterations. After detecting this dislocation using a shadow filter as described by Mader et al. [154], the estimated system distance is reinitialized, leading to a decrease of the regularization parameter. In the last diagram of Fig. 13.21 the real system distance is shown. The maximal achievable system distance is around -20 dB, since the implemented filter order ($N = 256$) was distinctively smaller than the simulated echo pathlength ($\tilde{N} = 2048$). The system distance has been computed as

$$\|\boldsymbol{h}_\Delta(n)\|^2 = \sum_{i=0}^{N-1}\left(h_i(n) - \hat{h}_i(n)\right)^2 + \sum_{i=N}^{\tilde{N}-1} h_i^2(n)\,. \tag{13.75}$$

Despite all approximations used in the derivation, the regularization control method works very well; all matrix inversions (projection order $p = 4$) were conveniently stabilized without degrading the adaptation speed.

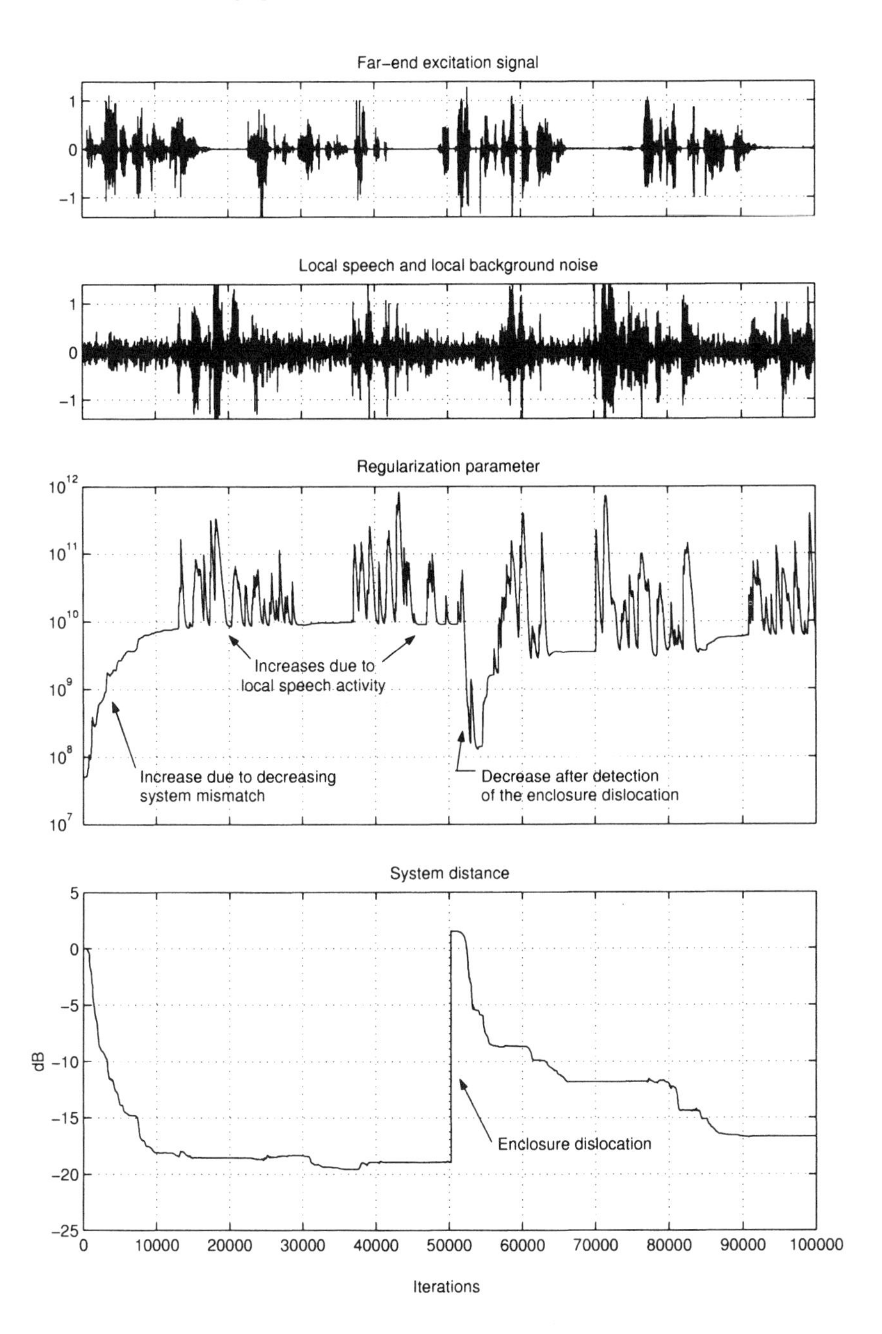

Fig. 13.21 Simulation example for a regularization control unit of an AP algorithm of order $p = 4$. Speech signals were used for excitation as well as for local distortion (see first two diagrams). Additionally, the microphone signal consists of background noise (car noise, 130 km/h). After 50,000 iterations the volume of the loudspeaker was increased by 3 dB. In the third diagram the regularization parameter (computed according to Eq. 13.74) is depicted. The lowest diagram shows the system distance (with permission from [169]).

14

Control of Noise and Echo Suppression Systems

In Chapter 10 we derived algorithms for suppressing background noise as well as residual echoes. During the derivation all necessary quantities such as the power spectral densities of the distorted speech signal $\widehat{S}_{ee}(\Omega)$ and of the background noise $\widehat{S}_{bb}(\Omega)$ were assumed to be known. The latter quantity is not directly measurable in a real scenario. Thus, we will introduce methods for estimating this quantity in this chapter.

Furthermore, several parameters for modifying the derived filter characteristics—such as overestimation and maximal attenuation—were introduced in Chapter 10. If these parameters are chosen adaptively in a time/frequency-selective manner, the quality of a residual echo and noise suppression system can be improved considerably. For this reason, schemes for adjusting the subband overestimation factors and the maximal spectral attenuations will be introduced in Section 14.3.

14.1 ESTIMATION OF SPECTRAL POWER DENSITY OF BACKGROUND NOISE

One of the key elements in a background noise suppression system is the estimation of the short-term power spectral density of the background noise $\widehat{S}_{bb}(\Omega, n)$. This quantity has to be estimated by utilizing the noisy speech signal $y(n)$, the output of a beamformer $u(n)$, or the output of an echo cancellation stage $e(n)$, respectively. Usually, the noise measurement is performed during speech pauses, which have to be detected properly. Estimation schemes based on this simple principle will be described briefly in the first part of this section.

In environments with only a low signal-to-noise ratio, such as speech recordings within a car traveling at high speed, the reliability of speech pause detection is rather low. Thus, a variety of noise estimation schemes that do not require speech pause detection have been proposed. These methods will be described in Sections 14.1.2 – 14.1.4.

14.1.1 Schemes with Speech Pause Detection

As described in Chapter 10, residual echo and noise suppression is usually performed in the frequency or subband domain. We will assume that a polyphase filterbank according to Section 9.1.3.2.3 has been applied to achieve the required frequency resolution. Furthermore, we will assume that the noise suppression is performed after subtracting the estimated echo components (this would be the case if we had the application of hands-free telephony in mind). The complex subband signals that enter the background noise estimation will be denoted with $e_\mu(n)$. If a DFT/IDFT structure is utilized, $e_\mu(n)$ can also be written as the short-term spectrum $E(\Omega_\mu, n)$ at time or frame index n and at the discrete frequencies Ω_μ.

If we assume the background noise $b(n)$ to be stationary, it is possible to estimate its power spectral density in periods without local and far-end speech by first-order IIR smoothing of the squared absolute values of the input subband signals:

$$\widehat{S}_{bb}(\Omega_\mu, n) = \begin{cases} \gamma\, \widehat{S}_{bb}(\Omega_\mu, n-1) + (1-\gamma)\, |e_\mu(n)|^2, & \text{during speech pauses,} \\ \widehat{S}_{bb}(\Omega_\mu, n-1), & \text{else.} \end{cases} \tag{14.1}$$

The time constant γ is chosen much larger than in short-term power estimations $0 \ll \gamma < 1$. For detecting speech pauses, a variety of methods exist. A very simple one is based on the comparison of the broadband short-term power with a threshold. Estimating the broadband power can be performed by summing the squared magnitudes of the subband signals. In order to lower the estimation variance, first-order IIR smoothing can be applied:

$$\hat{\sigma}_e^2(n) = \gamma_e(n)\, \hat{\sigma}_e^2(n-1) + \big(1 - \gamma_e(n)\big) \sum_{\mu=\mu_0}^{M-1} |e_\mu(n)|^2 \,. \tag{14.2}$$

The smoothing is performed such that an increase of the short-term power can be followed faster than a decrease. For this reason, the time constant $\gamma_e(n)$ is chosen as

$$\gamma_e(n) = \begin{cases} \gamma_{e,r}\,, & \text{if } \sum\limits_{\mu=\mu_0}^{M-1} |e_\mu(n)|^2 > \hat{\sigma}_e^2(n-1)\,, \\ \gamma_{e,f}\,, & \text{else}\,. \end{cases} \tag{14.3}$$

To obtain the desired temporal behavior the time constant for increasing short-term power is chosen smaller than the one for decreasing power:

$$0 < \gamma_{e,r} < \gamma_{e,f} < 1\,. \tag{14.4}$$

In many applications a very low signal-to-noise ratio can be expected at low frequencies. Car noise, for example, has a very large power spectral density below 500 Hz (see Section 3.1.2.2). On the other hand, the frequency range containing power of a speech signal starts above the pitch frequency, which is around 200 Hz (on average). For this reason the summations in Eqs. 14.2 and 14.3 start at index $\mu = \mu_0$ instead of $\mu = 0$. The parameter μ_0 is usually set such that only frequencies above 500 Hz are taken into account for speech activity detection.

Speech pauses ($t_{\text{VAD}}(n) = 0$) are detected if the short-term broadband power is below a threshold

$$t_{\text{VAD}}(n) = \begin{cases} 0, & \text{if } \hat{\sigma}_e^2(n) < \sigma_n^2, \\ 1, & \text{else}. \end{cases} \tag{14.5}$$

Instead of a fixed threshold, it is also possible to estimate the broadband background noise level with the method presented in Section 13.4.1.2 and to use this estimate—raised by a few decibels—as an adaptive threshold. Further enhancement can be achieved if the subband error signals $e_\mu(n)$ are weighted according to the expected power spectral density $\tilde{S}_{bb}(\Omega)$ of the background noise

$$\hat{\sigma}_e^2(n) = \gamma_e(n)\,\hat{\sigma}_e^2(n-1) + \big(1 - \gamma_e(n)\big) \sum_{\mu=0}^{M-1} a_\mu \left|e_\mu(n)\right|^2 \tag{14.6}$$

with

$$a_\mu = \frac{M}{\sum\limits_{i=0}^{M-1} \frac{1}{\tilde{S}_{bb}(\Omega_i)}} \, \frac{1}{\tilde{S}_{bb}(\Omega_\mu)}\,. \tag{14.7}$$

This weighting of the subband error signals represents a predictor error filter within the time domain where filter coefficients are matched to the expected autocorrelation properties of the background noise. The first fraction on the right side of Eq. 14.7 was introduced in order to satisfy

$$\sum_{\mu=0}^{M-1} a_\mu = M\,. \tag{14.8}$$

To show the differences of the estimation methods, three short-term power estimations were performed for a noisy speech signal that was recorded (sampling rate $f_s = 8000$ Hz) in a car driven at 170 km/h. The oral statements "I think it would be wonderful. There will be guests coming. She seldom listens to anybody." were made by a male person. Speech periods start after 0.5 second and end at around 5.4 seconds. Significantly large pauses (pauses between sentences) were made between 1.8 and 2.2 seconds and between 3.5 and 3.9 seconds. The signal is depicted in the first diagram of Fig. 14.1. It was analyzed with a filterbank having $M = 256$ channels and a subsampling factor $r = 64$ was utilized.

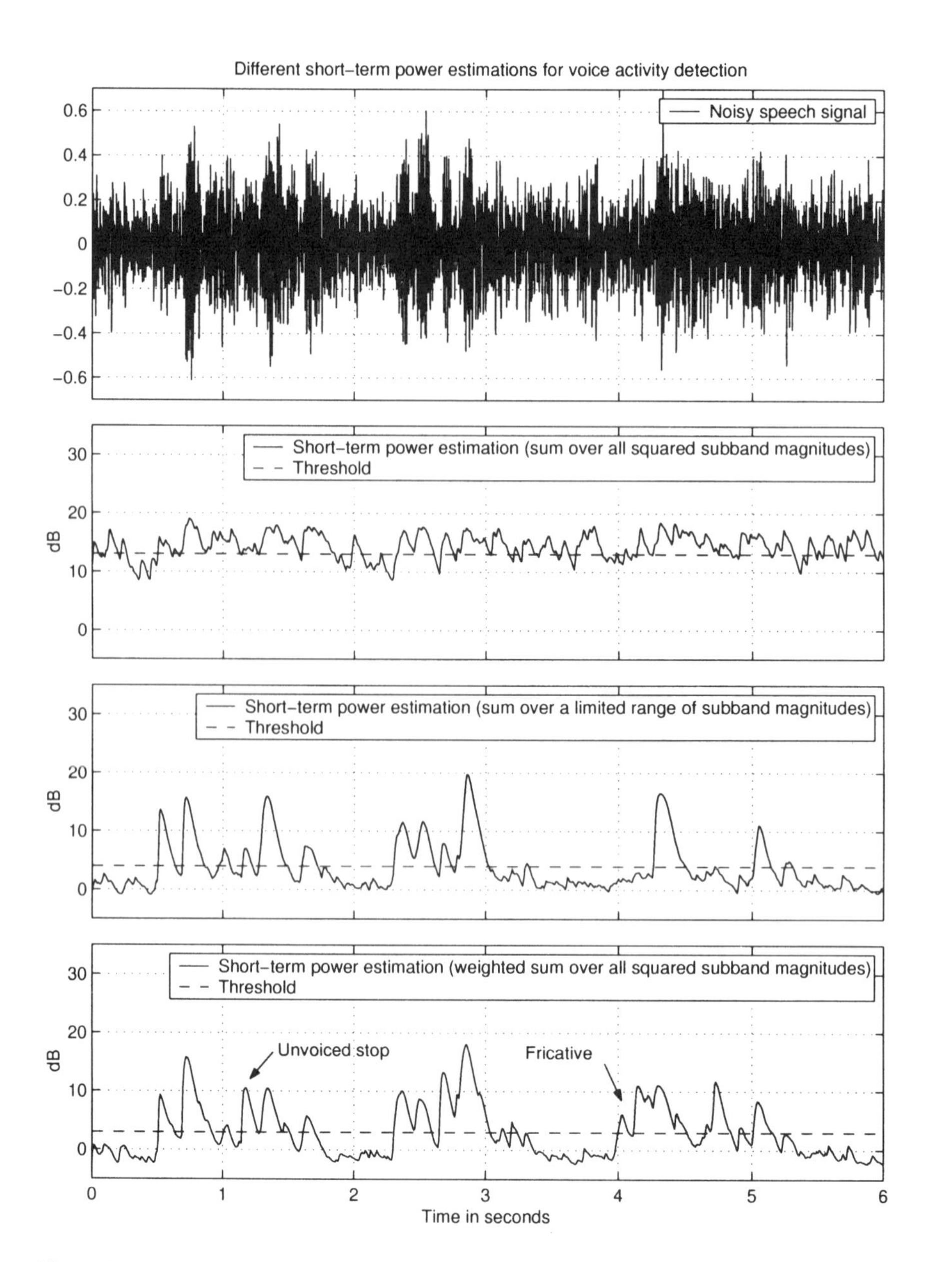

Fig. 14.1 Noisy speech signal and three different short-term power estimations. Three sentences were spoken by a male person in a car driven at 170 km/h. Speech periods start after 0.5 second and end at around 5.4 seconds. Significantly large pauses (pauses between sentences) were made between 1.8 and 2.2 seconds and between 3.5 and 3.9 seconds. The lower three diagrams show different short-term power estimations utilized for voice activity detection.

If a very basic power estimation according to Eq. 14.2 with $\mu_0 = 0$ is applied, it is nearly impossible to distinguish between speech and pause intervals (see second diagram of Fig. 14.1). If the starting index within Eq. 14.2 is set to $\mu_0 = 16$, which corresponds to a frequency of about 500 Hz, most of the subbands with very low signal-to-noise ratios were not considered for voice activity detection. With the resulting short-term power estimation—depicted in the third diagram of Fig. 14.1—a much better distinction between speech and pause intervals (compared to the broadband scheme) is possible. However, because of the large power of the background noise in the low- and medium-frequency subbands (500–1000 Hz), sounds that have most of their energy located at high frequencies, such as fricatives or unvoiced stops, cannot be detected properly. If spectral weighting according to Eqs. 14.6 and 14.7 is applied, this drawback of simple highpass filtering can be avoided. High-frequency sounds like the /t/ in *it* or the /ŝ/ in *she* can now be classified as speech.[1] A significant enhancement can be achieved even if the spectral envelope of the power spectral density $\tilde{S}_{bb}(\Omega)$ is known only approximately.

For each short-term power estimation an individual threshold was plotted in Fig. 14.1. Setting these thresholds in advance is often problematic or even impossible. For this reason it is preferable to analyze each short-term power estimation with the estimation method according to Section 13.4.1.2 and utilize the result (lifted by a few decibels) as a threshold.

A drawback of the method is that the envelope of the background noise power spectral density has to be known in advance. To overcome this problem, a two-stage voice activity detection was proposed [187]. In a first stage a highpass filtered power comparison was applied to obtain a rough classification into voiced and unvoiced segments. During signal intervals classified as "noise-only" ones, the power spectral density of the background noise was estimated by means of a low-order autoregressive model. In a second stage the coefficients of a predictor error filter were adjusted according to autocorrelation properties of the background noise. Furthermore, a hangover mechanism was applied in order to hold a once detected speech activity sequence for an additional short period of time.

Voice activity detection principles based on analysis of the short-term output power of predictor error filters are widely utilized, such as within the GSM full-rate speech codec [63, 65].

To get an impression of the performance of estimation schemes based on voice activity detectors, a simulation example is presented in Fig. 14.2. The noisy speech signal presented in Fig. 14.1 was utilized again. The voice activity detection was performed according to Eq. 14.6 with a fixed threshold of $\sigma_n^2 = 2$ (see lowest diagram in Fig. 14.1). The noisy speech signal was analyzed with a DFT of order $M = 256$. Neighboring frames overlap by 75%. The time constant γ in the first-order IIR smoothing according to Eq. 14.1 was chosen as $\gamma = 0.9$. In the second and third diagrams of Fig. 14.2 the short-term powers and the corresponding noise level

[1] Details of the phonetic description utilized here can be found in the book by Rabiner and Juang [193].

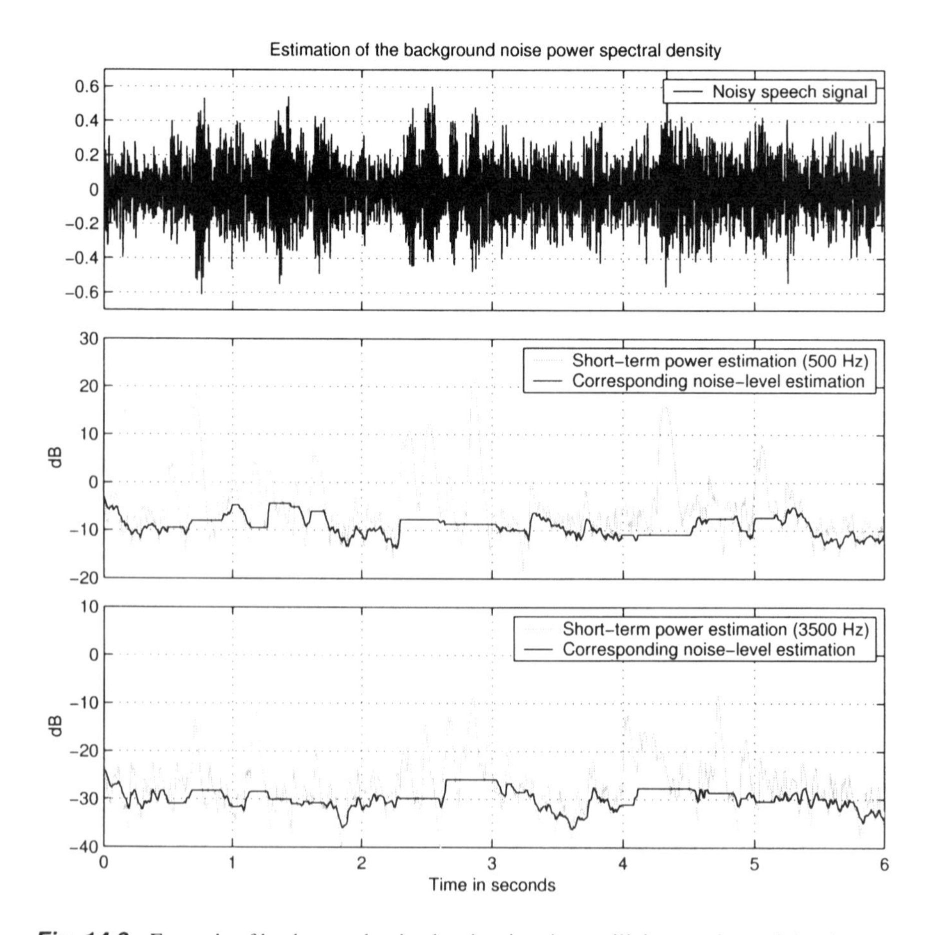

Fig. 14.2 Example of background noise-level estimations utilizing a voice activity detector.

estimates are depicted for a low-frequency subband (500 Hz) and a high-frequency subband (3500 Hz).

Besides power-based speech pause detection schemes, a variety of other methods has been proposed [189]:

- The *zero-crossing rate* can be monitored for distinguishing between speech and noise-only periods [121]. In this time-domain approach the number of neighboring sample pairs $\{e(n), e(n-1)\}$ with equal signs is compared to the number of pairs with different signs. The analysis is usually performed on short signal frames having a duration of about 20 ms. The zero-crossing criterion depends crucially on the type of noise. While white noise has a very large zero-crossing rate (larger than speech), typical car noises have very low zero-crossing rates (due to the lowpass characteristic).

- Another method for detecting speech pauses is to utilize the property that noise exhibits a lower fluctuation than does speech. For this reason, the fluctuations of the spectral envelope are monitored. This can be done by computing the *line spectral frequencies* [170], denoted by $\Omega_{\text{LSF},i}(n)$, from the coefficients of a low-order predictor error filter (see Chapter 6). Each line spectral frequency is smoothed, for example, by first-order IIR filtering. During voiced periods large fluctuations of the parameters $\Omega_{\text{LSF},i}(n)$ appear. Due to the stationary behavior of car noise, for instance, the line spectral frequencies do not change largely within noise-only periods. Thus, speech pauses can be detected by analyzing the sum of the differences between the current $\Omega_{\text{LSF},i}(n)$ and the smoothed line spectral frequencies $\overline{\Omega_{\text{LSF},i}(n)}$:

$$\Delta_{\text{LSF}}(n) = \sum_i \left(\Omega_{\text{LSF},i}(n) - \overline{\Omega_{\text{LSF},i}(n)} \right)^2 . \tag{14.9}$$

  Whenever $\Delta_{\text{LSF}}(n)$ is smaller than a threshold Δ_0, speech pauses are detected and the estimation of the power spectral density of the background noise is updated.

Even with sophisticated speech pause detections the principle has several drawbacks. Usually, the classification of signal frames into voiced ones and ones containing only background noise is done in a broadband manner. During voiced periods a signal consists mainly of frequency components at the pitch frequency and its harmonics. Thus, the background noise level can be estimated—assuming sufficiently high frequency selectivity of the analysis filterbank—in between the fundamental frequency and its first harmonic, between the first and the second harmonics, and so forth. Furthermore, the classification into voiced and pause segments is rather complicated in case of very low signal to noise ratios. For this reason, several schemes that do not require an explicit voice activity (or pause) detection have been proposed. Examples of such schemes will be presented in the next few sections.

14.1.2 Minimum Statistics

One of these schemes is called *minimum statistics* [159]. This approach is based on the fact that speech does not occupy the entire time–frequency space even during voiced periods. This means that on one hand short (temporal) pauses appear between words and even between syllables. On the other hand, voiced intervals have spectral gaps between the harmonics of the pitch frequency. These spectral and temporal gaps can be exploited to estimate the power spectral density of the noise by means of tracking the minima within the subbands for limited blocks of length N_{MS}. Because of the temporal limitation on N_{MS} frames for the minimum search, it is possible to also track increasing noise levels. In order to obtain estimates with small error variance, temporal smoothing has to be applied before the minimum tracking can be performed. This is usually achieved by first-order IIR smoothing of the squared magnitudes of the subband input signals:

$$\overline{|e_\mu(n)|^2} = \gamma_e \, \overline{|e_\mu(n-1)|^2} + (1-\gamma_e) \, |e_\mu(n)|^2 . \tag{14.10}$$

One way to track the subband minima would be a direct search over the last N_{MS} output signals of the smoothing filters:

$$\widehat{S}_{bb}(\Omega_\mu, n) = \min\left\{\overline{|e_\mu(n)|^2}, \ldots, \overline{|e_\mu(n - N_{\mathrm{MS}})|^2}\right\}. \tag{14.11}$$

However, this direct approach would require a large memory (N_{MS} memory words for each subband), and also the processing power required for the minimum search would be very high. For this reason, the search memory is split into a small number, usually 4–8, of subblocks having a smaller size of only N_{B} frames (see Fig. 14.3).

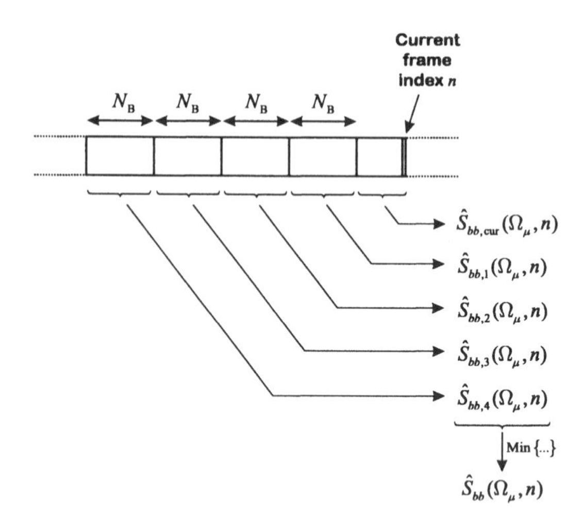

Fig. 14.3 Basic structure of a minimum statistic.

For each subblock, the minimum is stored within the variables $\widehat{S}_{bb,i}(\Omega_\mu, n)$. The background noise level within subband μ is estimated by tracking the minimum of the subblocks $\widehat{S}_{bb,i}(\Omega_\mu, n)$ and of the current block $\widehat{S}_{bb,\mathrm{cur}}(\Omega_\mu, n)$. The latter is computed by tracking the minimum of the smoothed short-term power:

$$\widehat{S}_{bb,\mathrm{cur}}(\Omega_\mu, n) = \min\left\{\widehat{S}_{bb,\mathrm{cur}}(\Omega_\mu, n-1), \overline{|e_\mu(n)|^2}\right\}, \quad \text{if } n \bmod N_{\mathrm{B}} \neq 0. \tag{14.12}$$

Whenever N_{B} new frames have entered the tracking process, the quantity is reinitialized with the current short-term power:

$$\widehat{S}_{bb,\mathrm{cur}}(\Omega_\mu, n) = \overline{|e_\mu(n)|^2}, \quad \text{if } n \bmod N_{\mathrm{B}} \equiv 0. \tag{14.13}$$

The minima of the previous blocks, which are stored within the variables $\widehat{S}_{bb,i}(\Omega_\mu, n)$, are updated also only every N_{B} frames. At these indices the minimum of the current block moves into $\widehat{S}_{bb,1}(\Omega_\mu, n)$, the minimum of subblock i will be passed to the subblock with index $i+1$ and the oldest minimum will be discarded:

$$\widehat{S}_{bb,1}(\Omega_\mu, n) = \begin{cases} \widehat{S}_{bb,\mathrm{cur}}(\Omega_\mu, n-1) & \text{if } n \bmod N_\mathrm{B} \equiv 0\,, \\ \widehat{S}_{bb,1}(\Omega_\mu, n-1), & \text{else}\,, \end{cases} \tag{14.14}$$

$$\widehat{S}_{bb,2}(\Omega_\mu, n) = \begin{cases} \widehat{S}_{bb,1}(\Omega_\mu, n-1) & \text{if } n \bmod N_\mathrm{B} \equiv 0\,, \\ \widehat{S}_{bb,2}(\Omega_\mu, n-1), & \text{else}\,, \end{cases} \tag{14.15}$$

$$\dots$$

$$\widehat{S}_{bb,4}(\Omega_\mu, n) = \begin{cases} \widehat{S}_{bb,3}(\Omega_\mu, n-1) & \text{if } n \bmod N_\mathrm{B} \equiv 0\,, \\ \widehat{S}_{bb,4}(\Omega_\mu, n-1), & \text{else}\,. \end{cases} \tag{14.16}$$

Because the minima, which are stored within the variables $\widehat{S}_{bb,i}(\Omega_\mu, n)$, have already been computed, the global minima $\widehat{S}_{bb}(\Omega_\mu, n)$ can be obtained in a very simple manner with only one compare instruction per frame and subband:

$$\widehat{S}_{bb}(\Omega_\mu, n) = \min\left\{\widehat{S}_{bb,\mathrm{cur}}(\Omega_\mu, n),\, \widehat{S}_{bb}(\Omega_\mu, n-1)\right\}, \quad \text{if } n \bmod N_\mathrm{B} \not\equiv 0\,. \tag{14.17}$$

Only when an old subblock is discarded and a new one is opened ($n \bmod N_\mathrm{B} \equiv 0$), the global minimum has to be updated according to

$$\widehat{S}_{bb}(\Omega_\mu, n) = \min\left\{\widehat{S}_{bb,1}(\Omega_\mu, n),\, \widehat{S}_{bb,2}(\Omega_\mu, n),\, \dots,\, \widehat{S}_{bb,i_{\max}}(\Omega_\mu, n)\right\}, \quad \text{if } n \bmod N_\mathrm{B} \equiv 0\,. \tag{14.18}$$

The estimation method described by Eqs. 14.12–14.18 is a very efficient way of computing a slightly modified version of Eq. 14.11. The modification is to change the fixed memory size N_MS by a slowly time varying one $N_\mathrm{MS}(n)$:

$$N_\mathrm{MS}(n) = i_{\max}\, N_\mathrm{B} + n \bmod N_\mathrm{B}\,. \tag{14.19}$$

To show the performance of this type of background noise estimation procedure a simulation is presented in Fig. 14.4. An input signal (sampling frequency = 8 kHz) consisting of speech and car noise is decomposed periodically using a 256-point FFT. A Hamming window is utilized and consecutive frames are overlapping by 75%.

Figure 14.4 depicts the short-term power $\overline{|e_\mu(n)|^2}$ and the corresponding background noise level estimation $\widehat{S}_{bb}(\Omega_\mu, n)$ of the subband covering the frequency range around 1250 Hz. For smoothing the squared magnitudes of the subband input signal, a time constant $\gamma_e = 0.6$ has been utilized. Six subblocks have been implemented to track the subband minima. Every $N_\mathrm{B} = 24$ frames, the minima were shifted by one block and the oldest block minimum was discarded. Assuming a frameshift of $r = 64$ samples and a sampling rate of $f_\mathrm{s} = 8000$ Hz, the total memory time (see Eq. 14.19) is between 1.15 and 1.34 seconds.

The memory size has been chosen rather small in this example—typically memory sizes of about 2 seconds or more are utilized. However, with this small length the

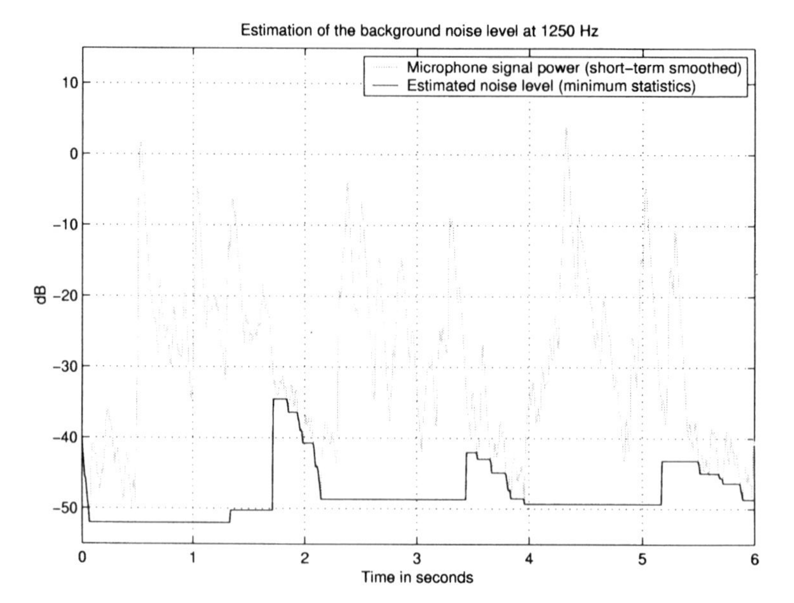

Fig. 14.4 Short-term power estimation and corresponding background noise-level estimation based on minimum statistics.

advantages and disadvantages of this procedure clearly turn out. First, the method works well even without any speech pause detection. The minima of the subband short-term powers are tracked very well (see Fig. 14.4). On the other hand, with this basic version of minimum statistics the estimates of the noise levels will be biased (caused by tracking the minima and not the average power of the noise). Furthermore, an increase of the estimated noise level is realized by a sudden increase. This appears whenever the global minimum leaves the last block $S_{bb,i_{\max}}(\Omega_\mu, n)$. This is rather unproblematic for noise suppression units but audible whenever the estimation is utilized within a comfort noise injection unit.

To overcome these drawbacks, several extensions to the basic version of the minimum statistic have been proposed. The extensions consist of utilizing an adaptive time constant $\gamma_e \rightarrow \gamma_e(n)$ within Eq. 14.10 and computing a bias compensation. The interested reader is referred to Martin's article [161].

14.1.3 Scheme with Fast and Slow Estimators

One of the simplest schemes—even simpler than minimum statistics—for estimating the power spectral density of the background noise without speech pause detection is first-order IIR smoothing with different time constants for increasing and decreasing amplitudes:

$$\widehat{S}_{bb}(\Omega_\mu, n) = \gamma_\mu(n)\, \widehat{S}_{bb}(\Omega_\mu, n-1) + \big(1 - \gamma_\mu(n)\big)\, |e_\mu(n)|^2 \tag{14.20}$$

with

$$\gamma_\mu(n) = \begin{cases} \gamma_\mathrm{r}, & \text{if } |e_\mu(n)|^2 > \widehat{S}_{bb}(\Omega_\mu, n-1), \\ \gamma_\mathrm{f}, & \text{else.} \end{cases} \tag{14.21}$$

The time constant for rising signal edges γ_r is chosen much larger than the one for falling signal edges γ_f:

$$1 > \gamma_\mathrm{r} > \gamma_\mathrm{f} > 0\,. \tag{14.22}$$

With this setup the estimator follows a decrease of the short-term power very quickly. For this reason, the estimation is decreased to the subband background noise level whenever a speech pause appears. During speech activity the estimation is increased very slowly. Thus, the success of this scheme depends on the average duration of speech activity periods. The closer γ_r is chosen to one, the longer the speech activity period can be. However, if the increasing behavior is too slow, it also takes very long before an increase of the background noise level can be followed. For a sampling rate of $f_\mathrm{s} = 8000$ Hz a time constant pair

$$\gamma_\mathrm{r} = 0.9995 \tag{14.23}$$

$$\gamma_\mathrm{r} = 0.9 \tag{14.24}$$

might be a good choice. A simulation example with these values is presented in Section 14.1.5.

14.1.4 Extensions to Simple Estimation Schemes

Besides the two basic estimation schemes presented earlier, a variety of extensions and modifications has been suggested. Two of them will be discussed in the following.

14.1.4.1 Minimum Tracking

A combination of minimum tracking and IIR smoothing has been presented [51]. As in the case of minimum statistics, first a smoothed version of the input power is computed according to

$$\overline{|e_\mu(n)|^2} = \gamma_e\, \overline{|e_\mu(n-1)|^2} + (1-\gamma_e)\, |e_\mu(n)|^2\,. \tag{14.25}$$

In a second stage the estimation of the short-term power spectral density $\widehat{S}_{bb}(\Omega_\mu, n)$ is performed. Whenever the short-term power $\overline{|e_\mu(n)|^2}$ is smaller than $\widehat{S}_{bb}(\Omega_\mu, n-1)$, the estimation follows this decrease immediately:

$$\widehat{S}_{bb}(\Omega_\mu, n) = \overline{|e_\mu(n)|^2}, \ \text{if } S_{bb}(\Omega_\mu, n-1) > \overline{|e_\mu(n)|^2}\,. \tag{14.26}$$

In the other case ($S_{bb}(\Omega_\mu, n-1) \leq \overline{|e_\mu(n)|^2}$) IIR smoothing with a large time constant is applied. In contrast to the smoothing method presented in Section 14.1.3, a weighted difference $\Delta_e(\Omega_\mu, n)$ of the current and the previous short-term power is smoothed:

$$\Delta_e(\Omega_\mu, n) = \max\left\{\frac{1}{1-\beta}\left(\overline{|e_\mu(n)|^2} - \beta\,\overline{|e_\mu(n-1)|^2}\right),\, 0\right\}. \tag{14.27}$$

The maximum operator is applied to avoid negative outputs. The smoothing process is performed using a first-order IIR filter:

$$\begin{aligned}\widehat{S}_{bb}(\Omega_\mu, n) &= \gamma_{\mathrm{mt}}\,\widehat{S}_{bb}(\Omega_\mu, n-1) + (1-\gamma_{\mathrm{mt}})\Delta_e(\Omega_\mu, n), \\ &\qquad \text{if } S_{bb}(\Omega_\mu, n-1) \leq \overline{|e_\mu(n)|^2}.\end{aligned} \tag{14.28}$$

Because of smoothing $\Delta_e(\Omega_\mu, n)$ instead of $\overline{|e_\mu(n)|^2}$, it is possible to adjust the power spectral density of the background noise more quickly to a new noise scenario. Typical parameter selections for a sampling rate of $f_s = 8000$ Hz are $\gamma_e = 0.6$, $\beta = 0.95$, and $\gamma_{\mathrm{mt}} = 0.995$. An example for this type of estimation is presented in Section 14.1.5.

14.1.4.2 Smoothing along the Frequency Axis

Methods that track the subband minima show the drawback that close to zero amplitude outliers introduce a large bias. To overcome this problem temporal smoothing is applied within the schemes presented until now (see Eqs. 14.10 and 14.25). In order to reduce these outliers, considerably large smoothing constants need to be implemented. In this case, however, the estimated short-term power does not decrease to the background noise level during short speech pauses as they appear in between words or syllables.

Another possibility to reduce outliers is to apply—in addition to temporal filtering—smoothing along the frequency axis. This could be achieved by first-order IIR filtering in positive and negative frequency directions. Smoothing in positive frequency direction can be described by

$$\overline{|\breve{e}_\mu(n)|^2} = \begin{cases} \overline{|e_\mu(n)|^2}, & \text{if } \mu = 0, \\ \gamma_{\mathrm{f}}\,\overline{|\breve{e}_{\mu-1}(n)|^2} + (1-\gamma_{\mathrm{f}})\,\overline{|e_\mu(n)|^2}, & \text{if } \mu = 1, \ldots, M-1. \end{cases} \tag{14.29}$$

To avoid a biased estimation, IIR filtering with the same time constant γ_{f} should be applied in the opposite direction (again along the frequency axis):[2]

$$\overline{|\tilde{e}_\mu(n)|^2} = \begin{cases} \overline{|\breve{e}_\mu(n)|^2}, & \text{if } \mu = M-1, \\ \gamma_{\mathrm{f}}\,\overline{|\tilde{e}_{\mu+1}(n)|^2} + (1-\gamma_{\mathrm{f}})\,\overline{|\breve{e}_\mu(n)|^2}, & \text{if } \mu = 0, \ldots, M-2. \end{cases} \tag{14.30}$$

To show the effect of frequency smoothing, the output signals of the short-term temporal smoothing $\overline{|e_\mu(n)|^2}$ and the resulting frequency smoothed signals $\overline{|\tilde{e}_\mu(n)|^2}$ (within a noise-only frame) are depicted in Fig. 14.5. The outliers (e.g., around

[2]Note that the results of the frequency smoothing $\overline{|\tilde{e}_\mu(n)|^2}$ are not copied into the memory of the *temporal* IIR filters. In this case the estimation process would become much too slow.

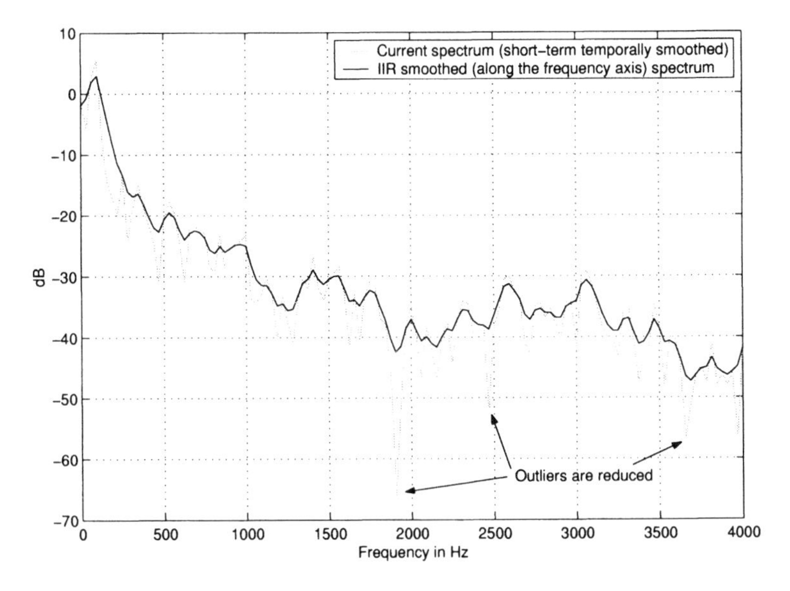

Fig. 14.5 Short-term smoothed subband powers and corresponding frequency-smoothed counterparts.

1850 Hz) can be reduced significantly, leading to a reduction of the estimation variance within the background noise-level estimation process. The larger the time constant γ_{f} is chosen, the larger will be this reduction. However, this is true only for spectrally flat (white) noise spectra. In case of colored noise, the smoothing constant γ_{f} should be chosen such that the spectral envelope can be followed without significant bias.

To visualize the effect of smoothing along the frequency axis in more detail three short-term power estimations within one subband (located around 500 Hz) are depicted in Fig. 14.6. The upper diagram shows the subband amplitudes without any smoothing. If the minimum of the amplitudes will be tracked directly, a large bias would occur (due to the low-amplitude outliers). The second diagram shows the output of a short-term power estimation according to Eq. 14.10 or 14.25 with a time constant[3] $\gamma_e = 0.5$. Due to the temporal smoothing, both types of outliers—negative and positive ones—can be reduced. However, if smoothing along the frequency axis is applied in addition to the time-domain filtering (depicted in the lowest diagram of Fig. 14.6), an even smoother behavior can be achieved.

The resulting subband quantities $\overline{|\tilde{e}_\mu(n)|^2}$ can be utilized for estimating the power spectral density of the background noise by applying the broadband estimation scheme presented in Section 13.4.1.2 for each subband:

$$\widehat{S}_{bb}(\Omega_\mu, n) = \min\left\{\overline{|\tilde{e}_\mu(n)|^2},\ \widehat{S}_{bb}(\Omega_\mu, n-1)\right\}\,(1+\epsilon)\,. \qquad (14.31)$$

[3]The sampling rate was $f_{\mathrm{s}} = 8000$ Hz, and a subsampling rate $r = 64$ was chosen.

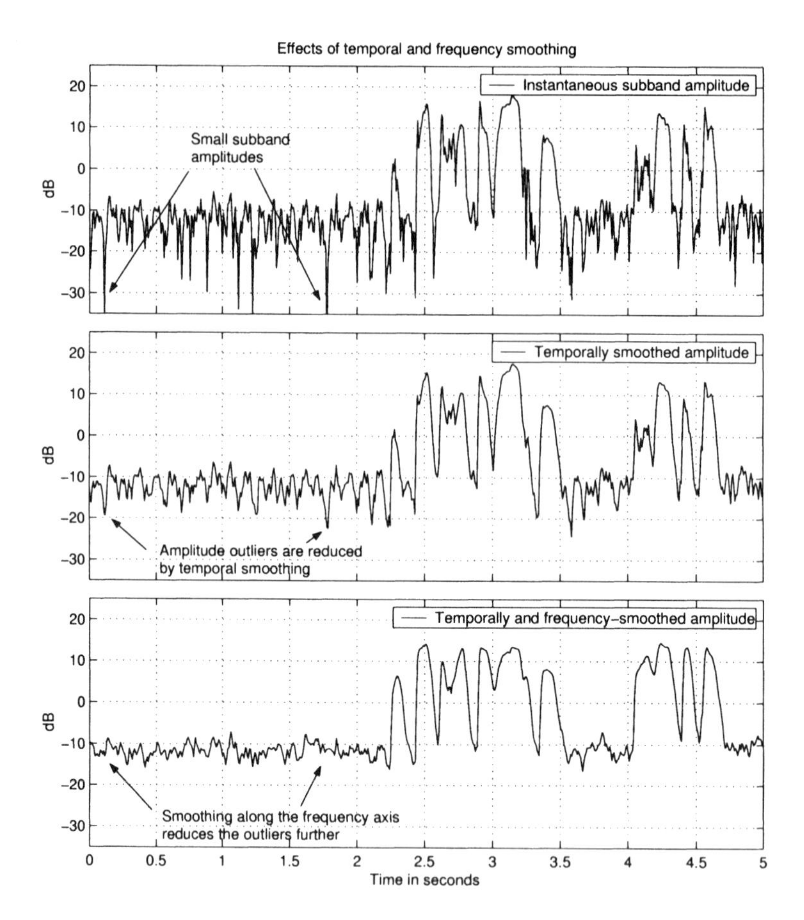

Fig. 14.6 Effects of temporal filtering and smoothing along the frequency direction. In the upper diagram the instantaneous amplitude of one of the subbands (around 500 Hz) of a noisy speech signal is depicted. If temporal IIR smoothing with a time constant $\gamma_e = 0.5$ is applied, the second graph is obtained. Further smoothing along the frequency axis can reduce outlier amplitudes even more (depicted in the lowest diagram).

For a sampling rate $f_s = 8000$ Hz and a subsample rate $r = 64$, the parameter ϵ is chosen usually from the interval [0.001, 0.01] (depending on the desired speed for increasing the estimation). Note that this method is applicable only because of the time and frequency smoothing. Without it a very large bias would appear. However, frequency smoothing can be applied only if the long-term power spectral density of the background noise has no deep notches. Fortunately, most types of car and office noises exhibit a broadband envelope without notches. An example for the noise estimation according to Eq. 14.31 will be presented in the next section.

14.1.5 Concluding Remarks

Before we finish the description of schemes for estimating the power spectral density of the background noise, some concluding remarks are given. When we compare the different approaches (see Fig. 14.7), it is not possible to name the best scheme. Which method is most appropriate for a certain application depends crucially on the boundary conditions of that specific application.

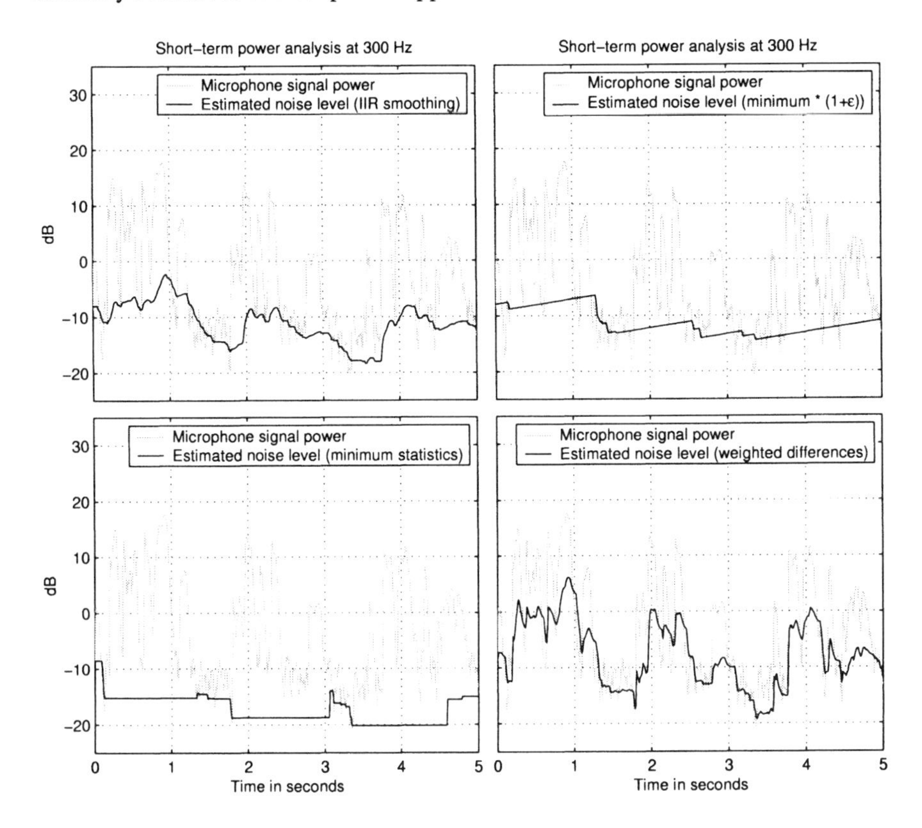

Fig. 14.7 Comparison of different noise estimation schemes. Four of the estimation schemes (upper left—scheme according to Section 14.1.3; upper right—frequency smoothing according to Section 14.1.4.2; lower left—minimum statistics according to Section 14.1.2; lower right—minimum tracker according to Section 14.1.4.1) were utilized to estimate the power of the background noise in a subband located around 300 Hz. The parameters of each method were adjusted as defined in the corresponding sections.

If $\widehat{S}_{bb}(\Omega_\mu, n)$ needs to be estimated for the purpose of noise reduction for an automotive application, it is rather important that the estimate can follow changes of the background noise level (e.g., caused by a change of the road surface, or after entering or leaving a tunnel) rather quickly. Increases of the subband noise power estimates during transitions from noise-only to speech can be tolerated up to a certain extent, because the attenuation during speech periods is rather small and the noise is masked by the speech components. However, if the estimation is utilized for a comfort noise injection, it is rather important that $\widehat{S}_{bb}(\Omega_\mu, n)$ changes slowly because power modulations within intervals of comfort noise injection are very disturbing for the listener.

Estimating the background noise level is a very important subunit within a speech processing system. Thus, the estimation method should be chosen carefully and often the standard methods as presented in this chapter need to be extended to fulfill the requirements of the specific application.

14.2 MUSICAL NOISE

During speech pauses the input signal of a noise suppression system consists only of background noise. Thus, the "theoretical" signal-to-noise ratio

$$SNR(\Omega) = \frac{S_{ss}(\Omega)}{S_{bb}(\Omega)} = \frac{S_{ee}(\Omega) - S_{bb}(\Omega)}{S_{bb}(\Omega)} \tag{14.32}$$

is zero and the "theoretical" attenuation of a noise suppression system should be infinitely large. In practice, the power spectral densities are estimated within each frame with one of the methods presented above. As we have seen in the previous sections, the estimation of $S_{ee}(\Omega)$ is able to change much faster than the estimation of $S_{bb}(\Omega)$. As a result, the attenuation factors—computed according to one of the filter characteristics presented in Section 10.1.2—are fluctuating randomly.

In Fig. 14.8 the estimated background noise level and the estimated signal level are depicted at a certain frequency (around 1000 Hz). The analyzed signal is background noise measured in a car driven at about 100 km/h. The estimated power spectral density of the distorted input signal $\widehat{S}_{ee}(\Omega, n)$ fluctuates much faster than does the background noise estimation $\widehat{S}_{bb}(\Omega, n)$. For the estimation, minimum statistics have been applied. In order to compensate the bias caused by the estimation method, the subband minima were raised by 4 dB. If the attenuation factors are computed according to the Wiener characteristic (see lower part of Fig. 14.8), random openings of the attenuation can be observed.

Without any modification, most of the filter characteristics presented in Section 10.1.2 transform broadband background noise into a signal consisting of short-duration tones with randomly distributed frequencies. This remaining signal is called *musical noise* or *musical tones* [232]. Each filter characteristic presented in Section 10.1.2 shows a different sensibility to this phenomenon. In the lowest two diagrams of Fig. 14.8 the Wiener characteristic and the one according to Ephraim and Malah [61, 62] are depicted. As we can see, the attenuation fluctuates much larger within the Wiener approach and produces more musical noise than the method proposed by Ephraim and Malah.

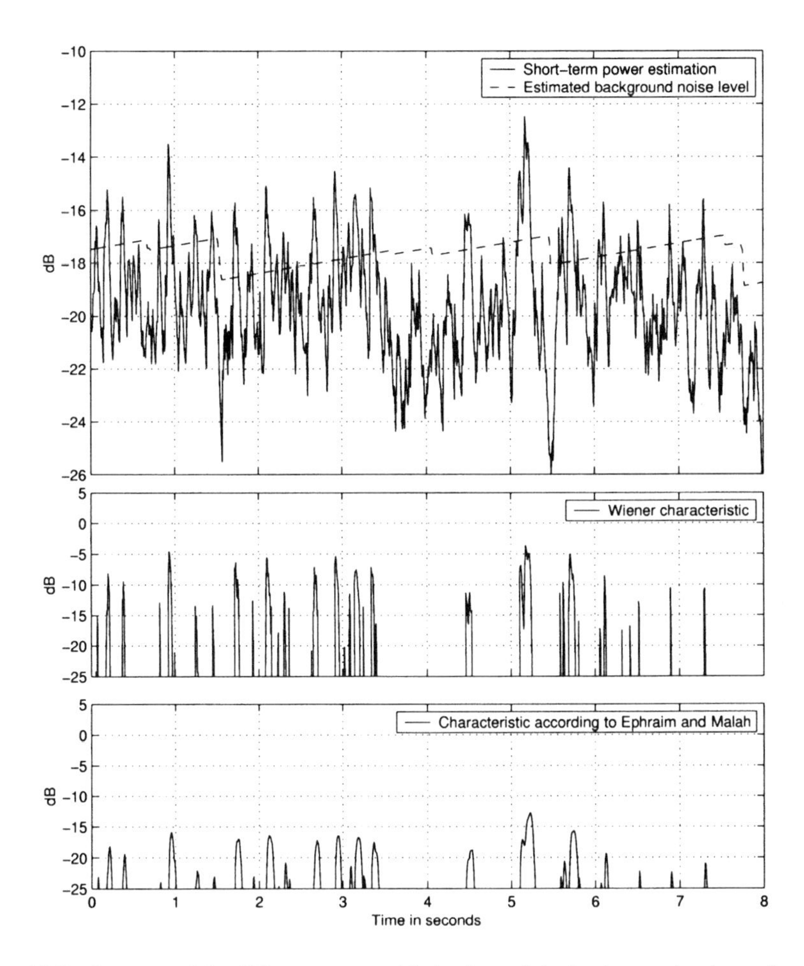

Fig. 14.8 Because of the different temporal behaviors of the background noise estimation and the short-term power of the distorted subband signals random openings of most filter characteristics (center diagram—Wiener characteristic; lowest diagram—characteristic according to [61]) can be observed. The remaining signal is called *musical noise*.

14.3 CONTROL OF FILTER CHARACTERISTICS

To reduce the effects of musical noise several approaches have been proposed. As explained in Chapter 10, the most common filter modifications are the introduction of an attenuation limit—called *spectral floor*—and the application of a multiplicative overestimation for the estimated subband noise levels.

If the attenuation factors in the lower two diagrams of Fig. 14.8 would be limited to -12 dB the fluctuation within the Ephraim–Malah characteristic would disappear completely. However, even a limitation of -12 dB would not be sufficient for the Wiener approach.

In practice a tradeoff between speech quality and residual noise level has to be found. If the overestimation parameter and the spectral floor are adjusted such that no musical noise appears, the speech quality is often degraded. An improvement can be achieved if the parameters of the filter characteristics are adjusted according to the current state (speech or pause) of the system [145, 185]. Examples for such adaptively controlled characteristics are presented in the next two sections.

14.3.1 Overestimation Factor

A very simple but also very effective method to reduce musical noise is to apply a recursion on the attenuation factors $H(e^{j\Omega_\mu}, n)$. The old attenuation factor $H(e^{j\Omega_\mu}, n-1)$ in subband μ is coupled back (inversely) into the current computation of the attenuation [145]. This recursion can be interpreted as an adaptive overestimation factor:

$$\beta \rightarrow \frac{\tilde{\beta}}{H(e^{j\Omega_\mu}, n-1)} . \tag{14.33}$$

The recursion can be applied to all of the filter characteristics presented in Section 10.1.2. In case of the Wiener characteristic, for instance, the modification can be denoted as

$$H\left(e^{j\Omega_\mu}, n\right) = \max\left\{H_{\text{min}}, 1 - \frac{\tilde{\beta}}{H(e^{j\Omega_\mu}, n-1)} \frac{\widehat{S}_{bb}\left(\Omega_\mu, n\right)}{\widehat{S}_{ee}\left(\Omega_\mu, n\right)}\right\} . \tag{14.34}$$

One result of the recursion is that the attenuation is fixed to the value H_{min} if the signal-to-noise ratio is rather low, or to be precise, if the input-to-noise ratio

$$INR(\Omega_\mu, n) = \frac{\hat{S}_{ee}\left(\Omega_\mu, n\right)}{\hat{S}_{bb}\left(\Omega_\mu, n\right)} \tag{14.35}$$

is below a threshold. The switch over to the maximum attenuation H_{min} appears suddenly. Thus, the filter characteristic consists of two parts:

- A part with constant attenuation
- Another part with a filter characteristic closely related to one of the standard attenuation curves

Due to the recursion of the attenuation factor, the filter characteristic results in a hysteresis. Fig. 14.9 shows the resulting gain curves of the standard Wiener characteristic and its recursive counterpart. Both curves have been calculated for a first slowly increasing and later slowly decreasing signal-to-noise ratio.

It is important to note that the recursive filter characteristic depends on the limitation to a maximum attenuation. For typical values

$$H_{\text{min}} \in [0.1,\, 0.5] \triangleq [-20\,\text{dB},\, -6\,\text{dB}] \tag{14.36}$$

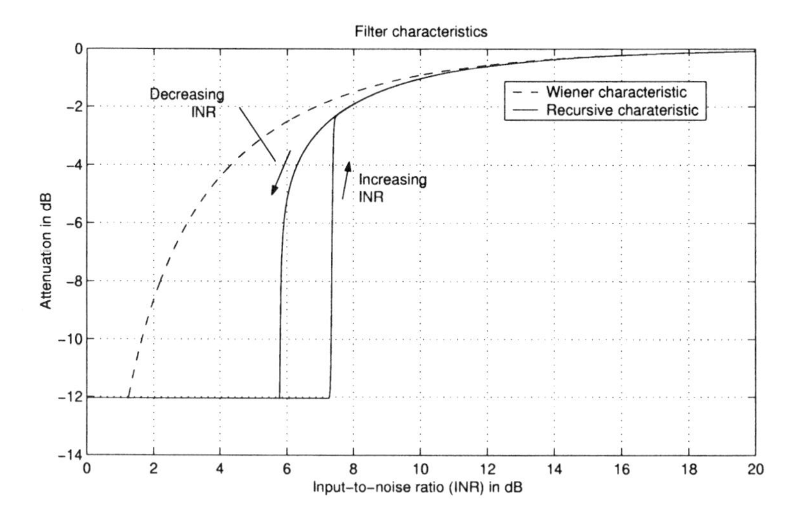

Fig. 14.9 Standard Wiener characteristic and its recursive counterpart. Both curves have been calculated for an initially slowly increasing and later slowly decreasing input-to-noise ratio (INR). No additional overestimation ($\beta = \tilde{\beta} = 1$) was applied, and the maximum attenuation was set to $H_{\text{min}} = 0.25$.

the characteristic is able to follow the onset of speech components with only a short delay. Without the limitation the filter characteristic would never leave the once reached value $H(e^{j\Omega_\mu}, n) = 0$. The disadvantage of this method is that signal components with only very low signal-to-noise ratios are attenuated. Nevertheless, for medium and even low signal-to-noise ratios the recursion is able to reduce musical noise properly.

14.3.2 Spectral Floor

When adjusting the spectral floor H_{min}, which is the maximum allowed subband attenuation, two different demands should be fulfilled:

- During speech periods, the quality of the enhanced signal is of special importance. On the other hand, the amount of residual noise is of only minor importance in such periods (the speech components usually mask most of the residual noise). To obtain optimal quality, only little attenuation should be inserted. Thus, a high spectral floor should be utilized.

- In speech pauses the residual noise is not masked. Thus, a larger amount of attenuation is desired and a lower spectral floor should be applied.

If both demands should be fulfilled at the same time, the spectral floor needs to be adjusted in a time/frequency-selective manner:

$$H_{\text{min}} \longrightarrow H_{\text{min}}\left(e^{j\Omega_\mu}, n\right) . \tag{14.37}$$

Puder [185] proposed a simple but effective approach. First, preliminary attenuation factors are computed for each subband by applying the Wiener characteristic according to Section 10.1.2.1:

$$H_{\mathrm{pre}}\left(e^{j\Omega_\mu}, n\right) = \max\left\{\left(1 - \beta_{\mathrm{pre}} \frac{\widehat{S}_{bb}\left(\Omega_\mu, n\right)}{\widehat{S}_{ee}\left(\Omega_\mu, n\right)}\right), 0\right\}. \tag{14.38}$$

A fixed overestimation parameter (e.g., $\beta_{\mathrm{pre}} = 4$) can be utilized. The power spectral density $\widehat{S}_{ee}(\Omega_\mu, n)$ is estimated in most systems simply by the instantaneous power $|e_\mu(n)|^2$.

Before the preliminary attenuation factors are mapped via a simple characteristic on the subband spectral floor values, smoothing along both the time and frequency axes is applied. The time-domain smoothing is performed by utilizing first-order IIR filters with a moderate time constant:

$$\tilde{H}_{\mathrm{pre}}\left(e^{j\Omega_\mu}, n\right) = \gamma\, \tilde{H}_{\mathrm{pre}}\left(e^{j\Omega_\mu}, n-1\right) + (1-\gamma)\, H_{\mathrm{pre}}\left(e^{j\Omega_\mu}, n\right). \tag{14.39}$$

For a frameshift of 64 samples and a sampling rate $f_s = 8000$ Hz, the time constant can be chosen as $\gamma = 0.5$. After temporal filtering smoothing along the frequency axis is applied. A simple 3-tap FIR filter can be applied:

$$\bar{H}_{\mathrm{pre}}\left(e^{j\Omega_\mu}, n\right) = \begin{cases} \frac{1}{2} \sum\limits_{i=0}^{1} \tilde{H}_{\mathrm{pre}}\left(e^{j\Omega_i}, n\right), & \text{if } \mu = 0, \\ \frac{1}{3} \sum\limits_{i=\mu-1}^{\mu+1} \tilde{H}_{\mathrm{pre}}\left(e^{j\Omega_i}, n\right), & \text{if } \mu \in \{1, ..., M-2\}, \\ \frac{1}{2} \sum\limits_{i=M-2}^{M-1} \tilde{H}_{\mathrm{pre}}\left(e^{j\Omega_i}, n\right), & \text{if } \mu = M-1. \end{cases} \tag{14.40}$$

Mapping of the smoothed preliminary subband attenuations $\bar{H}_{\mathrm{pre}}\left(e^{j\Omega_\mu}, n\right)$ on the subband attenuation limits $H_{\mathrm{min}}\left(e^{j\Omega_\mu}, n\right)$ is performed with a two-stage characteristic. In order to prevent large fluctuations of the subband limits, a smooth (linear) transition between the two stages is inserted (see Fig. 14.10):

$$H_{\mathrm{min}}\left(e^{j\Omega_\mu}, n\right) = \begin{cases} 0.5, & \text{if } \bar{H}_{\mathrm{pre}}\left(e^{j\Omega_\mu}, n\right) > 0.3, \\ 0.12, & \text{if } \bar{H}_{\mathrm{pre}}\left(e^{j\Omega_\mu}, n\right) < 0.05, \\ 0.044 + 1.52\, H_{\mathrm{pre}}\left(e^{j\Omega_\mu}, n\right), & \text{else}. \end{cases} \tag{14.41}$$

To visualize this method, the result of a time–frequency analysis of a noisy speech signal is depicted in the upper diagram of Fig. 14.11. The second diagram shows the preliminary subband attenuation factors $H_{\mathrm{pre}}(e^{j\Omega_\mu}, n)$ before applying any kind of smoothing. The lowest diagram shows the resulting subband spectral floor parameters. Within speech pauses a maximum attenuation of 0.12 is allowed, corresponding

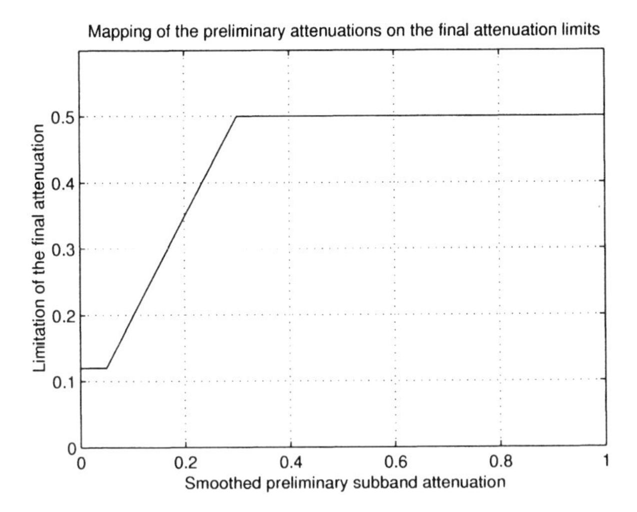

Fig. 14.10 Mapping function of the preliminary subband attenuation factors $\bar{H}_{\text{pre}}(e^{j\Omega_\mu}, n)$ on the subband spectral floor parameters according to Puder [185].

to -18 dB. Within periods of speech activity the attenuation limits are increased to 0.5, which corresponds to -6 dB. However, in between neighboring harmonics of the pitch frequency a large attenuation is still possible. The maximal and minimal attenuation limitations utilized in this example are a typical choice for scenarios in which large signal-to-noise ratios (car moving at low speed or offices with only moderate noise levels) can be expected.

In case of high-noise scenarios, such as telephone calls made out of cars at high speed, higher spectral floor parameters (-6 dB during speech periods and ranging from -8 to -10 dB during noise-only periods) should be utilized. Furthermore, the mapping function depends on the utilized overestimation factors as well as on the filter characteristic itself. Thus, a large amount of (dependent) parameters need to be optimized. Even if this is a time-consuming task, the overall quality of a noise suppression system can be improved significantly in case of a properly designed adaptive control of the subband spectral floor parameters.

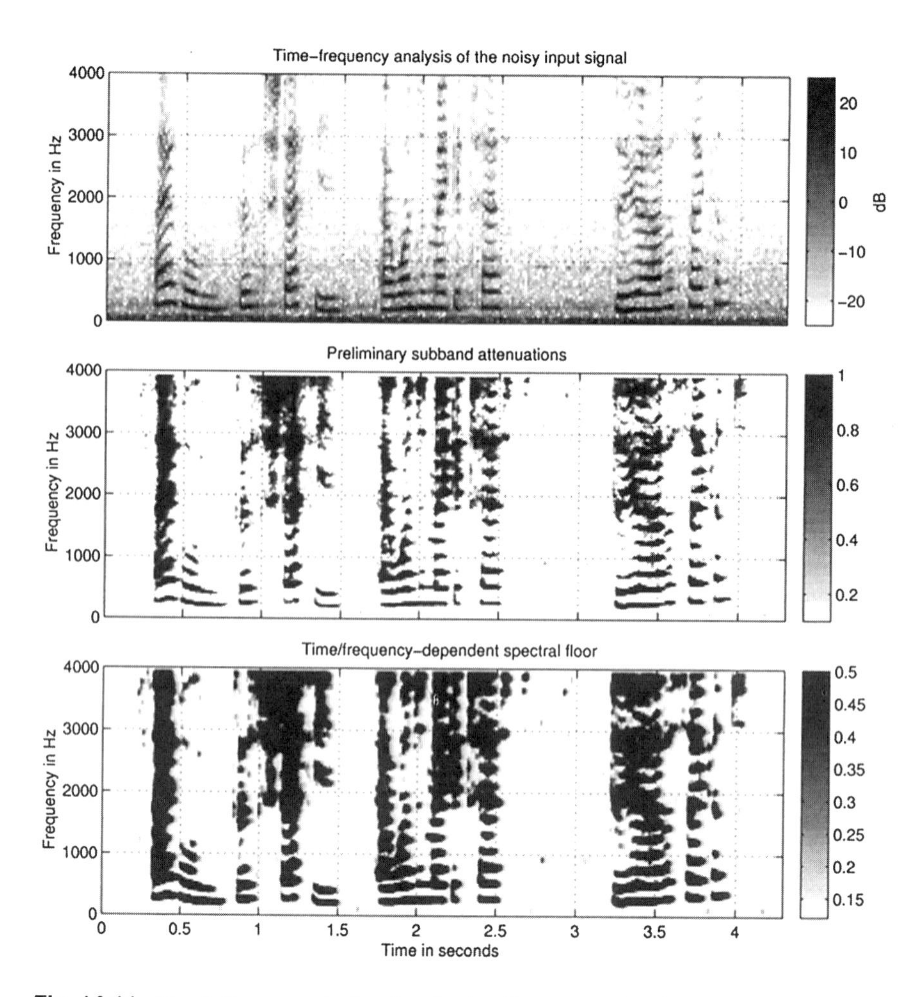

Fig. 14.11 Simulation example showing the time/frequency-selective adjustment of the subband spectral floor parameters. The upper diagram shows a time–frequency analysis of the noisy input signal. The preliminary subband attenuation factors are depicted in the second diagram. The resulting spectral floor parameters are depicted in the lowest diagram.

15

Control for Beamforming

In contrast to adaptive filters utilized for echo cancellation, the task of an adaptive beamformer is not to perform a system identification. As introduced in Section 11.3.2, an adaptive beamformer should adapt its filter coefficients such that signals coming from another than the steering direction are attenuated as much as possible. On the other hand, signals from the steering direction should be able to pass the beamformer without degradation. The latter point was taken into account by introducing the side condition (see Eq. 11.64 for the time-domain counterpart)

$$\sum_{l=0}^{L-1} G_{\mathrm{F},l}\left(e^{j\Omega}, n\right) = e^{j\Omega k_0} \tag{15.1}$$

in addition to a standard minimization of the beamformer output power:[1]

$$\mathrm{E}\left\{ |u(n)|^2 \right\} \longrightarrow \min \, . \tag{15.2}$$

When designing adaptive algorithms for beamforming, the constraint can be fulfilled either by using only a subspace of gradient based filter updates (see Section 11.4) or by introducing so-called *blocking matrices* (see Section 11.4.1), which are used as a preprocessor of the filter input signals. With both of these approaches the constraint of the minimization problem can be fulfilled. However, it is still not possible to adapt

[1]The signal $u(n)$ is denoting the (time domain) output signal of the beamformer. $G_{\mathrm{F},l}(e^{j\Omega}, n)$ is the frequency response of the adaptive filters utilized within the lth microphone path and k_0 is an appropriately chosen constant. For further details on the notation, see Section 11.3.2.

the coefficients of the beamformer filters without adaptation control. Before we analyze the reasons for this, we will present an example of a noncontrolled beamformer. Two microphones were mounted on the rear-view mirror inside a car. The distance between the microphones was $d = 10$ cm. The output signals of the microphones were processed with a subband domain beamformer in generalized sidelobe structure (see Section 11.4.1). The delay compensation filters have been designed such that the beamformer steers toward the driver of the car. The filters were adapted using the NLMS algorithm with a fixed stepsize $\mu(n) = 0.03$. During the first 5 seconds the passenger sitting beside the driver speaks; afterward the driver is active (also for about 5 seconds).

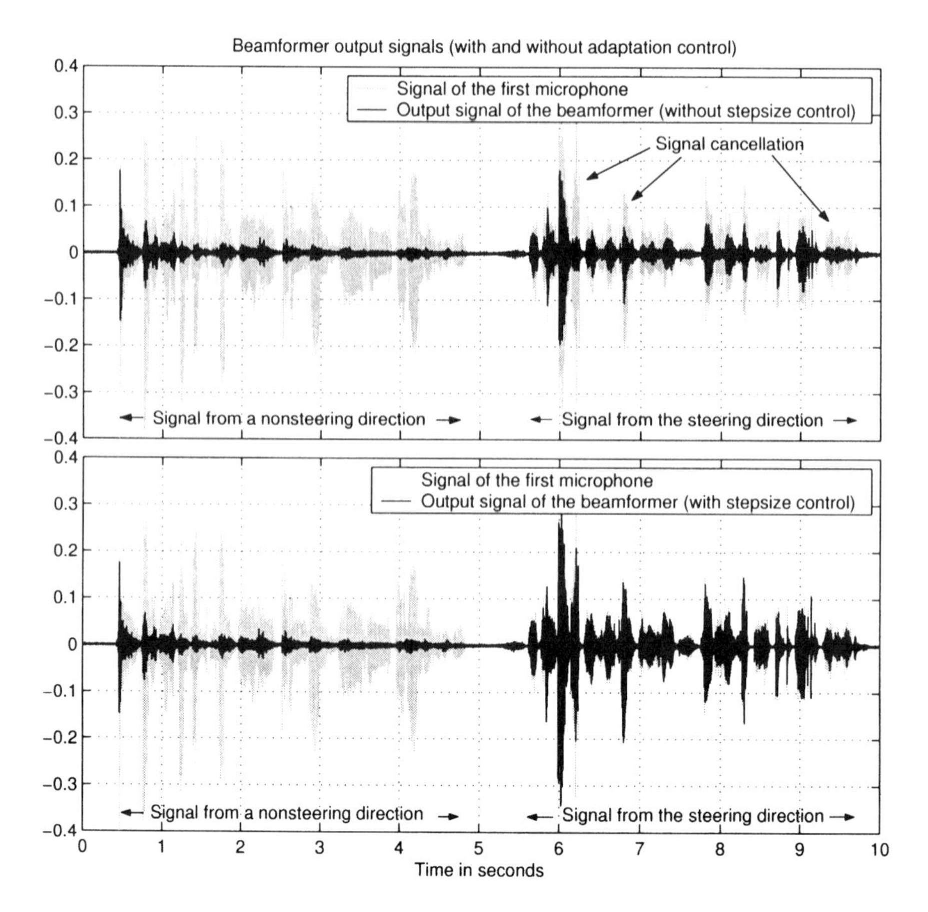

Fig. 15.1 Input and output signals of beamformers with both fixed and adaptive control parameters.

In the upper diagram of Fig. 15.1 the signal of the first microphone as well as the output signal of the (noncontrolled) beamformer are depicted. During the first 5 seconds the (undesired) signal of the passenger can be reduced considerably. However, during the second 5 seconds—when the driver speaks—the (desired) signal is also suppressed by a large amount. This happens even if the delay compensation filters are designed such that signals arriving from the direction pointing to the driver are delay aligned and thus cannot pass the blocking matrix.

This undesired behavior, called *signal cancellation* [239], is caused by multipath propagation within the car. The signals of the driver do not only arrive from the steering direction. Due to reflections on the boundary of the car cabin (windscreen, roof, etc.), the driver's signal also arrives from other directions, which can pass the blocking matrix. Without a control mechanism the adaptive filters are adjusted such that these reflections compensate for the direct signals. As a result, the desired signal is attenuated, as we can see in the right half of the upper diagram in Fig. 15.1.

The solution to this problem is to introduce a robust adaptation control. The control mechanisms should adjust the stepsize or the regularization parameter such that the adaptation of the filters is performed only in intervals without speech from the desired direction. In this chapter we will describe such procedures. The result of an adaptive beamformer exhibiting an appropriate stepsize control are depicted in the lower diagram of Fig. 15.1. In this case the suppression of the passenger's signal is barely affected. However, during periods of single talk from the steering direction no signal cancellation occurs. Before describing the details of the utilized stepsize control in Section 15.2, we will introduce into some further problems that appear when realizing a beamformer in a real scenario.

15.1 PRACTICAL PROBLEMS

When implementing a beamformer on real hardware and in an reverberant environment such as a car cabin or a small office, several problems occur that were not considered during the derivation of the algorithms in Chapter 11. However, as we have seen in the example above, these effects can degrade the performance of beamformers considerably. Besides multipath propagation, sensor imperfections and the far-field assumption, which was made during the design of the blocking matrix, also need to be reviewed in more detail.

15.1.1 Far-Field Assumptions

In most basic descriptions of adaptive beamformers the *far-field assumption* is made. This means that the microphones are assumed to be far away from the desired source. The term *far away* means in this case that the intersensor distance is relatively small compared to the distance between the microphones and the source (see Section 11.1).

In real applications the microphones are usually placed as close as possible to the speaker in order to obtain large signal-to-noise ratios. Thus, the far-field assumption is often violated. To show the extent of this violation, the measurement points of a diffuse field distance estimation performed within a car are depicted in Fig. 15.2.

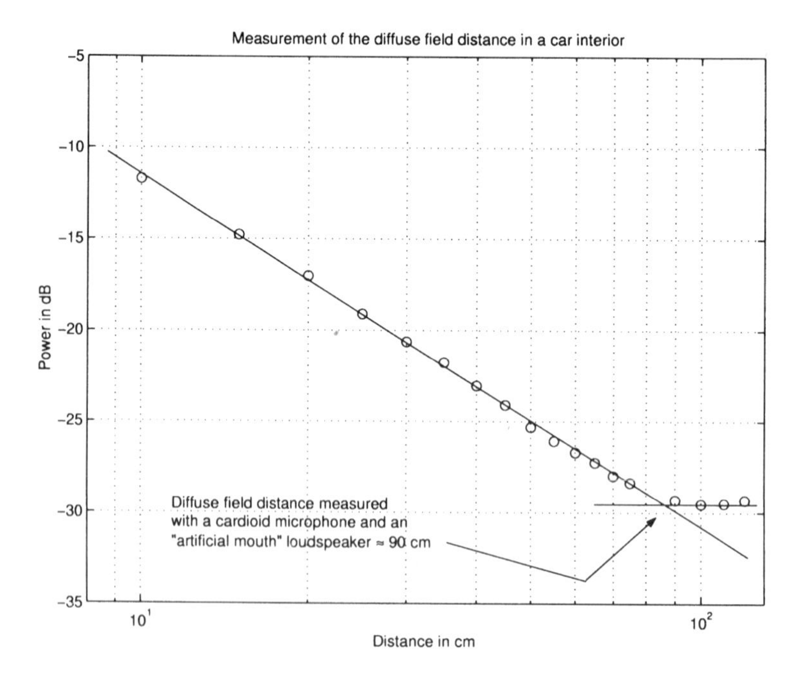

Fig. 15.2 Measurement of the diffuse field distance in a car.

A microphone was mounted in the middle of the driver's sun visor, and a loudspeaker was placed at several distances, ranging from 10 to 120 cm. The loudspeaker always emits the same signal with equal amplification for all measurements. In Fig. 15.2 the average power picked up by the microphone is depicted for each measurement. According to the linear wave equation for point sources (see Section 11.1), the power is decreased by 6 dB whenever the distance between the microphone and the loudspeaker is doubled. This is true up to a distance (in this example) of about 90 cm (called the *diffuse field distance*). Afterward the reflections on the boundaries of the enclosure (windscreen, roof, etc.) compensate for the loss within the direct path and the receiving power changes only marginally with increasing distance.

As a consequence of this effect, the delay compensation unit of beamformers should be extended by a gain compensation unit whenever the distance between the sensors and the source is smaller than the diffuse field distance of the enclosure. An example of a simple adaptive gain compensation method is presented by Aubauer and Leckschat [6].

15.1.2 Sensor Imperfections

Microphones are generally considered to be perfect point sensors with ideal and equal properties. This assumption is quite unrealistic. In most hands-free systems installed in cars, cardioid microphones with tolerances concerning the transfer function of about 2 dB or even more are utilized. Buck and Hacker [94] analyzed a set of 83 microphones from the same fabrication charge. The transfer functions and the group delay was measured in an anechoic chamber for several target directions. The upper and the lower diagrams of Fig. 15.3 show the mean, the minimum, and the maximum of the absolute value (top) and the phase (bottom) of the measured transfer functions for the endfire direction. In the right two diagrams the corresponding standard deviations (as a function of frequency) are depicted.

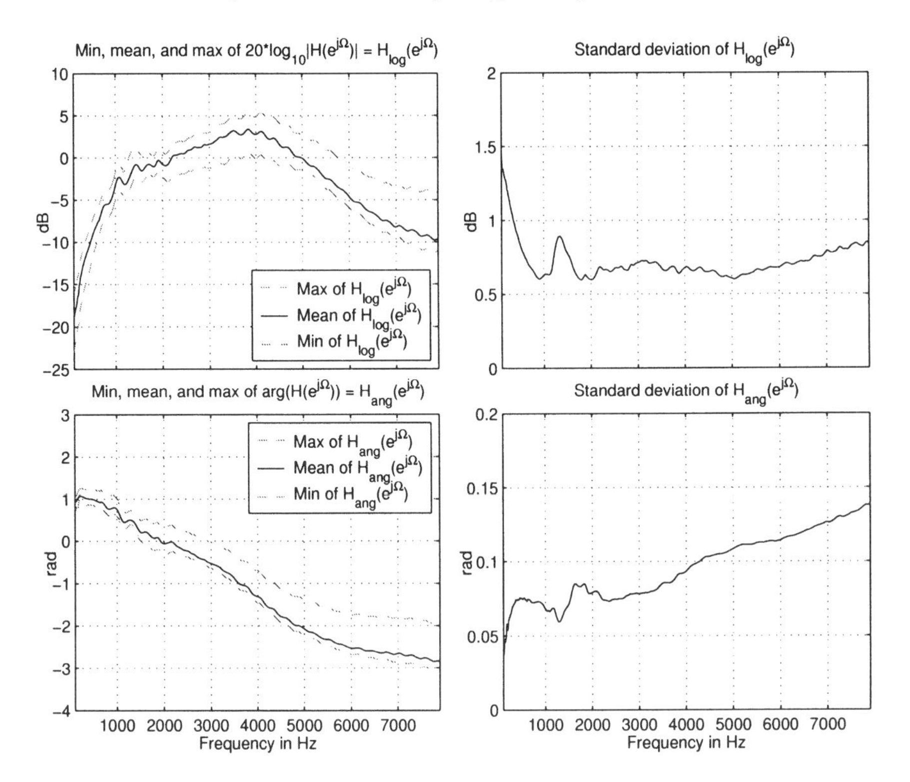

Fig. 15.3 Tolerances of microphones. The upper two diagrams show the minimum, the mean, and the maximum as well as the standard deviation of the logarithmic absolute values of the transfer functions measured at 83 microphones from the same fabrication charge. In the lower two diagrams the minimum, the mean, the maximum, and the standard deviation of the measured phases are depicted.

The measured deviations show that the assumption of ideal and equal sensors is far away from reality. In case of utilizing the generalized sidelobe structure (see Section 11.4.1), signals from the steering direction can pass the blocking matrix. As a result, signal cancellation—as presented in Fig. 15.1—can also appear without multipath propagation but with intersensor differences.

A solution to this problem would be to calibrate the microphones or to use preselected microphones. With the latter approach only those microphones with small intersensor differences will be combined in an array. Microphone calibration is usually performed in an unechoic chamber with precalibrated measurement microphones and loudspeakers. This is very costly and time-consuming. Furthermore, calibration is possible for only one direction, and because the calibration is performed only once (before installing the microphones) the effects of aging will not be covered.

A different solution is to calibrate the microphones during operation by means of adaptive filters. This could be achieved by using one of the microphones of a beamformer as a reference and trying to equalize the others. Fig. 15.4 shows such an adaptive calibration inserted into the generalized sidelobe structure (compare with Fig. 11.13). The delay in the reference microphone channel is introduced in order to allow also the compensation of negative delays. Details about this method can be found in papers by Van Copernolle and colleagues [43, 44].

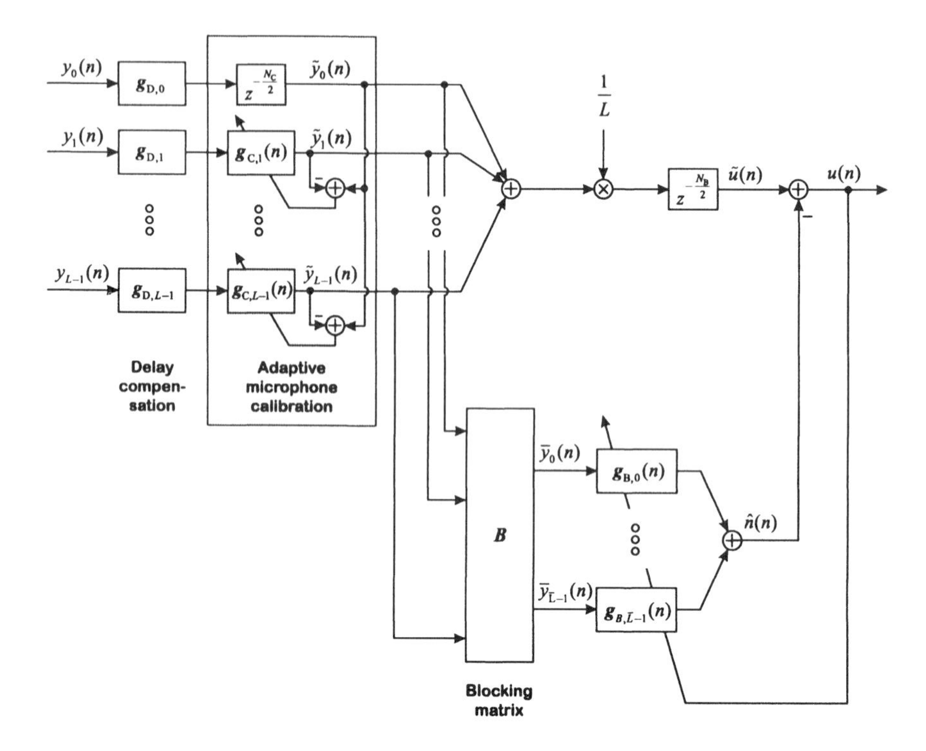

Fig. 15.4 Structure of the generalized sidelobe structure with additional microphone calibration stage.

The drawback of using one microphone as a reference is that one depends strongly on this specific microphone. More robust solutions have been presented [28] where the basic structure of Fig. 15.4 is extended in several ways. One possibility of reducing the dependence on one of the microphones is to first compute a preliminary beamformer output with the noncalibrated microphones. The output signal of this beamformer can be utilized as a reference signal in the calibration process. After calibration, the enhanced microphone signals are utilized again for obtaining the final beamformer output signal.

15.2 STEPSIZE CONTROL

Finally, a basic method for controlling the stepsize in order to avoid signal cancellation will be presented. Because the structure according to Griffiths and Jim [89] is a very popular one for realizing adaptive beamformers, we will show how the required information for adjusting the stepsize can be extracted in this structure.

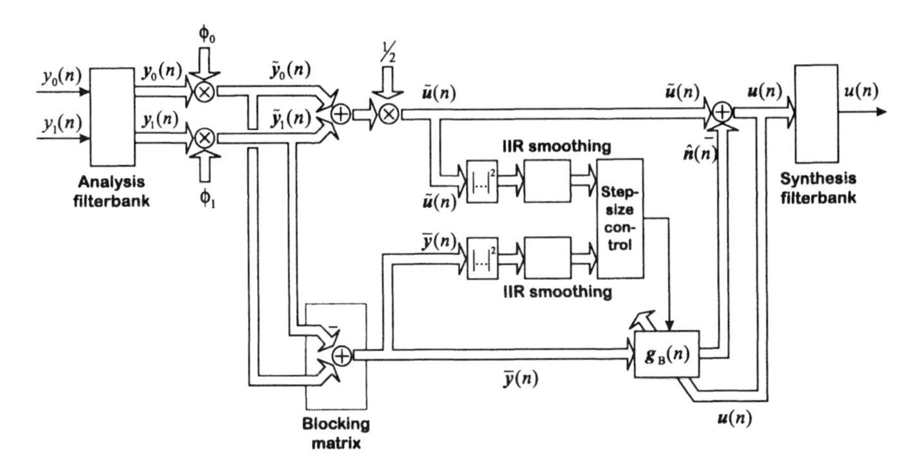

Fig. 15.5 Stepsize control for the generalized sidelobe structure within the subband domain.

In order to reduce the computational complexity, subband processing is applied. For this reason, overlapping blocks of the microphone signals are transformed periodically via a DFT-modulated polyphase filterbank into the subband domain. The basic structure of the system is depicted in Fig. 15.5 for the two-microphone case. We will denote the resulting vectors by

$$\boldsymbol{y}_i(n) = \left[y_{i,0}(n),\, y_{i,1}(n),\, \ldots,\, y_{i,M-1}(n)\right]^{\mathrm{T}}. \tag{15.3}$$

The elements $y_{i,\mu}(n)$ of the vector $\boldsymbol{y}_i(n)$ correspond to the subband μ of the ith microphone signal at time index n. In a first stage the delays of the subband signals are compensated for a predefined direction (steering direction of the array). In automotive applications the locations of the microphone array and the average seat positions are known a priori, which means that no estimation of the source direction is required. The delay compensation can be achieved within the subband domain by multiplication with appropriate exponential terms:

$$\tilde{\boldsymbol{y}}_i(n) = \boldsymbol{y}_i(n) \otimes \boldsymbol{\phi}_i\,. \tag{15.4}$$

The sign $\otimes$ denotes elementwise multiplication of two vectors. If the required delay for the ith microphone is denoted by $\tilde{\tau}_i$ and the sampling rate by f_s, the vectors $\boldsymbol{\phi}_i$

are defined as follows:

$$\boldsymbol{\phi}_i = \left[1,\, e^{-j\frac{2\pi}{M}\tilde{\tau}_i f_\text{s}},\, e^{-j\frac{2\pi}{M}\tilde{\tau}_i\, 2\, f_\text{s}},\, \ldots,\, e^{-j\frac{2\pi}{M}\tilde{\tau}_i\, [M-1)\, f_\text{s}}\right]^\text{T}. \tag{15.5}$$

A first signal enhancement can be achieved by summing the delay-compensated microphone signals:

$$\tilde{\boldsymbol{u}}(n) = \frac{1}{L}\sum_{i=0}^{L-1} \tilde{\boldsymbol{y}}_i(n)\,. \tag{15.6}$$

Further enhancement can be obtained by subtracting an adaptively generated noise estimate $\hat{\boldsymbol{n}}(n)$ from the so-called delay-and-sum signal $\tilde{\boldsymbol{u}}(n)$:

$$\boldsymbol{u}(n) = \tilde{\boldsymbol{u}}(n) - \hat{\boldsymbol{n}}(n)\,. \tag{15.7}$$

In order to apply unconstrained adaptive algorithms, so-called blocking matrices are utilized. In the simplest case of only two microphones ($L = 2$), one can simply use the difference between the delay-compensated signals:

$$\bar{\boldsymbol{y}}(n)\big|_{L=2} = \tilde{\boldsymbol{y}}_0(n) - \tilde{\boldsymbol{y}}_1(n)\,. \tag{15.8}$$

By filtering this input signal with an adaptive filter $\boldsymbol{g}_\text{B}(n)$, the noise estimate is generated:

$$\hat{\boldsymbol{n}}(n)\big|_{L=2} = \boldsymbol{g}_\text{B}^\text{H}(n)\,\bar{\boldsymbol{y}}(n)\,. \tag{15.9}$$

The filter $\boldsymbol{g}_\text{B}(n)$ can be adapted using one of the standard algorithms, such as NLMS or affine projection (see Chapter 7). If more than two microphones are used, one adaptive filter is applied for each blocking signal.

In order to detect speech activity from a predefined source direction, short-term powers in each subband μ of the beamformer output and of the blocking signal are estimated by first-order IIR smoothing of the squared magnitudes [117]:

$$p_{\text{s},\mu}(n) = \beta\, p_{\text{s},\mu}(n-1) + (1-\beta)\,|u_\mu(n)|^2, \tag{15.10}$$

$$p_{\text{b},\mu}(n) = \beta\, p_{\text{b},\mu}(n-1) + (1-\beta)\,|\bar{y}_\mu(n)|^2. \tag{15.11}$$

The choice of the time constant β depends on the subsampling rate and the sampling frequency. Usually β is chosen from the interval $0.4 < \beta < 0.8$. The ratio $r_\mu(n)$ of the short-term powers is computed individually for each subband:

$$r_\mu(n) = \frac{p_{\text{s},\mu}(n)}{p_{\text{b},\mu}(n)}. \tag{15.12}$$

The resulting power ratio as a function of frequency can vary widely (caused by the characteristics of the blocking beamformer). To derive a common range, $r_\mu(n)$ has to be normalized by a moving average $\bar{r}_\mu(n)$. The normalization value is obtained by slowly increasing or decreasing the preceding value according to

$$\bar{r}_\mu(n) = \begin{cases} \bar{r}_\mu(n-1)\,\Delta_{\text{r,inc}}, & \text{if}\, r_\mu(n) > \bar{r}_\mu(n-1), \\ \bar{r}_\mu(n-1)\,\Delta_{\text{r,dec}}, & \text{else.} \end{cases} \tag{15.13}$$

The adjustment constants are to be chosen in the range of $0 \ll \Delta_{\text{r,dec}} < 1 < \Delta_{\text{r,inc}}$. The normalized power ratios can be computed as

$$\tilde{r}_\mu(n) = \frac{r_\mu(n)}{\bar{r}_\mu(n)} \,. \tag{15.14}$$

Figure 15.6 shows an example of such a normalized power ratio based on a section of a hands-free telephone call recorded in a car. After a short interval of only background noise, the far-end speaker is active. After 9 seconds the local speaker—the driver—starts speaking, too (double talk). At the end of the measurement only the local speaker is active. The values of $\tilde{r}_\mu(n)$ have been averaged over all subbands

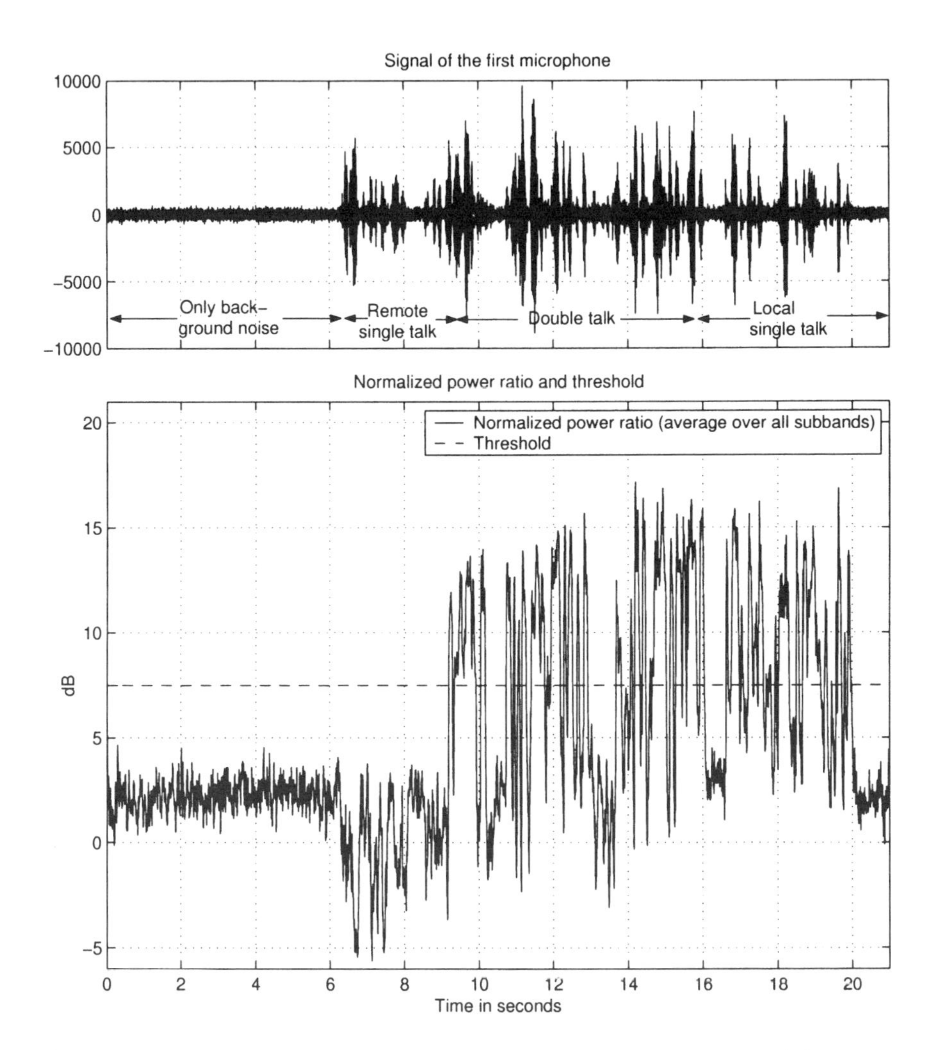

Fig. 15.6 Input signal of the first microphone and normalized power ratio $\tilde{r}_\mu(n)$ averaged over all subbands of a hands-free telephone call recorded in a car.

before plotting. It is obvious that a distinction between speech activity of the driver and the remote speaker can be made from the resulting values of $\tilde{r}_\mu(n)$, for example, by applying a threshold decision.

In order to avoid signal cancellation, the beamformer is adapted only if the averaged normalized power ratio is below a threshold (see the example in Fig. 15.1). The adaptive filter of the two-microphone beamformer was adapted with the NLMS algorithm. The stepsize was adjusted according to the averaged normalized power ratio:

$$\mu(n) = \begin{cases} 0.03, & \text{if } \frac{1}{M} \sum_{\mu=0}^{M-1} \tilde{r}_\mu(n) < 2.4\,, \\ 0, & \text{else}\,. \end{cases} \tag{15.15}$$

Note that the threshold (here 2.4) should be chosen according to the number of microphones. By comparing the output signals of the noncontrolled and controlled beamformers, it is obvious that the signal cancellation phenomenon can be reduced considerably by appropriate adaptation control.

16

Implementation Issues

Most algorithms used for adaptive filtering are derived using the assumption of infinite precision. Testing these algorithms with simulations or real-time implementations reveals that results derived under this assumption are no longer valid. Therefore, differences between the expected behavior and results obtained with finite-precision arithmetic may force modifications of the specific algorithm.

There are several reasons for deviations from the expected results [40, 144]. Most of them are due to the finite precision that is inherent in electronic calculations. The following reviews three different error sources.

16.1 QUANTIZATION ERRORS

First, all values used in electronic calculations are quantized. The quantization occurs during analog-to-digital conversion. The associated quantization error can be decreased by increasing the resolution of the quantizer.

For sinusoidal input signals, each additional bit results in a gain of 6 dB for the output signal-to-noise ratio. This results in a signal-to-noise ratio of 96 dB in the case of a 16-bit analog to digital conversion as it is most widely used in signal processing applications. For linear quantization, the relative quantization is large for small values to be quantized and decreases with increasing input amplitude. One should therefore ensure that both the excitation and the microphone signal are accurately scaled. An automatic gain control (AGC) may be necessary.

16.2 NUMBER REPRESENTATION ERRORS

The method chosen to represent a number in the binary system can be a second source of problems. Fixed-point representations lead to errors that are in a constant range, regardless of the magnitude of the quantized value. Therefore, storing small numbers in fixed-point precision results in substantial relative errors.

With 16 bits, a signal-to-noise ratio of 96 dB may be an appropriate quantization for audio signals. Nevertheless, there may be other applications that require a higher precision. This is particularly true for variables that can have a large dynamic range, like control variables, vector norms, or filter taps.

The necessary dynamic range can be provided by floating-point representations. Many modern digital signal processors (DSPs) use floating-point precision according to the IEEE P754 [139] standard (total of 32 bits: 1 bit for sign, 11 bits for exponent, and 23 bits for mantissa). Since the quantization error now occurs only in the mantissa, it increases with the quantizer input, which makes the relative quantization error constant over the entire dynamic range.

16.3 ARITHMETICAL ERRORS

Also, depending on the arithmetic used, different types of errors may occur. Whereas additions in fixed-point arithmetic do not produce errors (assuming no overflow condition), additions in floating-point arithmetic result in roundoff errors, unless both summands have the same exponent. On the other hand, errors arising from multiplications are usually smaller for floating-point rather than for fixed-point arithmetic.

Another type of error occurs in summations if the sum exceeds the permitted range. This event (overflow) is handled differently, depending on the processor. Some processors assign the result to the largest representable positive or to the largest negative number, respectively (saturation). Another way to handle overflows is to set the result to zero (zero saturation). In the worst case, the overflow condition is not handled at all (wrapping).

16.4 FIXED POINT VERSUS FLOATING POINT

The majority of today's DSPs provide fixed-point arithmetic and quantize data with 16-bit resolution. As mentioned previously in this chapter, low cost is a major criterion for manufacturers. Therefore, fixed-point DSPs are preferred to floating-point DSPs for consumer products.

Quantization with 16 bits may be well suited for raw speech data. However, for squared values of speech signals and for room impulse response filter taps, quantization with 16 bits does not suffice. Especially for the application in acoustic echo cancellation, the quantization of the filter taps noticeably limits the achievable echo attenuation.

16.5 QUANTIZATION OF FILTER TAPS

Two effects enlarge the residual echo of adaptive filters if implemented with fixed-point arithmetic. First, the result of the quantization error of each filter tap can be seen as an additional noise term that degrades the achievable residual echo if no stepsize control is used. However, this effect does not play the most important role for 16-bit quantization, unless the number of taps is huge. Second, if the update term on the right of the NLMS equation (Eq. 7.19) is smaller than the quantization interval, the corresponding filter tap freezes. This "stopping" phenomenon occurs especially if $\mu(n)$ is very small and conflicts with the rule that one can decrease the excess mean squared error by decreasing the stepsize $\mu(n)$. With fixed-point arithmetic, it can be advantageous to leave $\mu(n)$ at a higher value [35]. Fig. 16.1 shows the convergence of the system distance for fixed-point arithmetic with different word length and for floating-point arithmetic.

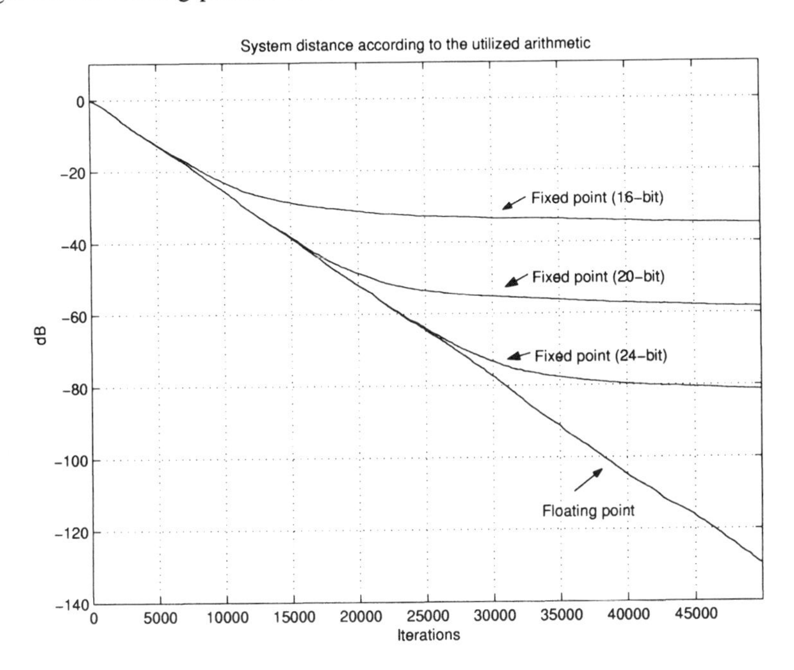

Fig. 16.1 Convergence of the system distance as a function of the arithmetic used. Filter length = LEM system length: 2000 taps. Excitation: white Gaussian noise.

Freezing coefficients can be easily observed when examining a specific algorithm that uses typical properties of room impulse responses to speed up the convergence of high order adaptive filters [156, 157]. The update equation of the *exponentially weighted stepsize* (ES)–NLMS algorithm is as follows:

$$\hat{\boldsymbol{h}}(n+1) = \hat{\boldsymbol{h}}(n) + \mu(n)\,\boldsymbol{A}\,\frac{e^*(n)}{\|\boldsymbol{x}(n)\|^2}\,\boldsymbol{x}(n)\,, \tag{16.1}$$

where $\boldsymbol{A}$ is a diagonal matrix that assigns exponentially decreasing stepsizes to the coefficients:

$$\boldsymbol{A} = \operatorname{diag}\left[a_0,\, a_1,\, \ldots,\, a_{N-1}\right] \,, \tag{16.2}$$

where $a_i = \alpha\, \gamma^{\,i}$ for $0 \leq i < N$ and γ is an exponential attenuation ratio leading to the same logarithmic decay as the impulse response ($0 < \gamma < 1$).

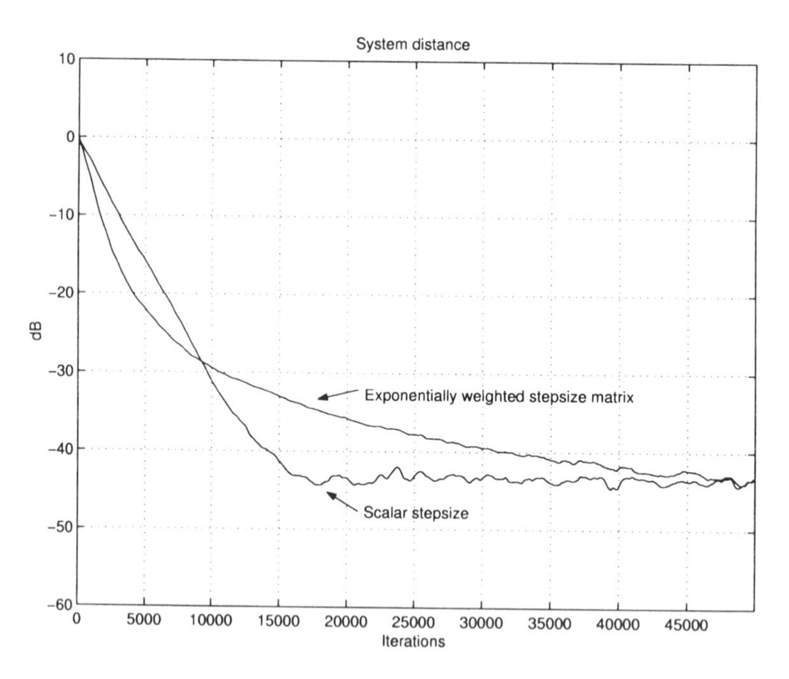

Fig. 16.2 Convergence of the system distance for the NLMS algorithm with scalar stepsize and for exponentially weighted stepsize. Filter length = 1200 taps, LEM system length: 2000 taps. Excitation: white Gaussian noise. Stepsize matrix: $a_i = 0.997^i$.

Figure 16.2 compares the convergence of the system distance for an NLMS algorithm with scalar stepsize and one using exponentially decaying stepsizes. The freezing of filter coefficients in case of scalar stepsize becomes clearly visible. In case of decaying stepsizes the slowdown of the convergence speed with increasing number of iterations is caused by a property of the NLMS algorithm; during startup the system distance error is distributed proportional to the envelope of the LEMS impulse response. With an increasing number of iterations, the algorithm distributes the error equally over all filter coefficients (see Fig. 16.3). Therefore, a decaying stepsize is no longer matched to the problem at this phase of the adaptation.

The convergence behavior can be improved by using the exponential decay of room impulse responses in a different way. Figure 16.4 shows the magnitude of the room impulse response and its quantized envelope. Since adaptive algorithms identify the room impulse response $\boldsymbol{h}(n)$ with the filter $\hat{\boldsymbol{h}}(n)$, the coefficients at the

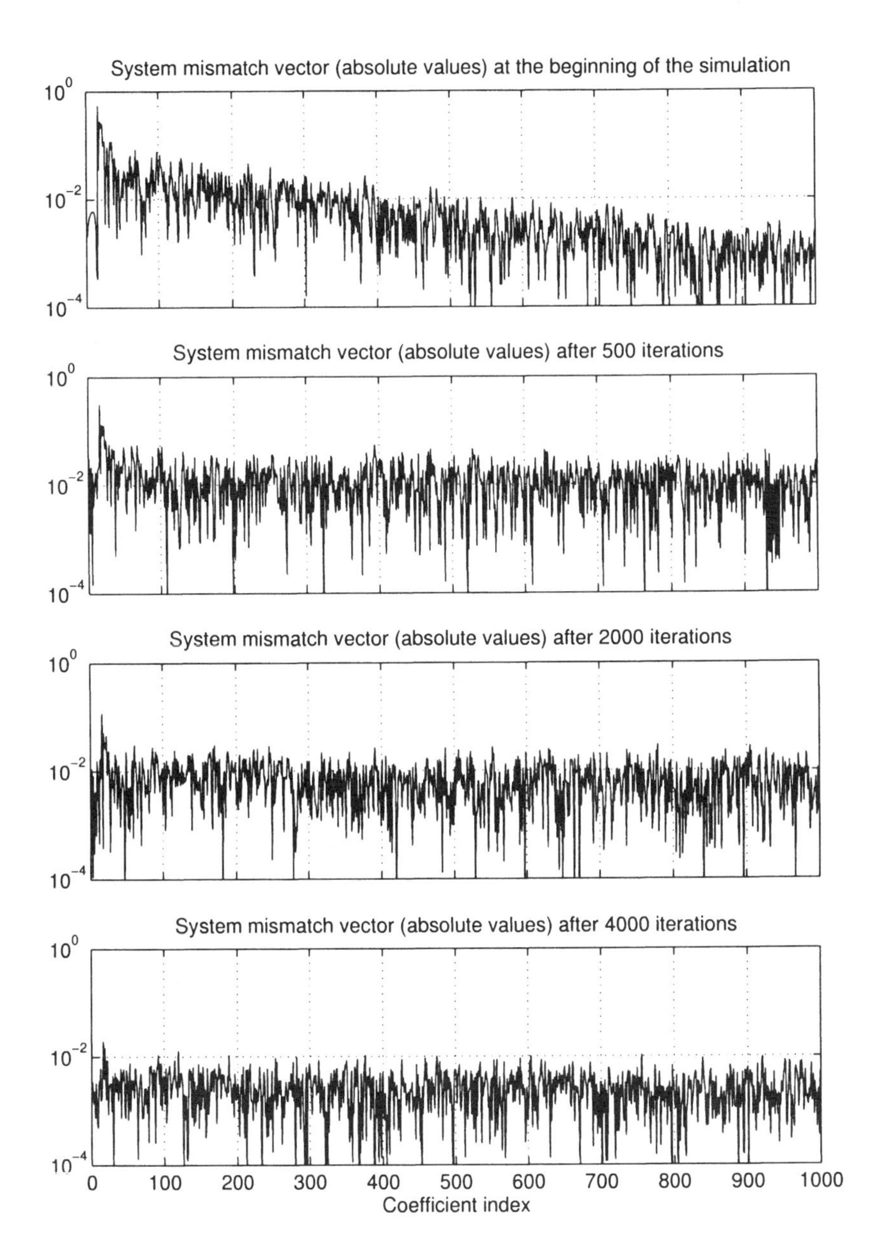

Fig. 16.3 System mismatch vector at the beginning of the adaptation (top) and after 500 (second diagram), 2000 (third diagram), and 4000 (bottom) iterations [201].

tail of $\hat{\boldsymbol{h}}(n)$ make use of only a fraction of the available bits. The bits above the dashed line in Fig. 16.4 remain unused. This "waste" of bits does not occur for an envelope that is approximately constant.

To achieve the desired behavior, one can introduce a scaled filter vector $\hat{\boldsymbol{h}}_s(n)$. Its relation to the filter vector $\hat{\boldsymbol{h}}(n)$ can be described by using a diagonal scaling matrix $\boldsymbol{B}$:

$$\hat{\boldsymbol{h}}(n) = \boldsymbol{B}\,\hat{\boldsymbol{h}}_s(n). \tag{16.3}$$

Using this equation, the error calculation and the filter update can be written as

$$e_s(n) = y(n) - \left[\boldsymbol{B}\,\hat{\boldsymbol{h}}_s(n)\right]^{\mathrm{H}} \boldsymbol{x}(n), \tag{16.4}$$

$$\hat{\boldsymbol{h}}_s(n+1) = \hat{\boldsymbol{h}}_s(n) + \mu(n)\frac{e_s^*(n)}{\|\boldsymbol{x}(n)\|^2}\,\boldsymbol{B}^{-1}\,\boldsymbol{x}(n), \tag{16.5}$$

where $e_s(n)$ is the error signal obtained using scaled filter coefficients.

Assuming infinite precision, the scaled filter update equation and the conventional NLMS equation are equivalent, despite the fact that the error signal $e(n)$ and the error used in filter scaling, $e_s(n)$, can differ during adaptation. By using the diagonal scaling matrix $\boldsymbol{B}$, the filter vector $\hat{\boldsymbol{h}}_s(n)$ adapts to a pseudo-room-impulse response.

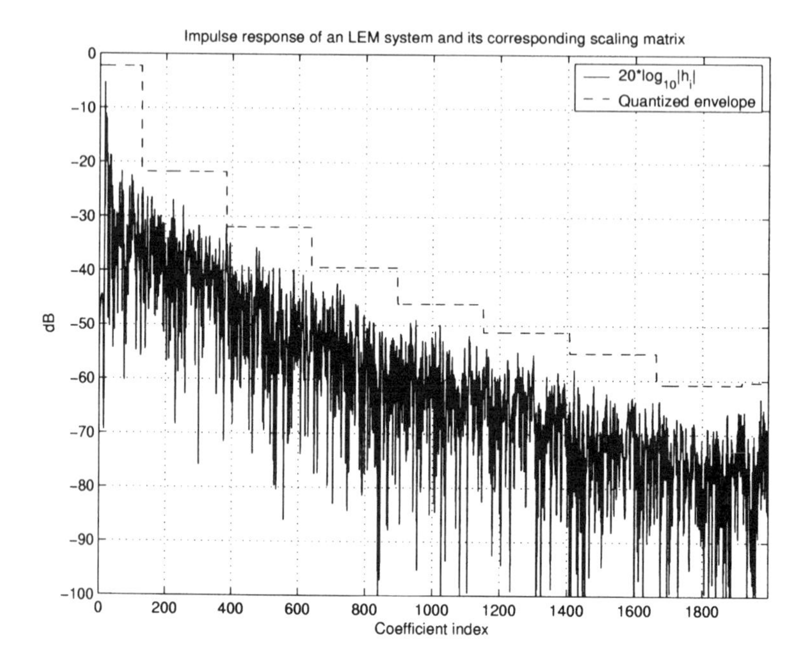

Fig. 16.4 Logarithmic representation of the absolute values of the impulse response of an LEM system and its quantized envelope (dashed line).

Using Eqs. 16.4 and 16.5 improves the performance of the learning curve with fixed-point coefficients. Additional information about filter scaling can be found in Schertler's paper [199].

As can be seen in Fig. 16.1, the quantization problem is considerably reduced by applying 24-bit fixed-point arithmetic. Consequently, 24-bit DSPs should be preferred in acoustic echo control applications. They offer a reasonable compromise between undesirable word length effects and implementation costs.

Part V

Outlook and Appendixes

17

Outlook

It is not difficult to foresee that future research and development in the area of acoustic echo cancellation and noise suppression will not have to take into account processing power and memory size restrictions. Powerful digital signal processing chips will enter even low-priced consumer products.

This has a number of consequences; for instance, the implementation of sophisticated procedures on ordinary (office) PCs will be possible. This will make it easier to test new ideas or modifications of existing procedures in real time and in real environments. The performance of future systems will depend only on the quality of the algorithms implemented and approach limits given solely by the environment they have to work in. The achievable performance will no longer be limited by the restricted capabilities of the affordable hardware.

Nevertheless, efforts will continue to reduce processing demands. Methods applying signal processing in blocks or in subbands will retain their importance whereever the signal delay inherent in them is tolerable. Since users gave up their reluctance—because of the wiring cost—to allow more than one microphone, multimicrophone systems will be applied more frequently. Therefore, the development of improved array processing algorithms for echo cancellation and noise reduction will continue to be a focus for researchers.

Further development of echo and noise control systems will strengthen the inclusion of methods from all areas of digital signal processing such as speech enhancement, blind source separation, neural networks, and nonlinear adaptive filters, to mention just a few.

In this book the authors tried to show that the technology at least for single-channel acoustic echo and noise control systems is available. Their performance is satisfactory, especially if compared to solutions in other voice processing areas. The fact that echo control systems have not yet entered the market on a large scale seems to be not a technical but a marketing problem—a customer who buys a high-quality echo suppressing system pays for the comfort of her or his communication partner. Using a poor system only effects the partner at the far end, who usually is too polite to complain.

Existing echo and noise control systems always motivate customers to ask for improvements. Therefore, researchers and developers will be kept busy. The attempt to satisfy customer demands is like trying to walk up to a red light at the end of a street. This, however, turns out to be the tail light of a moving road roller.

The authors of this book are strongly convinced that acoustic echo and noise control will continue to be one of the most interesting problems in digital signal processing.

Appendix A Subband Impulse Responses

The decaying properties of sound fields can be affected by the type of material that has been used in an enclosure. Porous materials such as textile fibers (carpets, curtains, etc.) or plasterboards (ceiling suspensions) absorb high-frequency sound energy more rapidly than low-frequency energy. Therefore, after splitting the excitation signal into subbands the same amount of echo reduction in high-frequency bands can be achieved with shorter filters than in the low-frequency bands.

A.1 CONSEQUENCES FOR SUBBAND ECHO CANCELLATION

For visualization of these effects, a time–frequency analysis of a room impulse response is depicted in Fig. A.1. The impulse response was measured in an office with a reverberation time of about 300 ms. Basically the office is the one that was presented in Section 3.2. The impulse response h_i was multiplied by a Hamming window with 256 coefficients to extract a part of it. The resulting vector was transformed via a DFT into the frequency domain. This procedure was repeated with windows shifted by 5 coefficients each. Finally the logarithm of the modulus of the transformed vectors was computed. All resulting vectors were grouped in a matrix

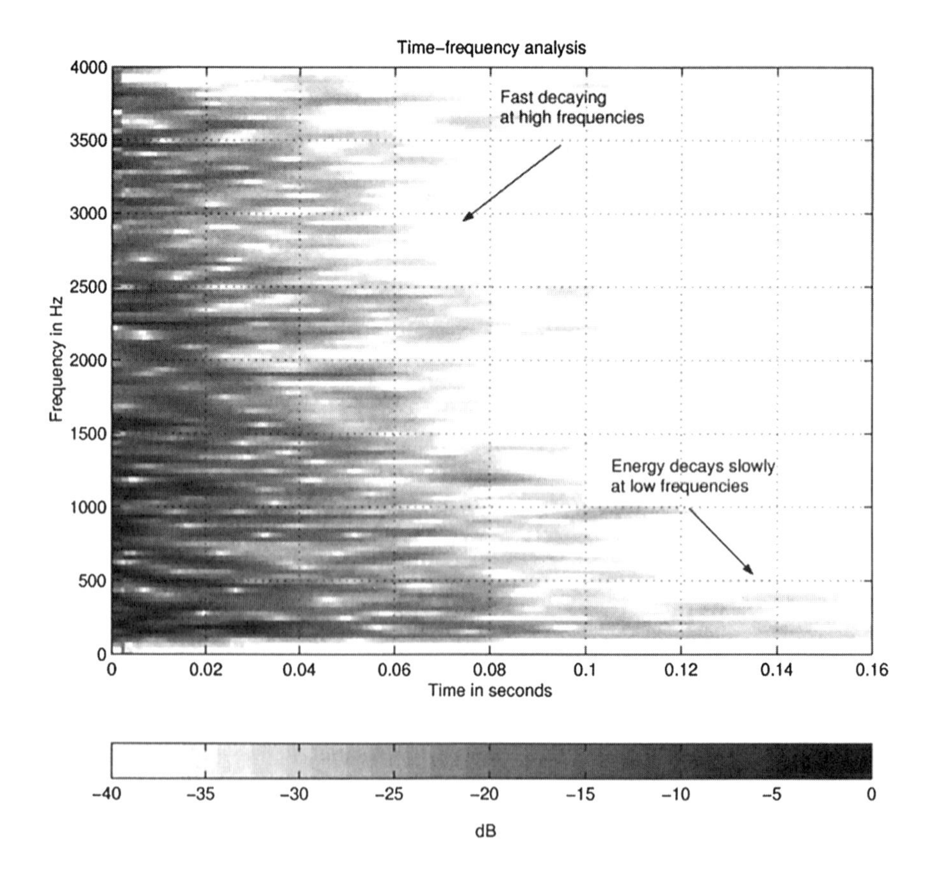

Fig. A.1 Time–frequency analysis of an LEM system impulse response. In most LEM systems the energy in the high frequencies is absorbed faster than the energy located in the low frequencies (see also Fig. 9.30).

and plotted via a grayscale image. The colour scale is selected in the same manner as in time–frequency analyses of speech signals (Section 3.1.1). The slower decay of the sound energy at lower frequencies is clearly visible.

Echo cancellation performed in subbands offers the possibility of allocating filters different orders within different subbands. The time–frequency analysis presented in Fig. A.1 motivates the usage of longer echo cancellation filters in the low-frequency subbands than in the high-frequency ones. In order to choose the filter orders in an appropriate manner, the fullband impulse response of an LEM system will be transformed into a set of subband impulse responses. Details about the choice of the filter orders can be found in Section 9.1.3.4.2. The transformation is performed independently of the filterbank system. In contrast to the results reported by Petraglia and coworkers [181, 182],[1] we will get infinitely long subband impulse responses even if the fullband impulse response has FIR character. In order to achieve optimal echo reduction, the adaptive subband filters for echo cancellation have to be placed in such a manner that also noncausal parts of the subband LEM impulse responses can be canceled. This can be achieved by using different analysis filterbanks for the microphone and the excitation channel or by introducing an artificial delay into the microphone or the loudspeaker path. Details about both methods can be found in Section 9.1.3.

A.2 TRANSFORMATION

In this section we discuss the transformation of a fullband impulse response into an equivalent set of subband impulse responses. When neglecting all local signals, the impulse response of an LEM system models the linear relationship of the remote speaker's signal $x(n)$ and the microphone signal $y(n)$, which is equal to the echo signal if no local signals are present[2] (see Fig. 3.13):

$$d(n) = y(n) \big|_{s(n)=0,\, b(n)=0} . \tag{A.1}$$

It will be assumed that the system does not change, resulting in a time-invariant impulse response $h_i(n) = h_i$. Within the subbands, the signals are bandpass-filtered and downsampled.

In the following we will investigate the relationship of the subband signals $x_\mu(n)$ and $d_\mu(n)$ in relation to the impulse response of the broadband system h_i. Note that the subscript μ will indicate subband quantities. In case of impulse responses all quantities with only one index (e.g., h_i) address broadband filter coefficients; for instance, quantities with two indices (e.g., $h_{\mu,i}$) address subband filter coefficients.

The terms $\breve{x}_\mu(t)$ and $\breve{d}_\mu(t)$ are continuous-time signals that were obtained from their discrete-time counterparts. Bows above signal expressions will indicate time continuity as follows, where the signal $\breve{x}_\mu(t)$ is generated by interpolation[3] of the samples of subband μ [174]:

$$\breve{x}_\mu(t) = \sum_{m=-\infty}^{\infty} x_\mu(m) \, \frac{\sin\left(\pi\left(\frac{t}{rT_0} - m\right)\right)}{\pi\left(\frac{t}{rT_0} - m\right)} \, e^{j\frac{2\pi\mu}{M}\left(\frac{t}{T_0} - mr\right)} . \tag{A.2}$$

The subband signal $x_\mu(n)$ is interpolated first by a sum of weighted Dirac impulses

[1] In [181, 182] *adaptive cross-filters* are permitted. Cross-filters are able to model aliasing components of adjacent subbands. We will not consider this alternative because the use of adaptive cross-filters would double the computational complexity as well as the memory requirements.

[2] For simplicity, the signals $b(n)$, $d(n)$, $x(n)$, and $y(n)$ as well as the impulse response $h_i(n)$ are assumed to be real.

[3] For $x = 0$ the term $\frac{\sin(x)}{x}$ is defined as 1.

$$\sum_{n=-\infty}^{\infty} x_\mu(n)\, \delta_{t-nrT_0}\,. \tag{A.3}$$

For suppression of imaging components, an ideal bandpass filter with center frequency $\mu f_s/M$ and (single-sided) bandwidth $f_s/2r$ is applied. The (continuous-time) impulse response of the bandpass is given by

$$h_{\text{BP, id}, t}\left(\mu \frac{f_s}{M}\right) = \frac{\sin\left(\frac{\pi t}{rT_0}\right)}{\frac{\pi t}{rT_0}}\, e^{j\frac{2\pi\mu t}{MT_0}}\,. \tag{A.4}$$

The sampling instants of the subband signal are $t_n = nrT_0$, where T_0 denotes the inverse of the sampling rate f_s and r is the subsampling rate. At the sampling times the time-continuous excitation signal and its sampled version are equal:

$$\begin{aligned} \breve{x}_\mu(nrT_0) &= \sum_{m=-\infty}^{\infty} x_\mu(m)\, \frac{\sin\left(\pi\,(n-m)\right)}{\pi\,(n-m)}\, e^{j\frac{2\pi\mu}{M}(nr-mr)} \\ &= x_\mu(n)\,. \end{aligned} \tag{A.5}$$

The sampling instants in the broadband domain are $t = nT_0$. At these times the interpolated signals $\breve{x}_\mu(t)$ are given by

$$\breve{x}_\mu(nT_0) = \sum_{m=-\infty}^{\infty} x_\mu(m)\, \frac{\sin\left(\pi\left(\frac{n}{r}-m\right)\right)}{\pi\left(\frac{n}{r}-m\right)}\, e^{j\frac{2\pi\mu}{M}(n-mr)}\,. \tag{A.6}$$

The relationship of the signals $\breve{x}_\mu(nT_0)$ and $\breve{d}_\mu(nT_0)$ can be expressed in terms of the discrete impulse response h_i of the broadband system:

$$\begin{aligned} \breve{d}_\mu(nT_0) &= \sum_{i=-\infty}^{\infty} h_i\, \breve{x}_\mu\left((n-i)\,T_0\right) \\ &= \sum_{i=-\infty}^{\infty} h_i \sum_{m=-\infty}^{\infty} x_\mu(m)\, \frac{\sin\left(\pi\left(\frac{n-i}{r}-m\right)\right)}{\pi\left(\frac{n-i}{r}-m\right)}\, e^{j\frac{2\pi\mu}{M}(n-i-mr)}\,. \end{aligned} \tag{A.7}$$

Because of the ideal interpolation, the samples $d_\mu(n)$ are equal to the values of the signal $\breve{d}_\mu(t)$ at the times $t = nT_0$:

$$\begin{aligned} d_\mu(n) &= \breve{d}_\mu(nrT_0) \\ &= \sum_{i=-\infty}^{\infty} h_i \sum_{m=-\infty}^{\infty} x_\mu(m)\, \frac{\sin\left(\pi\left(\frac{nr-i}{r}-m\right)\right)}{\pi\left(\frac{nr-i}{r}-m\right)}\, e^{j\frac{2\pi\mu}{M}(nr-i-mr)} \\ &= \sum_{m=-\infty}^{\infty} x_\mu(m) \sum_{i=-\infty}^{\infty} h_i\, \frac{\sin\left(\pi\left(n-m-\frac{i}{r}\right)\right)}{\pi\left(n-m-\frac{i}{r}\right)}\, e^{j\frac{2\pi\mu}{M}((n-m)r-i)} \\ &= \sum_{m=-\infty}^{\infty} x_\mu(m)\, h_{\mu,\,n-m}\,. \end{aligned} \tag{A.8}$$

The relationship between the sampled subband signals $x_\mu(n)$ and $d_\mu(n)$ can be described via the impulse response $h_{\mu,\,i}$. This subband impulse response

$$h_{\mu,\,i} = \sum_{m=-\infty}^{\infty} h_m \, \frac{\sin\left(\pi\left(i - \frac{m}{r}\right)\right)}{\pi\left(i - \frac{m}{r}\right)} \, e^{j\frac{2\pi\mu}{M}(ir-m)} \tag{A.9}$$

consists of a superposition of $\sin(x)/x$ and $e^{j\frac{2\mu r}{M}x}$ functions that are weighted with the coefficients of the impulse response of the fullband system h_i. The superposition is equal to a convolution of h_i with an ideal bandpass filter with center frequency $\mu\pi/M$ and (double-sided) bandwidth π/M.

According to the method presented above, the fullband impulse response of a fullband system h_i as been transformed into an equivalent set of subband impulse responses $h_{\mu,\,i}$. The decomposition was performed for $M = 16$ subbands and a subsampling rate $r = 12$. Figure A.2 shows the impulse response of the fullband system as well as the impulse response of the first ($\mu = 0$) subband. The noncausal part of the subband impulse response $h_{0,\,i}$ is clearly visible (even if the fullband impulse response was causal).

If the impulse response h_i would have nonzero values only at multiples of the subsampling rate $i = mr$, the resulting subband impulse responses would be causal. In this case the values h_{ir} would appear at the subband indices i, only weighted with the exponential term $e^{j\frac{2\pi\mu}{M}(ir-m)}$.

Generally, fullband impulse responses also have nonzero values at indices that are not multiples of the subsampling rate. The subband impulse responses $h_{\mu,\,i}$ have an infinite number of coefficients. The real and imaginary parts of the second subband impulse response $h_{1,\,i}$ are presented in Fig. A.3.

A.3 CONCLUDING REMARKS

Finally, we refer to the fact that the transformation of a fullband impulse response into an equivalent set of subband impulse responses has no unique solution. The general approach of setting the output of the fullband system $d(n)$ equal to the output the cascaded system, consisting of an analysis filterbank, subband filters, and a synthesis filterbank, has an infinite number of solutions [134]. When using this approach, it is also possible that one subband compensates the errors of another. In this case one gets a solution that is signal-dependent.

In our derivation we assume only that the prototype lowpass filter of the filterbanks has sufficient stopband attenuation and that the analysis synthesis system has nearly perfect reconstruction. If these demands are fulfilled, the set of subband impulse responses computed according to the previous section models the fullband system with high accuracy and in a signal-independent manner. Furthermore, in each subband a real system identification is achieved.

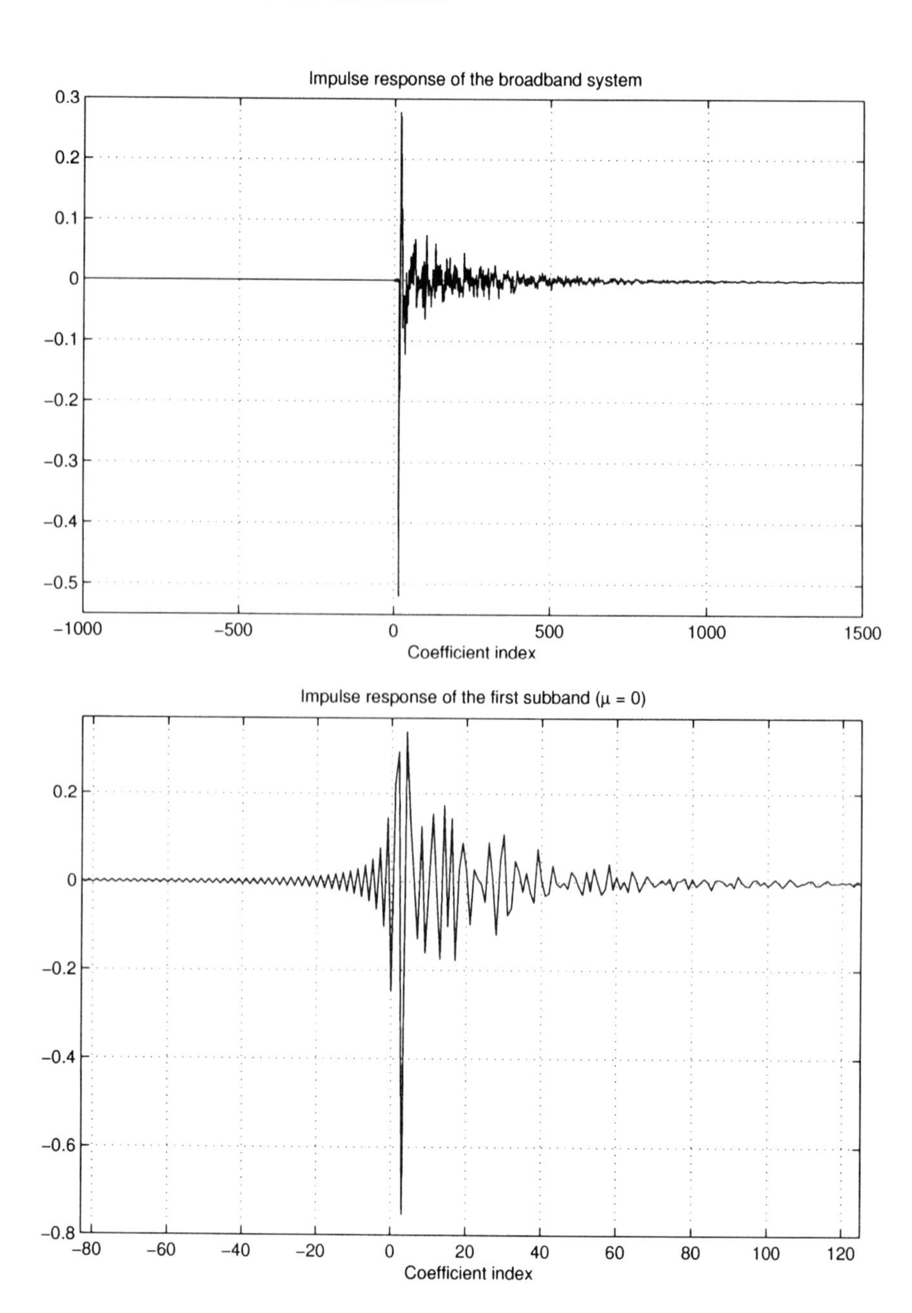

Fig. A.2 Impulse response measured in an office and corresponding first subband impulse response. The decomposition was performed for $M = 16$ channels and a subsampling rate $r = 12$. Discrete impulse response are defined only at discrete points. Thus, they are usually plotted by simple dots or circles. For the reason of clarity we have connected these points by straight lines.

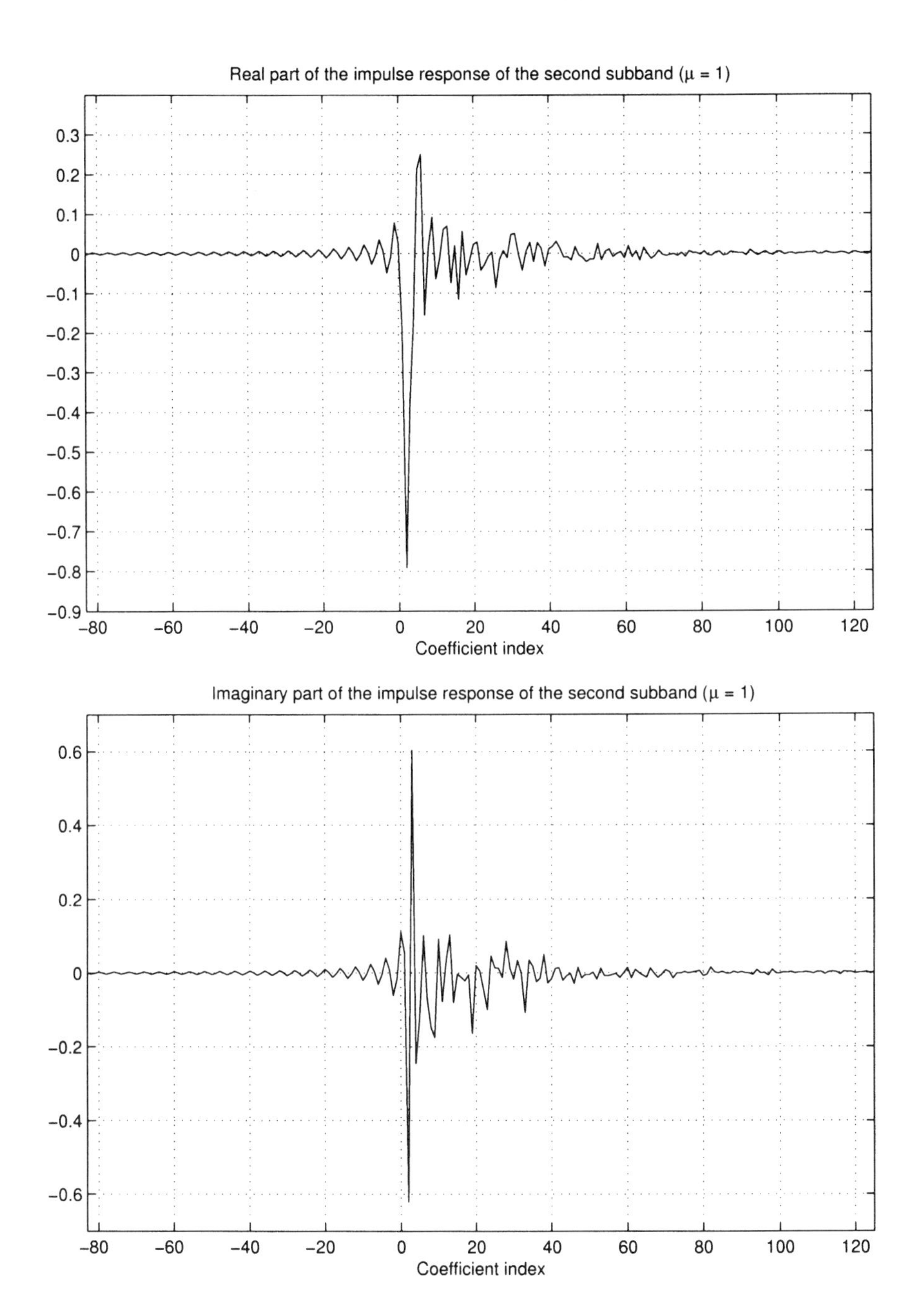

Fig. A.3 Second ($\mu = 1$) subband impulse response of an LEM system. As in Fig. A.2, the decomposition was applied for $M = 16$ subbands and a subsampling rate $r = 12$. Discrete impulse response are defined only at discrete points. Thus, they are usually plotted by simple dots or circles. For the reason of clarity we have connected these points by straight lines.

Appendix B
Filterbank Design

In this section the design of prototype lowpass filters for oversampled, DFT-modulated polyphase filterbanks is described. For this type of filterbank the aliasing terms can be kept small, making them well suited for the application of echo and noise control. Among a variety of design procedures[1] we have chosen the one presented in [231, 236]. The design process can be divided into four parts:

- First, conditions for a perfect reconstruction of the incoming signal after passing the analysis–synthesis system are derived. During this derivation it will be assumed that no subsampling is applied; accordingly, the aliasing terms can be neglected.

- We will see that the conditions for perfect reconstruction specify only the convolution of the analysis and synthesis prototype impulse responses $\psi_i = g_i * \tilde{g}_i$ (and not the filter coefficients themselves). The restrictions can be fulfilled with a simple product of two functions $\psi_{\mathrm{F},i}$ and $\psi_{\mathrm{B},i}$. The first one satisfies the reconstruction condition, while the second one can be designed to minimize aliasing components.

[1] See [230] for details as well as the references cited there.

- After deriving both functions, it is necessary to split the product $\psi_{\mathrm{F},i} \cdot \psi_{\mathrm{B},i}$ into its components g_i and $\tilde{g}_i$. To avoid an explicit computation of the zeros in the z domain, Boite's and Leich's method [17] is utilized. Here truncated power series of the complex cepstrum are used, leading to a maximum phase analysis and a minimum phase synthesis prototype lowpass filter.

- According to several approximations within the design process, it is necessary to verify whether the conditions for perfect reconstruction and neglecting aliasing terms are still fulfilled. If this is not the case, the design procedure should be restarted with modified parameters.

Figure B.1 shows the flowchart of the entire design procedure. In most cases only a few $(2, \cdots, 5)$ restarts are necessary until the final conditions are met.

During this section we will design—as an example—prototype filters for a filterbank with $M = 16$ channels. The boundary conditions are

- At most $N_{\mathrm{LP}} = 128$ coefficients should be used.

- The aliasing terms should be smaller than -40 dB.

- The ripple of the frequency response[2] of the entire filterbank system should not exceed 0.1 dB.

- The group delay should not vary by more than 0.5 samples.

B.1 CONDITIONS FOR APPROXIMATELY PERFECT RECONSTRUCTION

The demands on the absolute value of the frequency response of the analysis and the synthesis filter (e.g., 60 dB stopband attenuation and 0.5 dB passband ripple) allow to use the same filter for both stages. This may change if additional restrictions for the phase are introduced. It is desirable, for instance, that the entire system—consisting of an analysis and a synthesis part—has a linear phase response. In this case the group delay

$$\tau_{\mathrm{all}}(\Omega) = \frac{d\,\varphi_{\mathrm{all}}(\Omega)}{d\Omega} \tag{B.1}$$

would be independent of Ω and therefore equal for all frequencies. The next few lines will show how this could be—at least approximately—achieved. For this reason we set the impulse response of the prototype filter of the synthesis stage equal to the mirrored and shifted impulse response of the analysis filterbank:

$$\tilde{g}_i = g_{N_{\mathrm{LP}}-i}\,. \tag{B.2}$$

[2]For multirate systems the term *frequency response* is not explicitly defined. A more precise formulation can be found in Section B.4.

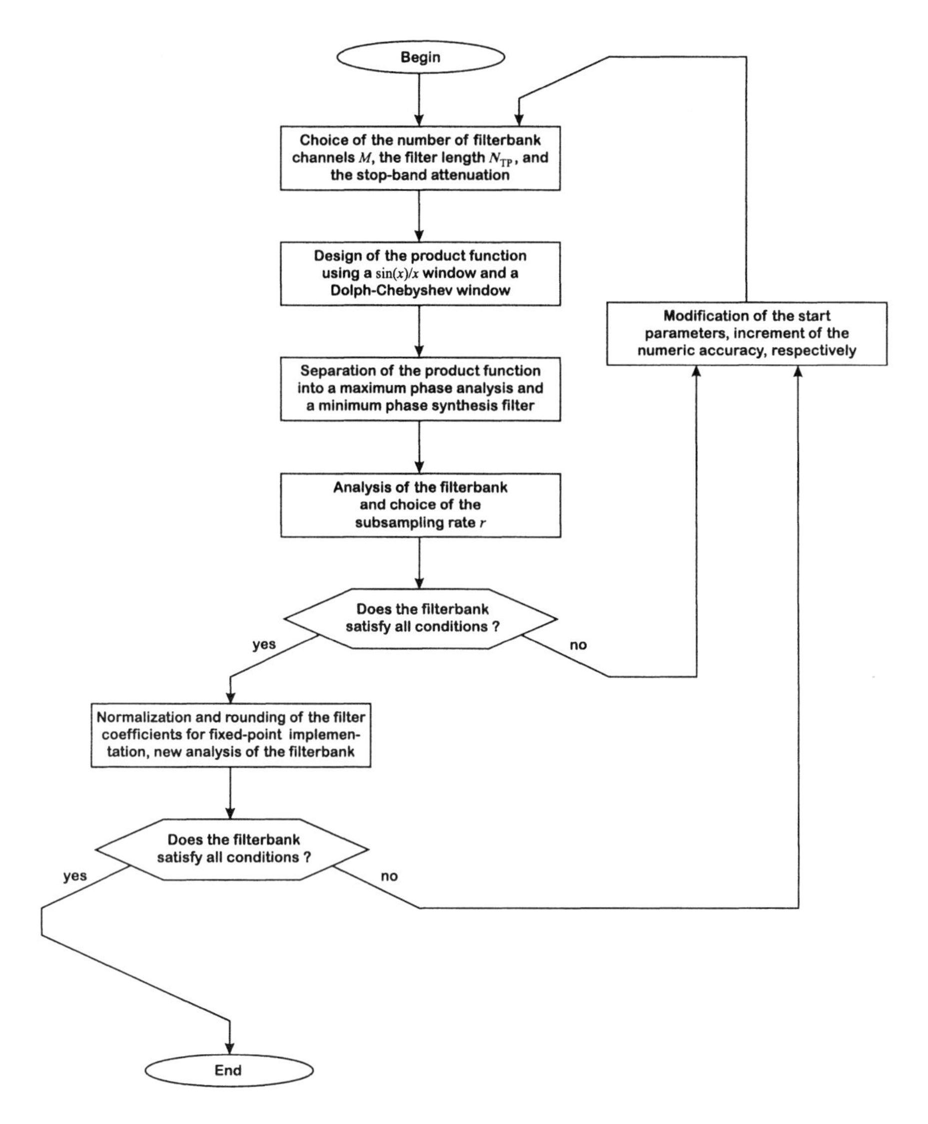

Fig. B.1 Structure of the filter design. To achieve optimal performance (e.g., minimal aliasing for a given subsampling rate r), typically only a few $(2, \cdots, 5)$ iterations are necessary.

Furthermore it is assumed that the impulse response of the analysis filter has FIR character $i \in \{0, \ldots, N_{\text{LP}} - 1\}$. In the frequency domain we obtain the following for the analysis response:

$$G(e^{j\Omega}) = \sum_{i=0}^{N_{\text{LP}}-1} g_i \, e^{-j\Omega i} \, . \tag{B.3}$$

For the frequency response of the synthesis filter, we get (assuming real coefficients g_i)[3]

$$\begin{aligned}
\tilde{G}\left(e^{j\Omega}\right) &= \sum_{i=-\infty}^{\infty} \tilde{g}_i \, e^{-j\Omega i} = \sum_{i=-\infty}^{\infty} g_{N_{\mathrm{LP}}-i} \, e^{-j\Omega i} \\
&= \sum_{i=-\infty}^{\infty} g_i \, e^{-j\Omega(N_{\mathrm{LP}}-i)} = e^{-j\Omega N_{\mathrm{LP}}} \sum_{i=0}^{N_{\mathrm{LP}}-1} g_i \, e^{j\Omega i} \\
&= e^{-j\Omega N_{\mathrm{LP}}} \, G^*\left(e^{j\Omega}\right) . \qquad \text{(B.4)}
\end{aligned}$$

Due to restriction B.2, both filters can use the same set of coefficients. With respect to hardware implementations, this means that a smaller amount of fixed memory (ROM) is required. If the analysis and the synthesis lowpass are connected in series, we get

$$\begin{aligned}
G\left(e^{j\Omega}\right) \, \tilde{G}\left(e^{j\Omega}\right) &= G\left(e^{j\Omega}\right) \, G^*\left(e^{j\Omega}\right) \, e^{-j\Omega N_{\mathrm{LP}}} \\
&= \left|G\left(e^{j\Omega}\right)\right|^2 \, e^{-j\Omega N_{\mathrm{LP}}} . \qquad \text{(B.5)}
\end{aligned}$$

The analysis filter can also be expressed in terms of the synthesis filter. In this case we obtain

$$\begin{aligned}
G\left(e^{j\Omega}\right) \, \tilde{G}\left(e^{j\Omega}\right) &= \tilde{G}^*\left(e^{j\Omega}\right) \, e^{-j\Omega N_{\mathrm{LP}}} \, \tilde{G}\left(e^{j\Omega}\right) \\
&= \left|\tilde{G}\left(e^{j\Omega}\right)\right|^2 \, e^{-j\Omega N_{\mathrm{LP}}} . \qquad \text{(B.6)}
\end{aligned}$$

$|G(e^{j\Omega})|^2$ and $|\tilde{G}(e^{j\Omega})|^2$, respectively, are real quantities. For this reason only the exponential term is responsible for the overall phase. The phase is linear in Ω, resulting in a constant group delay. The linear phase, which follows from the restriction $\tilde{g}_n = g_{\mathrm{LP}-n}$, is now clearly visible.

The entire filterbank system—consisting of M channels—should not distort the input signal (except for a delay of n_0 samples). This could be expressed by the following equation (the derivation follows):

$$\frac{1}{r} \sum_{\mu=0}^{M-1} G\left(e^{j\left(\Omega-\frac{2\pi\mu}{M}\right)}\right) \tilde{G}\left(e^{j\left(\Omega-\frac{2\pi\mu}{M}\right)}\right) = e^{-j\Omega n_0} . \qquad \text{(B.7)}$$

The condition above guarantees a perfect reconstruction if no subsampling was applied ($r = 1$). If the subsampling rate r would be chosen larger than one, aliasing terms will arise. These terms are not mentioned in Eq. B.7. In other applications, such as speech coding, it is often tried to design filterbanks in such a way that the aliasing terms compensate mutually. The compensation is perfect only for

[3]We will address only real input signals here. For cascaded filterbanks, the design procedure has to be repeated with complex convolutions.

through-connected filterbank systems. For the application of echo cancellation, this design would not be well suited. The cancellation of the aliasing terms in critically subsampled ($r = M$) filterbanks is performed within the synthesis stage of the filterbank. This means that after the analysis stage aliasing is present and distorts the convergence of the adaptive subband filters. For this reason we will mention here only oversampled ($r < M$) filterbanks.[4]

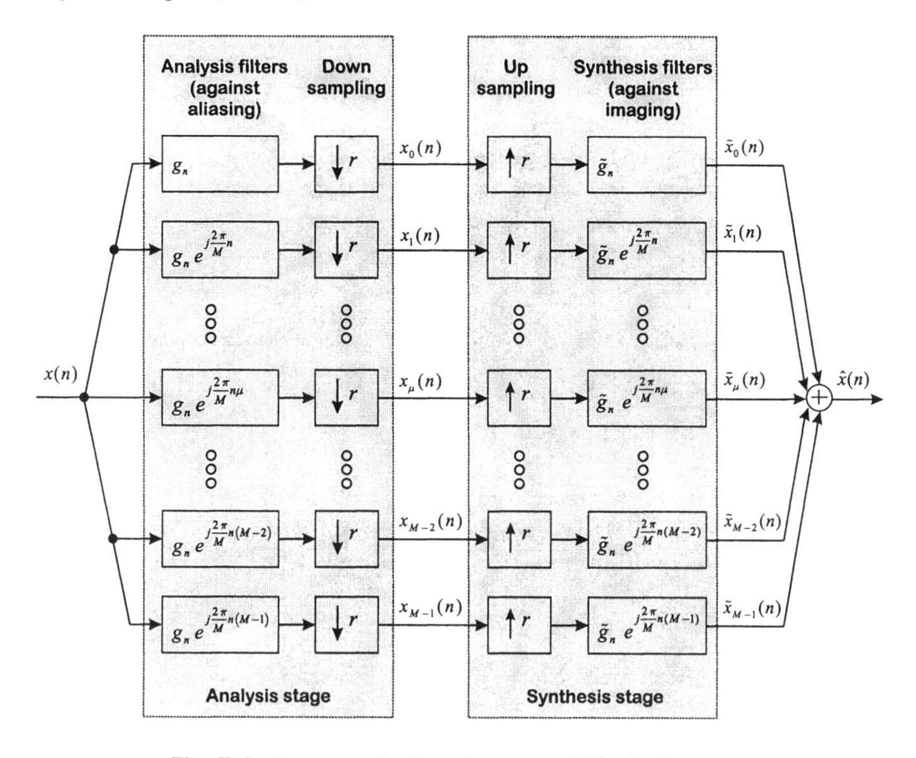

Fig. B.2 Structure of a through-connected filterbank system.

In case of a through-connected filterbank system (see Fig. B.2) we obtain for the spectrum of the subband signals $\tilde{x}_\mu(n)$:

$$\begin{aligned}
\tilde{X}_\mu(e^{j\Omega}) &= X_\mu(e^{j\Omega r})\ \tilde{G}\Big(e^{j(\Omega-\frac{2\pi}{M}\mu)}\Big) \\
&= \frac{1}{r}\sum_{m=0}^{r-1} X\Big(e^{j(\Omega-\frac{2\pi}{r}m)}\Big)\ G\Big(e^{j(\Omega-\frac{2\pi}{r}m-\frac{2\pi}{M}\mu)}\Big)\ \tilde{G}\Big(e^{j(\Omega-\frac{2\pi}{M}\mu)}\Big)\,.
\end{aligned} \tag{B.8}$$

[4]Another possibility is the use of so-called cross-filters [181, 183]. In this case the filterbank is designed to generate the main subband signals as well as the main aliasing components. Adaptive filters are adjusted for both types of signals. The disadvantage of this method is its high computational complexity (due to the additional aliasing filters).

The signals $\tilde{x}_\mu(n)$ consist of a linear part and aliasing components:

$$\begin{aligned}\tilde{X}_\mu\left(e^{j\Omega}\right) &= \frac{1}{r}\,X\left(e^{j\Omega}\right)\,\underbrace{G\left(e^{j\left(\Omega-\frac{2\pi}{M}\mu\right)}\right)\,\tilde{G}\left(e^{j\left(\Omega-\frac{2\pi}{M}\mu\right)}\right)}_{\text{linear transfer function}} \\ &+ \frac{1}{r}\sum_{m=1}^{r-1} X\left(e^{j\left(\Omega-\frac{2\pi}{r}m\right)}\right)\,\underbrace{G\left(e^{j\left(\Omega-\frac{2\pi}{r}m-\frac{2\pi}{M}\mu\right)}\right)\,\tilde{G}\left(e^{j\left(\Omega-\frac{2\pi}{M}\mu\right)}\right)}_{\text{aliasing components}}.\end{aligned} \tag{B.9}$$

The aliasing components should be kept as small as possible. For oversampled filterbanks it is possible—as we will see below—to minimize them, and we can approximate the spectra of the subband output signals as

$$\tilde{X}_\mu\left(e^{j\Omega}\right) \approx \frac{1}{r}\,X\left(e^{j\Omega}\right)\,G\left(e^{j\left(\Omega-\frac{2\pi}{M}\mu\right)}\right)\,\tilde{G}\left(e^{j\left(\Omega-\frac{2\pi}{M}\mu\right)}\right). \tag{B.10}$$

This approximation motivates us to use a design condition according to Eq. B.7. Inserting the restrictions that follow from $\tilde{g}_i = g_{N_{\mathrm{LP}}-i}$ into Eq. B.7 leads to

$$\begin{aligned}&\frac{1}{r}\sum_{\mu=0}^{M-1} G\left(e^{j\left(\Omega-\frac{2\pi\mu}{M}\right)}\right)\,\tilde{G}\left(e^{j\left(\Omega-\frac{2\pi\mu}{M}\right)}\right) \\ &\qquad = \frac{1}{r}\sum_{\mu=0}^{M-1}\left|\tilde{G}\left(e^{j\left(\Omega-\frac{2\pi\mu}{M}\right)}\right)\right|^2 e^{-jN_{\mathrm{LP}}\left(\Omega-\frac{2\pi\mu}{M}\right)} = e^{-j\Omega n_0}.\end{aligned} \tag{B.11}$$

If we demand N_{LP} and n_0 to be equal and to be a multiple of M

$$N_{\mathrm{LP}} = n_0 = k\,M \qquad \text{with} \qquad k \in \mathbf{N}, \tag{B.12}$$

we obtain

$$\frac{1}{r}\sum_{\mu=0}^{M-1}\left|G\left(e^{j\left(\Omega-\frac{2\pi\mu}{M}\right)}\right)\right|^2 = \frac{1}{r}\sum_{\mu=0}^{M-1}\left|\tilde{G}\left(e^{j\left(\Omega-\frac{2\pi\mu}{M}\right)}\right)\right|^2 = 1. \tag{B.13}$$

Transforming Eq. B.13 into the time domain yields

$$\frac{1}{r}\,\psi_i\sum_{\mu=0}^{M-1} e^{j\frac{2\pi\mu i}{M}} = \delta_{K,i}. \tag{B.14}$$

The term $\delta_{K,i}$ stands for the Kronecker function, and ψ_i abbreviates the following (equivalent) convolutions:

$$\begin{aligned}\psi_i &= g_i * g_{-i} = \sum_{\kappa=-\infty}^{\infty} g_\kappa\, g_{\kappa-i} \\ &= \tilde{g}_i * \tilde{g}_{-i} = \sum_{\kappa=-\infty}^{\infty} \tilde{g}_\kappa\, \tilde{g}_{\kappa-i}.\end{aligned} \tag{B.15}$$

The remaining sum—a finite geometric series—in Eq. B.14 can be simplified to

$$\sum_{\mu=0}^{M-1} e^{j\frac{2\pi\mu i}{M}} = \begin{cases} M & : \; i \in \{0, \pm M, \pm 2M, ...\} \\ 0 & : \; \text{else} \end{cases} \quad . \tag{B.16}$$

Finally, the result of the convolution ψ_i has to obey the following restriction:

$$\psi_i = \begin{cases} \frac{r}{M} & : \; i = 0, \\ 0 & : \; i \in \{\pm M, \pm 2M, ...\}, \\ \text{arbitrary} & : \; \text{else.} \end{cases} \tag{B.17}$$

B.2 FILTER DESIGN USING A PRODUCT APPROACH

The conditions for (approximately) perfect reconstruction can be fulfilled by designing the series ψ_i as a product of two subseries:

$$\psi_i = \psi_{\text{B},i} \, \psi_{\text{F},i} \, . \tag{B.18}$$

With $\psi_{\text{B},i}$ is a series that satisfies condition B.17 is denoted. Therefore, $\psi_{\text{B},i}$ has to be zero at $i \in \{\pm M, \pm 2M, ...\}$. If the function ψ_i is different from zero at $i = 0$, condition B.17 can be fulfilled easily by proper scaling. One possible series[5] that satisfies these properties is

$$\begin{aligned} \psi_{\text{B},i} &= \frac{1}{M} \,\text{sinc}\left(\frac{i\pi}{M}\right) \\ &= \begin{cases} \dfrac{1}{M} & : \; \text{for } i = 0, \\ \dfrac{1}{M} \dfrac{\sin\left(\frac{i\pi}{M}\right)}{\frac{i\pi}{M}} & : \; \text{else.} \end{cases} \end{aligned} \tag{B.19}$$

By transforming $\psi_{\text{B},i}$ into the frequency domain, we obtain

$$\Psi_{\text{B}}\left(e^{j\Omega}\right) = \begin{cases} 1 & : \; 0 \leq |\Omega| < \dfrac{\pi}{M}, \\ 0 & : \; \dfrac{\pi}{M} \leq |\Omega| \leq \pi. \end{cases} \tag{B.20}$$

This ideal lowpass filter has the following property:

$$\Psi_{\text{B}}\left(e^{j\Omega}\right) \, \Psi_{\text{B}}^*\left(e^{j\Omega}\right) = \Psi_{\text{B}}\left(e^{j\Omega}\right) \, . \tag{B.21}$$

[5] It is also possible to use a variety of other functions, such as weighted powers of the sinc(x) function. For the reason of simplicity we will mention here only the sinc function of first order. Furthermore, in some texts the factor π is omitted within the definition of the sinc function. We will use the definition $\text{sinc}(x) = \frac{\sin(x)}{x}$ (see Eq. B.19) here.

Using this property when adding shifted versions of $|\Psi_{\rm B}(e^{j\Omega})|^2$, we obtain

$$\sum_{\mu=0}^{M-1} \left|\Psi_{\rm B}\left(e^{j\left(\Omega-\frac{2\pi\mu}{M}\right)}\right)\right|^2 = 1\,. \tag{B.22}$$

By comparing this result with Eq. B.13, we see that the series $\psi_{{\rm B},n}$ (appropriate weighting assumed) is able to satisfy the filterbank restrictions without the second series $\psi_{{\rm F},i}$. Furthermore, the subsampling rate r could be chosen ideally to $r = M$, leading to a minimal computational complexity for any processing within the subsampled domain. However, the impulse response of the analysis and the synthesis lowpass filter would have infinite order and would be noncausal. Therefore, this direct solution is not applicable.

By multiplying with the second series $\psi_{{\rm F},i}$ the ideal lowpass can be truncated. Due to the equivalence of multiplication in the time domain and convolution in the frequency domain, the limitation of the ideal lowpass in the time domain leads to a widening of the passband in the frequency domain. For this reason, the spectrum of $\psi_{{\rm F},i}$ should have lowpass character with maximally narrow passband. In order to keep aliasing errors minimal, the stopband attenuation should be as large as possible. Finally we have to mention that the resulting function ψ_i corresponds in the frequency domain to a squared magnitude:

$$\begin{aligned}
\Psi\left(e^{j\Omega}\right) &= \sum_{m=-\infty}^{\infty} \psi_m\, e^{-j\Omega m} \\
&= \sum_{m=-\infty}^{\infty} \sum_{\kappa=-\infty}^{\infty} g_\kappa\, g_{\kappa-m}\, e^{-j\Omega m} \\
&= \sum_{\kappa=-\infty}^{\infty} g_\kappa \sum_{m=-\infty}^{\infty} g_{\kappa-m}\, e^{-j\Omega(m-\kappa)}\, e^{-j\Omega\kappa} \\
&= \sum_{\kappa=-\infty}^{\infty} g_\kappa\, e^{-j\Omega\kappa} \sum_{l=-\infty}^{\infty} g_l\, e^{-j(-\Omega)l} \\
&= G\left(e^{j\Omega}\right)\, G\left(e^{-j\Omega}\right) \\
&= \left|G\left(e^{j\Omega}\right)\right|^2 .
\end{aligned} \tag{B.23}$$

Therefore, the following condition has to be fulfilled:

$$\Psi\left(e^{j\Omega}\right) \geq 0\,, \quad \text{for all } \Omega\,. \tag{B.24}$$

Due to the three design criteria mentioned above, we chose $\psi_{{\rm F},i}$ as a modified Dolph–Chebyshev[6] lowpass filter:

$$\psi_{{\rm F},i} = K_{\rm F}\,\delta_{{\rm K},i} + \tilde{\psi}_{{\rm F},i}\,. \tag{B.25}$$

[6] In European textbooks on filter design the spelling or transliteration "Tschebyscheff" can also be found.

The term $K_{\mathrm{F}}\, \delta_{\mathrm{K},i}$ was added in order to guarantee the fulfillment of restriction B.24. The second term in Eq. B.25 denotes a Dolph–Chebyshev lowpass filter. This filter type has the property to minimize the width of the transition band at a given amount of stopband attenuation [53]. The frequency response of the filter is denoted by

$$\tilde{\Psi}_{\mathrm{F}}(e^{j\Omega}) = \begin{cases} K \, \cosh\left((N_{\mathrm{F}} - 1)\, \mathrm{arcosh}\left(\dfrac{\cos\left(\frac{\Omega}{2}\right)}{\cos\left(\frac{\Omega_{\mathrm{g}}}{2}\right)} \right) \right) & : \quad 0 \leq |\Omega| < \Omega_{\mathrm{g}}\,, \\ K \, \cos\left((N_{\mathrm{F}} - 1)\, \mathrm{arccos}\left(\dfrac{\cos\left(\frac{\Omega}{2}\right)}{\cos\left(\frac{\Omega_{\mathrm{g}}}{2}\right)} \right) \right) & : \quad \Omega_{\mathrm{g}} \leq |\Omega| \leq \pi\,. \end{cases} \tag{B.26}$$

The constant K is used for normalization purposes and should be chosen in such a way that condition B.17 is satisfied for $i = 0$. The parameters N_{F} and Ω_{g} denote the order and the cutoff frequency of the filter, respectively. In Fig. B.3 the frequency response of a Dolph–Chebyshev lowpass filter of order 255 and with a stopband attenuation of about -140 dB is depicted.

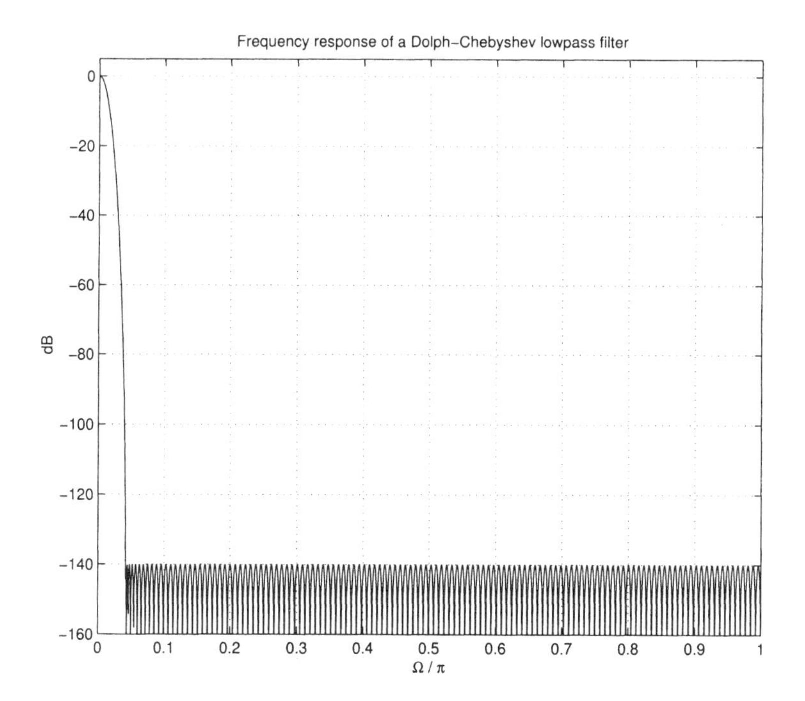

Fig. B.3 Frequency response of a Dolph–Chebyshev lowpass filter.

A specific property of Eq. B.26 is that the piecewise defined function can be transformed into a power series of $e^{j\Omega}$. By setting

$$\frac{\cos\left(\frac{\Omega}{2}\right)}{\cos\left(\frac{\Omega_g}{2}\right)} = \frac{1}{2\,\cos\left(\frac{\Omega_g}{2}\right)}\left(e^{j\frac{\Omega}{2}} + e^{-j\frac{\Omega}{2}}\right) = u\,, \tag{B.27}$$

the frequency response can be derived recursively. Using the start polynomials

$$\begin{aligned} T_0(u) &= 1, \\ T_1(u) &= u, \\ T_2(u) &= 2u^2 - 1 \end{aligned} \tag{B.28}$$

as well as Chebyshev's recursion

$$T_{k+1}(u) = 2uT_k(u) - T_{k-1}(u) \tag{B.29}$$

it is possible to rewrite Eq. B.26 as a Chebyshev polynomial of order N_F:

$$\tilde{\Psi}_F\left(e^{j\Omega}\right) = T_{N_F-1}\left(\frac{\cos\left(\frac{\Omega}{2}\right)}{\cos\left(\frac{\Omega_g}{2}\right)}\right). \tag{B.30}$$

Chebyshev polynomials have a further characteristic that motivates the choice of the argument[7] $e^{j\Omega}$ on the left side of Eq. B.26. Polynomials of even order $T_{2i}(u)$ only contain nonzero coefficients at even powers of u and odd polynomials only at odd powers of u:

$$T_k(u) = \begin{cases} t_0 + t_2\,u^2 + t_4\,u^4 + \cdots + t_k\,u^k & : \text{ if } k \text{ is even}, \\ t_1\,u + t_3\,u^3 + t_5\,u^5 + \cdots + t_k\,u^k & : \text{ else}. \end{cases} \tag{B.31}$$

In Fig. B.4 the graphs of the first six Chebyshev polynomials are depicted. The series ψ_i was generated by convolution of g_i und g_{-i}. For this reason the length of the window function has to be chosen as

$$N_F = 2N_{LP} - 1\,. \tag{B.32}$$

The resulting Chebyshev window should be of even order und should contain only powers of

$$u^2 = \frac{1}{4\,\cos^2\left(\frac{\Omega_g}{2}\right)}\left(e^{j\Omega} + 2 + e^{-j\Omega}\right). \tag{B.33}$$

The frequency response of the Chebyshev lowpass filter can be transformed into a finite power series:

$$\tilde{\Psi}_F\left(e^{j\Omega}\right) = \sum_{n=-N_{LP}+1}^{N_{LP}-1} \psi_{F,n}\; e^{j\Omega n}\,. \tag{B.34}$$

[7]By transforming the arguments of the cos function, one gets at first only powers of $e^{\,j\Omega/2}$.

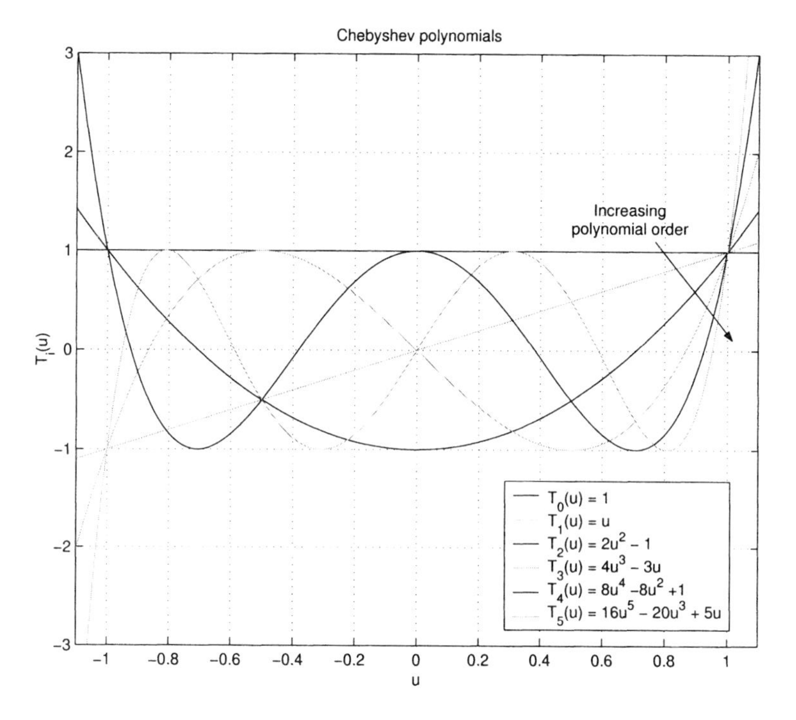

Fig. B.4 Graph of the first six Chebyshev polynomials.

Because of this constraint, the coefficients $\tilde{\psi}_{\mathrm{F},n}$ can be computed in a straightforward way by utilizing an inverse DFT. Therefore, the closed-form solution from Eq. B.26 is sampled at equidistant frequency supporting points and the resulting vector is transformed into the time domain via an inverse DFT.

Because of the modification term $K_\mathrm{F}\,\delta_{\mathrm{K},i}$ (see Eq. B.25), the frequency response of the modified Dolph–Chebyshev lowpass filter is raised by a constant value K_F:

$$\Psi_\mathrm{F}\big(e^{j\Omega}\big) = \begin{cases} K_\mathrm{F} + K\,\cosh\left((N_\mathrm{F}-1)\,\mathrm{arcosh}\left(\dfrac{\cos\left(\frac{\Omega}{2}\right)}{\cos\left(\frac{\Omega_\mathrm{g}}{2}\right)}\right)\right) & :\ 0 \le |\Omega| < \Omega_\mathrm{g}, \\ K_\mathrm{F} + K\,\cos\left((N_\mathrm{F}-1)\arccos\left(\dfrac{\cos\left(\frac{\Omega}{2}\right)}{\cos\left(\frac{\Omega_\mathrm{g}}{2}\right)}\right)\right) & :\ \Omega_\mathrm{g} \le |\Omega| \le \pi. \end{cases} \tag{B.35}$$

If $K_\mathrm{F} = K$ is chosen, the entire window function will always be larger than or equal to zero:

$$\Psi_\mathrm{F}\big(e^{j\Omega}\big) \ge 0\,. \tag{B.36}$$

The same is true for the frequency response of the ideal lowpass filter $\Psi_{\mathrm{B}}(e^{j\Omega})$. Because a convolution of two nonnegative functions again generates a nonnegative function, the frequency response $\Psi(e^{j\Omega})$ is also nonnegative:

$$\Psi\left(e^{j\Omega}\right) \geq 0\,. \tag{B.37}$$

If the constant K_{F} was chosen as mentioned above, the stopband attenuation would be reduced by 6 dB. For the resulting analysis and synthesis lowpass filters, this would cause a degradation by 3 dB. Close examinations show that condition B.24 can be fulfilled with values K_{F} even smaller than K. In this case the stopband attenuation would be reduced only marginally.

The condition concerning the absolute value of the convolution of $\Psi_{\mathrm{F}}(e^{j\Omega})$ and $\Psi_{\mathrm{B}}(e^{j\Omega})$ is only of importance within the stopband. In this area $\Psi_{\mathrm{F}}(e^{j\Omega})$ is a cosine function with amplitude K, raised by the positive constant K_{F}. During the convolution the Chebyshev window will be integrated over a few periods (depending on the width of the rectangular window $\Psi_{\mathrm{B}}(e^{j\Omega})$) of the cosine function. In order to obtain a nonnegative result, it is not necessary to move the cosine function completely into the positive range. It is sufficient that during the integration over the integer number of periods of the cosine function (which are included in the rectangle window) a value is obtained which is larger than the integral over one negative half of one cosine period. This is possible with quite small values $0 < K_{\mathrm{F}} << K$. In this case the stopband attenuation can be increased by a few decibels. This "fine tuning" of the filter design is done as follows:

- Start with the choice $K_{\mathrm{F}} = K$ and compute the minimum of $\Psi(e^{j\Omega})$ and the stopband attenuation.
- Halve the constant K_{F} and repeat the procedure.
- As soon as the minimum becomes negative or the stopband attenuation is reduced by more than 5 dB, stop the iteration and use the previous or the actual value of K_{F}, respectively.

On one hand, the window function $\psi_{\mathrm{F},i}$ determines the length of the final lowpass filter, which is $N_{\mathrm{LP}} = (N_{\mathrm{F}} + 1)/2$. On the other hand, the final stopband attenuation will also be dominated by the window function. The filter design is based on the specification of $g_i * g_{-i}$. For this reason, the attenuation of the Dolph–Chebyshev filter should be chosen twice as large as the desired stopband attenuation of the final analysis and synthesis filters.

The finite and real series $\tilde{\psi}_{\mathrm{F},i}$ is obtained by applying an inverse DFT to a vector of N_{F} equidistant chosen frequency supporting points of the frequency response $\tilde{\Psi}_{\mathrm{F}}(e^{j\Omega})$. By adding the weighted Kronecker function, the series $\psi_{\mathrm{F},i}$ is generated. Finally the multiplication with $\psi_{\mathrm{B},\,i}$ results in the filter autocorrelation function ψ_i. This finite and real series fulfills the design criteria summarized by Eq. B.17. The design procedure has to be repeated with different cutoff frequencies of the Dolph–Chebyshev filter until $\Psi(e^{j\Omega})$ achieves the desired properties in terms of transition bandwidth and stopband attenuation.

Both the primary series $\psi_{B,i}$ and the window function $\psi_{F,i}$ are even and real series. For this reason, ψ_i is real and even, too. These properties can be transformed into the z domain. If we compute the zeros of $\Psi(z)$, we will obtain only pairs of complex and corresponding conjugate complex zeros (resulting from the real-valued time-domain function). Furthermore, the symmetry in the time domain leads also to the following property in the z domain: a zero at $z = z_0$ appears always together with a zero at $z = z_0^{-1}$ (see Fig. B.5).

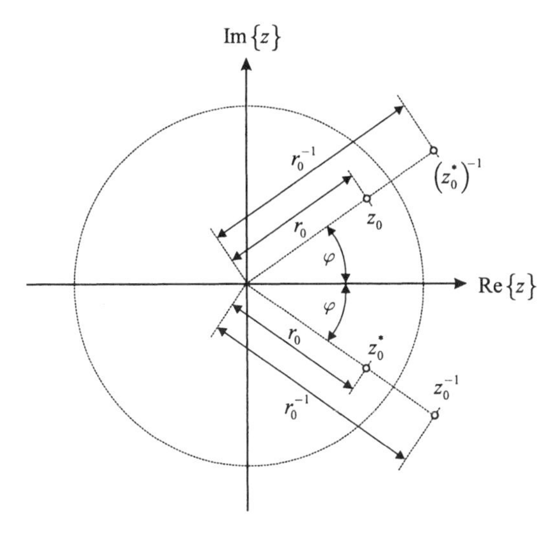

Fig. B.5 Quadruples of zeros. Due to the real and symmetric time-domain series, quadruples of zeros always appear at $z_0 = r_0\, e^{j\varphi}$, z_0^*, z_0^{-1}, and $(z_0^*)^{-1}$ (respectively pairs at z_0 and z_0^{-1}, if $\varphi = 0$).

There are several ways to compute the impulse response of the analysis filter g_i from the zeros of $\Psi(z)$. For each quadruple of zeros two possibilities exist. One of the zeros and its conjugate can be assigned to the analysis filter. In this case the zeros with the inverse magnitude belong to the synthesis filter. The second possibility is just the inverse assignment. In the long run, one has $2^{N_{\mathrm{LP}}/2}$ possible choices for the assignments.

For the application of acoustic echo cancellation, it is desirable to have a maximum phase analysis and a minimum phase synthesis filterbank (see Section 9.1.3.3). Therefore, all zeros within the unit circle have to be assigned to the synthesis filter and all zeros outside the unit circle should belong to the analysis filter. For this assignment a method for separating the zeros presented by Boite and Leich [17] is utilized. Special about this method is that it is not necessary to explicitly compute all zeros of $\Psi(z)$.

B.3 DESIGN OF PROTOTYPE LOWPASS FILTERS

For computing the impulse response of the analysis and the synthesis lowpass we start with frequency response $\Psi(e^{j\Omega})$, which is defined as

$$\Psi\big(e^{j\Omega}\big) = \tilde{G}\big(e^{j\Omega}\big)\ \tilde{G}^*\big(e^{j\Omega}\big) = \Big|\tilde{G}\big(e^{j\Omega}\big)\Big|^2 . \tag{B.38}$$

This means that up until now only the absolute value of the Fourier transform of $\tilde{g}_i$ is determined. The synthesis filter should have minimum phase. Therefore, it is possible to compute the phase out of the magnitude of $\tilde{G}\big(e^{j\Omega}\big)$. The direct way of solving this problem would be a polynomial factorization of $\Psi(z)$. By collecting all zeros inside the unit circle, we would obtain $\tilde{G}(z)$. Because of the finite precision of digital computers and the large order (here 254), such a computation of zeros would be a very challenging and difficult problem. For that reason, Boite and Leich [17] suggest the development of a power series of the logarithm of $\tilde{G}(e^{j\Omega})$. In the logarithmic frequency domain absolute value and phase are separated additively:

$$\begin{aligned} \ln\Big(\tilde{G}\big(e^{j\Omega}\big)\Big) &= \ln\Big(\Big|\tilde{G}\big(e^{j\Omega}\big)\Big|\ e^{j\arg\{\tilde{G}(e^{j\Omega})\}}\Big) \\ &= \ln\Big(\Big|\tilde{G}\big(e^{j\Omega}\big)\Big|\Big) + j\ \arg\Big\{\tilde{G}\big(e^{j\Omega}\big)\Big\}. \end{aligned} \tag{B.39}$$

The real part of this function is already known:

$$\mathrm{Re}\Big\{\ln\Big(\tilde{G}\big(e^{j\Omega}\big)\Big)\Big\} = \ln\Big(\sqrt{\Psi\big(e^{j\Omega}\big)}\Big). \tag{B.40}$$

The development of a power series of the logarithmic frequency response leads to the so-called *complex cepstrum* $\bar{g}_i$ [174]:

$$\ln\Big(\tilde{G}\big(e^{j\Omega}\big)\Big) = \sum_{i=0}^{\infty} \bar{g}_i\, e^{-j\Omega i} . \tag{B.41}$$

Herewith we exploited the property, that minimum phase frequency responses result in causal cepstral series [17]. The summation can be rearranged as follows:

$$\begin{aligned} \sum_{i=0}^{\infty} \bar{g}_i\, e^{-j\Omega i} &= \sum_{l=0}^{\infty} \sum_{i=0}^{N_\mathrm{B}} \bar{g}_{lN_\mathrm{B}+i}\, e^{-j\Omega(lN_\mathrm{B}+i)} \\ &= \sum_{i=0}^{N_\mathrm{B}} \sum_{l=0}^{\infty} \bar{g}_{lN_\mathrm{B}+i}\, e^{-j\Omega(lN_\mathrm{B}+i)} . \end{aligned} \tag{B.42}$$

By setting

$$\breve{g}_i = \sum_{l=0}^{\infty} \bar{g}_{lN_\mathrm{B}+i} \tag{B.43}$$

and computing the logarithmic frequency response only at discrete frequencies

$$\Omega = \frac{2\pi}{N_{\mathrm{B}}} k \quad \text{with} \quad k \in \{0, \ldots, N_{\mathrm{B}} - 1\} , \tag{B.44}$$

we obtain

$$\begin{aligned} \ln\left(\tilde{G}\left(e^{j\frac{2\pi}{N_{\mathrm{B}}}k}\right)\right) &= \sum_{i=0}^{N_{\mathrm{B}}} \breve{g}_i \, e^{-j\frac{2\pi}{N_{\mathrm{B}}}ki} \\ &= \mathrm{DFT}\left\{\breve{g}_{ln,i}\right\} \\ &= \alpha_{\mathrm{B},i} \; + \; j\,\beta_{\mathrm{B},i} \,. \end{aligned} \tag{B.45}$$

It can be shown [17], that for sufficiently large N_{B} the coefficients $\bar{g}_i$ can be approximated by

$$\bar{g}_i \approx 0, \quad \text{for} \quad i \geq \frac{N_{\mathrm{B}}}{2}. \tag{B.46}$$

In this case the DFT can be applied directly to the series $\bar{g}_i$ (instead of $\breve{g}_i$):

$$\begin{aligned} \mathrm{DFT}\left\{\bar{g}_i\right\} &\approx \mathrm{DFT}\left\{\breve{g}_i\right\} \\ &= \alpha_{\mathrm{B},i} \; + \; j\,\beta_{\mathrm{B},i} \,. \end{aligned} \tag{B.47}$$

The real part $\alpha_{\mathrm{B},i}$ and the imaginary part $\beta_{\mathrm{B},i}$ correspond approximately to the (known) logarithmic absolute value and the (unknown) phase, respectively, of the synthesis lowpass filter:

$$\alpha_{\mathrm{B},i} \approx \ln\left(\left|\tilde{G}\left(e^{j\frac{2\pi}{N_{\mathrm{B}}}i}\right)\right|\right), \tag{B.48}$$

$$\beta_{\mathrm{B},i} \approx \arg\left\{\tilde{G}\left(e^{j\frac{2\pi}{N_{\mathrm{B}}}i}\right)\right\}. \tag{B.49}$$

The real part can be computed by transforming only the even part of the sequence $\bar{g}_i$. Likewise, the imaginary portion is computed by applying a DFT to the odd part of $\bar{g}_i$:

$$\left\{\alpha_{\mathrm{B},i}\right\} \approx \mathrm{DFT}\left\{\bar{g}_{\mathrm{even},i}\right\}, \tag{B.50}$$

$$\left\{\beta_{\mathrm{B},i}\right\} \approx \mathrm{DFT}\left\{\bar{g}_{\mathrm{odd},i}\right\}. \tag{B.51}$$

The even and the odd parts of the series $\bar{g}_i$ are defined as follows:

$$\bar{g}_{\mathrm{even},i} = \frac{1}{2}\left(\bar{g}_i + \bar{g}_{N_{\mathrm{B}}-i}\right), \tag{B.52}$$

$$\bar{g}_{\mathrm{odd},i} = \frac{1}{2}\left(\bar{g}_i - \bar{g}_{N_{\mathrm{B}}-i}\right), \tag{B.53}$$

$$\text{with} \qquad i \in \{0, \ldots, N_{\mathrm{B}} - 1\} .$$

By applying an inverse DFT to the logarithmic magnitude sequence $\alpha_{B,i}$, the even sequence $\bar{g}_{\text{even},i}$ can be computed. Solving the Eqs. B.52 and B.53 (and using the approximation B.46) leads to the odd sequence:

$$\bar{g}_{\text{odd},i} \approx \begin{cases} \bar{g}_{\text{even},i} & : \quad 0 < i < \frac{N_B}{2}, \\ -\bar{g}_{\text{even},i} & : \quad \frac{N_B}{2} < i < N_B, \\ 0 & : \quad i = 0, i = \frac{N_B}{2}. \end{cases} \tag{B.54}$$

Again, by applying a DFT the phase of the synthesis lowpass can be computed. Now the filter is completely defined. What remains is to investigate the influence of approximation B.46. For that reason, the squared magnitude of the synthesis frequency response is compared with the function $\Psi(e^{j\Omega})$. In the case that the variations remain under a predefined threshold the computed filter will be utilized as the synthesis prototype. The analysis filter can be computed by shifting and inverse addressing

$$g_i = \tilde{g}_{N_{LP}-i} \,. \tag{B.55}$$

In the other case, the constant N_B has to be increased and the assignment of the zeros has to be repeated. After having performed several designs the choice $N_B = 8\,N_{LP}$ has proved itself as a good starting value. If this choice is not sufficient N_B should be doubled for the next design iteration.

Figure B.6 shows the impulse response g_i and the absolute value of the frequency response of an analysis prototype filter designed with the method described above. The length of the filter was $N_{LP} = 128$. The design was done for a filterbank with $M = 16$ channels.

B.4 ANALYSIS OF PROTOTYPE FILTERS AND THE FILTERBANK SYSTEM

Quantization of the filter coefficients and numerical inaccuracies can affect the design process negatively. For this reason, the initial conditions for approximately perfect reconstruction (Eqs. B.13 and B.17) and the desired lowpass characteristics (stopband attenuations, etc.) should be verified at the end of the design procedure.

The condition for a linear phase of the entire system

$$\tilde{g}_i = g_{N_{LP}-i} \tag{B.56}$$

is not affected by quantization. Therefore, no further verification is necessary.

Due to numerical effects, the frequency response of the analysis–synthesis system (without subsampling)

$$G_{\text{ent}}\left(e^{j\Omega}\right)\Big|_{r=1} = \sum_{\mu=0}^{M-1} G\left(e^{j\left(\Omega-\frac{2\pi\mu}{M}\right)}\right) \tilde{G}\left(e^{j\left(\Omega-\frac{2\pi\mu}{M}\right)}\right) \tag{B.57}$$

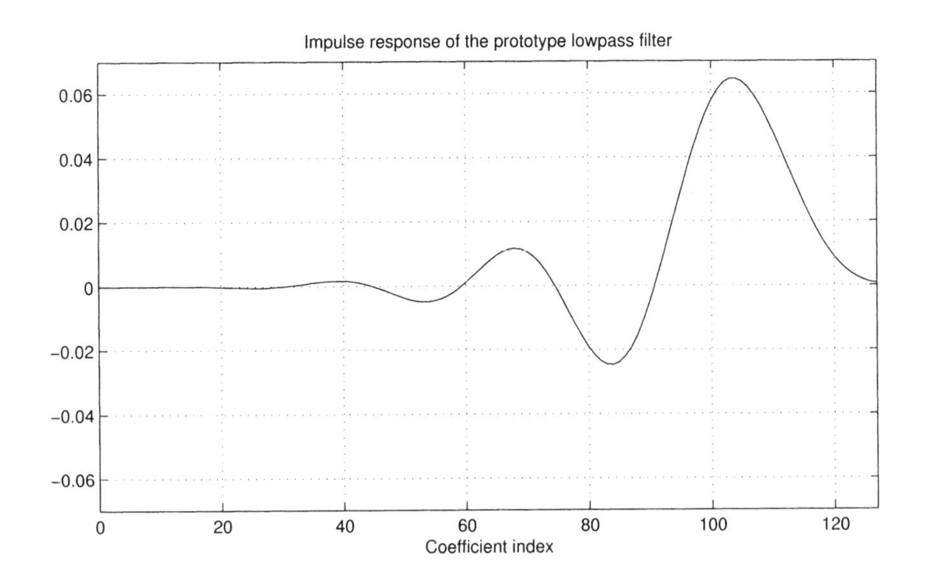

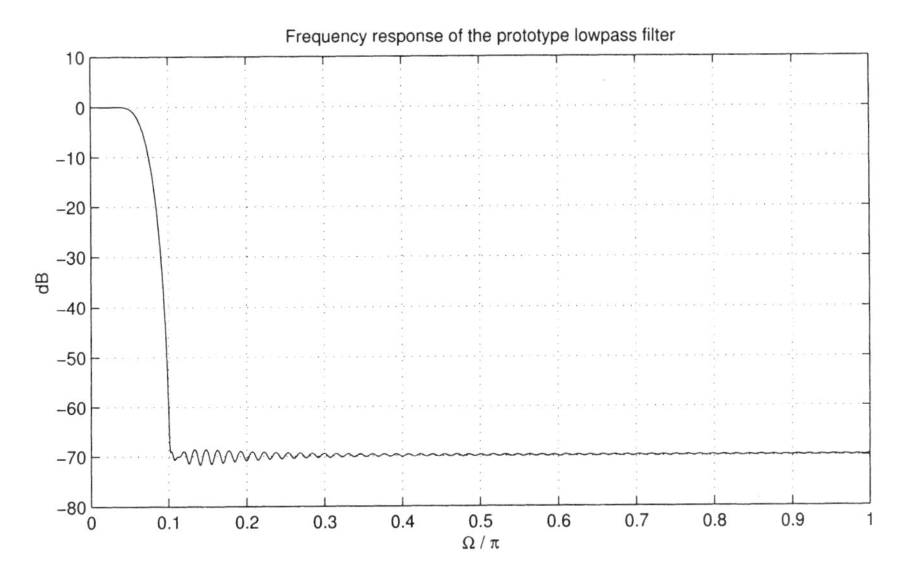

Fig. B.6 Impulse and frequency response of the designed analysis prototype lowpass filter. The filter is designed for a filterbank with $M = 16$ bands. Its length was chosen to $N_{\mathrm{LP}} = 128$.

differs from the ideal response

$$G_{\text{ent, ideal}}\left(e^{j\Omega}\right) = e^{-j\Omega N_{\mathrm{LP}}}\,. \tag{B.58}$$

In our design example, the derivations after 16-bit quantization and appropriate scaling are smaller than 0.06 dB (see Fig. B.7). Therefore, the prototype lowpass filter is well designed also for fixed-point implementations.

However, the subsampling rate r has not been specified yet. If this parameter cannot be chosen sufficiently large, the whole design process has to be restarted (with different starting parameters). For finding the maximum subsampling rate, the aliasing components, caused by down- and upsampling, have to be analyzed. For the spectra of the subsampled signals $x_\mu(n)$ (see Fig. B.2), we obtain

$$X_\mu\left(e^{j\Omega}\right) = \frac{1}{r}\sum_{m=0}^{r-1} X\left(e^{j\left(\frac{\Omega}{r}-\frac{2\pi}{r}m\right)}\right) G\left(e^{j\left(\frac{\Omega}{r}-\frac{2\pi}{r}m-\frac{2\pi}{M}\mu\right)}\right). \tag{B.59}$$

For the spectrum of the reconstructed signal $\hat{x}(n)$ (assuming a through connection of the analysis and the synthesis stage), we have already derived

$$\hat{X}\left(e^{j\Omega}\right) = \frac{1}{r}\sum_{\mu=0}^{M-1} \tilde{G}\left(e^{j\left(\Omega-\frac{2\pi}{M}\mu\right)}\right) \sum_{m=0}^{r-1} G\left(e^{j\left(\Omega-\frac{2\pi}{r}m-\frac{2\pi}{M}\mu\right)}\right) X\left(e^{j\left(\Omega-\frac{2\pi}{r}m\right)}\right). \tag{B.60}$$

For analyzing the transmission characteristics of the entire filterbank system the *bi-frequency response* according to [47, 195] can be utilized. Another analysis would be to use the periodic time variance of the system [229]. In this case the filterbank system can be described by a set of r impulse responses. The Fourier transforms of these responses will differ only marginally in our case. For this reason, we will analyze only the response of the system to an impulse at time index $n = 0$

$$x(n) = \delta_{\text{K}}(n). \tag{B.61}$$

In the frequency domain we obtain

$$X\left(e^{j\Omega}\right) = X\left(e^{j\left(\Omega-\frac{2\pi}{r}m\right)}\right) = 1. \tag{B.62}$$

The response of the filterbank system (for this special input signal) is given by

$$G_{\text{ent}}\left(e^{j\Omega}\right)\Big|_{X(e^{j\Omega})=1} = \frac{1}{r}\sum_{\mu=0}^{M-1} \tilde{G}\left(e^{j\left(\Omega-\frac{2\pi}{M}\mu\right)}\right) \sum_{m=0}^{r-1} G\left(e^{j\left(\Omega-\frac{2\pi}{r}m-\frac{2\pi}{M}\mu\right)}\right). \tag{B.63}$$

For evaluation of the designed system, the absolute value of the frequency response $G_{\text{ent}}\left(e^{j\Omega}\right)$ and the group delay are depicted in Fig. B.7. The lowpass impulse response presented in Fig. B.6 was employed. All 128 coefficients were rounded according to a 16-bit fixed-point arithmetic. The filterbank consisted of $M = 16$ channels, and a subsampling rate $r = 12$ was used. The output signal of the through-connected system was attenuated or amplified maximally by 0.052 dB. The

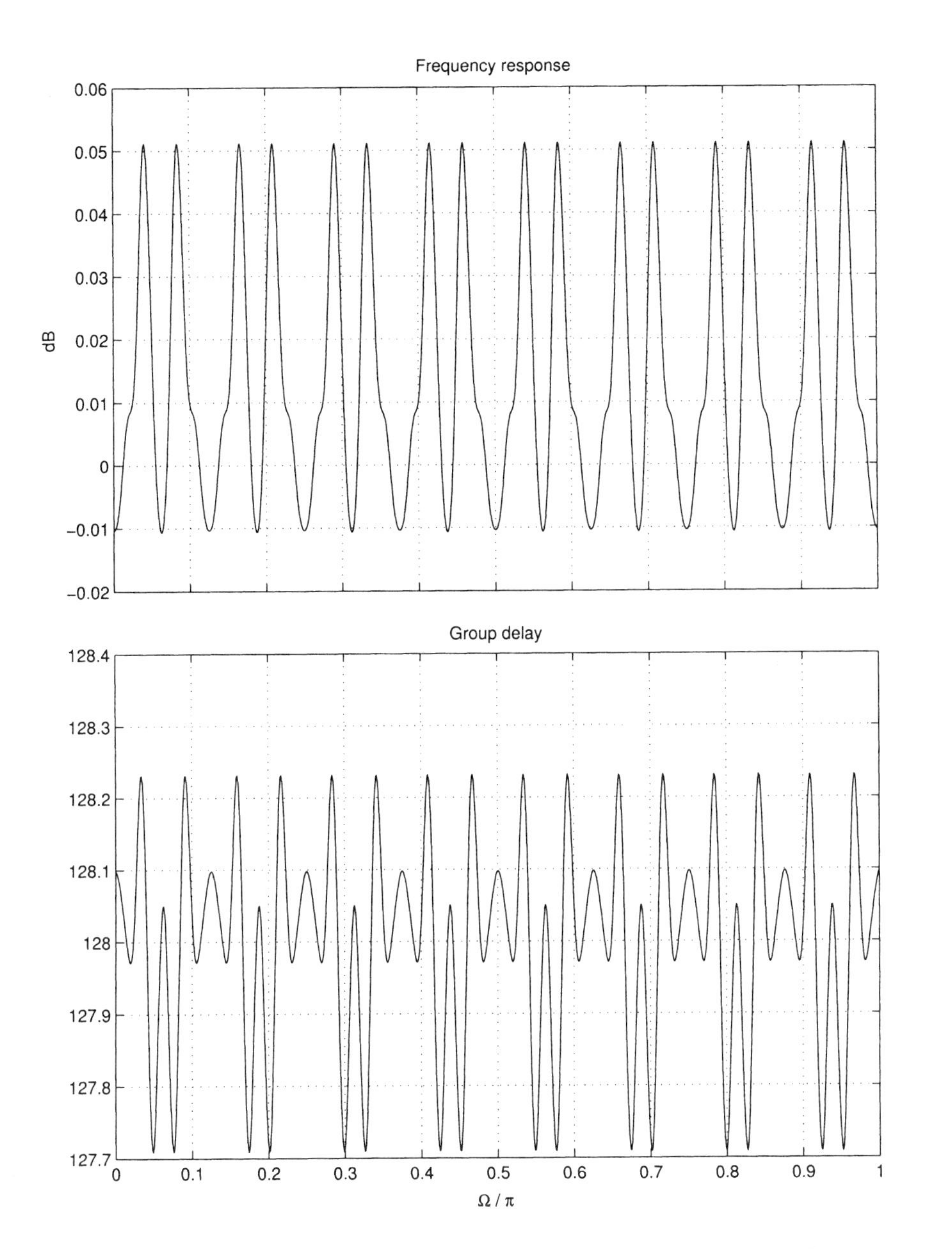

Fig. B.7 Transmission properties of the through-connected filterbank system. The upper diagram shows the logarithmic absolute value of the frequency response $G_{\text{ent}}(e^{j\Omega})$. The maximum deviation from the ideal response $G_{\text{ent,ideal}}(e^{j\Omega}) = 1$ is about 0.05 dB. In the lower diagram the group delay of the filterbank system is depicted. The maximum deviation is about 0.3 samples. The lowpass filter depicted in Fig. B.6 was utilized. The filterbank consists of $M = 16$ channels, and a subsampling rate $r = 12$ was chosen.

maximum deviation of the group delay from its average value (128 samples) was about 0.3 sample.

With these values a nearly distortion-free subband decomposition and a succeeding synthesis is possible. Nevertheless, for echo cancellation purposes the size of the aliasing components within the subband domain are of special importance. Aliasing terms cannot be modeled by a single linear filter within each subband. For this reason, aliasing is reducing the adaptation speed as well as the final misadjustment. In subband μ this type of distortion can be described by

$$A_\mu\left(e^{j\Omega}\right) = \frac{1}{r}\sum_{n=1}^{r-1} G\left(e^{j\left(\Omega - \frac{2\pi\mu}{M} - \frac{2\pi n}{r}\right)}\right). \tag{B.64}$$

After upsampling, these terms will be filtered with the synthesis bandpasses. For the (fullband) aliasing term $A(e^{j\Omega})$ we finally obtain:

$$\begin{aligned} A\left(e^{j\Omega}\right) &= \sum_{\mu=0}^{M-1} \tilde{G}\left(e^{j\left(\Omega - \frac{2\pi\mu}{M}\right)}\right) A_\mu\left(e^{j\Omega}\right) \\ &= \frac{1}{r}\sum_{\mu=0}^{M-1} \tilde{G}\left(e^{j\left(\Omega - \frac{2\pi\mu}{M}\right)}\right) \sum_{n=1}^{r-1} G\left(e^{j\left(\Omega - \frac{2\pi\mu}{M} - \frac{2\pi n}{r}\right)}\right). \end{aligned} \tag{B.65}$$

Figure B.8 shows the absolute value of $A(e^{j\Omega})$ for three different subsampling rates ($r = 11, 12, 13$). As in the last examples, the prototype impulse response depicted in the upper part of Fig. B.6 was used. For small subsampling rates ($r \leq 11$) the aliasing distortion stays at a small level (here $\leq$ –55 dB). This advantage has to be compensated by an increased amount of processing power and memory for the signal processing in the subband area.

If the subsampling rate is chosen too large ($r \geq 13$) the aliasing components become quite large (here $\geq$ –25 dB), especially at the band edges. Echo cancellation filter (if only a single filter in each subband is used) are not able to model these nonlinear distortions. A decreased final misadjustment arises. In practical systems for echo and noise control aliasing up to –40 dB can be tolerated, which results in a choice of $r = 12$ (see center diagram of Fig. B.8). With this choice all requirements, stated at the beginning of this section, can be fulfilled.

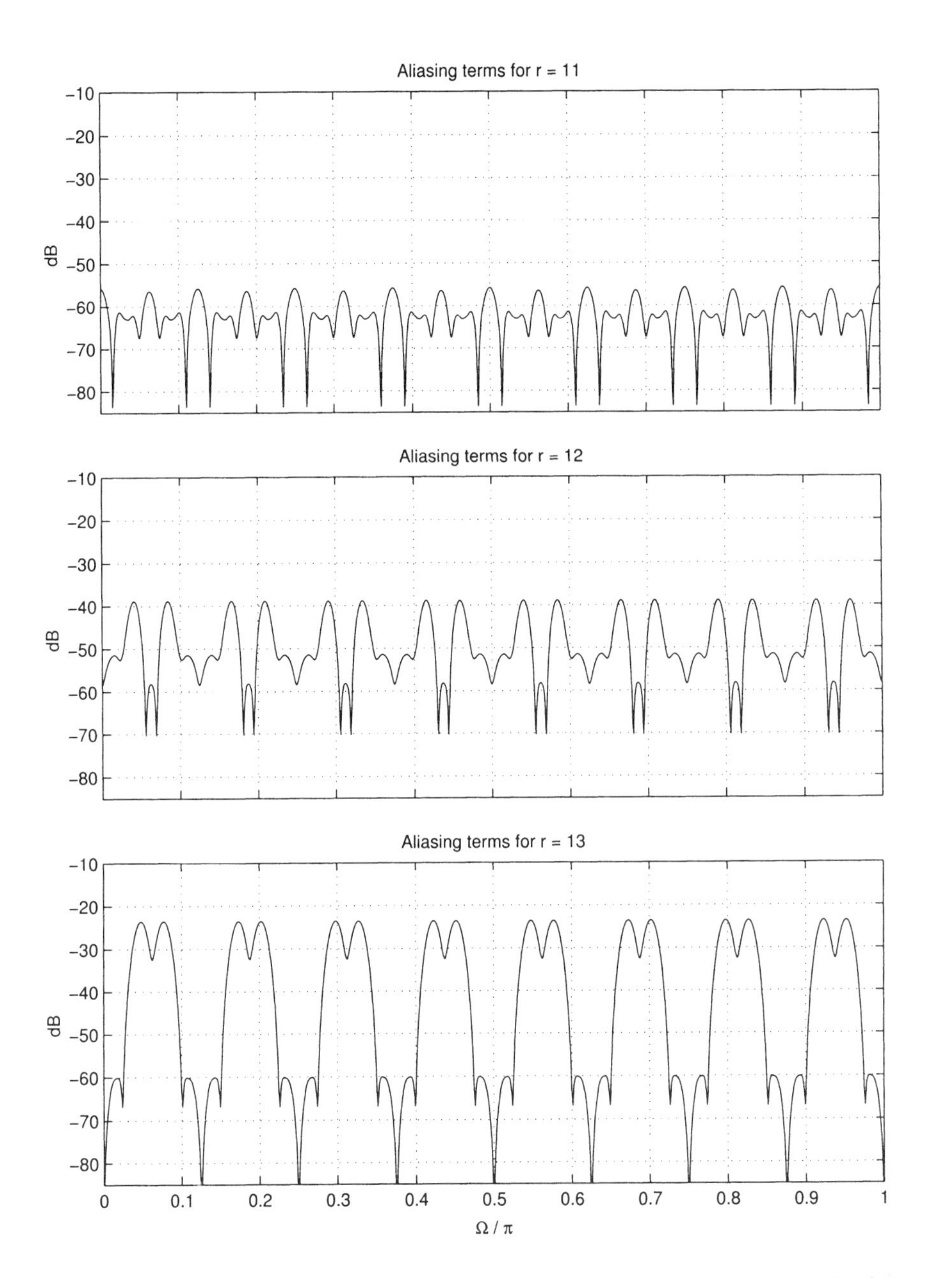

Fig. B.8 Aliasing components of the through-connected filterbank system. The absolute value of the aliasing spectrum is depicted for different subsampling rates (top $r = 11$, center $r = 12$, bottom $r = 13$).

References

1. The New Bell Telephone, *Sci. Am.* **37**(1), 1(1877).

2. T. Aboulnasr, K. Mayyas: Selective coefficient update of gradient-based adaptive algorithms, *Proc. ICASSP '97,* **3**, 1929–1932, Munich, Germany, 1997.

3. T. Aboulnasr, K Mayyas: MSE analysis of the M-Max NLMS adaptive algorithm, *Proc. ICASSP '98,* **3**, 1669–1672, Washington, DC, 1998.

4. M. Abramowitz, I. A. Stegun: *Handbook of Mathematical Functions,* New York, NY: Dover Publications, 1974.

5. B. S. Atal, L. R. Rabiner: A pattern recognition approach to voiced-unvoiced-silence classifiaction with applications to speech recognition, *IEEE Trans. Acoust. Speech Signal Process.,* **24**(3), 201–212, 1976.

6. R. Aubauer, D. Leckschat: Optimized second-order gradient microphone for hands-free speech recordings in cars, *Speech Communication,* **34**(1,2), 13–23, 2001.

7. C. Beaugeant, V. Turbin, P. Scalart, A. Gilloire: New optimal filtering approaches for hands-free telecommunication terminals, *Signal Process.,* **64**(1), 33–47, 1998.

8. M. G. Bellanger: *Adaptive Digital Filters,* 2nd ed., New York, NY: Marcel Dekker, 2001.

9. A. Benallal, A. Gilloire: A new method to stabilize fast RLS algorithms in transversal adaptive filters, *Proc. ICASSP '88,* 1373–1376, New York, NY, 1988.

10. J. Benesty, F. Amand, A. Gilloire, Y. Grenier: Adaptive filtering algorithms for stereophonic acoustic echo cancellation, *Proc. ICASSP '95,* **5**, 3099–3102, Detroit, MI, 1995.

11. J. Benesty, D. R. Morgan, M. M. Sondhi: A better understanding and an improved solution to the problems of stereophonic acoustic echo Cancellation, *Proc. ICASSP '97,* **1**, 303–306, Munich, Germany, 1997.

12. J. Benesty, D. R. Morgan, M. M. Sondhi: A Better Understanding and an Improved Solution to the Specific Problems of Stereophonic Acoustic echo Cancellation, *IEEE Trans. Speech Audio Process.,* **T-SA-6**(2), 156–165, 1998.

13. J. Benesty, D. R. Morgan, J. H. Cho: A new class of double-talk detectors based on cross correlation, *IEEE Trans. Speech Audio Process.,* **T-SA-8**(2), 168–172, 2000.

14. J. Benesty, T. Gänsler, D. R. Morgan, M. M. Sondhi, S. L. Gay: *Advances in Network and Acoustic Echo Cancellation,* Berlin, Germany: Springer, 2001.

15. D. A. Berkley, O. M. M. Mitchell: Seeking the ideal in "hands-free" telephony, *Bell Lab. Rec.,* **52**, 318–325, 1974.

16. J. Bitzer, K. U. Simmer: Superdirective microphone arrays, in M. Brandstein, D. Ward (eds.), *Microphone Arrays – Signal Processing Techniques and Applications,* Berlin, Germany: Springer, 2001, 19–38.

17. R. Boite, H. Leich: A new procedure for the design of high-order minimum-phase FIR-digital or CCD filters, *Signal Process.,* **3**(2), 101–108, 1980.

18. S. F. Boll: Suppression of acoustic noise in speech using spectral subtraction, *IEEE Trans. Acoust. Speech Signal Process.,* **27**(2), 113–120, 1979.

19. J. Botto, G. Moustakides: Stabilizing the fast Kalman algorithm, *IEEE Trans. Acoust. Speech Signal Process.,* **37**(9), 1342–1348, 1989.

20. R. Le Bouquin Jeannes, P. Scalart, G. Faucon, C. Beaugeant: Combined noise and echo reduction in hands-free systems: a survey, *IEEE Trans. Speech Audio Process.,* **T-SA-9**(8), 808–820, 2001.

21. K. Brammer, G. Siffling: *Kalman-Bucy Filters*, Artech House, Norwood, MA, 1989.

22. M. Brandstein, D. Ward (Eds.): *Microphone Arrays – Signal Processing Techniques and Applications,* Berlin, Germany: Springer, 2001.

23. H. Brehm, W. Stammler: Description and generation of sperically invariant speech-model signals, *Signal Process.*, **12**(2), 119–141, 1987.

24. C. Breining: Applying a neural network for stepsize control in echo cancellation, *Proc. IWAENC '97*, 41–44, London, UK, 1997.

25. C. Breining, P. Dreiseitel, E. Hänsler, A. Mader, B. Nitsch, H. Puder, T. Schertler, G. Schmidt, J. Tilp: Acoustic echo control, *IEEE Signal Process. Mag.*, **16**(4), 42–69, 1999.

26. C. Breining: A Robust fuzzy logic-based step-gain control for adaptive filters in acoustic echo Cancellation, *IEEE Trans. Speech Audio Process.*, **T-SA-9**(2), 162–167, 2001.

27. C. Breining: State detection for hands-free telephone sets by means of fuzzy LVQ and SOM, *Signal Process.*, **80**(7), 1361–1372, 2000.

28. M. Buck: Mehrkanalige Geräuschunterdrückungssysteme mit adaptiver Selbstkalibrierung, *Proc. ESSV '03*, 138–145, Karlsruhe, Germany, 2003 (in German).

29. K. M. Buckley: Spatial/spectral filtering with linearly constrained minimum variance beamformers, *IEEE Trans. Acoust. Speech Signal Process.*, **ASSP-35**(3), 249–266, 1987.

30. T. Burger, U. Schultheiss: A robust acoustic echo canceller for a hands-free voice-controlled telecommunication terminal, *Proc. EUROSPEECH '93*, **3**, 1809–1812, Berlin, Germany, 1993.

31. T. Burger: Practical application of adaptation control for NLMS-algorithms used for echo cancellation with speech signals, *Proc. IWAENC '95*, 87–90, Røros, Norway, 1995.

32. J. P. Campbell, T. E. Tremain: Voiced/unvoiced classification of speech with applications to the U.S. Government LPC-10e algorithm, *Proc. ICASSP '86*, **1**, 473–476, 1986.

33. O. Cappé: Elimination of the musical noise phenomenon, *IEEE Trans. Speech Audio Process.*, **T-SA-2**(2), 345–349, 1994.

34. G. Carayannis, D. Manolakis, N. Kalouptsidis: A fast sequential algorithm for the LS filtering and prediction, *IEEE Trans. Signal Process.*, **ASSP-32**, 1394–1402, 1983.

35. C. Caraiscos, B. Liu: A roundoff error analysis of the LMS adaptive algorithm, *IEEE Trans. Acoust. Speech Signal Process.*, **32**(1), 33–41, 1994.

36. J. Chang, J.R. Glover: The feedback adaptive line enhancer: a constrained IIR adaptive filter, *IEEE Trans. Signal Process.*, **41**(11), 3161–3166, 1993.

37. J. H. Chen, A. Gersho: Adaptive postfiltering for quality enhancement of coded speech, *IEEE Trans. Speech Audio Process.,* **T-SA-3**(1), 59–71, 1995.

38. C. K. Chui, G. Chen: *Kalman Filtering with Real-Time Applications*, 3rd ed., Berlin, Germany: Springer, 1999.

39. J. M. Cioffi, T. Kailath: Fast, recursive-least-squares transversal filters for adaptive filtering, *IEEE Trans. Acoust., Speech, Signal Process.,* **ASSP-32**, 304–337, 1984.

40. J. M. Cioffi: Limited-precision effects in adaptive filtering, *IEEE Trans. Circuits Systems,* **34**(7), 821–833, 1987.

41. W. F. Clemency, F. F. Romanow, A. F. Rose: The Bell System Speakerphone, *AIEE. Trans.,* **76**(I), 148–153, 1957.

42. W. F. Clemency, W. D. Goodale Jr.: Functional design of a voice–switched Speakerphone, *B.S.T.J.,* **40**, 649-668, 1961.

43. D. Van Copernolle: Switching adaptive filters for enhancing noisy and reverberant speech from microphone array recordings, *Proc. ICASSP '90,* **2**, 833–836, Albuquerque, NM, 1990.

44. D. Van Copernolle, S. Van Gerven: Beamforming with microphone arrays, in A.R. Figueiras–Vidal (ed.), *Digital Signal Processing to Telecommunications,* Cost 229, 1995, 107–131.

45. A. G. Constantinides, K.M. Knill, J.A. Chambers: A novel orthogonal set adaptive line enhancer tuned with fourth-order cumulants, *Proc. ICASSP '92,* **4**, 241–244, 1992.

46. H. Cox, R. Zeskind, M. Oven: Robust adaptive beamforming, *IEEE Trans. Acoust. Speech Signal Process.,* **ASSP-35**(10), 1365–1376, 1987.

47. R. E. Crochiere, L. R. Rabiner: *Multirate Digital Signal Processing,* Englewood Cliffs, NJ: Prentice Hall, 1983.

48. W. B. Davenport: An experimental study of speech wave probability distributions, *J. Acoust. Soc. Am.,* **24**(4), 390–399, 1952.

49. J. Deller, J. Hansen, J. Proakis: *Discrete-Time Processing of Speech Signals,* New York, NY: IEEE Press, 1993.

50. P. S. R. Diniz: *Adaptive Filtering – algorithms and Practical Implementations,* Boston, MA: Kluwer, 1997.

51. G. Doblinger: Computationally efficient speech enhancement by spectral minima tracking in subbands, *Proc. EUROSPEECH '95,* **2**, 1513–1516, Madrid, Spain, 1995.

52. M. Dörbecker: *Mehrkanalige Signalverarbeitung zur Verbesserung akustisch gestörter Sprachsignale am Beispiel elektonischer Hörhilfen,* Aachen, Germany: Verlag der Augustinus Buchhandlung, 10, 1998 (in German).

53. C. Dolph: A current distribution for broadside arrays which optimizes the relationship between beam width and sidelobe level, *Proc. IRE,* **35** 335–348, 1946.

54. S. Douglas: Adaptive filters employing partial updates, *IEEE Trans. Circuits Systems,* **44**(3), 209–216, 1997.

55. P. Dreiseitel, H. Puder: Speech enhancement for mobile telephony based on non-uniformly spaced frequency resolution, *Proc. EUSIPCO '98,* **2**, 965–968, Rhodos, Greece, 1998.

56. P. Dreiseitel: Quality measures for single-channel speech enhancement algorithms, *Proc. IWAENC '01,* 9–13, Darmstadt, Germany, 2001.

57. J. Durbin: The fitting of time series models, *Rev. Int. Stat. Inst.,* **28**, 233–244, 1960.

58. G. W. Elko: Superdirectional microphone arrays, in S. L. Gay, J. Benesty (eds.), *Acoustic Signal Processing for Telecommunication,* Boston, MA: Kluwer, 2000, 181–237.

59. G. W. Elko: Spatial coherence functions for differential microphones in isotropic noise fields, in M. Brandstein, D. Ward (eds.), *Microphone Arrays – Signal Processing Techniques and Applications,* Berlin, Germany: Springer, 2001.

60. G. Enzner, R. Martin, P. Vary: On spectral estimation of residual echo in handsfree telephony, *Proc. IWAENC '01,* 211–214, Darmstadt, Germany, 2001.

61. Y. Epharaim, D. Malah: Speech enhancement using a minimum mean-square error short-time spectral amplitude estimator, *IEEE Trans. Acoust. Speech Signal Process.,* **32**(6), 1109–1121, 1984.

62. Y. Epharaim, D. Malah: Speech enhancement using a minimum mean-square error log-spectral amplitude estimator, *IEEE Trans. Acoust. Speech Signal Process.,* **33**(2), 443–445, 1985.

63. ETS 300 580-6 (GSM 06.10 version 4.3.1): *Digital Cellular Telecommunication System (Phase 2), Full Rate Speech, Transcoding, Part 6: Voice Activity Detection (VAD) for Full Rate Speech Traffic Channels,* ETSI, France, 2nd edition, 1995.

64. ETS 300 903 (GSM 03.50): *Transmission Planning Aspects of the Speech Service in the GSM Public Land Mobile Network (PLMS) System,* ETSI, France, 1999.

65. ETS 300 961 (GSM 06.10): *Digital Cellular Telecommunication System (Phase 2+), Full Rate Speech,* ETSI, France, 3rd edition, 2000.

66. P. Eykhoff: *System Identification – Parameter and State Estimation,* New York, NY: Wiley, 1974.

67. H. Ezzaidi, I. Bourmeyster, J. Rouat: A new algorithm for double talk detection and separation in the context of digital mobile radio telephone, *Proc. ICASSP '97,* **3**, 1897–1900, Munich, Germany, 1997.

68. A. Feuer, E. Weinstein: Convergence analysis of LMS filters with uncorrelated Gaussian data, *IEEE Trans. Acoust. Speech Signal Process.,* **ASSP-33**(1), 222–230, 1985.

69. R. Frenzel: *Freisprechen in gestörter Umgebung*, Düsseldorf, Germany: Fortschr.-Ber. VDI-Reihe 10(228), VDI Verlag, 1992 (in German).

70. R. Frenzel, M. Hennecke: Using prewhitening and stepsize control to improve the performance of the LMS algorithm for acoustic echo compensation, *Proc. ISCAS '92,* **4**, 1930–1932, San Diego, CA, 1992.

71. O. L. Frost, III: An algorithm for linearily constrained adaptive array processing, *Proc. IEEE,* **60**(8), 926–935, 1972.

72. T. Gänsler, M. Hansson, C.-J. Ivarsson, G. Salomonsson: A double-talk detector based on coherence, *IEEE Trans. Commun.,* **COM-44**(11), 1421–1427, 1996.

73. T. Gänsler: A double-talk resistant subband echo canceller, *Signal Process.,* **65**(1), 89–101, 1998.

74. T. Gänsler, S. Gay, M. Sondhi, J. Benesty: Double-talk robust fast converging algorithms for network echo cancellation, *IEEE Trans. Speech Audio Process.,* **T-SA-8**(6), 656–663, 2000.

75. S. Gay, S. Travathia: The fast affine projection algorithm, *Proc. ICASSP '95,* **3**, 3023–3027, Detroit, MI, 1995.

76. S. Gay, J. Benesty (eds.): *Acoustic Signal Processing for Telecommunications,* Boston, MA: Kluwer, 2000.

77. S. Gay: Affine projection algorithms, in S. Haykin, B. Widrow (eds.), *Least-Mean-Square Adaptive filters,* New York, NY: Wiley, 2003, 241–291.

78. M. Ghanassi, B, Champagne: On the fixed–point implementation of a subband acoustic echo canceler based on a modified FAP algorithm, *Proc. IWAENC '99,* 128-131, Pocono Manor, NJ, 1999.

79. H. W. Gierlich: A measurement technique to determine the transfer characteristic of hands-free telephones, *Signal Process.,* **27**(3), 281–300, 1992.

80. A. Gilloire: State of the art in acoustic echo cancellation, in A.R. Figueiras, D. Docampo (eds.), *Adaptive algorithms: Applications and Non Classical Schemes,* Universidad de Vigo, Spain, 20–31, 1991.

81. A. Gilloire, E. Moulines, D. Slock, P. Duhamel: State of the art in acoustic echo cancellation, in A.R. Figueiras–Vidal (ed.), *Digital Signal Processing in Telecommunications,* Springer, London, UK, 1996, 45–91.

82. A. Gilloire: Current issues in stereophonic and multi-channel acoustic echo cancellation, *Proc. IWAENC '97,* 5–8, London, UK, 1997.

83. A. Gilloire, V. Turbin: Using auditory properties to improve the behaviour of stereophonic acoustic echo cancellation, *Proc. ICASSP '98,* **6**, 3681–3684, Washington, DC, 1998.

84. A. A. Giordano, F. M. Hsu: *Least Square Estimation with Application to Digital Signal Processing,* New York, NY: Wiley, 1985.

85. A. Gilloire, P. Scalart, C. Lamblin, C. Mokbel, S. Proust: Innovative speech processing for mobile terminals: an annotated bibliography, *Signal Process.,* **80**(7), 1149–1166, 2000.

86. G. Glentis, K. Berberidis, S. Theodoridis: Efficient least squares adaptive algorithms for FIR transversal filtering: a unified view, *IEEE Signal Process. Mag.,* **16**(4), 13–41, 1999.

87. A. H. Gray, J. D. Markel: Distance measures for speech processing, *IEEE Trans. Acoust. Speech Signal Process.,* **ASSP-24**(5), 380–391, 1976.

88. L. J. Griffiths: Frequency tracking using an adaptive oscillator, in M. Kunt, F. De Coulon (eds.), *Signal Processing: Theory and Applications,* Amsterdam, The Netherlands: North-Holland Publishing Co., 1980.

89. L. J. Griffiths, C. W. Jim: An alternative approach to linearly constrained adaptive beamforming, *IEEE Tans. Antennas and Propagation,* **AP-30**(1), 24–34, 1982.

90. T. Gülzow, A. Engelsberg, U. Heute: Comparision of a discrete wavelet transformation and a nonuniform polyphase filterbank applied to spectral–subtraction speech enhancement, *Signal Process.,* **64**(1), 5–19, 1998.

91. S. Gustafsson, R. Martin, P. Vary: Combined acoustic echo control and noise reduction for hands-free telephony, *Signal Process.,* **64**(1), 21–32, 1998.

92. S. Gustafsson, P. Jax: Combined residual echo and noise reduction: a novel psychoacoustically motivated algorithm, *Proc. EUSIPCO '98,* **2**, 961–964, Rhodos, Greece, 1998.

93. S. Gustafsson, P. Jax, A. Kamphausen, P. Vary: A postfilter for echo and noise reduction avoiding the problem of musical tones, *Proc. ICASSP '99,* **2**, 873–876, Phoenix, AR, 1999.

94. R. Hacker: *Einsatzbedingungen für Mikrofon-Arrays in der Praxis,* master's thesis, Universtity of Ulm, Germany, 2002 (in German).

95. E. Hänsler: The hands-free telephone problem – an annotated bibliography, *Signal Process.,* **27**(3), 259–271, 1992.

96. E. Hänsler: The hands-free telephone problem – an annotated bibliography update, *Annales des Télécommunications,* **49**, 360–367, 1994.

97. E. Hänsler: The hands-free telephone problem – a second annotated bibliography update, *Proc. IWAENC '95,* 107–114, Røros, Norway, 1995.

98. E. Hänsler, G. Schmidt: Hands-free telephones – joint control of echo cancellation and post filtering, *Signal Process.,* **80**(11), 2295–2305, 2000.

99. E. Hänsler: *Statistische Signale – Grundlagen und Anwendungen,* Berlin, Germany: Springer, 2001 (in German).

100. E. Hänsler, G. Schmidt: Single–channel acoustic echo cancellation, in J. Benesty, Y. Huang (eds.), *Adaptive Signal Processing*, Berlin, Germany: Springer, 2003, 59–93.

101. E. Hänsler: Acoustic echo cancellation, in J. Proakis (ed.), *Encyclopedia of Telecommunications*, New York, NY: Wiley, 2003, 1–15.

102. E. Hänsler, G. Schmidt: Control of LMS-type adaptive filters, in S. Haykin, B. Widrow (eds.), *Least-Mean-Square adaptive Filters,* New York, NY: Wiley, 2003, 175–240.

103. V. Hamacher: Comparison of advanced monaural and binaural noise reduction algorithms for hearing aids, *Proc. ICASSP '02,* **4**, 4008–4011, Orlando, FL, 2002.

104. Y. Haneda, S. Makino, J. Kojima, S. Shimauchi: Implementation and evaluation of an acoustic echo canceler using duo-filter control system, *Proc. EUSIPCO '96,* **2**, 1115–1118, Trieste, Italy, 1996.

105. J. H. L. Hanson: Morphological constrained feature enhancement with adaptive cepstral compensation (MCE-ACC) for speech recognition in noise and Lombard effect, *IEEE Trans. Speech Audio Process.,* **T-SA-2**(4), 598–614, 1994.

106. D. Hardmann: Noise and voice quality in VoIP environments, in G.M. Davis (ed.), *Noise reduction in Speech Applications,* Boca Raton, FL: CRC Press, 2002, 277–304.

107. T. Haulick: Beamformer applications in cars, *Fifth International Workshop on Microphone Array Systems,* Erlangen, Germany, 2003.

108. T. Haulick, G. Schmidt: Signalverarbeitungskomponenten zur Verbesserung der Kommunikation in Fahrzeuginnenräumen, *Proc. ESSV '03,* 130–137, Karlsruhe, Germany, 2003 (in German).

109. M. H. Hayes: *Statistical Digital Signal Processing and Modeling,* New York, NY: Wiley, 1996.

110. S. Haykin (ed.): *Kalman Filtering and Neural Networks,* New York, NY: Wiley, 2001.

111. S. Haykin: *Adaptive Filter Theory,* 4th ed., Englewood Cliffs, NJ: Prentice Hall, 2002.

112. D. A. Heide, G. S. Kang: Speech enhancement for bandlimited speech, *Proc. ICASSP '98,* **1**, 393–396, Washington, DC, 1998.

113. P. Heitkämper, M. Walker: Adaptive gain control and echo cancellation for hands-free telephone systems, *Proc. EUROSPEECH '93,* **3**, 107–1080, Berlin, Germany, 1993.

114. P. Heitkämper: An adaptation control for acoustic echo cancellers, *IEEE Signal Process. Letters,* **4**(6), 170–172, 1997.

115. W. Hess: *Pitch Determination of Speech Signals,* Berlin, Germany: Springer, 1983.

116. M. L. Honig, D. G. Messerschmitt: *Adaptive Filters: Structures, Algorithms, and Applications,* Boston, MA: Kluwer, 1984.

117. O. Hoshuyama, B. Begasse, A. Sugiyama, A. Hirano: A realtime robust adaptive microphone array, *Proc. ICASSP '98,* **6**, 3605–3608, Washington, DC, 1998.

118. O. Hoshuyama, A. Sugiyama, A. Hirano: A robust adaptive beamformer for microphone arrays with a blocking matrix using constrained adaptive filters, *IEEE Trans. Signal Process.,* **47**(10), 2677–2684, 1999.

119. T. Hoya, Y. Lore, J. A. Chamber, P. A. Naylor: Application of the leaky extended LMS (XLMS) algorithm in stereophonic acoustic echo cancellation, *Signal Process.,* **64**(1), 87–91, 1997.

120. ITU-T Recommendation G.167: *General Characteristics of International Telephone Connections and International Telephone Circuits – Acoustic echo Controllers,* Helsinki, Finland, 1993.

121. ITU-T Recommendation G.729: *Annex B: A Silence Compression Scheme for G.729 Optimized for Terminals Conforming to Rec. V.70,* Geneva, Switzerland, 1996.

122. ITU-T Recommendation P.50: *Artifical Voices,* Geneva, Switzerland, 1999.

123. ITU-T Recommendation P.501: *Test Signals for Use in Telephonometry,* Geneva, Switzerland, 2000.

124. ITU-T Recommendation P.581: *Use of Head And Torso Simulator (HATS) for Hands-Free Terminal Testing,* Geneva, Switzerland, 2000.

125. K. Jaervinen, J. Vainio, et al.: Adaptive postfiltering for quality enhancement of coded speech, *Proc. ICASSP '97,* **2**, 771–774, Munich, Germany, 1997.

126. N. S. Jayant, P. Noll: *Digital Coding of Waveforms,* Englewood Cliffs, NJ: Prentice Hall, 1984.

127. D. H. Johnson, D. E. Dudgeon: *Array Signal Processing – Concepts and Techniques,* Englewood Cliffs, NJ: Prentice Hall, 1993.

128. W. B. Joyce: Sabine's reverberation time and ergodic auditoriums, *J. Acoust. Soc. Am.,* **58**(3), 643–655, 1975.

129. Y. Joncour, A. Sugiyama, A. Hirano: DSP implementations and performance evaluation of a stereo echo canceller with pre-processing, *Proc. EUSIPCO '98,* **2**, 981–984, Rhodos, Greece, 1998.

130. M. Kappelan, B. Strauss, P. Vary: Flexible nonuniform filter banks using allpass transformation of multiple order, *Proc. EUROSPEECH '96,* **3**, 1745–1748, Trieste, Italy, 1996.

131. R. E. Kalman: A new approach to linear filtering and prediction problems, *Trans. ASME, J. Basic Eng.,* **82**, 35–45, 1960.

132. R. E. Kalman , R. S. Bucy: New results in linear filtering and prediction theory, *Trans. ASME, J. Basic Eng.,* **83**, 95–108, 1961.

133. N. Kalouptsidis, S. Theodoridis (eds.): *Adaptive Systems Identification and Signal Processing Algorithms,* Englewood Cliffs, NJ: Prentice Hall, 1993.

134. W. Kellermann: *Zur Nachbildung physikalischer Systeme durch parallelisierte digitale Ersatzsysteme im Hinblick auf die Kompensation akustischer Echos,* Düsseldorf, Germany: Fortschr.-Ber. VDI-Reihe 10(102), VDI Verlag, 1989 (in German).

135. D. E. Knuth: *The Art of Computer Programming – Sorting and Searching,* vol. 3, 2nd ed., Boston, MA: Addison-Wesley, 1998.

136. H. Kuttruff: Sound in enclosures, in M. J. Crocker (ed.), *Encyclopedia of Acoustics,* New York, NY: Wiley, 1997.

137. T. Kwan, K. Martin: Adaptive detection and enhancement of multiple sinusoids Uuing a cascade IIR filter, *IEEE Trans. Circuits Systems,* **36**(7), 937–947, 1998.

138. T. I. Laakso, V. Välimäki, M. Karjalainen, U. K. Laine: Splitting the unit delay – tools for fractional delay filter design, *IEEE Signal Process. Mag.,* **13**(1), 30–60, 1996.

139. P. Lapsley, J. Bier, A. Shoham, E. A. Lee: *DSP Processor Fundamentals – Architectures and Features,* New York, NY: IEEE Press, 1997.

140. S. J. Leese: Echo cancellation, in G. M. Davis (ed.), *Noise Reduction in Speech Applications,* Boca Raton, FL: CRC Press, 2002, 199–216.

141. N. Levinson: The Wiener RMS error criterion in filter design and prediction, *J. Math. Phys.,* **25**, 261–268, 1947.

142. A. P. Liavas, P. A. Regalia: Acoustic echo cancellation: do IIR filters offer better modelling capabilities than their FIR counterparts? *IEEE Trans. Signal Process.,* **46**(9), 2499–2504, 1998.

143. D. Lippuner: *Model-Based Step-Size Control for Adaptive filters,* Diss. ETH Zurich, No. 14461, Konstanz, Germany: Hartung-Gorre Verlag, 2002.

144. B. Liu: Effect of finite word length on the accuracy of Ddigital filters – a review, *IEEE Trans. Circuit Theory,* **18**, 6, 670–667, 1971.

145. K. Linhard, T. Haulick: Spectral noise subtraction with recursive gain curves, *Proc. ICSLP '98,* **4**, 1479–1482, Sydney, Australia, 1998.

146. L. Ljung, M. Morf, D. Falconer: Fast Cclculation of gain matrices for recursive estimation schemes, *Int. J. Control,* **27**, 1–19, 1978.

147. S. Ljung, L. Ljung: Error propagation properties of recursive LS adaptation algorithms, *Automatica,* **21**, 157–167, 1985.

148. E. Lleida, E. Masgrau, A. Ortega: Acoustic echo and noise reduction for car cabin communication, *Proc. EUROSPEECH '01,* **3**, 1585–1588, Aalborg, Denmark, 2001.

149. E. Lombard: Le signe de l'elevation de la voix, *Ann. Maladies Oreille, Larynx, Nez. Pharynx,* **37**, 101–119, 1911 (in French).

150. R. W. Lucky, H. R. Rudin: Generalized automatic equalization for communication channels, *Proc. IEEE,* **54**(3), 439–440, 1966.

151. D. Luenberger: *Optimization by Vector Space methods,* New York, NY: Wiley, 1969.

152. O. Macchi: *Adaptive Processing: The Least Mean Squares Approach with Applications in Transmission,* New York, NY: Wiley, 1994.

153. A. Mader: Automatic optimization for a control of hands-free telephone sets, *Proc. IWAENC '99,* 120–123, Pocono Manor, NJ, 1999.

154. A. Mader, H. Puder, G. Schmidt: Step-size control for acoustic echo cancellation filters – an overview, *Signal Process.,* **80**(9), 1697–1719, 2000.

155. A. Mader: Control for acoustic echo cancellers applying an RBF network, *Proc. EUSIPCO '00,* **3**, 1847–1850, Tampere, Finland, 2000.

156. S. Makino, Y. Kaneda: Exponentially weighted step-Size projection algorithm for acoustic echo cancellers, *IECE Trans. Fund.,* **E75-A**(11), 1500–1507, 1992.

157. S. Makino, Y. Kaneda, N. Koizumi: Exponentially weighted step-size NLMS adaptive filter based on the statistics of a room impulse response, *IEEE Trans. Speech and Audio Process.,* **1**, 101–108, 1993.

158. S. Makino, K. Strauss, S. Shimauchi, Y. Haneda: Subband stereo echo canceller using the projection algorithm with fast convergence to the true echo path, *Proc. ICASSP '97,* **1**, 299–302, Munich, Germany, 1997.

159. R. Martin: An Efficient algorithm to estimate the instantaneous SNR of speech signals, *Proc. EUROSPEECH '93,* **3**, 1093–1096, Berlin, Germany, 1993.

160. R. Martin, P. Vary: Combined acoustic echo control and noise reduction for hands-free telephony – state of the art and perspectives, *Proc. EUSIPCO '96,* **2**, 1107–1110, Trieste, Italy, 1996.

161. R. Martin: Noise power spectral density estimation based on optimal smoothing and minimum statistics, *IEEE Trans. Speech Audio Process.,* **T-SA-9**(5), 504–512, 2001.

162. R. Martin: Speech enhancement using MMSE short-time spectral estimation with Gamma distributed speech priors, *Proc. ICASSP '02,* **1**, 253–256, Orlando, FL, 2002.

163. J. Marx: *Akustische Aspekte der Echokompensation in Freisprecheinrichtungen,* Düsseldorf, Germany: Fortschr.-Ber. VDI-Reihe 10(400), VDI Verlag, 1996 (in German).

164. M. Mboup, M. Bonnet: On the adequatness of IIR adaptive filtering for acoustic echo cancellation, *Proc. EUSIPCO '92,* **1**, 111–114, Bruxells, Belgium, 1995.

165. W. Mikhael, F. Wu: Fast algorithms for block FIR adaptive digital filtering, *IEEE Trans. Circuits Systems,* **CAS-34**(10), 1152–1160, 1987.

166. Z. J. Mou, P. Duhamel: Fast FIR filtering: algorithms and implementations, *Signal Process.,* **13**(4), 377–384, 1987.

167. B. Mulgrew, C. F. Cowan: *Adaptive Filters and Equalisers,* Boston, MA: Kluwer, 1988.

168. V. Myllylä: Robust fast affine projection algorithm for acoustic echo cancellation, *Proc. IWAENC '01,* 143–146, Darmstadt, Germany, 2001.

169. V. Myllylä, G. Schmidt: Pseudo optimal regularization for affine projection algorithms, *Proc. ICASSP '02,* **2**, 1917–1920, Orlando, FL, 2002.

170. H. C. Ng, S. H. Leung, C. W. Tsang: Scalar quantization of LSF parameters using vector measurement, *IEEE Electron. Letters,* **34**(10) 961–963, 1998.

171. B. Nitsch: The partitioned exact frequency domain block NLMS algorithm, a mathematically exact version of the NLMS algorithm working in the frequency domain, *AEÜ Int. J. Electron. Commun.,* **52**, 293–301, 1998.

172. K. Ochiai, T. Araseki, T. Ogihara: Echo canceler with two echo path models, *IEEE Trans. Commun.,* **COM-25**(6), 589–595, 1977.

173. S. Oh, D. Linabarger, B. Priest, B. Raghothaman: A fast affine projection algorithm for an acoustic echo canceller using a fixed-point DSP processor, *Proc. ICASSP '97,* **3**, 2345–2348, Munich, Germany, 1997.

174. A. V. Oppenheim, R. W. Schafer: *Discrete-Time Signal Processing,* Englewood Cliffs, NJ: Prentice Hall, 1989.

175. A. Ortega, E. Lleida, E. Masgrau, F. Gallego: Cabin car communication system to improve communication inside a car, *Proc. ICASSP '02,* **4**, 3836–3839, Orlando, FL, 2002.

176. K. Ozeki, T. Umeda: An adaptive filtering algorithm using an orthogonal projection to an affine subspace and its properties, *Electron. Commun. Jpn.,* **67-A**(5), 19–27, 1984.

177. T. Painter, A. Spanias: A review of algorithms for perceptual coding of digital audio, *Proc. Internat. Conf. Digital Signal Process. (DSP)*, 179–205, Santorini, Greece, 1997.

178. A. Papoulis, S. U. Pillai: *Probability, Random Variables and Stochastic Processes,* 4th ed., New York, NY: McGraw-Hill, 2002.

179. T. W. Parks, C. S. Burrus: *Digital filter Design,* New York, NY: Wiley, 1987.

180. G. Pays, J. M. Person: Modèle de laboratoire d'un poste téléphonique à hautparleur, *FASE,* **75**, 88–102, Paris, France, 1975 (in French).

181. M. Petraglia, R. Alves: A new adaptive subband structure with critical sampling, *Proc. ICASSP '99,* **4**, 1861–1864, Phoenix, AR, 1999.

182. Petraglia, M., Alves, R., Diniz, P.: Convergence analysis of a new subband adaptive structure with critical sampling, *Proc. ISCAS '99,* **3**, 134–137, Orlano, FL, 1999.

183. M. Petraglia, R. Alves, P. Diniz: New structures for adaptive filtering in subbands with critical sampling, *IEEE Trans. Signal Process.*, **48**(12), 3316–3327, 2000.

184. I. Pitas: Fast algorithms for running Oordering and max/min calculation, *IEEE Trans. Circuits Systems,* **36**(6), 795–804, 1989.

185. H. Puder: Single channel noise reduction using time-frequency dependent voice activity detection, *Proc. IWAENC '99,* 68–71, Pocono Manor, NJ, 1999.

186. H. Puder, P. Dreiseitel: Implementation of a hands-free car phone with echo cancellation and noise-dependent loss control, *Proc. ICASSP '00,* **6**, 3622–3625, Istanbul, Turkey, 2000.

187. H. Puder, O. Soffke: An approach for an optimized voice-activity detector for noisy speech signals, *Proc. EUSIPCO '02,* **1**, 243–246, Toulouse, France, 2002.

188. H. Puder: Kalman-filters in subbands for noise reduction with enhanced pitch-adaptive speech model estimation, *Europ. Trans. on Telecommunication,* **13**, 139–148. 2002.

189. H. Puder: *Geräuschreduktionsverfahren mit modellbasierten Ansätzen für Freisprecheinrichtungen in Kraftfahrzeugen,* Düsseldorf, Germany: Fortschr.-Ber. VDI-Reihe 10(721), VDI Verlag, 2003 (in German).

190. H. Quast, O. Schreiner. M. R. Schroeder: Robust pitch tracking in the car environment, *Proc. ICASSP '02,* **1**, 353–356, Orlando, FL, 2002.

191. T. F. Quatieri: *Discrete-Time Speech Signal Processing,* Englewood Cliffs, NJ: Prentice Hall, 2002.

192. L. Rabiner, R. W. Schafer: *Digital Processing of Speech Signals,* Englewood Cliffs, NJ: Prentice Hall, 1978.

193. L. Rabiner, B. H. Juang: *Fundamentals of Speech Recognition,* Englewood Cliffs, NJ: Prentice Hall, 1993.

194. Z. Ren, H. Schütze: A stabilized fast transversal filter algorithm for recursive least square adaptive filtering, *Signal Process.,* **39**(3), 235–246, 1994.

195. R. Reng, H. W. Schüssler: Measurement of aliasing distortions and quantization noise in multirate systems, *Proc. ISCAS '92,* **5**, 2328–2331, San Diego, CA, 1992.

196. D. L. Richards, Statistical properties of speech signals, *Proc. IEE,* **111**(5), 941–949, 1964.

197. T. Schertler, P. Heitkämper: Erhöhung der Stabilitätsgrenze von Saalbeschallungsanlagen, *Proc. DAGA '95,* **1**, 331–334, Saarbrücken, Germany, 1995 (in German).

198. T. Schertler, G. Schmidt: Implementation of a low-cost acoustic echo canceller, *Proc. IWAENC '97,* 49–52, London, UK, 1997.

199. T. Schertler: Cancelation of acoustic echos with exponentially weighted step-size and fixed point architecture, *Proc. 32nd Annual Asilomar Conference on Signals, Systems, and Computers,* **1**, 399–403, 1998.

200. T. Schertler: Selective block update of NLMS-type algorithms, *Proc. ICASSP '98,* **3**, 1717–1720, Washington, DC, 1998.

201. T. Schertler: *Kostengünstige Realisierung von Verfahren zur Kompensation akustischer Echos,* Düsseldorf, Germany: Fortschr.-Ber. VDI-Reihe 10(619), VDI Verlag, 2000 (in German).

202. V. Schless: *Automatische Erkennung von gestörten Sprachsignalen,* Aachen, Germany: Shaker, 1999 (in German).

203. D. Schlichthärle: *Digital Filter – Basics and Design,* Berlin, Germany: Springer, 2000.

204. G. Schmidt: Step-size control in subband echo cancellation Systems, *Proc. IWAENC '99,* 116–119, Pocono Manor, NJ, 1999.

205. M. R. Schroeder: Improvement of acustic-feedback stability by frequency shifting, *J. Acoust. Soc. Am.,* **36**(9), 1718–1724, 1964.

206. M. R. Schroeder: Period histogram and product spectrum: new methods for fundamental frequency measurements, *J. Acoust. Soc. Am.,* **43**(4), 829–834, 1968.

207. M. R. Schroeder: *Computer Speech – Recognition, Compression, Synthesis,* Berlin, Germany: Springer, 1999.

208. K. Shenoi: *Digital Signal Processing in Telecommunications,* Englewood Cliffs, NJ: Prentice Hall, 1995.

209. B. G. Sherlock: Windowed discrete Fourier transform for shifting data, *Signal Process.,* **74**(2), 169–177, 1999.

210. S. Shimauchi, S. Makino: Stereo projection echo canceller with true echo path estimation, *Proc. ICASSP '95,* **5**, 3059–3062, Detroit, MI, 1995.

211. J. Shynk: Frequency-domain and multirate adaptive filtering, *IEEE Signal Process. Mag.,* **9**(1), 14–37, 1992.

212. Siemens AG: *Für das ganze Spektrum des Hörens,* prospectus, 2001 (in German).

213. Siemens AG: *Triano—revolutionäre Technik für besseres Hören,* prospectus, 2002 (in German).

214. D. T. M. Slock, T. Kailath: Numerically stable fast RLS transversal filters, *Proc. ICASSP '88,* 1365–1368, New York, NY, 1988

215. D. T. M. Slock, T. Kailath: Fast transversal RLS algorithms, in N. Kalouptsidis, S. Theodoridis (eds.), *Adaptive System Identification and Signal Processing Algorithms,* Englewood Cliffs, NJ: Prentice Hall, 1993.

216. W.-J. Song, M.-S. Park: A complementary pair LMS algorithm for adaptive filtering, *Proc. ICASSP '97,* **3**, 2261–2264, Munich, Germany, 1997.

217. M. M. Sondhi: An adaptive echo canceller, *B.S.T.J.,* **46**, 497–511, 1967.

218. M. M. Sondhi, D. R. Morgan: Stereophonic acoustic echo cancellation – An overview of the fundamental problem, *IEEE Signal Process. Letters,* **2**(8), 148–151, 1995.

219. H. W. Sorensen (ed.): *Kalman filtering: Theory and Application,* IEEE Press, Piscataway, NJ, 1985.

220. A. Spanias: Speech coding – a tutorial review, *Proc. IEEE,* **82**(10), 1541–1582, 1994.

221. A. Sugiyama, Y. Joncour, A. Hirano: A stereo echo canceller with correct echo-path identification based on an input-sliding technique, *IEEE Trans. Signal Process.,* **49**(1), 2577–2587, 2001.

222. A. Sugiyama, T. P. Hua, M. Kato, M. Serizawa: Noise suppression with synthesis windowing and pseudo noise injection, *Proc. ICASSP '02,* **1**, 545–548, Orlando, FL, 2002.

223. M. Tanaka, Y. Kaneda, S. Makino, J. Kojima: Fast projection algorithm and its step-size control, *Proc. ICASSP '95,* **2**, 945–948, Detroit, MI, 1995.

224. J. Tilp: Single-channel noise reduction with pitch-adaptive post-filtering, *Proc. EUSIPCO '00,* **1**, 101–104, Tampere, Finland, 2000.

225. H. L. Van Trees, *Detection, Estimation, and Modulation Theory,* Part I, New York, NY: Wiley, 2001.

226. J. R. Treichler, C. R. Johnson, M. G. Rarimor: *Theory and Design of Adaptive filters,* New York, NY: Wiley, 1987.

227. J. R. Treichler, C. R. Johnson: *Theory and Design of Adaptive Filters,* Englewood Cliffs, NJ: Prentice Hall, 2001.

228. V. Turbin, A. Gilloire, P. Scalart: Comparison of three post-filtering algorithms for residual acoustic echo reduction, *Proc. ICASSP '97,* **1**, 307–310, Munich, Germany, 1997.

229. P. P. Vaidyanathan: Mulitrate digital filter banks, polyphase networks, and applications: a tutorial, *Proc. IEEE,* **78**(1), 56–93, 1990.

230. P. P. Vaidyanathan: *Mulitrate Systems and filter Banks,* Englewood Cliffs, NJ: Prentice Hall, 1992.

231. P. Vary, G. Wackersreuther: A unified approach to digital polyphase filter banks, *AEÜ Int. J. Electron. Commun.,* **37**, 29–34, 1983.

232. P. Vary: Noise suppression by spectral magnitude estimation – mechanism and theoretical limits, *Signal Process.,* **8**(4), 387–400, 1985.

233. P. Vary, U. Heute, W. Hess: *Digitale Sprachsignalverarbeitung,* Stuttgart, Germany: Teubner, 1998 (in German).

234. B. D. Van Veen, K. M. Buckley: Beamforming: A versatile approach to spatial filtering, *IEEE ASSP Mag.,* **5**(2), 4–24, 1988.

235. B. D. Van Veen: An analysis of several paritally adaptive beamformer designs, *IEEE Trans. Acoust. Speech Signal Process.,* **ASSP-37**(2), 192–203, 1989.

236. G. Wackersreuther: On the design of filters for ideal QMF and polyphase filter banks, *AEÜ Int. J. Electron. Commun.,* **39**, 123–130, 1985.

237. B. Widrow, M. E. Hoff Jr.: Adaptive switching circuits, *IRE WESCON Conv. Rec.* **IV**, 96–104, 1960.

238. B. Widrow, J. R. Glover, F. M. McCool, J. Kaunitz, C. S. Williams, R. H. Hearn, J. R. Zeidler, E. Dong, R. C. Goodlin: Adaptive noise cancellation: principles and applications, *Proc. IEEE,* **63**(12), 1962–1718, 1975.

239. B. Widrow, K. Duvall, R. Gooch, W. Newman: Signal cancellation phenomena in adaptive antenna arrays, *IEEE Tans. Antennas and Propagation,* **30**(3), 469–478, 1982.

240. B. Widrow, S. Stearns: *Adaptive Signal Processing,* Englewood Cliffs, NJ: Prentice Hall, 1985.

241. N. Wiener: *Extrapolation, Interpolation, and Smoothing of Stationary Time Series, with Engineering Applications,* MIT Press, Cambridge, MA, 1949 (originally published as confidential report in 1942).

242. D. Wolf: Entropy of speech signals and rate distortion function, *Proc. 1974 Zurich Seminar on Digital Communications,* **B 8**, 1–3, 1974.

243. D. Wolf: Statistical models of speech signals, *NTG Fachberichte,* **65**, 1–9, VDE-Verlag, Berlin, Germany, 1978.

244. G. Wustmann: Erzeugung von Pseudo-Zufallsfolgen mit binären Schieberegistern, *Elektronik* **18**, Arbeitsblatt no. 153, 79–81, 1982 (in German).

245. S. Yamamoto, S. Kitayama, J. Tamura, H. Ishigami: An adaptive echo canceller with linear predictor, *Trans. IECE Jpn.,* **62**(12), 851–857, 1979.

246. S. Yamamoto, S. Kitayama: An adaptive echo canceller with variable step gain method, *Trans. IECE Jpn.,* **E65**(1), 1–8, 1982.

247. H. Yasukawa, S. Shimada: An acoustic echo canceller using subband sampling and decorrelation methods, *IEEE Trans. Signal Process.,* **41**(2), 926–930, 1993.

248. H. Ye, B.-X. Wu: A new double-talk detection algorithm based on the orthogonality theorem, *IEEE Trans. Commun.,* **COM-39**(11), 1542–1545, 1991.

249. U. Zölzer (ed.): *DAFX Digital Audio Effects,* New York, NY: Wiley, 2002.

250. E. Zwicker, H. Fastl: *Psychoacoustics – Facts and Models,* 2nd ed., Berlin, Germany: Springer, 1999.

Index